Methods in Cell Biology

VOLUME 69
Methods in Cell–Matrix Adhesion

ASCB

Series Editors

Leslie Wilson
Department of Biological Sciences
University of California, Santa Barbara
Santa Barbara, California

Paul Matsudaira
Whitehead Institute for Biomedical Research and
Department of Biology
Massachusetts Institute of Technology
Cambridge, Massachusetts

Methods in Cell Biology

Prepared under the auspices of the American Society for Cell Biology

VOLUME 69
Methods in Cell–Matrix Adhesion

Edited by

Josephine C. Adams

Department of Cell Biology
Lerner Research Institute
Cleveland Clinic Foundation
Cleveland, Ohio

ACADEMIC PRESS
An imprint of Elsevier Science

Amsterdam   Boston   London   New York   Oxford   Paris
San Diego   San Francisco   Singapore   Sydney   Tokyo

Cover image (paperback only): Collage of images include FITC-phallodin staining of actin organization in a cell spatially constrained on a square fibronectin-coated island (courtesy of Prof. D. Ingber, Harvard University, reproduced by permission of Elsevier), indirect immunofluorescent staining of cells within a fibrin-fibronectin matrix (courtesy of Prof. J. Schwarzbauer, Princeton University), TRITC-phallodin staining of actin in the spikes and filopodia produced by a syndecan-1-expressing cell, and indirect immuno-fluorescent staining of the bPS integrin subunit in clusters of Drosophila cells (from Adams' laboratory notebooks).

This book is printed on acid-free paper. ∞

Academic Press
An imprint of Elsevier Science
525 B Street, Suite 1900, San Diego, California 92101-4495, USA
http://www.academicpress.com

Academic Press
84 Theobalds Road, London WC1X 8RR, UK
http://www.academicpress.com

International Standard Book Number: 0-12-544168-1 (case)
International Standard Book Number: 0-12-044142-X (pb)

PRINTED IN THE UNITED STATES OF AMERICA
02   03   04   05   06   07   MM   9   8   7   6   5   4   3   2   1

CONTENTS

3. Expression of Recombinant Matrix Components Using Baculoviruses

Deane F. Mosher, Kristin G. Huwiler, Tina M. Misenheimer, and Douglas S. Annis

4. Heparan Sulfate–Growth Factor Interactions

Alan C. Rapraeger

5. Analysis of Basement Membrane Self-Assembly and Cellular Interactions
with Native and Recombinant Glycoproteins

Peter D. Yurchenco, Sergei Smirnov, and Todd Mathus

6. Preparation and Analysis of Synthetic Multicomponent Extracellular Matrix

Kim S. Midwood, Iwona Wierzbicka-Patynowski, and Jean E. Schwarzbauer

12. Intracellular Coupling of Adhesion Receptors: Molecular
Proximity Measurements

Maddy Parsons and Tony Ng

PART IV   Functional Applications of Cell–Matrix Adhesion in Molecular Cell Biology

13. Functional Analysis of Cell Adhesion: Quantitation of
Cell–Matrix Attachment

Steven K. Akiyama

14. Measurements of Glycosaminoglycan-Based Cell Interactions

J. Kevin Langford and Ralph D. Sanderson

15. Applications of Adhesion Molecule Gene Knockout Cell Lines

Jordan A. Kreidberg

PART V   General Information

CONTRIBUTORS

Numbers in parentheses indicate the pages on which the authors' contributions begin.

Steven K. Akiyama (281), Laboratory of Molecular Carcinogenesis, National Institute of Environmental Health Sciences, National Institutes of Health, Research Triangle Park, North Carolina 27709

Douglas S. Annis (69), Department of Medicine, University of Wisconsin, Madison, Wisconsin 53706

Hannelore Asmussen (341), Department of Cell Biology, University of Virginia, Charlottesville, Virginia 22908

Karen A. Beningo (325), Department of Physiology, University of Massachusetts Medical School, Worcester, Massachusetts 01605

Fedor Berditchevski (223), CRC Institute for Cancer Studies, The University of Birmingham, Edgbaston, Birmingham, B15 2TA United Kingdom

Paul Bornstein (7), Departments of Biochemistry and Medicine, University of Washington, Seattle, Washington 98195

William G. Carter (27), Basic Sciences, Fred Hutchinson Cancer Research Center, Seattle, Washington 98109; and Department of Pathobiology, University of Washington, Seattle, Washington 98195

Dennis O. Clegg (163), Department of Molecular, Cellular, and Developmental Biology, University of California, Santa Barbara, California 93106

Johannes A. Eble (223), Institute of Physiological Chemistry and Pathobiochemistry, Universität Münster, 48149 Münster, Germany

Susana G. Gil (27), Basic Sciences, Fred Hutchinson Cancer Research Center, Seattle, Washington 98109

Mark Ginsberg (209), The Scripps Research Institute, La Jolla, California 92037

Helen G. Hansma (163), Department of Physics, University of California, Santa Barbara, California 93106

Alan F. Horwitz (341), Department of Cell Biology, University of Virginia, Charlottesville, Virginia 22908

Kristin G. Huwiler (69), Department of Medicine, University of Wisconsin, Madison, Wisconsin 53706

Donald Ingber (385), Departments of Pathology and Surgery, Children's Hospital and Harvard Medical School, Boston, Massachusetts 02115

Renato V. Iozzo (53), Department of Pathology, Anatomy and Cell Biology, Thomas Jefferson University, Philadelphia, Pennsylvania 19107

Efrosini Kokkoli (163), College of Engineering, University of California, Santa Barbara, California 93106

Jordan A. Kreidberg (309), Department of Medicine, Division of Nephrology, Children's Hospital and Department of Pediatrics, Harvard Medical School, Boston, Massachusetts 02115

J. Kevin Langford (297), Department of Pathology, Arkansas Cancer Research Center, University of Arkansas for Medical Sciences, Little Rock, Arkansas 72205

Philip LeDuc (385), Departments of Pathology and Surgery, Children's Hospital and Harvard Medical School, Boston, Massachusetts 02115

Chun-Min Lo (325), Department of Physics, Cleveland State University, Cleveland, Ohio 44115

Roy R. Lobb (17), Biogen Inc., Cambridge, Massachusetts 02142

Todd Mathus (111), Department of Pathology and Laboratory Medicine, Robert Wood Johnson Medical School, Piscataway, New Jersey 08854

Kim S. Midwood (145), Department of Molecular Biology, Princeton University, Princeton, New Jersey 08544

Cindy Miranti (359), Laboratory of Integrin Signaling and Tumorigenesis, Van Andel Research Institute, Grand Rapids, Michigan 49503

Tina M. Misenheimer (69), Department of Medicine, University of Wisconsin, Madison, Wisconsin 53706

Deane F. Mosher (69), Department of Medicine, University of Wisconsin, Madison, Wisconsin 53706

H. G. Munshi (195), Department of Medicine, Division of Hematology/Oncology, Northwestern University Medical School, Chicago, Illinois 60611

Shin-ichi Murase (341), Department of Cell Biology, University of Virginia, Charlottesville, Virginia 22908

Tony Ng (261), Richard Dimbleby Department of Cancer Research/Cancer Research UK Labs, The Rayne Institute, St. Thomas Hospital, London SE1 7EH, United Kingdom

Emin Oroudjev (163), Department of Physics, University of California, Santa Barbara, California 93106

Emanuele Ostuni (385), Departments of Chemistry and Chemical Biology, Harvard University, Cambridge, Massachusetts 02138

Maddy Parsons (261), Richard Dimbleby Department of Cancer Research/Cancer Research UK Labs, The Rayne Institute, St. Thomas Hospital, London SE1 7EH, United Kingdom

Joe W. Ramos (209), Rutgers, The State University of New Jersey, Piscataway, New Jersey 08854

Alan C. Rapraeger (83), Department of Pathology and Laboratory Medicine, University of Wisconsin-Madison, Madison, Wisconsin 53706

Ralph D. Sanderson (297), Departments of Pathology and Anatomy, Arkansas Cancer Research Center, University of Arkansas for Medical Sciences, Little Rock, Arkansas 72205

Martin A. Schwartz (13), The Scripps Research Institute, Department of Vascular Biology, La Jolla, California 92037

Jean E. Schwarzbauer (145), Department of Molecular Biology, Princeton University, Princeton, New Jersey 08544

Randy O. Sigle (27), Basic Sciences, Fred Hutchinson Cancer Research Center, Seattle, Washington 98109

Sergei Smirnov (111), Department of Pathology and Laboratory Medicine, Robert Wood Johnson Medical School, Piscataway, New Jersey 08854

Jeffrey W. Smith (249), The Burnham Institute, La Jolla, California 92037

M. Sharon Stack (195), Department of Cell and Molecular Biology, Northwestern University Medical School, Chicago, Illinois 60611

Charles H. Streuli (403), School of Biological Sciences, University of Manchester, Manchester M13 9PT, United Kingdom

Matthew Tirrell (163), College of Engineering, University of California, Santa Barbara, California 93106

Yu-Li Wang (325), Department of Physiology, University of Massachusetts Medical School, Worcester, Massachusetts 01605

Harriet Watkin (403), School of Biological Sciences, University of Manchester, Manchester M13 9PT, United Kingdom

Donna J. Webb (341), Department of Cell Biology, University of Virginia, Charlottesville, Virginia 22908

John M. Whitelock (53), CSIRO Molecular Science, North Ryde 21113, Sydney, Australia

George Whitesides (385), Departments of Chemistry and Chemical Biology, Harvard University, Cambridge, Massachusetts 02138

Iwona Wierzbicka-Patynowski (145), Department of Molecular Biology, Princeton University, Princeton, New Jersey 08544

Peter D. Yurchenco (111), Department of Pathology and Laboratory Medicine, Robert Wood Johnson Medical School, Piscataway, New Jersey 08854

ACKNOWLEDGMENTS

This book has taken shape through the efforts of many people. I thank all the contributing authors for their willingness to share their expertise and for their enthusiasm, time, and care in writing the chapters. I thank Mica Haley for very effective editorial support at Academic Press and to Alexia Zaromytidou at University College London for valuable help with collating chapters and bringing together and checking all the general information in Part V. I thank Bill Tawil, Suneel Apte, Nina Kureishy, Raymond Monk, Soren Prag, Vasileia Sapountzi and Alexia Zaromytidou for contributions of website addresses. The cover of the softbound version of the book is a collage featuring images from the chapters of Leduc *et al.* (reproduced with permission of Elsevier), Westwood *et al.,* and Adams' laboratory notebooks. I especially thank Paul Matsudaira and Leslie Wilson for their invitation to develop this book.

PART I

Preface and Perspectives on Cell–Matrix Adhesion

Preface

Josephine C. Adams

Department of Cell Biology
Lerner Research Institute
The Cleveland Clinic Foundation
Cleveland, Ohio 44195

The responses of cells to extracellular matrix (ECM) have long been a source of fascination for experimentalists. Effects of solid surfaces on cell shape, locomotion, and growth were noted during the earliest days of tissue culture. As one example, the experiments of Harrison memorably compared the behavior of embryo cells in suspension or explanted onto clotted plasma, threads of spiders webs, or glass coverslips. The cells only locomoted when in contact with a solid surface and their spreading and movement behavior was different on the three surfaces; in particular, those plated on spider web became elongated and extended processes that conformed to the geometry of the threads (Harrison, 1914). From these results, Harrison concluded that the nature of the surface could regulate the form and arrangement of cells and that a solid surface was essential for cell locomotion. Cell responses to extracellular matrix, the determinants of specificity, and the molecular processes that underlie these responses remain a major focus of interest for modern biologists.

A second source of fascination lies in the structure and properties of ECM molecules themselves. ECM molecules compel attention because of their very large size, both as single polypeptides and as oligomers, and because of the elegance with which repeated modular units are used in different combinations to build up the polypeptides. The modules have been a magnet for structural biologists (Campbell and Downing, 1994; Timpl *et al.*, 2000; Liddington, 2001). Tracking a module through otherwise structurally distinct proteins can give a unique window on the workings of evolution (Engel *et al.*, 1994). Measurements of the biophysical properties of modules and protein units provide first steps toward precise understanding of the mechanical behavior of single ECM molecules and complex extracellular matrices (Oberhauser *et al.*, 1998; Ohashi *et al.*, 1999).

To work with the large matrix molecules, even with the advantages of recombinant protein expression technology, poses experimental challenges. To then understand and control the buildup of fibrillar or sheetlike matrix networks from the interaction and assembly of many molecular species adds further complexity. Harrison remarked in his paper that early in the season it was difficult to find a suitable number of spiders for his experiments: a modern equivalent of this problem has been the need to obtain appropriate recombinant proteins in sufficient quantities to carry forward accessible, well-controlled, and well-defined functional experiments. The chapters on methods for purification or recombinant expression of matrix glycoproteins and proteoglycans describe expert approaches that have dealt effectively with the spider problem.

All these questions and challenges are of more than academic interest because of the essential roles of cell–matrix adhesions for cell and tissue organization, homeostasis, and repair throughout life (for example, Springer, 1995; Hynes, 1999). Chapters in this book describe methods of considerable sophistication and precision for control of the adhesive materials, the mechanical properties and geometry of the adhesive substrata, and the properties of test cell populations. The knowledge arising from use of these experimental systems has broad applications in the alleviation of chronic diseases, ageing, control of angiogenesis, development of artificial organs, and all forms of tissue engineering (for example, Endelman *et al.,* 1997; Guthiel *et al.,* 2000).

Because of the success of the genome projects, there is now available a vast amount of information for grouping proteins according to their primary sequence similarity or the presence of recognizable protein modules. For *C. elegans* and *D. melanogaster,* the complete set of matrix adhesion molecules that can be recognized by sequence comparison is now known (The *C. elegans* Sequencing Consortium, 1998; Adams *et al.,* 2000). This has enabled a first view of the recognizable primordial components of cell–matrix adhesions (Hutter *et al.,* 2000; Hynes and Xhao, 2000). The comparison of protein modules and the comparison of genomes both offer extraordinarily unifying views of biology. It is fortunate to be a life scientist at this time. Yet definition of protein function beyond these pointers, or to demonstrate that a novel protein is a functional adhesion molecule, continues to rely heavily on experiments conducted on a molecule-by-molecule basis at cellular level. For this reason, I believe there is a continuing need for methods books such as this which contain integrated guidance on the specialized methods that address a particular area of cell function. It is also clear that, in general, cell biological research is moving from a focus on molecules to a coherent understanding of functional systems within cells. The process of cell–matrix adhesion is one such system has many points of entry for researchers.

Many excellent reviews have discussed the state-of-the-art knowledge of cell–matrix adhesion molecules and their functions at cell and organismal level. Instead of repeating these in a synopsis form, I invited three eminent, long-term experts to write personal perspectives on the development of the field and to speculate about areas they think could be of special future significance. Paul Bornstein surveys the history of matrix biology research and describes new developments with matricellular molecules. Martin Schwartz discusses how a cross-fertilization of ideas and technologies could bring about an understanding of the role of complex ECM structures in the spatial organization of signaling networks. Roy Lobb describes how advances in understanding integrin–ligand interactions are being translated into new targets for drug design in the fast-growing area of biotherapeutics. I thank these authors for generously contributing these essays.

The test of a methods book lies not in debate but in action. The methods brought together here distill many years of research experience in the authors' laboratories. As I read the contributed chapters, I wanted to abandon the text and run to the lab to try them and to start mixing new approaches. My hope is that, whether searching for details of tried and trusted protocols or dipping in for new connections, both novices and afficionados will be energized and inspired to further innovations in cell–matrix adhesion research.

References

Adams, M. D. *et al.* (2000). The genome sequence of *Drosophila melanogaster*. *Science* **287**, 2185–2195.

Campbell, I. D., and Downing, A. K. (1994). Building protein structure and function from modular units. *Trends Biotechnol.* **12**, 168–172.

The *C. elegans* Sequencing Consortium. (1998). The *C. elegans* genome. *Science* **282**, 2012.

Engel, J., Etimov, V. P., and Maurer, P. (1994). Domain organisations of extracellular matrix proteins and their evolution. *Develop.* S35–S42.

Engleman, V. W., Nickols, G. A., Ross, F. P., Horton, M. A., Griggs, D. W., Settle, S. L., Ruminski, P. G., and Teitelbaum, S. L. (1997). A peptidomimetic antagonist of the alpha v beta 3 integrin inhibits bone resorption *in vitro* and prevent osteoporosis *in vivo*. *J. Clin. Invest.* **99**, 2284–2292.

Gutheil, J. C., Campbell, T. N., Pierce, P. R., Watkins, J. D., Huse, W. D., Bodkin, D. J., and Cheresh, D. A. (2000). Targeted antianglogenic therapy for cancer using Vitaxin: A humanized monoclonal antibody to the integrin alphavbeta3. *Clin. Cancer Res.* **6**, 3056–3061.

Harrison, R. G. (1914). The reaction of embryonic cells to solid structures. *J. Expt. Zool.* **17**, 521–544.

Hutter, H., Vogel, B. E., Plenefisch, J. D., Norris, C. R., Proenca, R. B., Spieth, J., Guo, C., Mastwal, S., Zhu, X., Scheel, J., and Hedgecock, E. M. (2000). Conservation and novelty in the evolution of cell adhesion and extracellular matrix genes. *Science* **288**, 989–994.

Hynes, R. O. (1999). Cell adhesion: old and new questions. *Trends Cell Biol.* **9**, M33–M37.

Hynes, R. O., and Zhao, Q. (2000). The evolution of cell adhesion. *J. Cell Biol.* **150**, F89–96.

Liddington, R. C. (2001). Mapping out the basement membrane. *Nature Struc. Biol.* **8**, 573–574.

Oberhauser, A. F., Marszalek, P. E., Erickson, H. P., and Fernandez, J. M. (1998). The molecular elasticity of the extracellular matrix protein tenascin. *Nature* **393**, 181–185.

Ohashi, T., Kiehart, D. P., and Erickson, H. P. (1999). Dynamics and elasticity of the fibronectin matrix in living cell culture visualized by fibronectin-green fluorescent protein. *Proc. Natl. Acad. Sci. USA* **96**, 2153–2158.

Springer, T. A. (1994). Traffic signals for lymphocyte recirculation and leukocyte emigration: the multistep paradigm. *Cell* **76**, 301–314.

Timpl, R., Tisi, D., Talts, J. F., Andac, Z., Sasaki, T., and Hohenester, E. (2000). Structure and function of laminin LG modules. *Matrix Biol.* **19**, 309–317.

Cell–Matrix Interactions: The View from the Outside

Paul Bornstein

Departments of Biochemistry and Medicine
University of Washington
Seattle, Washington 98195

The term "fibronectin" was first used about a quarter of a century ago (Kuusela *et al.*, 1976; Keski-Oja *et al.*, 1976) to refer to an extracellular protein that had previously been known by many different names (cell surface protein, fibroblast surface antigen, LETS protein, galactoprotein *a*, Z protein) and that was recognized even at that time to play a major role in the adhesion of many different cells to extracellular matrices (Hynes, 1990). Shortly thereafter, but prior to the discovery by several laboratories of integrin adhesive receptors (see Hynes, 1987, 1992, for reviews), my laboratory proposed a model for cell–matrix interactions that invoked an "external protein meshwork," composed of fibronectin, collagens, and other proteins. This meshwork was thought to interact with transmembrane proteins to influence cytoskeletal assembly and function (Bornstein *et al.*, 1978). According to the fluid-mosaic model of the plasma membrane (Singer and Nicholson, 1972), the transmembrane proteins were "integral" proteins and the external protein meshwork was composed of extracellular "peripheral" proteins. The external-protein-meshwork model accommodated the possibility that intracellular changes could lead to alterations or even disassembly of the meshwork, as was observed in transformed and neoplastic cells (Bornstein *et al.*, 1978). These studies were performed prior to an appreciation that engagement of cell-surface integrin receptors could lead to covalent modification of intracellular proteins by phosphorylation/dephosphorylation, and the activation of intracellular signal transduction pathways (Clark and Brugge, 1995). Thus, changes in the assembly of protein complexes in our model were thought to be generated largely by changes in protein conformation and protein–protein interactions (Bornstein *et al.*, 1978).

The notion that a cell can control its cell-surface properties seems relatively straight-forward, but the reverse, namely that changes at the cell surface can modulate cell function, was less apparent, except perhaps in the context of the mechanism of action of growth factors and cytokines. The term "dynamic reciprocity" was initially introduced to suggest that secreted matrix macromolecules could modulate the properties and functions of the cells that produce them (Bornstein *et al.,* 1982). Subsequently, the term was applied more specifically by Bissell and co-workers to the reciprocal functions of epithelial cells, and the extracellular matrix (ECM) which they produce, in the differentiation of the mammary gland (Bissell *et al.,* 1982; Roskelley *et al.,* 1995; Boudreau *et al.,* 1995; Boudreau and Bissell, 1998). These investigators showed that the differentiation of mammary epithelial cells (MEC) *in vitro,* as judged by polarization of cells, formation of acini, and expression of β-casein, required a deformable substratum and the presence of laminin in basement membranes (Streuli and Bissell, 1990; Streuli *et al.,* 1991, 1995). A deformable substrate permitted cell rounding, and laminin served to interact with and cluster β1-integrin receptors (Roskelley *et al.,* 1994).

Although the interactions of fibronectin, laminin-1, and other laminins with cell-surface receptors, and their consequences for cell function, are perhaps the best-studied of cell–matrix interactions, it is likely that most, if not all, ECM proteins and proteoglycans can induce intracellular signals that modulate cellular behavior. Thus, as examples, fibrillar collagens (e.g., types I and III) and their fragments interact with $\alpha 1\beta 1$ and $\alpha 2\beta 1$ integrins (Pozzi *et al.,* 1998; Carragher *et al.,* 1999), and several collagen types bind one or both of two discoidin tyrosine kinase receptors (DDR1 and DDR2; Vogel, 1999). These interactions have diverse consequences for cell differentiation, proliferation, and function. It is of interest that mice that lack DDR1 function have a serious defect in the development of the mammary gland (Vogel *et al.,* 2001). Syndecans exist both as transmembrane heparan sulfate proteoglycans and as shed ectodomains. Together, these proteoglycans modulate cell adhesion, cytoskeletal organization, interaction of growth factors with their receptors, and other cellular functions (Bernfield *et al.,* 1999; Woods, 2001).

ECM proteins cover a spectrum, from those whose function is largely structural to proteins that play at best a minor structural role. Among the latter is a group of proteins that have been termed "matricellular" proteins to highlight their roles as extracellular modulators of cell function (Bornstein, 1995; Sage, 2001; Bornstein, 2001). A short list of these proteins includes thrombospondins (TSPs) 1 (Chen *et al.,* 2000) and 2 (Bornstein *et al.,* 2000), SPARC/osteonectin/BM-40 (Brekken and Sage, 2001), tenascin C (Jones and Jones, 2000), and osteopontin (Giachelli and Steitz, 2000; Denhardt *et al.,* 2001). Matricellular proteins typically interact with multiple cell-surface receptors; in the case of TSP1, interactions have been documented with six different integrins, and with syndecans, betaglycan, sulfatides, low-density lipoprotein receptor-related protein (LRP), CD36, and integrin-associated protein (IAP). Since no single cell type or local tissue environment is likely to display all of these receptors, the functions of TSP1 and other matricellular proteins are considered to be contextual. Since cytokines and hormones also function in the extracellular milieu and influence cell function by interactions with cell-surface receptors, how might one distinguish these effector molecules, conceptually, from matricellular proteins? A major distinction is that matricellular proteins, because of their size and propensity to bind to structural elements in the matrix, act locally rather than systemically. A possible exception is the TSP1 that is contained in platelet α granules. A second major difference is that a significant component of the function of matricellular proteins, and a factor in their tendency to function contextually, derives from their ability to bind, and sequester or activate, cytokines and proteases.

The ability of a matricellular protein to influence cell–matrix interactions indirectly is well illustrated by recent studies of the phenotype of the TSP2-null mouse. Dermal fibroblasts from a TSP2-null mouse show a marked defect in attachment to a wide variety of matrix proteins (Yang et al., 2000). The defect was shown to result from a twofold increase in matrix metalloproteinase 2 (MMP2) levels in the conditioned medium of the fibroblasts in the presence of normal MMP2 mRNA levels. Increased protease activity was responsible for the attachment defect, since normal attachment could be restored by treatment of cultured cells with TIMP2, an inhibitor of MMP2, or with a blocking antibody to MMP2 (Yang et al., 2000). It was subsequently shown that TSP2 bound MMP2 and that the complex was endocytosed by the LRP receptor (Yang et al., 2001). Thus, TSP2 functions to modulate cell–matrix interactions by regulating pericellular protease activity, and in its absence, steady-state levels of MMP2 rise. It is of interest that TSP1 production is unchanged in cultured TSP2-null fibroblasts, even though TSP1 binds MMP2 and is also recognized by the LRP receptor. An additional consequence of increased MMP2 levels in TSP2-null fibroblasts may be degradation of cell-surface tissue transglutaminase (tTG; N. Yan and P. Bornstein, unpublished observations). tTG functions as a co-receptor for $\beta 1$ and $\beta 3$ integrins and increases the affinity of binding of these receptors to fibronectin (Akimov et al., 2000). Thus, loss of tTG would contribute to the reduced attachment observed in these cells. Increased MMP2 levels have also been documented in the foreign body response in TSP2-null mice and might contribute to the increased angiogenesis and disordered collagen fibrillogenesis observed in these animals (Kyriakides et al., 2001)

Studies of other matricellular proteins, and of the phenotypes of mice deficient in these proteins, provide an unusual opportunity to further our understanding of the role that the ECM plays in cell–matrix interactions and in the regulation of cell function. TSP1 is capable of activating latent TGF-$\beta 1$ (Murphy-Ullrich and Poczatek, 2000), and the phenotype of the TSP1-null mouse is consistent with a mild TGF-$\beta 1$ deficiency (Crawford et al., 1998). TGF-$\beta 1$ is a potent inducer of expression of genes encoding fibronectin, collagens, and other matrix proteins, and therefore has an impact on the formation of the ECM. SPARC, when added to a variety of cells in culture, reduces focal adhesions and attachment to substrata and causes partial cell rounding (Brekken and Sage, 2001). The biochemical basis for these effects has not been established. SPARC also inhibits cell proliferation, possibly as a result of its ability to bind a number of growth factors, including PDGF-AB and BB, and VEGF-A (Brekken and Sage, 2001). Osteopontin is expressed primarily in response to inflammation and in mineralizing tissues and is thought to regulate the adhesion of macrophages and osteoclasts by interaction with one or more integrins (Giachelli and Steitz, 2000). Major aspects of the phenotypes of the SPARC-null mouse (cataracts) and the osteopontin-null mouse (compromised cell-mediated immunity and increased dystrophic calcification), while consistent with the results of studies in vitro, are still incompletely understood (Bradshaw and Sage, 2001; Giachelli and Steitz, 2000). Further studies of mice with disruptions in the genes encoding matricellular proteins are therefore likely to yield new insights into the complexly textured controls that the extracellular environment imposes on the cells that are also responsible for its synthesis.

Acknowledgments

Original studies from this laboratory were supported by grant AR45418 from the National Institutes of Health. I am grateful to Helene Sage and members of my laboratory for helpful discussions and a careful review of the manuscript.

References

Akimov, S. S., Krylov, D., Fleischman, L. F., and Belkin, A. M. (2000). Tissue transglutaminase is an integrin-binding adhesion coreceptor for fibronectin. *J. Cell Biol.* **148,** 825–838.

Bernfield, M., Gotte, M., Park, P. W., Reizes, O., Fitzgerald, M. L., Lincecum, J., and Zako, M. (1999). Functions of cell surface heparan sulfate proteoglycans. *Annu. Rev. Biochem.* **68,** 729–777.

Bissell, M. J., Hall, H. G., and Parry, G. (1982). How does the extracellular matrix direct gene expression? *J. Theoret. Biol.* **99,** 31–68.

Bornstein, P. (1995). Diversity of function is inherent in matricellular proteins: an appraisal of thrombospondin 1. *J. Cell Biol.* **130,** 503–506.

Bornstein, P. (2001). Thrombospondins as matricellular modulators of cell function. *J. Clin. Invest.* **107,** 929–934.

Bornstein, P., Duksin, D., Balian, G., Davidson, J. M., and Crouch, E. (1978). Organization of extracellular proteins on the connective tissue cell surface: Relevance to cell-matrix interactions *in vitro* and *in vivo. Ann. N.Y. Acad. Sci.* **312,** 93–105.

Bornstein, P., McPherson, J., and Sage, H. Synthesis and secretion of structural macromolecules by endothelial cells in culture. *In* "Pathobiology of the Endothelial Cell, P & S Biomedical Sciences Symposia," Vol. 6 (H. L. Nossel and H. J. Vogel, eds.), Academic Press, New York, pp. 215–228, 1982.

Bornstein, P., Armstrong, L. C., Hankenson, K. D., Kyriakides, T. R., and Yang, Z. (2000). Thrombospondin 2, a matricellular protein with diverse functions. *Matrix Biol.* **19,** 557–568.

Boudreau, N., and Bissell, M. J. (1998). Extracellular matrix signaling: integration of form and function in normal and malignant cells. *Curr. Opin. Cell Biol.* **10,** 640–646.

Boudreau, N., Myers, C., and Bissell, M. J. (1995). From laminin to lamin: regulation of tissue-specific gene expression by the ECM. *Trends Cell Biol.* **5,** 1–4.

Bradshaw, A. D., and Sage, E. H. (2001). SPARC, a matricellular protein that functions in cellular differentiation and tissue response to injury. *J. Clin. Invest.* **107,** 1049–1054.

Brekken, R. A., and Sage, E. H. (2001). SPARC, a matricellular protein: at the crossroads of cell–matrix communication. *Matrix Biol.* **19,** 815–827.

Carragher, N. O., Levkau, B., Ross, R., and Raines, E. W. (1999). Degraded collagen fragments promote rapid disassembly of smooth muscle focal adhesions that correlates with cleavage of pp125(FAK), paxillin, and talin. *J. Cell Biol.* **147,** 619–629.

Chen, H., Herndon, M. E., and Lawler, J. (2000). The cell biology of thrombospondin-1. *Matrix Biol.* **19,** 597–614.

Clark, E. A., and Brugge, J. S. (1995). Integrins and signal transduction pathways: the road taken. *Science* **268,** 233–239.

Crawford, S. E., Stellmach, V., Murphy-Ullrich, J. E., Ribeiro, S. M. F., Lawler, J., Hynes, R. O., Boivin, G. P., and Bouck, N. (1998). Thrombospondin-1 is a major activator of TGF-β1 in vivo. *Cell* **93,** 1159–1170.

Denhardt, D. T., Noda, M., O'Regan, A. W., Pavlin, D., and Berman, J. S. (2001). Osteopontin as a means to cope with environmental insults: regulation of inflammation, tissue remodeling, and cell survival. *J. Clin. Invest.* **107,** 1055–1061.

Giachelli, C. M., and Steitz, S. (2000). Osteopontin: a versatile regulator of inflammation and biomineralization. *Matrix Biol.* **19,** 615–622.

Hynes, R. O. (1987). Integrins: A family of cell surface receptors. *Cell* **48,** 549–554.

Hynes, R. O. (1990). "Fibronectins." Springer-Verlag, New York.

Hynes, R. O. (1992). Integrins: Versatility, modulation, and signaling in cell adhesion. *Cell* **69,** 11–25.

Jones, F. S., and Jones, P. L. (2000). The tenascin family of ECM glycoproteins: Structure, function, and regulation during embryonic development and tissue remodeling. *Dev. Dyn.* **218,** 235–259.

Keski-Oja, J., Mosher, D. F., and Vaheri, A. (1976). Cross-linking of a major fibroblast surface-associated glycoprotein (fibronectin) catalyzed by blood coagulation factor XIII. *Cell* **9,** 29–35.

Kuusela, P., Rouslahti, E., Engvall, E., and Vaheri, A. (1976). Immunological interspecies cross-reactions of fibroblast surface antigen (fibronectin). *Immunochemistry* **13,** 639–642.

Kyriakides, T. R., Zhu, Y-H., Yang, Z., Huynh, G., and Bornstein, P. (2001). Altered extracellular matrix remodeling and angiogenesis in sponge granulomas of TSP2-null mice. *Am. J. Pathol.* **159,** 1255–1262.

Murphy-Ullrich, J. E., and Poczatek, M. (2000). Activation of latent TGF-β by thrombospondin-1: mechanisms and physiology. *Cytokine Growth Factor Revi.* **11,** 59–69.

Pozzi, A., Wary, K. K., Giancotti, F. G., and Gardner, H. A. (1998). Integrin $\alpha 1 \beta 1$ mediates a unique collagen-dependent proliferation pathway in vivo. *J. Cell Biol.* **142,** 587–594.

Roskelley, C. D., Desprez, P. Y., and Bissell, M. J. (1994). Extracellular matrix-dependent tissue-specific gene expression in mammary epithelial cells requires both physical and biochemical signal transduction. *Proc. Natl. Acad. Sci. USA* **91,** 12378–12382.

Roskelley, C. D., Srebrow, A., and Bissell, M. J. (1995). A hierarchy of ECM-mediated signalling regulates tissue-specific gene expression. *Curr. Opin. Cell Biol.* **7,** 736–747.

Sage, E. H. (2001). Regulation of interactions between cells and extracellular matrix: a command performance on several stages. *J. Clin. Invest.* **107,** 781–783.

Singer, S. J., and Nicolson, G. L. (1972). The fluid mosaic model of the structure of cell membranes. *Science* **175,** 720–731.

Streuli, C. H., and Bissell, M. J. (1990). Expression of extracellular matrix components is regulated by substratum. *J. Cell Biol.* **110,** 1405–1415.

Streuli, C. H., Bailey, N., and Bissell, M. J. (1991). Control of mammary epithelial differentiation: Basement membrane induces tissue-specific gene expression in the absence of cell-cell interaction and morphological polarity. *J. Cell Biol.* **115,** 1383–1395.

Streuli, C. H., Schmidhauser, C., Bailey, N., Yurchenco, P., Skubitz, A. P. N., Roskelley, C., and Bissell, M. J. (1995). Laminin mediates tissue-specific gene expression in mammary epithelia. *J. Cell Biol.* **129,** 591–603.

Vogel, W. (1999). Discoidin domain receptors: structural relations and functional implications. *FASEB J.* **13,** S77–S82.

Vogel, W. F., Aszodi, A., Alves, F., and Pawson, T. (2001). Discoidin domain receptor 1 tyrosine kinase has an essential role in mammary gland development. *Mol. Cell Biol.* **21,** 2906–2917.

Woods, A. (2001). Syndecans: transmembrane modulators of adhesion and matrix assembly. *J. Clin. Invest.* **107,** 935–941.

Yang, Z., Kyriakides, T. R., and Bornstein, P. (2000). Matricellular proteins as modulators of cell-matrix interactions: the adhesive defect in thrombospondin 2-null fibroblasts is a consequence of increased levels of matrix metalloproteinase-2. *Mol. Biol. Cell* **11,** 3353–3364, 2000.

Yang, Z., Strickland, D. K., and Bornstein, P. (2001). Extracellular MMP2 levels are regulated by the LRP scavenger receptor and thrombospondin 2. *J. Biol. Chem.* **276,** 8403–8408.

Matrix and Meaning

Martin A. Schwartz

The Scripps Research Institute
Department of Vascular Biology
La Jolla, California 92037

Integrin signaling was discovered and has generally been studied by stimulating suspended cells with homogenous artificial extracellular matrices (ECM). For example, cells have been treated with soluble fibrinogen or anti-integrin antibodies, or plated on glass or plastic coated with fibronectin or collagen to stimulate integrin-dependent pathways (Schwartz *et al.,* 1989; Kornberg *et al.,* 1991; Guan *et al.,* 1991; Ferrell and Martin, 1989). This approach has been very useful, facilitating the identification and mapping of many pathways (Giancotti and Ruoslahti, 1999). This body of work has also contributed significantly to our understanding of signaling networks, as integrin signals are tightly interweaved with signals from other kinds of receptors (Schwartz and Baron, 1999). Nevertheless, the time has come to discuss the limitations of this approach and how and why we should move to the next phase of investigation.

It may be useful to first briefly summarize the basic principles that have emerged from studies with simplified matrices. First, integrin signals are intimately tied to organization of the actin cytoskeleton. Integrins contribute to assembly of cytoskeletal structures via both physical protein–protein interactions and via regulation of signaling pathways such as Rho family GTPases (Schoenwaelder and Burridge, 1999). Conversely, assembly of cytoskeletal structures such as focal adhesions and focal complexes strongly influences integrin signaling, and disruption of these structures (for example with cytochalasin or Rho inhibitors) strongly inhibits many integrin-dependent signals (Schoenwaelder and Burridge, 1999). Second, receptor clustering is a major stimulus for signaling (Schwartz, 1992). Occupancy obviously contributes as well, but the potent effects of clustering alone argue that spatial organization of receptors in the membrane is critical for signal transduction. Third, most integrins share a common set of core signaling pathways that promote cytoskeletal organization, growth, and survival. Additionally, different integrins (or subsets thereof) are able to regulate unique pathways that can give rise to effects on cell growth, survival, and gene expression that are specific to individual ECM proteins (Schwartz, 2001). Fourth, integrin-mediated adhesion is required for transmission of

13

many signals initiated by other receptors (e.g., growth factor, cytokine or antigen receptors) (Schwartz and Baron, 1999). These effects may involve cytoskeletal organization but in many cases are clearly due to signaling events (i.e., activation of kinases or GTPases) triggered by the integrins.

Despite a growing appreciation that signaling pathways are organized into complex networks, there is a surprising paucity of detailed information about how information is conveyed (Pawson and Saxton, 1999). Different stimuli that induce distinct responses flow into a limited set of common pathways. Even worse, distinct cytoplasmic pathways elicit very similar responses at the level of gene expression or function (Fambrough *et al.,* 1999; Tallquist *et al.,* 2000), leading biologists to argue that quantitative, spatial, and temporal aspects are critical. Examples of all three of these can be produced, supporting the importance of these aspects of signaling. But these are surprisingly few, leaving the origin of specificity unknown in most cases.

The ECM field occupies a parallel universe in which cell function is potently regulated by adhesion to ECM. In addition to the global effects of ECM in controlling growth and survival, distinct regulatory effects of different ECM proteins have been cataloged in many systems (Adams and Watt, 1993). Intriguing effects of matrix assembly on cell responses have also been identified (Bourdoulous *et al.,* 1998; Sechler and Schwarzbauer, 1998). As discussed earlier signaling pathways regulated by integrins have been extensively characterized, and other ECM receptors such as transmembrane proteoglycans also exert regulatory effects via cytoplasmic signaling pathways (Rapraeger, 2001).

A related area that has advanced considerably concerns structural studies of the ECM. For example, most ECM proteins are large and contain multiple domains. Significant progress has been made toward resolving the three-dimensional structure of individual domains (Leahy, 1997) and toward the assembly of simpler matrices (Schwarzbauer and Sechler, 1999; Yurchenco and O'Rear, 1995). The cell surface receptors for some multidomain ECM proteins (fibronectin or thrombospondin-1, for example) have been identified and include integrins, proteoglycans, and other transmembrane proteins (Rapraeger, 2001; Brown and Frazier, 2001; Damsky and Werb, 1992). These receptors show distinct effects on cell function and interesting cases of combinatorial effects, suggesting crosstalk among different receptors for single ECM proteins (Saoncella *et al.,* 1999; Zhou and Brown, 1993; Tunggal *et al.,* 2000). Although much remains to be accomplished, these areas are proceeding nicely.

But I cannot avoid the conclusion that there is a large gap in our knowledge. It appears to me that workers in the ECM field who appreciate the complexity and structure of three-dimensional matrices have little commerce with the signaling field, despite the recognition that signaling networks show complex spatial organization critical to signal transduction and cellular responses. Conversely, those of us working on signaling have not begun to seriously consider how complex assemblies of ECM proteins contribute to organization of signaling networks.

This idea is scarcely novel. I recall discussions in the late 1980s about how multidomain ECM proteins might bring about co-clustering of different receptors, leading to generation of specific signals. These ideas appeared in reviews (Damsky and Werb, 1992) and were commonly invoked at conferences, especially when beer was served. What is new is the means to carry out the program. I believe that the signaling field has advanced to the point that well designed experiments can be carried out to probe the effects of different matrices or differently organized matrices on cell signaling. Techniques to localize signaling events within cells are available, for example, using anti-phospho epitope antibodies to stain cells (Sells *et al.*, 2000; Sieg *et al.*, 2000) or fluorescence-based *in vivo* assays (Kraynov *et al.*, 2000; Mochizuki *et al.*, 2001).

This approach offers mutual benefits. Signaling studies will lend a functional side to structural studies of multidomain ECM proteins and their assembly into matrices. Conversely, analyses of organized ECM may provide critical information about the factors that govern spatial organization of signaling networks. This synergistic approach may even contribute to the elucidation of how the spatial and temporal characteristics of signaling networks are critical to their function.

In short, we face the growing recognition that both ECM and signaling networks exhibit considerable spatial organization, but the relationship between these two has not been much explored. A major barrier to progress has been the absence of information and techniques to assay localized signaling, but recent advances have diminished this problem. The remaining barrier is the difficulty of crossing disciplinary boundaries, of learning new fields in detail and making new friends.

Acknowledgments

Writing this article was supported by grants from the US Public Health Service. I thank Bette Cessna for secretarial assistance.

References

Adams, J. C., and Watt, F. M. (1993). Regulation of development and differentiation by the extracellular matrix. *Development* **117,** 1183–1198.

Bourdoulous, S., Orend, G., Mackenna, D. A., Pasqualini, R., and Ruoslahti, E. (1998). Fibronectin matrix regulates activation of Rho and Cdc42 GTPases and cell cycle progression. *J. Cell Biol.* **143,** 267–276.

Brown, E. J., and Frazier, W. A. (2001). Integrin-associated protein (CD47) and its ligands. *Trends Cell Biol.* **11,** 130–135.

Damsky, C. H., and Werb, Z. (1992). Signal transduction by integrin receptors for extracellular matrix: cooperative processing of extracellular information. *Curr. Opin. Cell Biol.* **4,** 772–781.

Fambrough, D. *et al.* (1999). Diverse signaling pathways activated by growth factor receptors induce broadly overlapping, rather than independent, sets of genes. *Cell* **97,** 727–741.

Ferrell, G. E., and Martin, G. S. (1989). Tyrosine-specific phosphorylation is regulated by glycoprotein IIb/IIIa in platelets. *Proc. Natl. Acad. Sci. USA* **86,** 2234–2238.

Giancotti, F. G., and Roslahti, E. (1999). Integrin signaling. *Science* **285,** 1028–1032.

Guan, J.-L., Trevithick, J. E., and Hynes, R. O. (1991). Fibronectin/integrin interaction induces tyrosine phosphorylation of a 120-kD protein. *Cell Regul.* **2,** 951–964.

Kornberg, L. J. *et al.* (1991). Signal transduction by integrins: increased protein tyrosine phosphorylation caused by integrin clustering. *Proc. Natl Acad. Sci. USA* **88,** 8392–8396.

Kraynov, V. S., Chamberlain, C., Bokoch, G. M., Schwartz, M. A., Slaubaugh, S., and Hahn, K. M. (2000). Localized Rac activation dynamics visualized in living cells. *Science* **290,** 333–337.

Leahy, D. J. (1997). Implications of atomic-resolution structures for cell adhesion. *Annu. Rev. Cell Dev. Biol.* **13,** 363–393.

Mochizuki, N., Yamashita, A. S., Kurokawa, K., Ohbe, Y., Nagai, T., Miyawaki, A., and Matsuda, M. (2001). Spatio-temporal images of growth factor induced activation of Ras and Rap1. *Nature* **411,** 1065–1068.

Pawson, T., and Saxton, T. M. (1999). Signaling networks—Do all roads lead to the same genes? *Cell* **97,** 675–678.

Rapraeger, A. C. (2001). Molecular interactions of syndecans during development. *Semin. Cell Dev. Biol.* **12,** 107–116.

Saoncella, S. *et al.* (1999). Syndecan-4 signals cooperatively with integrins in a Rho-dependent manner in the assembly of focal adhesions and actin stress fibers. *Proc. Natl. Acad. Sci. USA* **96,** 2805–2810.

Schoenwaelder, S. M., and Burridge, K. (1999). Bidirectional signaling between the cytoskeleton and integrins. *Curr. Opin. Cell Biol.* **11,** 274–286.

Schwartz, M. A. (1992). Signaling by integrins. *Trends Cell Biol.* **2,** 304–308.

Schwartz, M. A. (2001). Integrins and cell proliferation: regulation of cyclin-dependent kinases via cytoplasmic signaling pathways. *J. Cell Sci.* **114,** 2553–2560.

Schwartz, M. A., and Baron, V. (1999). Interactions between mitogenic stimuli, or, a thousand and one connections. *Curr. Opin. Cell Biol.* **11,** 197–202.

Schwartz, M. A., Both, G., and Lechene, C. (1989). The effect of cell spreading on cytoplasmic pH in normal and transformed fibroblasts. *Proc. Natl. Acad. Sci. USA* **86,** 4525–4529.

Schwarzbauer, J. E., and Sechler, J. L. (1999). Fibronectin fibrillogenesis: a paradigm for extracellular matrix assembly. *Curr. Opin. Cell Biol.* **11,** 622–627.

Sechler, J. L., and Schwarzbauer, J. E. (1998). Control of cell cycle progression by fibronectin matrix architecture. *J. Biol. Chem.* **273,** 25533–25536.

Sells, M. A., Pfaff, A., and Chernoff, J. (2000). Temporal and spatial distribution of activated PAK1 in fibroblasts. *J. Cell Biol.* **151,** 1449–1457.

Sieg, D. J. *et al.* (2000). FAK integrates growth factor and integrin signals to promote cell migration. *Nat. Cell Biol.* **2,** 249–256.

Tallquist *et al.* (2000). Retention of PDGFR-β function in mice in the absence of phosphatidylinositol 3'-kinase and phospholipase Cγ signaling pathways. *Genes Dev.* **14,** 3179–3190.

Tunggal, P., Smyth, N., Paulsson, M., and Ott, M. C. (2000). Laminins: structure and genetic regulation. *Microsc. Res. Tech.* **51,** 214–227.

Yurchenco, P. D., and O'Rear, J. J. (1995). Basal lamina assembly. *Curr. Opin. Cell Biol.* **6,** 674–681.

Zhou, M., and Brown, E. J. (1993). Leukocyte response integrin and integrin-associated protein act as a signal transduction unit in generation of a phagocyte respiratory burst. *J. Exp. Med.* **172,** 1165–1174.

Cell–Matrix Adhesion Research and the Development of Biotherapeutics

Roy R. Lobb

Biogen, Inc.
Cambridge, Massachusetts 02142

The short history of cell/matrix research and its application to biotherapeutics has been dominated by integrin adhesion. Two examples illustrate how fundamental research into integrin/matrix interactions has provided important insights into disease, leading both to early drug candidates and to increasingly sophisticated approaches to drug development.

Integrin $\alpha V \beta 3$ recognizes multiple RGD-containing ECM proteins, including vitronectin, fibronectin, and osteopontin (Eliceiri and Cheresh, 1999). Early data suggested key roles for this integrin in both bone resorption (Ross *et al.*, 1993) and angiogenesis (Brooks *et al.*, 1994; Eliceiri and Cheresh, 1999). There was rapid translation of these early observations to a clinical drug candidate, a humanized form of the $\alpha V \beta 3$-specific mAb LM609 (Gutheil *et al.*, 2000), in parallel with significant efforts, still ongoing, by biopharmaceutical companies to develop small molecule antagonists (see, e.g., Engleman *et al.*, 1997).

Deletion of genes for αV or $\beta 3$ in mice, while confirming the central role of $\alpha V \beta 3$ in osteoclast function (Hodivala-Dilke *et al.*, 1999), produced only highly selective defects in the vascular tree (Bader *et al.*, 1998; Hodivala-Dilke *et al.*, 1999), raising questions about the broad relevance of this integrin to angiogenesis and its validity as a target for angiogenic diseases. However, several years of effort by multiple laboratories has contributed to a much more sophisticated understanding of endothelial cell $\alpha V \beta 3$ biology: $\alpha V \beta 3$-dependent angiogenesis is linked to certain growth factors, such as FGF, and not others (Eleceiri and Cheresh, 1999); the integrin $\alpha 5 \beta 1$/fibronectin interaction, which plays a central role in angiogenesis (Kim *et al.*, 2000a), is upstream of $\alpha V \beta 3$-dependent function in endothelium (Kim *et al.*, 2000b); MMP2 associates with $\alpha V \beta 3$ via its hemopexin domain, and modulates its angiogenic activity (Brooks *et al.*, 1998); pro-MMP2 is activated by the membrane-anchored protease MT1-MMP (Sato *et al.*, 1994), which can cluster with $\alpha V \beta 3$ and MMP2, accelerating locomotion and invasion (Deryugina *et al.*, 2001); the MMP2/$\alpha V \beta 3$ interaction can be inhibited by small molecules which inhibit neither the enzymatic activity of MMP2 nor the adhesive activity of $\alpha V\beta 3$, but which nevertheless can inhibit angiogenesis (Silletti *et al.,* 2001), and lack of $\alpha V\beta 3$ ligation leads to direct activation of cellular apoptosis through a novel caspase 8-dependent mechanism (Stupack *et al.,* 2001).

Our current understanding of $\alpha V\beta 3$ biology now provides multiple mAb and small molecule targets for drug development, including $\alpha V\beta 3$ itself, integrin $\alpha 5\beta 1$, MMP2, the interaction between MMP2 and $\alpha V\beta 3$, MT1-MMP, and multiple intracellular targets involved in integrin signaling and apoptosis. MAbs and small molecules directed to $\alpha V\beta 3$ itself will validate, if successful, the importance of this pathway in human pathology, as well as provide an impetus for second generation efforts aimed towards the other targets, with likely benefits to the patient in terms of convenience, efficacy, price, or safety.

The leukocyte integrin VLA-4 was found to mediate adhesion to both an alternately spliced form of fibronectin and the endothelial cell surface molecule VCAM-1 (reviewed in Lobb and Hemler, 1994). Studies with mAbs in animal models of autoimmune and allergic diseases suggested an important role for VLA-4 in human pathology (Lobb and Hemler, 1994). A humanized mAb to VLA-4 was moved into clinical trials, and following initial clinical promise (Tubridy *et al.,* 1999), has shown efficacy in large well-controlled phase II studies in both multiple sclerosis and Crohn's disease (Ghosh *et al.,* 2001). As with integrin $\alpha V\beta 3$, there was a rapid translation of early observations to a clinical drug candidate, and in parallel, initiation of significant efforts by biopharmaceutical companies to develop small molecule antagonists (see, e.g., Abraham *et al.,* 2000).

Our understanding of the biology of VLA-4 has also increased in sophistication in recent years. For example, fibronectin (Fn) can be expressed on surface of endothelium in functional form and, like VCAM-1, can bind leukocyte VLA-4 (Shih *et al.,* 1999); many new VLA-4 ligands have been described, including osteopontin (Bayless *et al.,* 1998), uPAR (Tarui *et al.,* 2001), transglutaminases (Takahashi *et al.,* 2000), and EDA-fibronectin (Liao *et al.,* 2000); the cytoplasmic tail of the α-4 chain associates directly and selectively with paxillin, providing a novel small-molecule target (Liu *et al.,* 1999); and T cell adhesion to VCAM-1 induces MMP expression (Yakubenko *et al.,* 2000). Finally, studies have shown that the VLA-4 adhesive interaction is subverted by multiple B cell malignancies, allowing their growth and resistance to chemotherapy within the bone marrow microenvironment, and suggesting VLA-4 blockade may be of therapeutic benefit in multiple myeloma, B-ALL, and related malignancies (Damiano *et al.,* 1999; Michigami *et al.,* 2000; Mudry *et al.,* 2000).

These two examples define a paradigm for drug development in this field. Insight into a molecule or pathway of cell–matrix adhesion suggests a plausible link to human pathology; several years of intensive research "validate" the target *in vivo,* usually through rodent experiments with blocking mAbs or insights from transgenic or knockout animals, or both; a mAb is moved into early clinical trials, to serve as a "proof of concept" for the pathway in humans and possibly as a viable product candidate in its own right; if appropriate as a small-molecule target, either peptidomimetic approaches using peptides isolated from matrix sequences as "leads" or high-throughput screening generate small-molecule drug candidates; and finally, intensive study by many groups leads to a broader and more subtle understanding of the pathway and its biology, defining new approaches to antagonism as well as novel clinical applications. Although this is illustrated with integrin/matrix interactions, this paradigm is applicable to almost any cell/matrix pathway.

Many novel cell/matrix pathways are under intensive study, and we can already predict that new approaches to drug development will occur from these studies in multiple fields. These include:

• Angiogenesis, where cell–matrix interactions are already having a particularly profound impact. Multiple novel angiogenic and antiangiogenic activities are associated with matrix molecules, their proteolytically derived matrix fragments, or secreted proteins that associate with matrix. Examples include thrombospondin (Good *et al.*, 1990), the C-terminal fragment of collagen XVIII, endostatin (O'Reilly *et al.*, 1997), fragments of the collagen IV NC1 domain (Maeshima *et al.*, 2000; Petitclerc *et al.*, 2000), and Cyr61 (Jedsadayanmata *et al.*, 1999).

• Oncology, where the host–tumor interface is receiving ever increasing attention as an important field of research (Park *et al.*, 2000; Matrisian *et al.*, 2001). Examples include appreciation of the role of cell adhesion in resistance to chemotherapy (St. Croix and Kerbel, 1997; Damiano *et al.*, 1999; Sethi *et al.*, 1999); the direct role in tumorigenesis of carcinoma-associated stromal cells (Olumi *et al.*, 1999; Zucker *et al.*, 2001); the role of proteolyzed matrix in tumor cell invasion (Giannelli *et al.*, 1997; Davis *et al.*, 2000); and the use of tumor-associated matrix for targeting and diagnostic imaging (Nilsson *et al.*, 2001, and references therein).

• Fibrosis, where the fibroblast/myofibroblast transition and resulting collagen secretion in response to TGF-β is regulated by matrix molecules (Serini *et al.*, 1998), as is TGF-β activation itself (Crawford *et al.*, 1998; Munger *et al.*, 1999).

• Immunology, where the interaction of leukocytes with matrix is proving central to immune responses. The collagen-binding integrins VLA-1 and VLA-2, once thought merely to allow cells to adhere to a major structural component of the ECM, are now understood to play important roles in the immune response (de Fougerolles *et al.*, 2000a). Moreover, adhesion through these integrins initiates major changes in gene expression, proliferative status, and sensitivity to apoptosis (de Fougerolles *et al.*, 2000b; Aoudjit and Vuori, 2000). Importantly, recent studies show that a significant portion of memory/effector T cells reside outside of lymphoid organs and in peripheral tissue, suggesting a major role for leukocyte/matrix interactions in the homeostatic control of the immune system (Masopust *et al.*, 2001; Reinhardt *et al.*, 2001).

Based on our current knowledge, it is reasonable to assume that aberrant or subverted cell–matrix interactions will underlie many human pathologies. Unfortunately the time frames for clinical development are long, often well over a decade, before research insights are translated into marketed drugs. Nevertheless, over the past decade the symbiotic interaction between academic and biopharmaceutical research has begun to validate these fundamental insights, through the generation of drug candidates now showing efficacy in clinical trials. While integrin–matrix interactions have been at the forefront of the biotheraputic approaches to date, our rapidly increasing and sophisticated understanding of basic molecular processes such as matrix assembly, the formation of supramolecular structures, proteoglycan structure and function, matrix-dependent signaling, and cell/matrix contact structures (Huang *et al.,* 1998; Woods and Couchman, 1998; Schwarzbauer and Sechler, 1999; Colognato and Yurchenko, 2000; Adams, 2001; Rapraeger, 2001), combined with access to the sequences of multiple genomes (Rubin *et al.,* 2000) and novel research tools allowing highly systematic approaches (St. Croix, *et al.,* 2000; de Fougerolles *et al.,* 2000b), suggest that we can expect a truly fundamental impact of cell–matrix research on human disease in the near future.

References

Abraham, W. M., Gill, A., Ahmed, A., Sielczak, M. W., Lauredo, I. T., Botinnikova, Y., Lin, K. C., Pepinsky B., Leone, D. R., Lobb, R. R., and Adams, S. P. (2000). A small-molecule, tight-binding inhibitor of the integrin alpha(4)beta(1) blocks antigen-induced airway responses and inflammation in experimental asthma in sheep. *Am. J. Respir. Crit. Care Med.* **162,** 603–611.

Adams, J. C. (2001). Cell-matrix contact structures. *Cell. Mol. Life Sci.* **58,** 371–392.

Aoudjit, F., and Vuori, K. (2000). Engagement of the alpha2beta1 integrin inhibits Fas ligand expression and activation-induced cell death in T cells in a focal adhesion kinase-dependent manner. *Blood* **95,** 2044–2051.

Bader, B. L., Rayburn, H., Crowley, D., and Hynes, R. O. (1998). Extensive vasculogenesis, angiogenesis, and organogenesis precede lethality in mice lacking all alpha v integrins. *Cell* **95,** 507–519.

Bayless, K. J., Meininger, G. A., Scholtz, J. M., and Davis, G. E. (1998). Osteopontin is a ligand for the alpha4beta1 integrin. *J. Cell Sci.* **111,** 1165–1174.

Brooks, P. C., Clark, R. A. F., and Cheresh, D. A. (1994). Requirement for vascular integrin $\alpha V\beta 3$ for angiogenesis. *Science* **264,** 569–571.

Brooks, P. C., Silletti, S., von Schalscha, T. L., Friedlander, M., and Cheresh, D. A. (1998). Disruption of angiogenesis by PEX, a noncatalytic metalloproteinase fragment with integrin binding activity. *Cell* **92,** 391–400.

Colognato, H., and Yurchenco, P. D. (2000). Form and function: the laminin family of heterotrimers. *Dev. Dyn.* **218,** 213–234.

Crawford, S. E., Stellmach, V., Murphy-Ullrich, J. E., Ribeiro, S. M., Lawler, J., Hynes, R. O., Boivin, G. P., and Bouck, N. (1998). Thrombospondin-1 is a major activator of TGF-beta1 in vivo. *Cell* **93,** 1159–1170.

Damiano, J. S., Cress, A. E., Hazlehurst, L. A., Shtil, A. A., and Dalton, W. S. (1999). Cell adhesion mediated drug resistance (CAM-DR): role of integrins and resistance to apoptosis in human myeloma cell lines. *Blood* **93,** 1658–1667.

Davis, G. E., Bayless, K. J., Davis, M. J., and Meininger, G. A. (2000). Regulation of tissue injury responses by the exposure of matricryptic sites within extracellular matrix molecules. *Am. J. Pathol.* **156,** 1489–1498.

de Fougerolles, A. R., Sprague, A. G., Nickerson-Nutter, C. L., Chi-Rosso, G., Rennert, P. D., Gardner, H., Gotwals, P. J., Lobb, R. R., and Koteliansky, V. E. (2000a). Regulation of inflammation by collagen-binding integrins alpha1beta1 and alpha2beta1 in models of hypersensitivity and arthritis. *J. Clin. Invest.* **105,** 721–729.

de Fougerolles, A. R., Chi-Rosso, G., Bajardi, A., Gotwals, P., Green, C. D., and Koteliansky, V. E. (2000b). Global expression analysis of extracellular matrix-integrin interactions in monocytes. *Immunity* **13,** 749–758.

Deryugina, E. I., Ratnikov, B., Monosov, E., Postnova, T. I., DiScipio, R., Smith, J. W., and Strongin, A. Y. (2001). MT1-MMP initiates activation of pro-MMP-2 and integrin alphavbeta3 promotes maturation of MMP-2 in breast carcinoma cells. *Exp. Cell Res.* **263,** 209-223.

Eliceiri, B. P., and Cheresh, D. A. (1999). The role of alpha-V integrins during angiogenesis: insights into potential mechanisms of action and clinical development. *J. Clin. Invest.* **103,** 1227–1230.

Engleman, V. W., Nickols, G. A., Ross, F. P., Horton, M. A., Griggs, D. W., Settle, S. L., Ruminski, P. G., and Teitelbaum, S. L. (1997). A peptidomimetic antagonist of the alpha(v)beta3 integrin inhibits bone resorption in vitro and prevents osteoporosis in vivo. *J. Clin. Invest.* **99,** 2284–2292.

Ghosh, S., Goldin, E., Malchow, H. A., Pounder, R. E., Rask-Madsen, J., Rutgeerts, P., Vyhnalek, P., Zadorova, Z., Palmer, T., and Donoghue, S. (2001). A randomised, placebo-controlled, pan-European study of a recombinant antibody to $\alpha 4$ integrin (Antegren^{TM}) in moderate to severely active Crohn's disease. *Gastroenterology* **120**(suppl. 2), A 682.

Giannelli, G., Falk-Marzillier, J., Schiraldi, O., Stetler-Stevenson, W. G., and Quaranta, V. (1997). Induction of cell migration by matrix metalloprotease-2 cleavage of laminin-5. *Science* **277,** 225–228.

Good, D. J., Polverini, P. J., Rastinejad, F., Le Beau, M. M., Lemons, R. S., Frazier, W. A., and Bouck, N. P. (1990). A tumor suppressor-dependent inhibitor of angiogenesis is immunologically and functionally indistinguishable from a fragment of thrombospondin. *Proc. Natl. Acad. Sci. USA* **87,** 6624–6628.

Gutheil, J. C., Campbell, T. N., Pierce, P. R., Watkins, J. D., Huse, W. D., Bodkin, D. J., and Cheresh, D. A. (2000). Targeted antiangiogenic therapy for cancer using Vitaxin: a humanized monoclonal antibody to the integrin alphavbeta3. *Clin. Cancer Res.* **6,** 3056–3061.

Hodivala-Dilke, K. M., McHugh, K. P., Tsakiris, D. A., Rayburn, H., Crowley, D., Ullman-Cullere, M., Ross, F. P., Coller, B. S., Teitelbaum, S., and Hynes, R. O. (1999). Beta3-integrin-deficient mice are a model for Glanzmann thrombasthenia showing placental defects and reduced survival. *J. Clin. Invest.* **103,** 229–238.

Huang, S., Chen, C. S., and Ingber, D. E. (1998). Control of cyclin D1, p27(Kip1), and cell cycle progression in human capillary endothelial cells by cell shape and cytoskeletal tension. *Mol. Biol. Cell* **9,** 3179–3193.

Jedsadayanmata, A., Chen, C. C., Kireeva, M. L., Lau, L. F., and Lam, S. C. (1999). Activation-dependent adhesion of human platelets to Cyr61 and Fisp12/mouse connective tissue growth factor is mediated through integrin alpha(IIb)beta(3). *J. Biol. Chem.* **274,** 24321–24327.

Kim, S., Bell, K., Mousa, S. A., and Varner, J. A. (2000a). Regulation of angiogenesis in vivo by ligation of integrin alpha5beta1 with the central cell-binding domain of fibronectin. *Am. J. Pathol.* **156,** 1345–1362.

Kim, S., Harris, M., and Varner, J. A. (2000b). Regulation of integrin alpha v beta 3-mediated endothelial cell migration and angiogenesis by integrin alpha5beta1 and protein kinase A. *J. Biol. Chem.* **275,** 33920–33928.

Liao, Y., Gotwals, P., Koteliansky, V., Sheppard, D., and Van de Water, L. (2000). The EIIIA (EDA) segment of fibronectin is a novel ligand for integrins a4b1 and a9b1. *Mol. Biol. Cell* **11,** 394A.

Liu, S., Thomas, S. M., Woodside, D. G., Rose, D. M., Kiosses, W. B., Pfaff, M., and Ginsberg, M. H. (1999). Binding of paxillin to alpha4 integrins modifies integrin-dependent biological responses. *Nature* **402,** 676–681.

Lobb, R. R., and Hemler, M. E. (1994). The pathophysiologic role of alpha 4 integrins in vivo. *J. Clin. Invest.* **94,** 1722–1728.

Maeshima, Y., Colorado, P. C., Torre, A., Holthaus, K. A., Grunkemeyer, J. A., Ericksen, M. B., Hopfer, H., Xiao, Y., Stillman, I. E., and Kalluri, R. (2000). Distinct antitumor properties of a type IV collagen domain derived from basement membrane. *J. Biol. Chem.* **275,** 21340–21348.

Masopust, D., Vezys, V., Marzo, A. L., and Lefrancois, L. (2001). Preferential localization of effector memory cells in nonlymphoid tissue. *Science* **291,** 2413–2417.

Matrisian, L. M., Cunha, G. R., and Mohla, S. (2001). Epithelial-stromal interactions and tumor progression: meeting summary and future directions. *Cancer Res.* **61,** 3844–3846.

Michigami, T., Shimizu, N., Williams, P. J., Niewolna, M., Dallas, S. L., Mundy, G. R., and Yoneda, T. (2000). Cell–cell contact between marrow stromal cells and myeloma cells via VCAM-1 and alpha(4)beta(1)-integrin enhances production of osteoclast-stimulating activity. *Blood* **96,** 1953–1960.

Mudry, R. E., Fortney, J. E., York, T., Hall, B. M., and Gibson, L. F. (2000). Stromal cells regulate survival of B-lineage leukemic cells during chemotherapy. *Blood* **96,** 1926–1932.

Munger, J. S., Huang, X., Kawakatsu, H., Griffiths, M. J., Dalton, S. L., Wu, J., Pittet, J. F., Kaminski, N., Garat, C., Matthay, M. A., Rifkin, D. B., and Sheppard, D. (1999). The integrin alpha v beta 6 binds and activates latent TGF beta 1: a mechanism for regulating pulmonary inflammation and fibrosis. *Cell* **96,** 319–328.

Nilsson, F., Kosmehl, H., Zardi, L., and Neri, D. (2001). Targeted delivery of tissue factor to the ED-B domain of fibronectin, a marker of angiogenesis, mediates the infarction of solid tumors in mice. *Cancer Res.* **61,** 711–716.

Olumi, A. F., Grossfeld, G. D., Hayward, S. W., Carroll, P. R., Tlsty, T. D., and Cunha, G. R. (1999). Carcinoma-associated fibroblasts direct tumor progression of initiated human prostatic epithelium. *Cancer Res.* **59,** 5002–5011.

O'Reilly, M. S., Boehm, T., Shing, Y., Fukai, N., Vasios, G., Lane, W. S., Flynn, E., Birkhead, J. R., Olsen, B. R., and Folkman, J. (1997). Endostatin: an endogenous inhibitor of angiogenesis and tumor growth. *Cell* **88,** 277–285.

Park, C. C., Bissell, M. J., and Barcellos-Hoff, M. H. (2000). The influence of the microenvironment on the malignant phenotype. *Mol. Med. Today* **6,** 324–329.

Petitclerc, E., Boutaud, A., Prestayko, A., Xu, J., Sado, Y., Ninomiya, Y., Sarras, M. P. Jr, Hudson, B. G., and Brooks, P. C. (2000). New functions for non-collagenous domains of human collagen type IV. Novel integrin ligands inhibiting angiogenesis and tumor growth in vivo. *J. Biol. Chem.* **275,** 8051–8061.

Rapraeger, A. C. (2001). Molecular interactions of syndecans during development. *Semin. Cell Dev. Biol.* **12,** 107–116.

Reinhardt, R. L., Khoruts, A., Merica, R., Zell, T., and Jenkins, M. K. (2001). Visualizing the generation of memory CD4 T cells in the whole body. *Nature* **410,** 101–105.

Ross, F. P., Alvarez, J. I., Chappel, J., Sander, D., Butler, W. T., Farach-Carson, W. C., Mintz, K. A., Robey, P. G., Teitelbaum, S. L., and Cheresh, D. A. (1993). Interactions between the bone matrix proteins osteopontin and bone sialoprotein and the osteoclast integrin aVb3 potentiate bone resorption. *J. Biol. Chem.* **268,** 9901–9907.

Rubin, G. M., Yandell, M. D., Wortman, J. R., Gabor Miklos, G. L., Nelson, C. R., Hariharan, I. K., Fortini, M. E., Li, P. W., Apweiler, R., Fleischmann, W., Cherry, J. M., Henikoff, S., Skupski, M. P., Misra, S., Ashburner, M., Birney, E., Boguski, M. S., Brody, T., Brokstein, P., Celniker, S. E., Chervitz, S. A., Coates, D., Cravchik, A., Gabrielian, A., Galle, R. F., Gelbart, W. M., George, R. A., Goldstein, L. S., Gong, F., Guan, P., Harris, N. L., Hay, B. A., Hoskins, R. A., Li, J., Li, Z., Hynes, R. O., Jones, S. J., Kuehl, P. M., Lemaitre, B., Littleton, J. T., Morrison, D. K., Mungall, C., O'Farrell, P. H., Pickeral, O. K., Shue, C., Vosshall, L. B., Zhang, J., Zhao, Q., Zheng, X. H., and Lewis, S. (2000). Comparative genomics of the eukaryotes. *Science* **287,** 2204–2215.

Sato, H., Takino, T., Okada, Y., Cao, J., Shinagawa, A., Yamamoto, E., and Seiki, M. (1994). A matrix metalloproteinase expressed on the surface of invasive tumour cells. *Nature* **370,** 61–65.

Schwarzbauer, J. E., and Sechler, J. L. (1999). Fibronectin fibrillogenesis: a paradigm for extracellular matrix assembly. *Curr. Opin. Cell Biol.* **11,** 622–627.

Serini, G., Bochaton-Piallat, M. L., Ropraz, P., Geinoz, A., Borsi, L., Zardi, L., and Gabbiani, G. (1998). The fibronectin domain ED-A is crucial for myofibroblastic phenotype induction by transforming growth factor-beta1. *J. Cell Biol.* **142,** 873–881.

Sethi, T., Rintoul, R. C., Moore, S. M., MacKinnon, A. C., Salter, D., Choo, C., Chilvers, E. R., Dransfield, I., Donnelly, S. C., Strieter, R., and Haslett, C. (1999). Extracellular matrix proteins protect small cell lung cancer cells against apoptosis: a mechanism for small cell lung cancer growth and drug resistance in vivo. *Nat. Med.* **5,** 662–668.

Shih, P. T., Elices, M. J., Fang, Z. T., Ugarova, T. P., Strahl, D., Territo, M. C., Frank, J. S., Kovach, N. L., Cabanas, C., Berliner, J. A., and Vora, D. K. (1999). Minimally modified low-density lipoprotein induces monocyte adhesion to endothelial connecting segment-1 by activating beta1 integrin. *J. Clin. Invest.* **103,** 613–625.

Silletti, S., Kessler, T., Goldberg, J., Boger, D. L., and Cheresh, D. A. (2001). Disruption of matrix metalloproteinase 2 binding to integrin alpha vbeta 3 by an organic molecule inhibits angiogenesis and tumor growth in vivo. *Proc. Natl. Acad. Sci. USA* **98,** 119–124.

St. Croix, B., and Kerbel, R. S. (1997). Cell adhesion and drug resistance in cancer. *Curr. Opin. Oncol.* **9,** 549–556.

St. Croix, B., Rago, C., Velculescu, V., Traverso, G., Romans, K. E., Montgomery, E., Lal, A., Riggins, G. J., Lengauer, C., Vogelstein, B., and Kinzler, K. W. (2000). Genes expressed in human tumor endothelium. *Science* **289,** 1197–1202.

Stupack, D. G., Puente, X. S., Boutsaboualoy, S., Storgard, C. M., and Cheresh, D. A. (2001). Apoptosis of adherent cells by recruitment of caspase-8 to unligated integrins. *J. Cell Biol.* **155,** 459–470.

Takahashi, H., Isobe, T., Horibe, S., Takagi, J., Yokosaki, Y., Sheppard, D., and Saito, Y. (2000). Tissue transglutaminase, coagulation factor XIII, and the pro-polypeptide of von Willebrand factor are all ligands for the integrins alpha 9beta 1 and alpha 4beta 1. *J. Biol. Chem.* **275,** 23589–23595.

Tarui, T., Mazar, A. P., Cines, D. B., and Takada, Y. (2001). Urokinase-type plasminogen activator receptor (CD87) is a ligand for integrins and mediates cell-cell interaction. *J. Biol. Chem.* **276,** 3983–3990.

Tubridy, N., Behan, P. O., Capildeo, R., Chaudhuri, A., Forbes, R., Hawkins, C. P., Hughes, R. A., Palace, J., Sharrack, B., Swingler, R., Young, C., Moseley, I. F., MacManus, D. G., Donoghue, S., and Miller, D. H. (1999). The effect of anti-alpha4 integrin antibody on brain lesion activity in MS. The UK Antegren Study Group. *Neurology* **53,** 466–472.

Woods, A., and Couchman, J. R. (1998). Syndecans: synergistic activators of cell adhesion. *Trends Cell Bio.* **8,** 189–192.

Yakubenko, V. P., Lobb, R. R., Plow, E. F., and Ugarova, T. P. (2000). Differential induction of gelatinase B (MMP-9) and gelatinase A (MMP-2) in T lymphocytes upon alpha(4)beta(1)-mediated adhesion to VCAM-1 and the CS-1 peptide of fibronectin. *Exp. Cell Res.* **260,** 73–84.

Zucker, S., Hymowitz, M., Rollo, E. E., Mann, R., Conner, C. E., Cao, J., Foda, H. D., Tompkins, D. C., and Toole, B. P. (2001). Tumorigenic potential of extracellular matrix metalloproteinase inducer. *Am. J. Pathol.* **158,** 1921–1928.

PART II

Matrix Methodologies

CHAPTER 1

Detection and Purification of Instructive Extracellular Matrix Components with Monoclonal Antibody Technologies

Susana G. Gil,*,† Randy O. Sigle,* and William G. Carter‡

*Basic Sciences
Fred Hutchinson Cancer Research Institute
Seattle, Washington 98109

‡Basic Sciences
Fred Hutchinson Cancer Research Institute
Seattle, Washington 98109
and
Department of Pathobiology
University of Washington
Seattle, Washington 98195

†Current address: Department of Medicine, Beth Israel Hospital, Harvard Institute of Medicine, Boston, Massachusetts 02215.

I. Introduction

We describe a technical approach that has been of value in the initial detection (discovery), identification, and purification of adhesive and instructive extracellular matrix (ECM) components, or adhesion receptors in whole cells. Instructive ECM is defined here as an adhesive ECM ligand that promotes a cell response in reporter cells that is not promoted by other adhesive ligands. The reporter cell and its response are defined by the investigator and can include adhesion, migration, cell signals, mRNA transcription, differentiation, or cell death. The approach utilizes *in vitro* assays and monoclonal antibodies (Mabs) to identify instructive ECM. The techniques were developed through trial and error and have been successful when other apparently more direct biochemical approaches were not successful. The approach utilizes three steps in detecting, identifying, and purifying the ECM components with instructive activity:

Detect instructive ECM component in complex ECM
↓
Develop Mabs that bind instructive ECM component
↓
Purification of ECM with Mabs that instructs reporter cells

First, we suggest assays for the detection of instructive activity in complex mixtures of ECM components that promote a selected response in reporter cells. Desirably, the selected response will be unique to the instructive ECM mixture and not duplicated by adhesion of the reporter cells to control ECM components. Second, we describe development of Mabs that bind the instructive ECM component within the complex ECM mixture. Either inhibitory or noninhibitory Mabs are useful in identifying the instructive ECM component. Third, we discuss use of the new Mabs for purification of the instructive ECM and analysis of the instructive pathway in the reporter cells.

We outline both the strengths and weakness of the Mab approach as utilized in this laboratory. We recommend additional reading and protocols relating to the production of Mabs (Delcommenne and Streuli, 1999; Galfre and Milstein, 1981; Oi and Herzenberg, 1980; Sanchez-Madrid and Springer, 1986; Taggart and Samloff, 1983) and the many excellent published techniques for purification of ECM components (Miller and Rhodes, 1982; Timpl and Aumailley, 1989; Timpl *et al.,* 1987). Methodologies for the expression of recombinant forms of ECM and basement membrane (BM) components are described elsewhere in this volume.

A. Why Purification of ECM Components Is a Problem

ECM components commonly have both structural and instructive functions for cells with which they interact. The individual components of the ECM are usually synthesized and secreted as soluble homo- or heteromultimers and then assembled into insoluble, higher order macromolecular complexes outside the cell. The structural role of ECM components necessitates the extracellular insolubility of the ECM. The macromolecular nature of insoluble ECM promotes multiple interactions with cells that are instructive but are usually of low affinity, requiring high multiplicity. The high repeat number of instructive ECM components facilitates the cross-linking of integrin adhesion receptors and other receptors required for cell signaling. Finally, individual components of ECM including laminins, collagens, fibronectins, matricellular proteins (thrombospondins, tenascins, SPARC, etc.), and proteoglycans are each large and heterogeneous, with many isoforms present in any one tissue or ECM compartment. The isoforms may result from different gene products, mRNA splice variants, or combinations of individual protein chains that are assembled into homo- and heteromultimers.

Classical biochemical purification approaches, when applied to ECM, usually involve disruption of the insoluble macromolecular structure in order to generate soluble components, complexes, or fragments. Solubilization utilizes detergents, chaotropes, pH extremes, chemical modification, or proteolytic cleavage as approaches to fragment macromolecular complexes (Miller and Rhodes, 1982; Timpl and Aumailley, 1989; Timpl *et al.,* 1987). The soluble fragments are frequently large and heterogeneous and/or modified by the solubilization methods. As a result, subsequent biochemical fractionation has insufficient resolving power to distinguish the many different isoforms of laminins, collagen, fibronectins, and other ECM components. Arguably, the single biggest problem with these solubilization and purification approaches is that the structural and instructive function of any one ECM component is due to both the number of copies in the complex and its interaction with other components in the complex, both of which are disrupted by the solubilization.

In contrast to biochemical approaches, monoclonal antibodies (Mabs) and molecular approaches (George and Hynes, 1994; Pytela *et al.,* 1994) attempt to block the structural or instructive roles of individual ECM components while still within the context of the insoluble ECM complex. The Mabs bind to, and identify, the individual ECM component and may inhibit one or more functions of the instructive ECM within the insoluble ECM complex. We have used the Mab approach for the identification and inhibition of adhesion receptors within the complex of the plasma membrane as well as for adhesion components within complex ECM. The Mabs provide tools for the affinity purification of the instructive ECM component and the identification of the subdomain within the instructive ECM component that possesses the instructive function. Thus, the Mab approach adds to the resolving power of biochemical approaches, particularly for distinguishing closely related homologous subunits of integrins or chains of laminin isoforms. On the negative side, the Mab approach can be labor intensive and is dependent on the immunogenicity of the instructive ECM component. Therefore, the approach is not foolproof, but it has been more successful in our hands than any other approach.

B. Example of the Mab Approach: Laminin 5 Instructs Reporter Keratinocytes to Adhere and Assemble Cell Junctions

Here, we will provide an example of the Mab approach that led to the detection, identification and characterization of laminin 5, an instructive ECM component for reporter keratinocytes in the epidermis. We sought to identify ECM components that mediate adhesion of keratinocytes both in quiescent normal epidermis and in activated migratory outgrowths in wounds or culture. For clarity, we will first outline current understanding of the changes in adhesion that occur as quiescent keratinocytes transit into activated migratory cells. We will then point out technical approaches that have contributed to the current understanding. Initial wound activation of quiescent epidermis results from multiple events including disruption of cell–matrix and cell–cell junctions, the influx of soluble factors, and stress responses, all of which change cell adhesion and signaling (Martin, 1997; Nguyen *et al.*, 2000b). Each of these wound events stimulates the transition of keratinocytes from quiescent anchorage on the basement membrane (BM; see A below) through to initial activation of keratinocytes at the wound margin (B). This initial activation generates leading keratinocytes in the outgrowth that migrate over exposed dermis and repair the BM (C):

(A)	*Wounding*	(B)	*Response*	(C)
Quiescent epidermis	$\rightarrow$	Initial activation	$\rightarrow$	Leading cells in outgrowth
(anchored)		(spreading)		(processive migration)

Steps A–C are outlined: (A) Quiescent basal keratinocytes anchor to laminin 5 in the BM via integrin $\alpha6\beta4$ in hemidesmosome (HD) cell junctions (Borradori and Sonnenberg, 1999; Nguyen *et al.*, 2000b). Integrin $\alpha3\beta1$ also interacts with laminin 5 in actin-associated junctions to control motility (Carter *et al.*, 1990b, 1991; DiPersio *et al.*, 1997; Kreidberg, 2000; Nguyen *et al.*, 2000b). (B) Keratinocytes at the wound edge adhere and spread via $\alpha6\beta4$ and $\alpha3\beta1$ on endogenous BM laminin 5 (Goldfinger *et al.*, 1999; Nguyen *et al.*, 2000b) and execute adhesion-dependent initial activation signals that generate the migratory epidermal outgrowth (Martin, 1997; Woodley *et al.*, 1999). (C) The epidermal outgrowth is composed of two subpopulations of keratinocytes, leading and following (Lampe *et al.*, 1998; Martin, 1997; Nguyen *et al.*, 2000a). Leading keratinocytes migrate over exposed dermal collagen via integrin $\alpha2\beta1$ (Wayner and Carter, 1987; Wayner *et al.*, 1988) and over fibronectin via integrin $\alpha5\beta1$ (Toda *et al.*, 1987) while expressing elevated levels of precursor laminin 5 and active RhoGTP required for collagen adhesion (Lampe *et al.*, 1998; Nguyen *et al.*, 2000a,b). Deposition of laminin 5 by the leading cells changes the matrix ligands, adhesion receptors, and signaling utilized by the following keratinocytes (Nguyen *et al.*, 2000a). The deposits of laminin 5 initiate the repair of the BM and reassembly of HDs (Goldfinger *et al.*, 1999; Ryan *et al.*, 1999) and promote gap junctional communication (Lampe *et al.*, 1998), all characteristic of more quiescent epithelium.

We indicated that ECM ligands "instruct" a response in reporter cells. This is a simplification for purposes of presentation. For example, adhesion of human foreskin keratinocytes (HFKs) to laminin 5 promotes assembly of hemidesmosomes. However, different ECM components can also "select" subpopulations of reporter cells from heterogeneous primary cells. For example, the epidermal outgrowth of a wound or primary cultures of HFKs are heterogeneous, containing leading, following, and quiescent keratinocytes. Leading keratinocytes can adhere to collagen and fibronectin or laminin 5, whereas confluent HFKs or quiescent HFKs isolated fresh from tissue adhere to laminin 5 but not dermal collagen or fibronectin (Nguyen *et al.*, 2001). However, both leading and

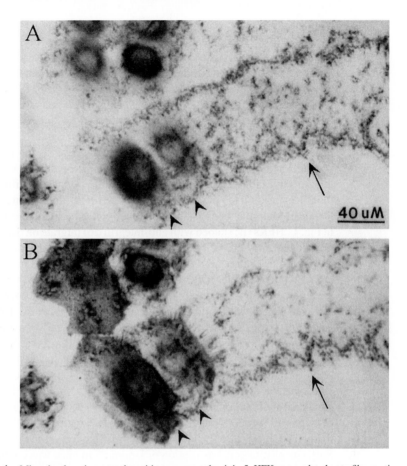

Fig. 1 Migrating keratinocytes depositing precursor laminin 5. HFKs were plated onto fibronectin-coated surfaces and allowed to adhere, spread, and migrate. Panel A: anti-laminin 5α3 chain (Mab P1E1). The migrating HFKs deposit a path of laminin 5 (marked with arrow) over the fibronectin surface. (B) Same field as (A); anti-integrin α6 (Mab G0H3). Two HFKs (marked with arrowheads) migrate on the fibronectin surface. Integrin a6b4 localizes to the trailing edge of the migrating HFKs and in retraction fibers and in footprints left on the substratum (arrow). The footprints of α6β4 and the deposited path of laminin 5 overlap at many points.

quiescent HFKs receive instructive signals from adhesion to ECM that are required for downstream responses including survival, proliferation, or differentiation. Thus, both instruction and selection result from adhesion to ECM components. However, differences between instruction and selection are important in evaluating mechanisms of communication between the matrix ligands and the downstream cell responses.

At the beginning of these studies, we had already used the Mab approach to identify two new adhesion receptors that were subsequently termed integrins α2β1, as a collagen receptor, and α3β1, as a promiscuous receptor or a receptor for an uncharacterized ligand (Takada *et al.*, 1988; Wayner and Carter, 1987; Wayner *et al.*, 1988). Integrins α2β1 and α3β1 are expressed in quiescent basal cells of epidermis (Wayner and Carter,

1987; Wayner *et al.*, 1988). Consistently, using primary HFKs as reporter cells, we found that $\alpha2\beta1$ localized in focal adhesions on collagen and adhesion could be blocked with anti-$\alpha2$ (P1H5) Mab. Surprisingly, $\alpha3\beta1$ localized in faint focal adhesions on all ligands tested, including collagen, fibronectin, and laminin 1 (Burgeson *et al.*, 1994), the only known laminin at that time (Carter *et al.*, 1990a,b, 1991). In explanation for the localization of $\alpha3\beta1$ in focal adhesions, HFKs were found to deposit a novel adhesive ligand onto all adhesive ligands tested. Further, HFKs deposited the instructive ECM ligand even onto BSA-blocked surfaces and then adhered and spread on the deposits (Carter *et al.*, 1991). This led to the assay described hereafter as the BSA adhesion assay that detects deposits of endogenous instructive ECM (Section II.B.2.b). Consistently, anti-integrin $\alpha6$ Mab (G0H3) partially blocked adhesion, and anti-integrin $\alpha3$ Mab (P1B5) blocked spreading, on the deposits of ECM. The BSA adhesion assay allowed for the selection of Mabs that bound and/or inhibited cell adhesion to the deposited ECM. As seen in Fig. 1, HFKs (B; stained with anti-integrin $\alpha6$) migrate on dermal ligands (fibronectin or collagen) and deposit trails of the new ligand (A stained with anti-laminin 5 $\alpha3$ chain). The Mabs also stained the BM of quiescent epidermis and provisional BM of wounds, colocalizing with the reporter keratinocytes. This confirmed that the deposited ECM and the reporter keratinocytes were in proximity to each other under physiological conditions. This new ligand was originally termed epiligrin (Carter *et al.*, 1991), kalinin (Rousselle *et al.*, 1991) or nicien (Verrando *et al.*, 1993) and eventually renamed laminin 5 (Burgeson *et al.*, 1994). The new Mabs were used to purify laminin 5 (see Section IV.B) and to characterize its function as an adhesive ligand for $\alpha6\beta4$ that instructs the assembly of HD-like stable anchoring contacts (SACs) (Carter *et al.*, 1990b) and as a ligand for adhesion and spreading via $\alpha3\beta1$ (Carter *et al.*, 1990b, 1991; Xia *et al.*, 1996). The physiological significance of laminin 5 was extended with a new adhesion assay on cryostat sections of epidermis (see Sections II.B.4 and IV.C): HFKs adhered to laminin 5 in the basement membrane of cryostat sections of epidermis (Nguyen *et al.*, 2000a; Ryan *et al.*, 1999). Subsequently, targeted disruption of laminin 5 in mice was confirmed to inhibit assembly of $\alpha6\beta4$ in mature HDs in epidermis (Ryan *et al.*, 1999) and replicate the epidermal blistering disease junctional epidermolysis bullosa gravis (Christiano and Uitto, 1996).

II. Methods for Detecting ECM Components That Instruct Cell Responses in Reporter Cells

Outlined next are relatively simple assays for detecting ECM that instructs a response of interest in a reporter cell. Preferably, these assays would be executed early in the studies to determine whether instructive ECM activity is present in complex ECM. The instructive ECM component may subsequently be identified by the more time-consuming Mab approaches. The assays utilize cells as reporters of instruction from ECM components. The cell response activated in the reporter cells may be adhesion, spreading, migration, proliferation, assembly of cell junctions, cell signals, cytoskeletal arrangements, transcriptional events, or cell death, to name just a few. In the assays and antibody screens that follow, adhesion is required but insufficient to generate specific cellular responses in

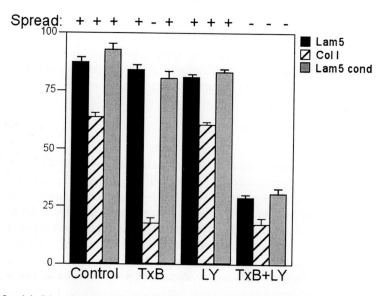

Fig. 2 Laminin 5 deposited over collagen substratum converts HFKs to toxin B-insensitive spreading. HFKs were untreated (control) or treated with toxin B (50 ng/ml; TxB), LY294002 (5 n*M*; LY), or the combination of toxin B and LY294002 (TxB + LY) for three hours. Cells were then trypsinized, treated with G0H3, and plated on exogenous laminin 5 (Lam5; black bars), collagen (Col I; striped bars), or laminin 5-conditioned collagen (Lam5 cond; gray bars). Cells were also scored as to whether they were spread (+) or not spread (−). (Reprinted from Nguyen *et al.*, 2000. *J. Biol. Chem.* **41**, 31896–31907, with permission of The American Society for Biochemistry and Molecular Biology.)

the reporter cells. For example, the reporter keratinocytes (Section I.B) can adhere and spread on laminin 5, collagen, or fibronectin and utilize distinct receptors for adhesions. (Fig. 2). However, only laminin 5 promotes assembly of hemidesmosomes or gap junctions. Therefore adhesion may be required for the desired reporter cell response, but only when combined with a unique signal from the instructive ECM component. For HFKs, adhesion and spreading on collagen or fibronectin require distinct cell signals and instruct different cell responses when compared to adhesion on laminin 5. Therefore dermal ligands provide important specificity controls for studies on laminin 5.

A. Reporter Cells and Cell Response

Presumably, the interested researcher may already have a specific reporter cell and cell response that he or she suspects is instructed by interaction with ECM components. It is necessary to establish a source for reporter cells, a cell response, and the instructive ECM. Reporter cells can be obtained fresh from tissue as suspended but uncultured (passage 0 or P0) cells, or as primary or secondary cultures (P1 or P2), or as cells with extended lifespans in culture (both cell lines and cell populations with extended lifespans). The reporter cells must interact with exogenous or endogenous instructive ECM to generate the cellular response of interest. In general, we have found that extended passage of primary cells alters the ability of the reporter cell to respond to instruction from ECM. The inability to respond to instruction from the ECM is even more profound in immortalized cell lines. Therefore, it may be undesirable to utilize cell lines, or even late passage diploid cells, when attempting to identify instructive ECM components. It is also desirable that the cellular response be easy to detect, so that it may be used in subsequent screens for Mabs.

B. Sources of Instructive ECM

It is necessary to establish a source for the macromolecular ECM that instructs reporter cells. The reporter cells may make their own instructive ECM. Instructive ECM may also be produced by another cell population, or in coculture with the reporter cells. Alternatively, instructive ECM may be obtainable directly from tissue as cryostat sections (see Section II.B.4). It should be reemphasized that solubilization and fractionation of insoluble tissue ECM can be time-consuming and unpredictable. Therefore, we will focus on conditioned culture media and extracts of cultured cells, or the tissue sections, as a source for the instructive ECM.

Reporter cells may synthesize, secrete, and/or deposit endogenous ECM components that dominate instructive signals from exogenous ECM (Nguyen *et al.*, 2000b). Endogenous ECM can be either a desirable benefit, or an undesirable problem, depending on how it affects the reporter cells and their response. If endogenous ECM is not the desired instructive ECM, its effects can be reduced or eliminated by utilizing short-term assays for the reporter cell response, or by addition of inhibitors of the endogenous ECM (antibodies or inhibitors of secretion such as brefeldin A, cycloheximide, or cycloheximide with monensin). In our own studies, the endogenous deposits of laminin 5 dominated other exogenous ligands. A wild-type HFK will deposit endogenous laminin 5 on any exogenous ligand with which it comes in contact. To eliminate the effects of endogenous deposited laminin 5 it was necessary to disrupt expression of laminin 5 in mice. After gene disruption, we obtained reporter keratinocytes from the knockout mice that could be instructed by exogenous collagen, fibronectin, or laminin 5 (Nguyen *et al.*, 2000b; Ryan *et al.*, 1999).

1. Testing of Known ECM Components for Instructive Activity

Prior to the preparation of new antibodies, it is prudent to evaluate known ECM components as candidates for the instructive ECM. This can be accomplished by coating virgin styrene surfaces (either petri dishes or beads [SM2 Biobeads, Bio-Rad Laboratories, Richmond CA]) with purified ECM components, washing to remove unbound materials, and blocking the surfaces with heat-denatured bovine serum albumin (HD-BSA), followed by interaction with the reporter cells. HD-BSA is prepared by dissolving bovine serum albumin (BSA) in calcium-free phosphate-buffered saline (PBS) (5 mg/ml; Fraction V BSA, Sigma A-4503). The BSA solution is heated to 80°C with stirring for 5 min, then rapidly cooled on ice. The faintly opalescent solution can be stored at 4°C after sterilization by filtration (Corning 500 ml, Bottle Top Filter 0.22 μM CA, #430521) but should not be frozen and thawed. Antibodies against individual ECM components can also be utilized as a trap for the ECM. The antibodies can be immobilized either

(i) as purified antibodies or (ii) as a sandwich in combination with purified secondary antibodies or protein G. Immobilized antibodies work well because they both trap and orient the ECM components on the substratum (Lampe *et al.,* 1998). This trap protocol may be utilized to test both inhibitory and noninhibitory Mabs to identify the instructional ECM component (Section III.C.2, "Secondary Screens for Mabs").

2. Cultured Reporter Cells That Produce Their Own Instructive ECM

a. Deposited ECM

Primary cultures of many adherent cells synthesize, secrete, and/or deposit ECM components. In the case of many epithelial cells (keratinocytes, esophageal and cervical epithelium) and cultured fibroblasts, the secreted or deposited ECM is characteristic of the ECM that is made during wound repair in tissue. This selective expression occurs in culture because culture systems are usually optimized for conditions similar to wound repair and proliferation, and not for the differentiation, morphogenesis, or cell death that are associated with development or histogenesis. For the epithelial cells mentioned, the secreted/deposited laminin 5 is also present in BM of quiescent tissue in a proteolytically processed form (Goldfinger *et al.,* 1998; Nguyen *et al.,* 2000a,b; Tsubota *et al.,* 2000). Surfaces coated with ECM, either glass coverslips to facilitate microscopy or culture dishes, are readily prepared by trypsinizing off cultured cells and using the residual ECM on the surface (Xia *et al.,* 1996). The cell layer can be removed by treatments with 1% Triton X-100 in PBS, 2 M urea/1 M NaCl, and 8 M urea (Carter, 1982b) or alternatively, by extraction with 20 mM NH$_4$OH (Robinson and Gospodarowicz, 1984). The ECM surface should be blocked with HD-BSA before use. It is important to verify that the instructional activity has been preserved in the ECM preparation. Thus, reporter cells that produce their own instructive ECM can be assayed as they attach to their endogenous ECM, or as they readhere to exogenous ECM.

b. Deposited ECM: BSA Adhesion Assay

We developed an *in vitro* assay to identify cells that deposit laminin 5 and subsequently adhere to the deposits. This assay, called the BSA adhesion assay, can also be used to determine whether soluble factors (either activator or inhibitors) regulate the deposition of ECM. The assay is presented in cartoon form in Fig. 3A. Cultures of HFKs are suspended by trypsin/EDTA digestion, then incubated on a nonadhesive surface prepared from virgin styrene that has been blocked with HD-BSA. The surface is nonadhesive, unless the cells express and deposit a matrix ligand onto the blocked surface that enables their adhesion to the deposits. The cells are incubated on the surface for 3–8 h to allow for deposition of endogenous ECM and subsequent cell adhesion. As seen in Fig. 3B, a subpopulation of keratinocytes adheres to the surface, and adhesion can be blocked with inhibitory Mabs against laminin 5 (C2-9), but not by noninhibitory anti-laminin 5 Mab (C2-5). Anti-integrin $\alpha6\beta4$ Mab (G0H3) also blocks adhesion, whereas Mabs against $\alpha3\beta1$ block cell spreading. Consistently, keratinocytes from an individual with homozygous null mutations in the integrin $\beta4$ gene do not adhere to the surface but surprisingly can still deposit laminin 5 (Frank and Carter, manuscript in preparation). Further, treatment of HFKs with transforming growth factor $\beta1$ (TGF$\beta1$), a transcriptional activator

A

DEPOSITION OF LAMININ 5 REGULATES ADHESION

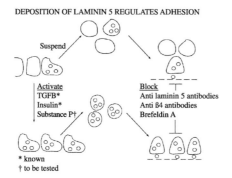

* known
† to be tested

B

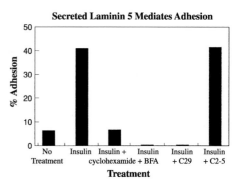

Fig. 3 BSA adhesion assay: Soluble factors stimulate deposition of laminin 5 and subsequent adhesion to the laminin 5 deposits. Panel A presents a diagrammatic explanation for the results presented in Panel B (Harper and Carter, unpublished results). HFKs in culture are a mixture of leading and following cells that differ in expression of laminin 5. When suspended and incubated on a nonadhesive surface, only leading cells that express laminin 5 can deposit it onto the nonadhesive surface and adhere. The adhesion is blocked by inhibitory anti-laminin 5 Mab (C2-9) but not noninhibitory (C2-5) Mab. Deposition and adhesion are also blocked by cycloheximide (an inhibitor of protein translation), or brefeldin A (BFA; an inhibitor of vesicle transport). Addition of TGFβ1 or insulin or other soluble factors that stimulate expression of laminin 5 also increases deposition and adhesion in this assay.

of laminin 5 (Ryan *et al.,* 1994), or insulin (Harper and Carter, unpublished results) increases the expression, and deposition of laminin 5 and adhesion to the deposits. Thus, the assay reports the effects of growth factors on the deposition of laminin 5 and subsequent adhesion to the deposits. The assay is scored by quantitating the number of labeled cells that have adhered. However, analysis can include quantitation of ECM deposits by ELISA (Gagnoux-Palacios *et al.,* 2001) or a quantitation of cell migration by time-lapse image analysis (Frank and Carter, manuscript in preparation).

c. Soluble ECM Components

Many cells in culture synthesize, secrete, and/or deposit ECM components. A subpopulation of ECM components is usually secreted as soluble glycoproteins as well as deposited on the substratum. The soluble secreted ECM is a convenient source of ECM material with possible instructional activity. The soluble secreted ECM can be passively adsorbed on plastic or glass surfaces, to provide a convenient instructional surface. To collect soluble ECM, cells should be grown in serum-free, defined culture medium. Alternatively, cells can be grown in serum, then shifted to serum-free culture conditions for 24 h or more. Conditioned culture medium is collected and stored in the presence of protease inhibitors (5 mM ethylenediaminetetraacetate [EDTA], 1 mM phenylmethylsulfonyl fluoride [PMSF], and 2 mM N-ethylmaleimide [NEM]). Concentration of secreted proteins in the conditioned media is achieved by salting out with $(NH_4)_2SO_4$ (50% saturation), or by passing conditioned media over a column of an immobilized lectin such as wheat germ agglutinin-agarose (Carter and Hakomori, 1976, 1979; Merkle and Cummings, 1987), or a column of heparin agarose (Farooqui, 1980). Bound glycoproteins elute from the wheat germ agglutinin column in a single peak when the column is washed with 400 mM N-acetylglucosamine. Heparin binding proteins can be eluted with a 2 M NaCl wash or fractionated with a salt gradient. The concentrated conditioned culture proteins can be adsorbed onto petri dishes (10 μg/ml PBS, 2 h to overnight). After adsorption, the wells are washed and blocked with 0.5% BSA in PBS for at least 30 min.

3. Coculture of Reporter Cells and Cells That Produce Instructive ECM

In contrast to the epithelium, primary cultures of endothellial cells (M. Ryan personal communication), or melanocytes (Gil and Carter, unpublished results), do not deposit components of ECM to which they adhere in quiescent tissue. Deposition of ECM characteristic of quiescent tissue may require special culture conditions optimized for the desired cell characteristics. For example, reporter cells may utilize ECM ligands synthesized and deposited by other cells. Type VII collagen is produced by mesenchymal cells that are in proximity to basal epithelial cells (Burgeson and Christiano, 1997; Marinkovich et al., 1993). Therefore assembly of the endogenous BM ligand may occur only in coculture of epidermis and mesenchyme. Alternatively, an ECM component, or the adhesion and signaling components necessary to interact with ECM components, may only be expressed or functional at a specific developmental stage, and not present in log-phase cultures. For example, laminin 10/11 that is present in the basement membrane of the quiescent epidermis is not synthesized in culture by primary HFKs. In wounds, it only appears in the BM after wound closure (Lampe et al., 1998). Presumably, the conditions of quiescent epidermis promote expression of laminin 10/11 by epidermal or mesenchymal cells. Establishing conditions for the expression of different ECM components may require considerable nonlinear effort (see Section II.B.4). Microarray technology may prove to be a useful tool for identifying suitable conditions for the expression of a particular ECM component (Perou et al., 1999). However, it should be emphasized that expression of mRNA for any one chain of a heteromeric ECM component such as laminin or collagen does not mean that the heteromultimeric protein will be translated, assembled, and secreted. Expression of heteromultimeric proteins should be confirmed by appropriate biochemical characterization at the protein level, not just mRNA levels.

4. Adhesion Assays on Cryostat Sections of Tissue

It may not be possible to find a source of instructive ECM characteristic of quiescent tissue in a culture system when the culture system does not duplicate the tissue environment. However, instructive ECM may be present in tissue. For the reasons discussed in Section I.A, we will not describe approaches for solubilization of instructive ECM components from tissue. Instead, we suggest an assay that has been useful for identification of adhesion or instructive components in cryostat sections of tissue. Further, this assay may be useful to determine if reporter cells actually adhere and respond to instructive ECM in tissue within their own physiological environment (see Section IV.C). The assay confirms that an identified ECM is instructive for the reporter cells in tissue. We developed an adhesion assay on cryostat sections of tissue to evaluate interactions of reporter epithelial cells via $\alpha6\beta4$ with laminin 5 (Figs. 4 and 5). It is based on a tissue adhesion assay utilized to detect lymphocyte interactions with endothelium on cryostat

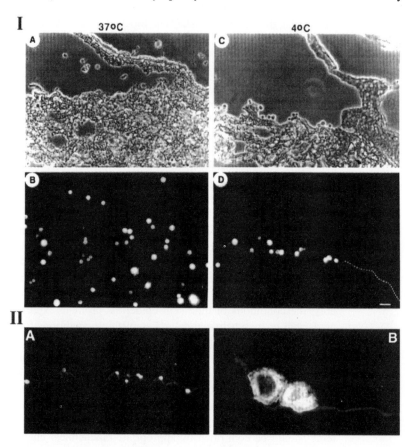

Fig. 4   HFKs anchor and align with laminin 5 in the BM of tissue sections. (*Panel I*) Cryostat sections of salt-split normal neonatal foreskin were utilized as substrate for cell adhesion. Normal HFKs were labeled with DiI and allowed to adhere to the skin sections at either 37°C (A and B, same field) or 4°C (C and D, same field). A and C show the phase contrast images whereas B and D show the DiI-fluorescent cells using a rhodamine filter. At 4°C, HFKs selectively adhere via integrins $\alpha6\beta4$ (Xia *et al.*, 1996) and decorate the BM zone in the floor of split skin. Dotted line in D identifies intact epidermal–dermal contact blocking access of HFKs to the BM. (*Panel II*) Cryostat sections of salt-split normal neonatal foreskin stained with anti-laminin 5 antibody (noninhibitory, C2-5) and fluorescein-labeled secondary antibody. Subsequently, the laminin 5-labeled tissue was used as substrate for cell adhesion. DiI-labeled HFKs were allowed to adhere to the tissue at 4°C for 30 min and the unbound cells were washed with PBS. The remaining adherent HFKs were fixed and photographed using fluorescein filters to visualize laminin 5 (fluorescent line) and the attached cells (fluorescent cells). Note that HFKs specifically attach to laminin 5 reactive basement membrane. (B) is a higher magnification field from (A).

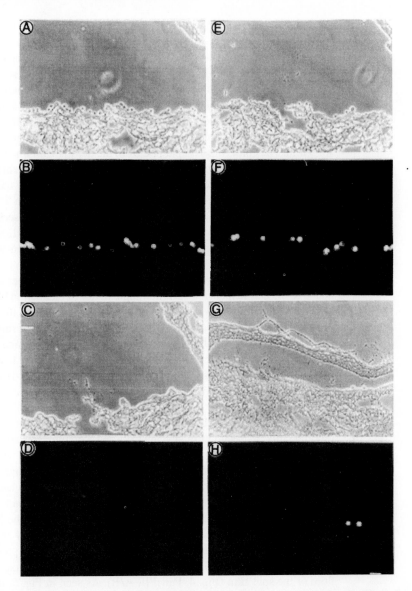

Fig. 5 Anchorage to laminin 5 in epidermis via $\alpha 6 \beta 4$ is defective in individuals with JEB-G. Cryostat sections of normal salt-split foreskin (A–F) and JEBG blistered skin (G and H) were utilized for the tissue adhesion assay. DiI-labeled HFKs were incubated with the tissues at 4°C for 30 min in the presence of a control (A and B; G and H, same field, respectively), an inhibitory anti-laminin 5 antibody (C2-9; E and F, same field) or 5 mM EDTA (C and D, same field). Compare the lack of cell adhesion in the JEBG tissue (G and H) with the high level of anchorage in the normal split skin (A and B). Note that inhibitory anti-laminin 5 Mab blocks cell adhesion to tissue 50% (E and F), indicating that laminin 5 is the primary ligand in this tissue. EDTA completely inhibits cell anchorage (C and D), suggesting a strong dependency on divalent cations for cell adhesion.

sections of tissue (Stamper and Woodruff, 1976). Subsequently, this tissue adhesion assay was utilized to screen for and select a Mab that inhibits adhesion of lymphocytes to endothelium (Gallatin *et al.,* 1983). This effort identified the lymphocyte homing receptor. Tissue with ECM confined to a small cross-sectional area, such as an epithelial BM, should be "salt-split" as described, to expose the basement membrane (Kurpakus *et al.,* 1991). Briefly, epithelial tissue is incubated for 2 h in a dissociation buffer containing sodium potassium phosphate (6 mM), NaCl (120 mM), EDTA (20 mM), KCl (3 mM), and PMSF (1 mM), then transferred to the same buffer containing 1 M KCl for 5 min. The epidermis is separated from the dermis with fine forceps, then embedded in OCT medium and snap frozen. Cryostat sections (8–10 μm) of tissue are cut and mounted on a flat surface, which can later form the bottom of the well for the adhesion assay. We found it convenient to mount sections on the top of a 35-mm petri dish lid and then use Scotch tape to form the walls of the well. The tissue section should be blocked for at least 30 min with 0.5% HD-BSA in PBS prior to addition of the suspension of reporter cells. The reporter cells can be fluorescently labeled while in suspension to aid visualization on the tissue sections (Calcein AM or DiI, 1 μg/ml PBS for 1–2 h; Molecular Probes). Activating or inhibitory antibodies that react with the cells should be added to the cell suspension several minutes prior to incubation with the tissue sections. Likewise, antibodies that target the ECM should be added to the tissue prior to adding the cells. The tissue section is covered with the suspension of reporter cells. Incubation conditions will vary according to the reporter cells, the tissue, and the ECM (see the next paragraph for an example of incubation conditions utilized for laminin 5). Following the incubation, unbound cells are gently rinsed away with PBS, and the bound cells are fixed for 20 min in 2% formaldehyde, 0.1 M sucrose, and 0.1 M cacodylate and prepared for viewing.

It is important to establish the specificity of this adhesion assay for the tissue and reporter cells. As seen in Fig. 4 (Panel I, A/B), keratinocytes adhere to both the epidermal BM of split epidermis and the dermal connective tissue when incubated at 37°C, conditions that allow both β1 and β4 integrins to bind ligand (Xia *et al.,* 1996). At 4°C (Fig. 4, Panel I, C/D), adhesion is restricted to the BM when β4 integrins, but not β1 integrins, are functional (Xia *et al.,* 1996). In controls (not shown), keratinoyctes from individuals with null mutations in β4 fail to adhere to the BM under these selective conditions, but do adhere at 37°C via α3β1. Selectivity in integrin β4 adhesive function can also be obtained by adhesion at 37°C in the presence of cytochalasin D (Ryan *et al.,* 1999). HFKs that adhere to the BM colocalize with laminin 5 in the floor of the split skin (Fig. 4, Panel II, A/B). As seen in Fig. 5, inhibitory anti-laminin 5 Mab (C2–9; 50% inhibition; Fig. 5 compare A/B to E/F) or EDTA (100% inhibition; Fig. 5 compare A/B to C/D) inhibited adhesion to BM in the split skin. Further, BM in skin from individuals with defects in laminin 5 (Fig. 5 G/H) or mice with targeted disruptions in laminin 5 (Ryan *et al.,* 1999) fail to adhere reporter keratinocytes to the BM when assayed for adhesion via integrin α6β4. Agitation or drugs can also be utilized as variables in the assay. At 37°C without agitation (Fig. 4, Panel I, A/B), HFKs adhere to dermal and BM ECM ligands. With agitation, adhesion is selective for the BM.

III. Preparation of Mabs for Identification of Instructive ECM

After the detection of instructive ECM activity within crude ECM, it is reasonable to attempt to identify and/or purify the instructive ECM component(s). We found it more productive to identify the instructive ECM component within the macromolecular ECM, before disruption of the complex during solubilization. Therefore our first choice for technical approaches is the production of Mabs that bind and inhibit the instructive activity within the complex of ECM. If the source of ECM components is soluble, for example in conditioned culture media from cultured cells, then either inhibitory or binding Mabs can be used to purify the instructive ECM component. ECM components in the cell layer may be solubilized by sequential extraction with 1% Triton X-100, followed by 2 M urea, then 8 M urea, or 8 M urea in the presence of reducing agents (Carter, 1982a; Carter and Hakomori, 1976, 1979). Once solubilized, the instructive ECM may be purified from the extracts as described later.

Preparation of Mabs requires significant time and energy. However, once prepared, the antibodies provide a means for identifying, purifying, and characterizing the instructive ECM component. Even if it is not possible to purify the ECM component in an instructionally active form, the Mab may still be useful for purification of a denatured form and providing amino acid sequence, or for screening cDNA expression libraries or for screening of expression libraries (Seed, 1995). All are routes to obtaining cDNA sequence information.

A. Preparation of Antigen and Immunization

It is necessary to generate sufficient antibody diversity in order to identify anti-ECM Mabs. This means that the cell fusion needs to generate a library of Mabs with many different specificities in order to identify useful Mabs that recognize the instructive ECM. Thus, 4–5 hybridoma colonies per well of the primary fusion plates is a minimum for these screens. Immunization of mice with human ECM and cellular components has been able to generate adequate antibody diversity. However, we have not been as successful in generating large antibody diversity when immunizing rats with mouse ECM components. As a result, making rat Mabs against mouse adhesion proteins has been difficult but still possible (Delcommenne and Streuli, 1999). We have had better success in generating the desired antibody diversity when immunizing either rats or mice with the complex ECM, or with whole cells, in comparison to purified proteins. Presumably, the diversity of Mabs benefits from an adjuvant effect of the macromolecular structures.

1. Soluble and Deposited ECM as Antigens

Cells are grown on both sides of sterile cellophane (Bio-Rad Laboratories, Richmond, CA) to allow deposition of ECM components. The mixture of adherent ECM and cells may be immunized directly or subject to sequential extraction with 1% Triton X-100 detergent in PBS, followed by 2 M urea in 500 mM NaCl to remove cytosol, cell membranes, and nuclear components. The residual protein on the cellophane is rich in ECM proteins.

The cellophane is ground with a mortar and pestle in liquid nitrogen, emulsified in Freund's incomplete with Ribi adjuvant (MPL+TDM emulsion R-700; Tel. 800-548-7424; Ribi, Hamilton, MO), then immunized IP or SC in mice. Alternatively, secreted proteins are concentrated (see Section II.B.2.c), then immunized in mice as a soluble antigen mixture.

2. Tissue as a Source of ECM Antigens

We have not immunized mice with human tissue or extracts of human tissue. However, we have immunized rats with mouse tissues rich in epithelial populations (intestine and lung). Mouse antigens do not appear to be particularly immunogenic in rats, but these tissues have worked equally as well as cultures of mouse keratinocytes, with regard to the number of hybridoma colonies produced and the antibody diversity.

B. Fusion and Preparation of Hybridomas

1. Fusion Techniques and Myeloma Fusion Partners

Monoclonal antibodies were produced by the methods of Oi and Herzenberg (1980) and Taggart and Samloff (1983). Briefly, either BALB/c or RBF/Dn mice (Jackson labs, Bar Harbour, ME) were immunized with the antigen mixtures discussed earlier. Immune spleens were removed and fused with either SP2/0 (ATCC # CRL 1588) or the NS-1/FOX-NY (ATCC # CRL 1581) myeloma cell lines. Viable heterokaryons were selected in RPMI 1640 supplemented with hypoxanthine/aminopterin/thymidine or adenine/aminopterine/thymidine. Heterokaryons were selected for the production of Mabs in staged screens as described later (Wayner and Carter, 1987). After screening, selected heterokaryons were cloned by limiting dilution with thymocytes as feeder cells.

2. Collection and Storage of Hybridoma Supernatants

Primary heterokaryons are plated out in 4 to 8 96-well culture plates. Depending on the types of screening assay to be performed on the hybridoma supernatants, it may be desirable to freeze the primary fusion plates after collecting antibody-containing culture supernatants. Freezing provides adequate time to perform the desired screen assays before the cultures overgrow the plates. It is best to freeze the cells in 96 tube clusters (Costar #4413, 1.2 ml, 8 strip). This allows individual hybridoma wells to be thawed after screening and cloned. For freezing of primary wells, culture media is replaced with 90% serum containing 10% v/v dimethyl sulfoxide, followed by slow freezing over 2–4 h in a −80°C freezer. Cells are stable at this temperature for at least 1 month. For longer storage, frozen cells should be transferred to liquid nitrogen.

In some cases, we have found difficulty in cloning out the desired hybridoma well because of low viability of the heterokaryons. This generates significant wasted effort in the cloning phases of antibody production. To reduce wasted time, we prepare replica plates of the primary hybridoma wells using 1/20 of the total primary well volume. Heterokaryons that survive in the replica plate are more stable than the primary cells during cloning by limiting dilution. Antibody supernatants from the primary or replica plates are saved in 96-tube clusters (Costar #4413, 1.2 ml, 8 strip). These antibody supernatants are the working stocks for the antibody screens and should be stored sterile at 4°C or frozen at −80°C.

C. Screening for Mabs That React with Instructive ECM

Screens of primary hybridoma wells are usually performed in stages. Each primary hybridoma well may contain multiple antibodies that must be selected or rejected. We frequently obtain 10 or more different antibodies in any one primary hybridoma well. Therefore, it is desirable to perform the easiest screens first. This reduces the numbers of candidate antibodies to a manageable quantity with the least effort. As the number of candidate antibodies is reduced, more difficult screens may be performed comfortably. Solid-phase assays that may be automated are performed first, and functional assays are performed last. It is important to understand that the nature of the screen determines the antibodies that are selected. For example, screening for antibodies that react with denatured antigens usually selects antibodies that immunoblot but may not immunoprecipitate native protein, or do not inhibit function of native proteins. Similarly, antibodies prepared against low molecular weight synthetic peptide antigens may cross-react with apparently unrelated protein antigens that share similar sequence epitopes, or may not react with the native protein because of steric factors. Realize that one of the major reasons for failure in selection of antibodies is that the researcher is unprepared for the large number of candidate Mabs to be tested and does not have the screens, or the time, available for reducing these numbers. Listed next are antibody screens that we have utilized and prefer for detection of ECM components or adhesion receptors.

1. Primary Screens for Mabs

a. Fluorescence Microscopy

Screening by fluorescence microscopy is one of the most informative and sensitive methods for detection of antibodies against instructive ECM components (Delcommenne and Streuli, 1999). Reasonably, 200–400 antibody supernatants can be examined in 1 day utilizing the following protocol. Cells that produce instructive ECM are grown on sterile 12-well glass slides (Cel-Line/Erie Scientific Co. 1-800-258-0834, HTC super cured #10-968, 7 mm wells). A drop (50 μl) of suspended cells is placed on each well of the slide and is cultured for sufficient time to allow for deposition of ECM (usually 24–48 h). Adherent cells and ECM are fixed (2% formaldehyde in 100 mM sodium cacodylate, 100 mM sucrose, 10 min) and permeabilized with Triton X-100 detergent (0.1% in PBS for 5 min). Fixed cells are blocked with HD-BSA (1% in PBS for 2 h). Fixed cells on slides can be stored for at least 1 month in block solution at 4°C. Each well on the slide is incubated with the antibody supernatants (10 μl diluted 1:4 in HD-BSA/PBS) for 2–4 h, washed in slide washers 2× with PBS containing 0.1% Triton X-100, followed by 2× with PBS alone, stained with fluorescent secondary antibody at appropriate dilutions and times, then examined by fluorescent microscopy. It is possible to detect antibodies reacting with deposited ECM even when other antibodies are reacting with the cells.

A weakness in this assay results from the large numbers of antibodies present in each primary hybridoma well. The staining patterns of the multiple Mabs overlap in the cells. This overlap usually does not interfere with detection of deposited ECM antibodies, but may not be resolved in other subcellular domains. Overlap in staining can be reduced by culturing primary hybridoma plates at lower cell densities. Unfortunately, this increases the total number of wells to be screened. As a side benefit of this screen, we record supernatants that stain subcellular structures that may be of interest to the laboratory (cell junctions, plasma membrane, etc.).

b. Screens by ELISA

Antibody screens by ELISA are probably the easiest and quickest, but also the least informative. The assay only reports whether an antibody will bind to the material in the well. There is no information regarding the subcellular location of the antigen. However, more than 1000 samples can be tested in just a few hours. Antigen can be prepared in a variety of ways and coated onto 96-well plates to correspond directly with the wells of the primary cell fusion. Deposited ECM is prepared by plating about 3×10^4 cells/well in 96-well flat-bottom tissue culture plates, and then growing the cells for 24 h while they deposit ECM. Cells are removed by trypsinization, or differential extraction (Section II.B.2.a) and the remaining ECM is fixed with 2% formaldehyde in PBS for 10 min. These plates can be stored frozen with PBS covering the wells. After thawing, the wells can be blocked with 0.5% HD-BSA prior to use. Soluble ECM components, which are secreted into serum-free conditioned culture medium can be immobilized onto polyvinyl chloride assay plates by incubating 100 μl of the medium in each well overnight at 4°C. The antigen density in the well can be increased by changing the medium every 2 h, or by using soluble ECM that has been concentrated from the conditioned medium (see Section II.B.2). These plates are best used fresh, but can be stored cold in the blocking solution for a time.

Diluted hybridoma culture supernatants (100 μl/well) are incubated with the immobilized ECM components for 1 h, then washed three times with PBS. Immobilized hybridoma Mabs are allowed to bind peroxidase-conjugated rabbit anti-mouse immunoglobin (DAKO P0260; diluted 1:5000 in HD BSA, 100 μl/well); the excess is washed free with PBS, then the bound antibody is detected with peroxidase substrate (100 μl/well; ABTS, Kirkegaard & Perry Labs #50-66-00). A blue-green color will develop in positive wells within a few minutes. The reaction can be stopped with 100 μl of 1% SDS and the color will be preserved. The reaction in the wells is quantitated in a plate reader at 405–410 nm. It is essential to run adequate controls to ensure that any positive reaction is due to a specific antibody that was generated in response to the immunization and is not present in the control wells or with the secondary antibody alone.

2. Secondary Screens for Mabs

Candidate antibody supernatants identified in the primary screens contain antibody(s) that react with deposited or secreted soluble ECM components. The candidate antibodies will be used in secondary screens to determine if antibodies can bind instructive ECM, and/or inhibit instructive ECM. The two following screen assays can be used to detect Mabs that bind ECM that promotes adhesion and/or spreading. It is also possible to adapt the same assay to detect an adhesion-dependent response of the reporter cells. For example, using HFKs as reporter cells, we might screen for ECM ligands that promote adhesion and spreading but are resistant to toxin B, an inhibitor of Rho-dependent cell spreading. HFK spreading on collagen and fibronectin, but not on laminin 5, is inhibited by toxin B (Nguyen *et al.,* 2000a). Alternatively, we might screen for adhesive ligands that promote gap junctional communication utilizing a rapid dye transfer assay (Lampe *et al.,* 1998).

a. BSA Adhesion Assay for Mabs

We utilize the BSA adhesion assay (Section II.B.2.b) to detect antibody in the hybridoma supernatants that inhibits deposition of ECM or inhibits adhesive function of the deposited ECM. Here it can be performed on a small scale employing the same slides utilized in the fluorescence microscopy screen. Wells on the 12-well glass slides (Cel-Line/Erie Scientific Co. 1-800-258-0834, HTC super cured #10-968, 7 mm wells) are blocked with 0.5% HD-BSA. A drop (30 μl) of suspended cells that produce ECM is mixed with antibody supernatants (10 μl undiluted), applied to each well on the blocked slide, and incubated for 4 h to overnight. Adhesion of the cells to the blocked surface is dependent on the deposition of an adhesive ligand. Adhesion and/or spreading may be affected by the presence of the anti-ECM antibodies. The slides are washed 4× in slide washers, fixed, and the numbers of adherent cells are quantitated.

b. Trapping of Instructive ECM with Mabs

Antibodies immobilized on the substratum can trap instructive ECM activity either from conditioned culture medium or directly from cells. If the immobilized anti-ECM antibody is inhibitory, it will block cell adhesion or the instructive activity of the ECM. If the immobilized antibody is not inhibitory, it will trap instructive ECM on the substratum either as soluble secreted ECM or deposited ECM.

Affinity-purified, rabbit secondary antibodies against mouse Ig (10 μg/ml PBS, 250 μl vol) is adsorbed to virgin styrene surfaces (24-well petri dishes, Falcon Cat. #351147) for 2 h. The surfaces are washed with PBS and blocked with 0.5% HD-BSA in PBS. Selected primary hybridoma culture supernatants (400 μl) are added to each well and incubated for 2 h to allow antibody to bind to the immobilized rabbit anti-mouse Ig. The wells are washed and reblocked with HD-BSA. Conditioned culture medium from cells expressing the instructive ECM is incubated with the immobilized Mabs to trap soluble instructive ECM, then washed and reblocked. In controls, it is necessary to perform this assay with or without the source of the instructive ECM, to eliminate possible interaction of the immobilized primary or secondary antibodies with the surface of the reporter cells. Suspended reporter cells are then incubated with the surfaces. Adhesion, spreading, or the specific cell response of the reporter cells can be evaluated. As an example, we performed this trap assay with conditioned culture medium of HFKs grown in ^{35}S-met (Fig. 6). We have then added unlabeled HFKs as reporter cells and recorded adhesion and spreading on the immobilized Mabs. The adherent cells and labeled ECM were disrupted

MW(X10^{-3})

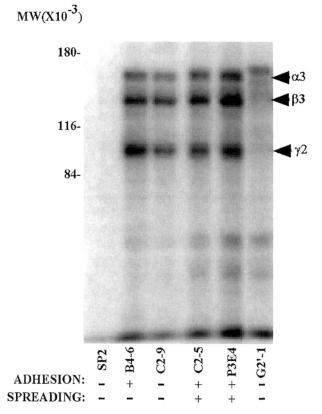

Fig. 6 Immobilized anti-laminin 5 Mabs trap laminin 5 to promote adhesion of HFKs. HFKs were metabolically labeled with ^{35}S-met (18 h, 50 μCi/ml in met-free KGM containing 100 ug/ml carrier BSA). The labeled conditioned culture medium containing approximately 40 different labeled proteins was added to surfaces of immobilized Mabs (Mabs were immobilized on surfaces as described in text). B4-6, C2-9, C2-5, and P3E4 are anti-laminin 5 Mabs. SP2 (irrelevant Mab) and G2'-1 (anti-thrombospondin 1) are controls. The trapped labeled proteins were incubated with unlabeled HFKs for 45 min to allow adhesion and spreading on immobilized ligands and recorded as indicated (− or +). Unattached cells were removed by washing. The adherent, labeled proteins were collected in SDS sample buffer, fractionated by SDS–PAGE under reducing conditions (8% gel), and detected by fluorography. Migration of labeled α3, β3, and γ2 chains of laminin 5 is indicated in the right margin. Migration of molecular weight (MW) standards (180, 116, etc.) is indicated in the left margin.

in SDS sample buffer and fractionated by sodium dodecyl sulfate–polyacrylamide gel electrophoresis (SDS–PAGE). Laminin 5 composed of labeled α3, β3, and γ2 chains bound to Mabs B4-6, C2-9, C2-5, and P3E4. Labeled thrombospondin (migrating just above the α3 chain of laminin 5) bound to immobilized Mab G2'-1. Nothing bound to negative control Mab SP2. Neither cell adhesion nor spreading occurred on SP2 or G2'-1. Adhesion occurred on laminin 5 trapped by Mabs B4-6, C2-5, and P3E4 but spreading only occurred on the latter two. B4-6, an anti-γ2 chain Mab, blocks spreading but not adhesion. Mab C2-9 binds the $\alpha3$ chain of laminin 5 but blocks cell adhesion and spreading activity.

D. Cloning of Selected Mabs

After primary or secondary antibody screens, the frozen hybridoma well producing the selected Mab can be thawed and cloned by limiting dilution. Recovery of frozen cells from the primary or replica wells is usually excellent (90% + viability).

IV. Purification of Instructive ECM with Mabs and Interactions with Reporter Cells

A. Initial Characterization of Instructive ECM with Mabs

1. Immunoprecipitation or Western Blotting of Instructive ECM

At this time, we usually determine if the hybridoma supernatants can immunoprecipitate and/or immunoblot antigen from cell extracts, or conditioned culture media. Immunoblotting with small volumes of hybridoma culture supernatants (120 to 200 μl of 1:10 dilution of hybridoma supernatant) is performed on the Miniblotter apparatus #45 or #25 (Immunetics, 145 Bishop Allen Dr., Cambridge, MA, tel. 617-492-5416). Immunoprecipitations are performed as previously described (Wayner and Carter, 1987) from cell extracts or conditioned culture media of cells labeled with ^{35}S-met.

2. Colocalization of Reporter Cells and Instructive ECM in Tissue

It is important to establish that the instructive activity of the identified ECM component is physiologically relevant for the reporter cell. Therefore, the instructive ECM and the reporter cells should colocalize by immunostaining in tissue under the appropriate conditions (quiescence, development, wound repair, etc.). For a complete discussion of Immunostaining procedures see Hoffstrom and Wayner (1994).

3. Adhesion of Reporter Cells to Instructive ECM in Tissue

It is desirable to establish that the instructive ECM interacts with the reporter cells within tissue and that the interaction instructs a physiologically meaningful response *in vitro* and *in vivo*. This may be evaluated with the tissue adhesion assay described in Section II.B.4. An example is given in Section IV.C.

B. Purification of Instructive ECM with Mabs

As an example of the affinity purification of instructive ECM, we will focus on laminin 5 (Carter *et al.*, 1991). Laminin 5 is a heterotrimeric glycoprotein containing $\alpha3$, $\beta3$, and $\gamma2$ chains. Chains $\beta3$ and $\gamma2$ are unique to laminin 5, while $\alpha3$ is also a subunit of laminins 6 ($\alpha3\beta1\gamma1$) and laminin 7 ($\alpha3\beta2\gamma1$). Therefore, affinity purification of laminin 5 utilizes Mabs reacting with either the $\beta3$ or $\gamma2$ chains, followed by purification of laminin 6 and laminin 7 from the laminin 5-depleted material. Mabs are affinity purified from hybridoma culture supernatants on protein G-Sepharose, then covalently coupled to Affigel-10 following the manufacturer's procedures (Bio-Rad Laboratories). Instructive ECM is purified from concentrated conditioned culture medium prepared and concentrated as described in Section II.B.2. The secreted proteins are applied to Mab affinity columns coupled in series to bind one or more different ECM components. The number and possibly the order of the columns can be varied after some consideration of the desired purifications. The first column is (i) gelatin Sepharose to remove fibronectin, (ii) followed by immobilized anti-$\gamma2$ laminin chain Mab (B4-6) to remove laminin 5, (iii) followed by anti-$\alpha3$ laminin chain Mab (C2-5) to remove laminin 6 and 7, (iv) followed by anti-thrombospondin (Mab G2'-1) to removed thrombospondin-1. The columns are then disconnected from each other, washed separately with PBS to remove unbound material, then eluted separately with 3 *M* KSCN. The eluted protein peaks are dialyzed against 25 m*M* borate buffer pH 8.5, aliquotted into plastic tubes, and frozen rapidly on dry ice. It should be emphasized that purified ECM proteins are frequently unstable and inclined to aggregate and will tend to precipitate out of solution on repeated freeze–thaw.

C. Characterization of Responses in Reporter Cell by Instructive ECM

Once the instructive ECM component has been isolated, assays should be performed to verify that both the reporter cells and the purified instructive ECM promote the desired cell response *in vitro* and possibly *in vivo*. As an example, we will describe studies with laminin 5. Laminin 5 promotes a cell response in reporter HFKs that is distinct from the responses promoted by dermal collagen or fibronectin, although they all induce adhesion and spreading (Nguyen *et al.*, 2000a,b). First, HFKs plated on antibody-immobilized laminin 5, or exposed to laminin 5 coated beads, assemble gap junctions to a greater extent than when exposed to collagen or fibronectin (Lampe *et al.*, 1998). Second, adhesion and spreading of HFKs on collagen via $\alpha2\beta1$ is dependent on RhoGTP, a regulator of the actin cytoskeleton (Barry *et al.*, 1997; Hotchin *et al.*, 1995). In contrast, adhesion and spreading of HFKs on laminin 5, via $\alpha6\beta4$ and $\alpha3\beta1$, is dependent on phosphoinositide 3-OH-kinase (PI3K) and is not inhibited by toxin B, a Rho inhibitor (Just *et al.*, 1994). Further, deposition of laminin 5 onto the collagen surface switches adhesion ligands, receptors, and signaling from collagen-dependent to laminin 5-dependent. In the third example, the tissue adhesion assay was able to distinguish between laminin 5 and laminin 10/11 ($\alpha5\beta1\gamma1/\alpha5\beta2\gamma1$) as ligands for $\alpha6\beta4$. In *in vitro* adhesion assays, laminin 10/11 purified with Mab 4C7 is an adhesive ligand for integrins $\alpha3\beta1$ and $\alpha6\beta4$ (Kikkawa *et al.*, 1998, 2000). Further, laminin 10/11, like laminin 5, is expressed in mouse and human epidermal BM (Pierce *et al.*, 1998). This suggests that laminin 10/11 in epidermal BM may interact with integrins $\alpha3\beta1$ and $\alpha6\beta4$ in keratinoyctes. However, keratinocytes adhere to cryostat sections of wild-type skin that contains laminin 5, but not to skin from laminin 5 null mice, at 4°C (Fig. 5) or at 37°C in the presence of cytochalasin D (Ryan *et al.*, 1999). These conditions are inhibitory for $\beta1$ integrins but not $\alpha6\beta4$.

Laminin 10/11 may be a ligand for $\alpha3\beta1$ since keratinocytes adhere to laminin 5 null skin at 37°C when $\alpha3\beta1$ is active, and the adhesion is inhibited with anti-$\alpha3\beta1$ (P1B5) (Ryan *et al.,* 1999). Based on the tissue adhesion assay, $\alpha6\beta4$ interacts with laminin 5, whereas $\alpha3\beta1$ may interact with both laminin 5 and laminin 10/11. Consistent with this interpretation, targeted disruption of laminin 5 (Ryan *et al.,* 1999), not laminin 10/11 (Miner *et al.,* 1998), disrupts assembly of mature $\alpha6\beta4$-containing hemidesmosmes.

Each of these examples of responses in reporter cells is dependent on cell adhesion to the substratum via a specific adhesive ECM component, in this case laminin 5, but not via other adhesive ECM ligands. Mabs that disrupt the instruction without disrupting adhesion will contribute to an understanding of the mechanism of instruction.

V. Summary

Historically, Mabs have been one of the most productive and reliable methods for the identification of adhesion receptors and adhesive ECM ligands. In large part, this is because Mabs can identify the function of the adhesion components within the context of the complex ECM or the cell surface. There are now many isoforms of laminin, collagen, and other ECM components that have been identified by molecular and Mab approaches. It is not clear when and where these isoforms are expressed at the protein level, nor what unique functions each ECM isoform may serve within the context of tissue. Undoubtedly, specific *in vitro* assays in combination with specific Mabs will help illuminate the instructive roles of ECM components for reporter cells within *in vitro* models and tissue. Delineation of cell responses to the instructive ECM will require additional high-resolution technologies including DNA microarrays and targeted disruption of ECM components.

Acknowledgment

This work was supported by the National Institutes of Health grants CA49259 (W.G.C).

References

Barry, S. T., Flinn, H. M., Humphries, M. J., Critchley, D. R., and Ridley, A. J. (1997). Requirement for Rho in integrin signalling. *Cell Adhesion Commun.* **4,** 387–398.

Borradori, L., and Sonnenberg, A. (1999). Structure and function of hemidesmosomes: more than simple adhesion complexes. *J. Invest. Dermatol.* **112,** 411–418.

Burgeson, R. E., and Christiano, A. M. (1997). The dermal–epidermal junction. *Curr. Opin. Cell Biol.* **9,** 651–658.

Burgeson, R. E., Chiquet, M., Deutzmann, R., Ekblom, P., Engel, J., Kleinman, H., Martin, G. R., Meneguzzi, G., Paulsson, M., and Sanes, J. (1994). A new nomenclature for the laminins. *Matrix Biol.* **14,** 209–211.

Carter, W. G. (1982a). The cooperative role of the transformation-sensitive glycoproteins, GP140 and fibronectin, in cell attachment and spreading. *J. Biol. Chem.* **257,** 3249–3257.

Carter, W. G. (1982b). Transformation-dependent alterations in glycoproteins of extracellular matrix of human fibroblasts. Characterization of GP250 and the collagen-like GP140. *J. Biol. Chem.* **257,** 13805–13815.

Carter, W. G., and Hakomori, S. (1976). Isolation and partial characterization of "galactoprotein a" (LETS) and "galactoprotein b" from hamster embryo fibroblasts. *Biochem. Biophys. Res. Commun.* **76,** 299–308.

Carter, W. G., and Hakomori, S. (1979). Isolation of galactoprotein a from hamster embryo fibroblasts and characterization of the carbohydrate unit. *Biochemistry* **18,** 730–738.

Carter, W. G., Kaur, P., Gil, S. G., Gahr, P. J., and Wayner, E. A. (1990b). Distinct functions for integrins $\alpha3\beta1$ in focal adhesions and $\alpha6\beta4$/bullous pemphigoid antigen in a new stable anchoring contact (SAC) of keratinocytes: relationship to hemidesmosomes. *J. Cell Biol.* **111,** 3141–3154.

Carter, W. G., Ryan, M. C., and Gahr, P. J. (1991). Epiligrin, a new cell adhesion ligand for integrin $\alpha3\beta1$ in epithelial basement membranes. *Cell* **65,** 599–610.

Carter, W. G., Wayner, E. A., Bouchard, T. S., and Kaur, P. (1990a). The role of integrins $\alpha2\beta1$ and $\alpha3\beta1$ in cell–cell and cell–substrate adhesion of human epidermal cells. *J. Cell Biol.* **110,** 1387–1404.

Christiano, A. M., and Uitto, J. (1996). Molecular complexity of the cutaneous basement membrane zone. Revelations from the paradigms of epidermolysis bullosa. *Exp. Dermatol.* **5,** 1–11.

Delcommenne, M., and Streuli, C. H. (1999). Production of rat monoclonal antibodies specific for mouse integrins. *Methods Mol. Biol.* **129,** 19–34.

DiPersio, C. M., Hodivala-Dilke, K. M., Jaenisch, R., Kreidberg, J. A., and Hynes, R. O. (1997). Alpha3beta1 integrin is required for normal development of the epidermal basement membrane. *J. Cell Biol.* **137,** 729–742.

Farooqui, A. A. (1980). Purification of enzymes by heparin-sepharose affinity chromatography. *J. Chromatogr.* **184,** 335–345.

Gagnoux-Palacios, L., Allegra, M., Spirito, F., Pommeret, O., Romero, C., Ortonne, J. P., and Meneguzzi, G. (2001). The short arm of the laminin gamma2 chain plays a pivotal role in the incorporation of laminin 5 into the extracellular matrix and in cell adhesion. *J. Cell Biol.* **153,** 835–850.

Galfre, G., and Milstein, C. (1981). Preparation of monoclonal antibodies: strategies and procedures. *Methods Enzymol.* **73,** 3–46.

Gallatin, W. M., Weissman, I. L., and Butcher, E. C. (1983). A cell-surface molecule involved in organ-specific homing of lymphocytes. *Nature* **304,** 30–34.

George, E. L., and Hynes, R. O. (1994). Gene targeting and generation of mutant mice for studies of cell–extracellular matrix interactions. *Methods Enzymol.* **245,** 386–420.

Goldfinger, L. E., Hopkinson, S. B., deHart, G. W., Collawn, S., Couchman, J. R., and Jones, J. C. (1999). The alpha3 laminin subunit, alpha6beta4 and alpha3beta1 integrin coordinately regulate wound healing in cultured epithelial cells and in the skin. *J. Cell Sci.* **112,** 2615–2629.

Goldfinger, L. E., Stack, M. S., and Jones, J. C. (1998). Processing of laminin-5 and its functional consequences: role of plasmin and tissue-type plasminogen activator. *J. Cell Biol.* **141,** 255–265.

Hoffstrom, B. G., and Wayner, E. A. (1994). Immunohistochemical techniques to study the extracellular matrix and its receptors. *Methods Enzymol.* **245,** 316–346.

Hotchin, N. A., Gandarillas, A., and Watt, F. M. (1995). Regulation of cell surface beta 1 integrin levels during keratinocyte terminal differentiation. *J. Cell Biol.* **128,** 1209–1219.

Just, I., Fritz, G., Aktories, K., Giry, M., Popoff, M. R., Boquet, P., Hegenbarth, S., and von Eichel-Streiber, C. (1994). *Clostridium difficile* toxin B acts on the GTP-binding protein Rho. *J. Biol. Chem.* **269,** 10706–10712.

Kikkawa, Y., Sanzen, N., Fujiwara, H., Sonnenberg, A., and Sekiguchi, K. (2000). Integrin binding specificity of laminin-10/11: laminin-10/11 are recognized by alpha 3 beta 1, alpha 6 beta 1 and alpha 6 beta 4 integrins. *J. Cell Sci.* **113,** 869–876.

Kikkawa, Y., Sanzen, N., and Sekiguchi, K. (1998). Isolation and characterization of laminin-10/11 secreted by human lung carcinoma cells—laminin-10/11 mediates cell adhesion through integrin alpha-3-beta-1. *J. Biol. Chem.* **273,** 15854–15859.

Kreidberg, J. A. (2000). Functions of alpha3beta1 integrin [In Process Citation]. *Curr. Opin. Cell Biol.* **12,** 548–553.

Kurpakus, M. A., Quaranta, V., and Jones, J. C. (1991). Surface relocation of alpha 6 beta 4 integrins and assembly of hemidesmosomes in an *in vitro* model of wound healing. *J. Cell Biol.* **115,** 1737–1750.

Lampe, P. D., Nguyen, B. P., Gil, S., Usui, M., Olerud, J., Takada, Y., and Carter, W. G. (1998). Cellular interaction of integrin a3b1 with laminin 5 promotes gap junctional communication. *J. Cell Biol.* **143,** 1735–1747.

Marinkovich, M. P., Keene, D. R., Rimberg, C. S., and Burgeson, R. E. (1993). Cellular origin of the dermal–epidermal basement membrane. *Developmental Dynam.* **197,** 255–267.

Martin, P. (1997). Wound healing—aiming for perfect skin regeneration. *Science* **276,** 75–81.

Merkle, R. K., and Cummings, R. D. (1987). Lectin affinity chromatography of glycopeptides. *Methods Enzymol.* **138,** 232–259.

Miller, E. J., and Rhodes, R. K. (1982). Preparation and characterization of the different types of collagen. *Methods Enzymol.* **82 Pt A,** 33–64.

Miner, J. H., Cunningham, J., and Sanes, J. R. (1998). Roles for laminin in embryogenesis: exencephaly, syndactyly, and placentopathy in mice lacking the laminin alpha5 chain. *J. Cell Biol.* **143,** 1713–1723.

Nguyen, B. P., Gil, S. G., and Carter, W. G. (2000a). Deposition of laminin 5 by keratinocytes regulates integrin adhesion and signaling. *J. Biol. Chem.* **275,** 31896–31907.

Nguyen, B. P., Gil, S. G., Ryan, M. C., and Carter, W. G. (2000b). Deposition of laminin 5 in epidermal woulds regulates integrin signaling and adhesion. *Curr. Opin. Cell Biol.* **12,** 554–562.

Nguyen, B. P., Ren, X.-D., Schwartz, M. A., and Carter, W. G. (2001). Ligation of integrin $\alpha3\beta1$ by laminin 5 at the wound edge activates Rho-dependent adhesion of leading keratinocytes on collagen. *J. Biol. Chem.* **276,** 43860–43870.

Oi, V. T., and Herzenberg, L. A. (1980). Immunoglobulin producing hybrid cell lines. *In* "Selected Methods in Cellular Immunology" (B. B. Mishell and S. M. Shiigi, eds.). W. H. Freeman and Co., San Francisco, pp. 351–373.

Perou, C. M., Jeffrey, S. S., van de Rijn, M., Rees, C. A., Eisen, M. B., Ross, D. T., Pergamenschikov, A., Williams, C. F., Zhu, S. X., Lee, J. C., Lashkari, D., Shalon, D., Brown, P. O., and Botstein, D. (1999). Distinctive gene expression patterns in human mammary epithelial cells and breast cancers. *Proc. Natl. Acad. Sci. USA* **96,** 9212–9217.

Pierce, R. A., Griffin, G. L., Mudd, M. S., Moxley, M. A., Longmore, W. J., Sanes, J. R., Miner, J. H., and Senior, R. M. (1998). Expression of laminin alpha3, alpha4, and alpha5 chains by alveolar epithelial cells and fibroblasts. *Am. J. Resp. Cell Mol. Biol.* **19,** 237–244.

Pytela, R., Suzuki, S., Breuss, J., Erle, D. J., and Sheppard, D. (1994). Polymerase chain reaction cloning with degenerate primers: homology-based identification of adhesion molecules. *Methods Enzymol.* **245,** 420–451.

Robinson, J., and Gospodarowicz, D. (1984). Effect of *p*-nitrophenyl-beta-D-xyloside on proteoglycan synthesis and extracellular matrix formation by bovine corneal endothelial cell cultures. *J. Biol. Chem.* **259,** 3818–3824.

Rousselle, P., Lunstrum, G. P., Keene, D. R., and Burgeson, R. E. (1991). Kalinin: an epithelium-specific basement membrane adhesion molecule that is a component of anchoring filaments. *J. Cell Biol.* **114,** 567–576.

Ryan, M. C., Lee, K., Miyashita, Y., and Carter, W. G. (1999). Targeted disruption of the LAMA3 gene in mice reveals abnormalities in survival and late stage differentiation of epithelial cells. *J. Cell Biol.* **145,** 1309–1323.

Ryan, M. C., Tizard, R., VonDevanter, D. R., and Carter, W. G. (1994). Cloning of the LamA3 gene encoding the $\alpha3$ chain of the adhesive ligand epiligrin: expression in wound repair. *J. Biol. Chem.* **269,** 22779–22787.

Sanchez-Madrid, F., and Springer, T. A. (1986). Production of Syrian and Armenian hamster monoclonal antibodies of defined specificity. *Methods Enzymol.* **121,** 239–244.

Seed, B. (1995). Developments in expression cloning. *Curr. Opin. Biotechnol.* **6,** 567–573.

Stamper, H. B., and Woodruff, J. J. (1976). Lymphocyte homing into lymph nodes: *in vitro* demonstration of the selective affinity of recirculating lymphocytes for high-endothelial venules. *J. Exp. Med.* **144,** 828–833.

Taggart, R. T., and Samloff, I. M. (1983). Stable antibody-producing murine hybridomas. *Science* **219,** 1228–1230.

Takada, Y., Wayner, E. A., Carter, W. G., and Hemler, M. E. (1988). Extracellular matrix receptors, ECMRII and ECMRI, for collagen and fibronectin correspond to VLA-2 and VLA-3 in the VLA family of heterodimers. *J. Cell. Biochem.* **37,** 385–393.

Timpl, R., and Aumailley, M. (1989). Biochemistry of basement membranes. *Adv. Nephrol. Necker Hosp.* **18,** 59–76.

Timpl, R., Paulsson, M., Dziadek, M., and Fujiwara, S. (1987). Basement membranes. *Methods Enzymol.* **145,** 363–391.

Toda, K., Tuan, T. L., Brown, P. J., and Grinnell, F. (1987). Fibronectin receptors of human keratinocytes and their expression during cell culture. *J. Cell Biol.* **105,** 3097–3104.

Tsubota, Y., Mizushima, H., Hirosaki, T., Higashi, S., Yasumitsu, H., and Miyazaki, K. (2000). Isolation and activity of proteolytic fragment of laminin-5 alpha3 chain. *Biochem. Biophys. Res. Commun.* **278,** 614–620.

Verrando, P., Schofield, O., Ishida-Yamamoto, A., Aberdam, D., Partouche, O., Eady, R. A., and Ortonne, J. P. (1993). Nicein (BM-600) in junctional epidermolysis bullosa: polyclonal antibodies provide new clues for pathogenic role. *J. Invest. Dermatol.* **101,** 738–743.

Wayner, E. A., and Carter, W. G. (1987). Identification of multiple cell adhesion receptors for collagen and fibronectin in human fibrosarcoma cells possessing unique alpha and common beta subunits. *J. Cell Biol.* **105,** 1873–1884.

Wayner, E. A., Carter, W. G., Piotrowicz, R. S., and Kunicki, T. J. (1988). The function of multiple extracellular matrix receptors in mediating cell adhesion to extracellular matrix: preparation of monoclonal antibodies to the fibronectin receptor that specifically inhibit cell adhesion to fibronectin and react with platelet glycoproteins Ic-IIa. *J. Cell Biol.* **107,** 1881–1891.

Woodley, D. T., O'Toole, E., Nadelman, C. M., and Li, W. (1999). Mechanisms of human keratinocyte migration: lessons for understanding the re-epithelialization of human skin wounds. *In* "Progress in Dermatology" (A. N. Moshell, ed.). Vol. 33, pp. 1–12.

Xia, Y., Gil, S. G., and Carter, W. G. (1996). Anchorage mediated by integrin alpha6beta4 to laminin 5 (epiligrin) regulates tyrosine phosphorylation of a membrane-associated 80-kD protein. *J. Cell Biol.* **132,** 727–740.

CHAPTER 2

Isolation and Purification of Proteoglycans

John M. Whitelock* and Renato V. Iozzo[†]

*CSIRO Molecular Science
North Ryde 21113, Sydney, Australia

†Department of Pathology, Anatomy and Cell Biology
Thomas Jefferson University
Philadelphia, Pennsylvania 19107

I. Overview and Scope

Proteoglycans belong to an expanding family of complex macromolecules that are primarily, but not exclusively, located at the cell membranes and within the extracellular matrices of most vascular and avascular tissues. They have been implicated in a variety of physiological and pathological processes, some of which are just beginning to be elucidated. By virtue of their large hydrodynamic size, high net charge, and heterogeneity, they have been difficult to study using standard biochemical methods. Proteoglycan core proteins themselves are highly diverse in structure and this, together with the highly variable glycosaminoglycan chains, make them central players in various signaling events. Proteoglycans, for example, interact with hyaluronan, lectins, and numerous growth factors and cell surface receptors. They modulate angiogenic cues and are involved in filtration, axonal guidance, and morphogenesis. Proteoglycans comprise a large family of molecules with more than 40 different genes, which is made more complicated by the ability of these genes to encode different isoforms generated by alternative splicing, as well as further complexities provided by posttranslational modifications.

Proteoglycans have been categorized and named depending on their distribution, biochemical structure, and, when possible, their function (Iozzo, 1998). They consist of a protein component to which is attached one or more glycosaminoglycan (GAG) side chains. The latter are covalently attached to the protein backbone, usually via serine residues, although keratan sulfate has been shown to be coupled to asparagine residues (Funderburgh, 2000). The major types of GAG chains that have been isolated and characterized include chondroitin sulfate (CS), dermatan sulfate (DS), keratan sulfate (KS), hyaluronic acid (HA), heparan sulfate (HS), and heparin (He). Each type of GAG chain has a distinctive polymeric structure in which the repeat disaccharides are linked by certain bonds. These chains endow the proteoglycans with unique biochemical characteristics, which are so important for several of their observed bioactivities (Esko and Lindahl, 2001; Gallagher, 2001; Iozzo, 2001a). They are usually long, unbranched polymeric chains, which have areas of highly charged density due to the substitution of sulfate groups at various positions on either the hexosamine or the hexuronic acid saccharide units. The other characteristic of proteoglycans is their relatively large molecular weight, which is due in part to the presence of these long GAG chains that attract and entrap water molecules. The latter increase their relative hydrodynamic size. Most of the purification methodologies used to isolate proteoglycans from biological systems have taken into account these two facets of their structure, with most procedures utilizing some form of anion exchange chromatography followed by separation on gel filtering matrices. Because there are many types of proteoglycans that have different structures, no two proteoglycans can be purified by exactly the same procedure. Over the years, there have been many reviews on this topic of proteoglycan purification (Vogel and Peters, 2001; Whitelock, 2001; Yanagishita, 2001; Yamaguchi, 2001). Given the vast and complex nature of the literature, we will present a simplified review of the methodology that would be suitable for the purification of most proteoglycans, and we will update it to include methods that have been developed for use on high-pressure chromatography systems such as FPLC. We will also provide a section describing the use of monoclonal antibodies to isolate the proteoglycan perlecan and discuss its advantages compared to other more standard purification protocols.

This approach can easily be adjusted to encompass other proteoglycans given the increasing availability of monoclonal antibodies directed to specific protein core epitopes. Throughout this chapter, we will mostly refer to review articles rather than to the original work where the methodology was first described. These references can be traced by referring to these reviews. Thus, we would like to acknowledge all those pioneers in the area of proteoglycan purification, a number of whom are still very active in the field today, for their contributions. Our chapter starts with a section on the extraction of proteoglycans from cultured cell lines and tissues and moves onto the use of chromatographic procedures used to isolate them. This section will include the use and advantages of immunopurification protocols, which will lead into a small section on the characterization of the isolated proteoglycans with antibodies (some of which are commercially available) as probes.

II. Extraction of Proteoglycans

Proteoglycans have been extracted from both tissues and cell cultures. A general flow chart that can be applied to the isolation and purification of various proteoglycan species is provided in Fig. 1. This figure presents a path that researchers can take to isolate their particular proteoglycan (Iozzo, 1989; Yanagishita, 2001). It is only meant to be a guide and in most cases will need to be modified for each proteoglycan species. As you work your way down this flow chart most of the steps will be described in more detail in the following sections. Variations on the theme for the extraction of proteoglycans from diverse sources have been recently reviewed (Iozzo, 2001b). Bulk extraction from tissue invariably involves the use of chaotropic dissociating agents such as urea or guanidine (Yanagishita *et al.*, 1987; Hascall *et al.*, 1994; Fischer *et al.*, 1996) and detergents to interrupt hydrophobic interactions of some of the membrane intercalated proteoglycans. Guanidine is an ionic dissociative agent and hence is not compatible with anion-exchange chromatography. In order to maintain the solubility of proteoglycans throughout the chromatographic procedure, the samples can be dialyzed against urea. Extraction of proteoglycans from the cultured cells themselves and/or the matrix produced by these cells also involves the use of either urea or guanidine and either the zwitterionic detergent 3-[(3-cholamidopropyl)dimethylammonio]1-propanesulfonate (CHAPS) or the nonionic detergent Triton X-100 (Yanagishita, 2001). The advantage of Triton X-100 is that it is more affordable than CHAPS to include in the buffers; however, it is more difficult to dialyze away as it forms large micelles. Proteoglycans have also been purified from the conditioned medium of cell cultures without the use of dissociating agents. However, it is important to remember that this may lead to the selection of proteoglycans that are more readily soluble in aqueous buffers. It is unknown if these are different from those that cells deposit in the cell-associated and matrix compartments.

Proteoglycans are susceptible to degradation by both proteases and glycosaminoglycan lyases. Proteases have been studied extensively over the years and this has led to an extensive range of inhibitors being available for inclusion in buffers to control their activity. The GAG-degrading enzymes from mammalian sources are poorly understood and most researchers use enzymes from bacterial sources to characterize their proteoglycans. Some of these include heparinase I, II, and III that degrade the HS-containing PGs, chondroitinase ABC and AC that degrade the chondroitin- and dermatan-containing PGs, and keratanase that degrades keratan sulfate-containing PGs. (All of these enzymes are available commercially from Seikagaku.) Some chemical inhibitors of the bacterial enzymes such as EDTA have been described but they are

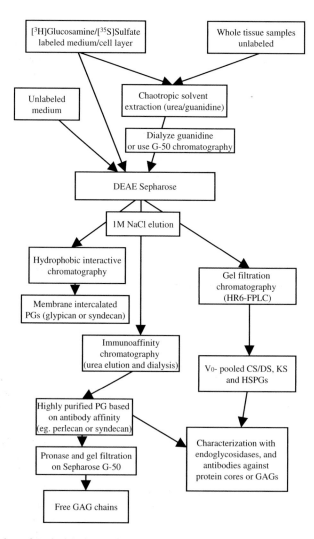

Fig. 1 Flow chart of methodologies used to isolate proteoglycans. This flow chart is designed to give an indication of some of the methods available that might be used to isolate a proteoglycan, together with suggested pathways to isolate and characterize that proteoglycan. It is meant as a guide only and may need to be modified for each particular proteoglycan species under investigation. It is also useful in the consideration of isolating free GAG chains from either a pool of proteoglycans or those that have been more highly purified.

not specific to this class of enzymes. The common protease inhibitors that are used in proteoglycan purification buffers include benzamidine hydrochloride and phenylmethylsulfonyl fluoride (PMSF) for serine proteases, disodium ethylenediamine tetraacetate (EDTA) for metalloproteases, and N-ethylmaleimide for thiol proteases. All of these are used at a final concentration of 10 mM except for PMSF, which can be used at lower concentrations. They can be stored as 100-fold concentrated stock solutions in aqueous solutions except for PMSF, which needs to be dissolved in ethanol and added just prior to use.

III. Radiolabeling of Proteoglycans for Ease of Purification

Prior to purification, cell cultures are often pulsed with radioactive Na_2SO_4 to label the sulfated GAG chains so that they can be followed throughout the procedure. It is difficult to do this with tissue samples and, in most cases, proteoglycans are extracted from tissue samples in a bulk fashion and their presence is monitored by chemical means that identify the reducing capacity of the attached GAG chains. With a few exceptions, organ cultures are of limited value primarily because of the hypoxic changes and the difficulty in maintaining a constant environment for proper and continuous radiolabeling. ^{3}H-Glucosamine can be used to label proteoglycan GAG chains but it is important to remember that this will also be incorporated into hyaluronic acid chains. Some researchers have used double-labeling procedures to enable an investigation of the degree of sulfation on the glycosaminoglycan chains. Glucosamine labeling has an advantage over $^{35}SO_4$ in that it has a relatively longer half-life. Although preferentially incorporated into HS, one has to keep in mind that glucosamine can be epimerized to galactosamine and, thus, be incorporated into CS- and DS-containing PGs. Efficient radiolabeling enables specific downstream structural analysis of GAG types to be performed (Turnbull et al., 1999; Merry et al., 1999). ^{35}S-Methionine or ^{35}S-cysteine can be utilized to label the protein component of the PGs. This is particularly useful in combination with immunoprecipitation procedures so that a particular proteoglycan species is isolated. Immunoprecipitation using antibodies that react with GAG stub epitopes can be useful, together with specific glycanases (i.e., glycosaminoglycan-degrading enzymes or lyases, described in Section II), to identify unknown proteoglycan molecules (Couchman and Tapanadechopone, 2001). In all cases, these radiolabels can be added to medium to give a final concentration between 10 and 100 μCi/ml and incubated for a periods of 6–72 h.

IV. Anion-Exchange Chromatography

The presence of the GAG chains on proteoglycans provides them with a relatively low pI. This property has allowed the successful use of anion-exchange resins to separate proteoglycans from complex biological milieus. It also means that the pH of the buffers can be lowered to around 6 with the result that most other proteins, which usually bind to these resins at neutral pH, will fail to bind, thereby improving the purity of the fractions. Most of the early work with anion-exchange columns involved the use DEAE cellulose-based resins (Hascall et al., 1994). This evolved into the use of DEAE agarose-based resins as the bed volume of these columns does not fluctuate with changes in temperature, pressure, or salt concentration. The use of weak anion-exchange resins has been expanded to include strong anion-exchange resins, such as Mono-Q resin, which is more compatible with higher-pressure chromatographic systems. Proteoglycans are eluted from these columns by increasing the salt concentration in the starting buffer. It is important to eliminate contaminating glycoproteins. This is in general achieved by using a gradient of NaCl, with most glycoproteins eluting at 0.3–0.5 M NaCl. Most proteoglycans are eluted by salt up to a concentration of 1 M as demonstrated by the lack of ^{35}S-labeled material in 2 M NaCl washes subsequent to a 1 M NaCl elution. However, it is conceivable that other proteoglycans from some biological sources may require greater concentrations. Heparin, which is one of the most highly charged GAG molecules of mammalian systems, is displaced from anion-exchange resins by salt concentrations in the range 1–1.5 M NaCl.

Chondroitin/dermatan sulfate-containing or HS-containing proteoglycans have been separated from each other using a gradient elution profile on anion-exchange columns (Benitz et al., 1990; Hascall et al., 1994; Whitelock et al., 1997). The CS/DS-containing molecules tend to have less charge density than the HS proteoglycans and so elute fractionally earlier when the salt gradient is shallow enough to allow proper separation. However, in some instances, CS-PGs or DS-PGs can elute at higher NaCl molarity than HS-PGs. It is advisable to characterize each batch of pooled fractions from these runs in order to maintain reproducibility as the elution times of each species can vary from run to run. It is also important to remember that this is a pooled fraction and hence contains many types of CS/DS and HS proteoglycan species.

Anion-exchange chromatography is a very useful first step that can be included in most proteoglycan purification protocols including those from tendon (Vogel and Peters, 2001), nervous tissue (Yamaguchi, 2001), mineralized tissues (Fedarko, 2001), and *Drosophila* (Staatz et al., 2001).

A. General Protocol for Conditioned Media, Cell Layers, or Matrices

1. After labeling, centrifuge the medium at 1500 g for 10 min to remove cell debris and store at −20° or −70°C (labeled CM).

2. The cells (approx. $5 \times 10^6/75$ cm^2 flask) can be removed from the flask leaving the extracellular matrix behind by first washing the cells with PBS, and then incubating the cell layers with a minimal volume (approx. 2 ml) of 0.02 N NH$_4$OH for 5 min. This can be repeated to ensure the complete removal of the cells. The matrix can then be extracted with either 6 M urea, 0.5% Triton X-100, 100 mM NaCl, 20 mM Tris, 0.02% NaN$_3$, pH 8.0 (with protease inhibitors), or guanidine in place of the urea and CHAPS in place of the Triton (labeled ECM).

3. Proteoglycans can be extracted from the cell layer and the ECM together with the same buffers described above. However, one has to be careful of contaminating DNA in the samples originating from the lysed cells. This can be removed by further digestion with 100 μg/ml DNAase (Sigma) for 4 h at 37°C (Labelled CL/ECM).

4. Protease inhibitors, benzamidine, ε-aminocaproic acid, EDTA, and PMSF are added to all buffers at a final concentration of 10 mM.

5. Prior to FPLC anion-exchange chromatography, solid urea is added to the CM fraction to give a final concentration of 6 M and it is filtered through a 0.45 μm nitrocellulose filter.

6. Fractions labeled CM, ECM, and CL/ECM can all be treated in a similar manner for anion-exchange chromatography. If guanidine is used, it must be dialyzed and the buffer exchanged to urea. Alternatively, buffer can be exchanged using Sepharose G-50 (Amersham Pharmacia) chromatography and disposable 10-ml pipettes as described before (Yanagishita, 2001). In the latter case, the V_0 is collected and further purified by anion-exchange chromatography.

7. Apply the samples to either a 1-ml or 10-ml (up to 100 ml on the 10-ml column) Mono-Q Sepharose column (Pharmacia) equilibrated with 6 M urea, 0.5% Triton X-100, 100 mM NaCl, 20 mM Tris, 0.02% NaN$_3$, pH 8.0 (with protease inhibitors). The flow rate for the 1-ml column is routinely 0.5 ml/min, whereas the flow rate for the 10-ml column is 1.0 ml/min.

8. The column is then washed with this buffer until the A_{280} nm baseline is reached. Bound molecules are subsequently eluted from the column using a linear gradient of NaCl from 300 mM to 1 M in start buffer (over 70 min), and collected in 1-ml (10-ml column) or 0.5-ml (1-ml column) fractions.

9. Add 50 to 100-μl aliquots of each fraction to 3 ml of scintillation fluid (Insta-gel; Packard) and count in a scintillation counter (LKB 1217 Rackbeta or equivalent).

10. The fractions containing radioactivity are finally pooled for further analysis.

B. Anion-Exchange Chromatography of Soluble Proteoglycans

Anion-exchange chromatography, in the absence of urea, can be used to isolate soluble proteoglycans in conditioned medium samples that have not been labeled using radioactive tracers.

1. Equilibrate a 100-ml bed volume of DEAE Sepharose Fast Flow (Amersham Pharmacia) with 250 mM NaCl, 20 mM Tris, 10 mM EDTA, pH 7.5, at 4°C.

2. Apply the conditioned medium (1.0–3.0 liter) to the column at a flow rate of 1–2 ml/min.

3. Wash the column with 250 mM NaCl, 20 mM Tris, 10 mM EDTA, pH 7.5, to obtain a baseline as monitored by ultraviolet at 280 nm.

4. Elute bound PGs with 1 M NaCl, 20 mM Tris, 10 mM EDTA, pH 7.5.

5. Wash DEAE Sepharose column with 2 M NaCl, 20 mM Tris, 10 mM EDTA, pH 7.5, and then 1 M NaCl, 20 mM Tris, 10 mM EDTA, pH 7.5, and store at 4°C.

V. Hydrophobic Interaction Chromatography

This form of chromatography is useful to separate proteoglycans on the basis of the hydrophobicity of their protein cores. In particular, membrane-intercalated proteoglycans are well suited to this methodology. These methods have been described in detail elsewhere (Yanagishita *et al.,* 1987; Hascall *et al.,* 1994; Midura *et al.,* 1990). One has to be careful with hydrophobic interaction chromatography, as molecules can be irreversibly bound to some matrices and it would be prudent to perform test runs with small samples to assess the suitability of this approach for your proteoglycan species.

VI. Gel Filtration Chromatography

This form of chromatography has been used in most proteoglycan purification protocols because of the other important characteristic of proteoglycans, their large hydrodynamic size relative to most other biological molecules. This method can be used to separate labeled proteoglycans from free label in conditioned media samples (Hascall *et al.,* 1994), as well as to separate out proteoglycans from other negatively charged molecules that coelute in anion-exchange columns. Initially, low-pressure matrices were used in size exclusion columns (Yanagishita *et al.,* 1987), but as matrices that could withstand both higher pressure and chaotropic buffers became available, the methodology was adapted to these systems. An example of this is the use of HR6 FPLC columns (Yanagishita, 2001; Whitelock *et al.,* 1997). These columns are useful in that they can separate whole proteoglycan populations eluted from anion-exchange resins, but they can also be used, in combination with specific glycosaminoglycan lyases, to characterize the nature and type of the proteoglycans (Whitelock *et al.,* 1997; Yanagishita, 2001). These columns are also useful for monitoring the degradation of proteoglycans by proteases (Poe *et al.,* 1992) or endoglycosidases (Maeda *et al.,* 1995), estimating the sizes of GAG chains that are attached to a protein core (Melrose and Ghosh, 1993), and separating whole proteoglycans from free GAG chains and full-length native proteoglycans from either truncated recombinant forms (Graham *et al.,* 1999) or proteoglycan fragments (Oda *et al.,* 1996). The HR6 FPLC columns allow for quick and reliable protocols that can be repeated throughout the working day enabling a fast turnaround in results. This system also enables a reliable comparison to be made among runs performed on the same day, since both the volume and timing of runs are very well controlled.

A. General Procedure

1. Equilibrate a HR6/30 Superose FPLC column (Pharmacia) with 6 M urea, 0.5% Triton X-100, 300 mM NaCl, 20 mM Tris, 0.02% NaN$_3$, pH 8.0 (with protease inhibitors). For estimating the size of GAG chains, equilibrate and run the column with 0.5 M CH$_3$COONa, 0.05% Triton-X 100, pH 7.0.

2. Apply the samples (via the sample loop) and chromatograph them at 0.4 ml/min, collecting 35 × 2-min fractions.

3. Assay each fraction for radioactive labeled proteoglycans or GAGs; or estimate the presence of a particular species using another suitable approach, such as dot or slot immunoblot with antibodies specific for a given protein core.

B. Glycosaminoglycanase Analysis to Investigate PG Type

1. Aliquot equal amounts of purified proteoglycans (usually based on radioactivity, cpm) into Eppendorf tubes.

2. Add either heparinase III (final concentration of 0.1 U/ml in PBS), or chondroitinase ABC (final concentration 0.5 U/ml in PBS) and incubate at 37°C for 6–8 h.

3. Apply the sample to the HR6/30 Superose FPLC column and perform the gel filtration chromatography as described in Section VI.A.

4. Analyse the peak profile and assess whether either or both of the enzyme treatments has shifted the peak to the void volume of the column. A shift to the right is indicative of degradation, which identifies the nature of the GAG chains attached to the proteoglycans present in the sample.

If a tissue or cellular extract is chromatographed through an anion-exchange column followed by gel filtration, it will contain many individual types of proteoglycans. Other chromatographic steps can be included to further purify the proteoglycan of choice. These can include hydrophobic interaction chromatography as mentioned earlier, rechromatographing samples over another gel filtration column (Castillo *et al.,* 1996), or using antibody affinity chromatography (Aviezer *et al.,* 1994; Whitelock, 2001).

VII. Immunoaffinity Chromatography

This form of chromatography is based upon the use of antibodies (mostly monoclonal) that are specific for a particular epitope present in the proteoglycan. This is usually on the protein core; however, antibodies that bind the GAG chains are becoming more widely available. The anti-HS monoclonal antibody 10E4 is a reliable and well-characterized antibody (David *et al.,* 1992) and is available commercially from Seikagaku. Other anti-HS monoclonal antibodies such as JM403 and single-chain variable regions generated by phage display technology are well characterized (Van Kuppevett *et al.,* 1998) but are not commercially available. The monoclonal anti-CS monoclonal antibody (CS-56; Avnur and Geiger, 1984) is commercially available from various suppliers including Sigma and ICN Flow laboratories. A major issue with this approach is the cost and availability of the appropriate antibodies. In-house developed monoclonal antibodies have a major advantage in this area but do require an investment of time to be produced and characterized. If this form of chromatography is to be used, the sample buffer should not contain either urea or guanidine, and the pH must be maintained between 7 and 8. It is also advisable to include NaCl at a concentration of 1 *M* to control for nonspecific ionic interactions between the isolated proteoglycan on the column and other molecules present in the mixture.

A. General Protocol

1. Conjugate the appropriate antibody to cyanogen bromide-activated Sepharose or other suitable matrix in the ratio of 5 mg of antibody per ml of matrix.

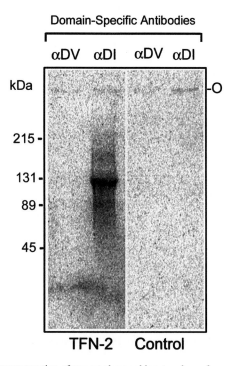

Fig. 2 Differential immunoseparation of truncated recombinant perlecan from native full-length material. TFN-2 is a cell line established from HEK-293 cells that had been transfected with a construct encoding the N-terminal Domain I of perlecan (Graham *et al.,* 1999). Media conditioned by these cells in the presence of 5 μCi/ml of $^{35}SO_4$, along with medium conditioned by the untransfected parental cells (Control), were immunoprecipitated with domain-specific antibodies directed toward either Domain V (mAb A74, αDV) or Domain I (mAb A71, αDI). The aim of this was to establish a purification protocol that would separate full-length endogenous perlecan from truncated recombinant forms. The media samples were first immunoprecipitated with mAb A74 and the supernatants from these incubations were subsequently incubated with mAb A71. Note the presence of the labeled truncated Domain I recombinant with an approximate molecular weight of 130 kDa (20 kDa protein plus ~110 kDa of GAG decoration) only in the stable transfected TNF-2 cells. The endogenous perlecan, which does not enter the gel, could be alternatively removed by an initial immunoaffinity chromatography purification step utilizing αDV mAb. Molecular weight markers (in kDa) are shown on the left margin. Origin (O).

2. Equilibrate the immunoaffinity column with 1 M NaCl, 20 mM Tris, 10 mM EDTA, pH 7.5.

3. Wash with 5 column volumes of 6 M urea in PBS and then 5 column volumes of 1 M NaCl, 20 mM Tris, 10 mM EDTA, pH 7.5.

4. This is repeated while monitoring the A_{280} nm baseline to ensure that antibody is not leaking from the column.

5. After the column is stable, apply the 1 M NaCl eluate from the DEAE Sepharose column directly, and cycle the eluate over the column overnight in a cold room at a slow flow rate.

6. Elute the proteoglycans by loading 1 column volume of 6 M urea in PBS, stop the flow, and let the column stand for 1 h. Resume flow and collect the eluate until an A_{280} nm baseline is achieved.

7. Wash the column with 1 M NaCl, 20 mM Tris, 10 mM EDTA, pH 7.5. A further peak may elute at this stage; this can be collected and pooled with the first peak.

8. Wash the column with 10 column volumes of 1 M NaCl, 20 mM Tris, 10 mM EDTA, pH 7.5, and store at 4°C.

9. Remove the urea and 1 M NaCl in the samples by dialysis against PBS, concentrate if required, and store aliquots at -70°C.

This method can be adapted to include a further purification step through a second immunoaffinity column, which has had another antibody conjugated to it that has different binding characteristics from the first. This is useful in separating full-length, endogenously produced proteoglycans from truncated recombinant forms. For example, it can be used to separate the perlecan produced naturally by the human embryonic kidney cells (HEK-293) from a truncated Domain I product, with which these cells have been transfected (Fig. 2). In the case of the data shown in Fig. 2, separation relies on the fact that the domain specificity of these antibodies had been previously characterized (Whitelock *et al.,* 1996). Antibody A74 binds the C-terminal Domain V of perlecan and so does not immunopurify the 130-kDa Domain I, whereas A71 binds the N-terminal Domain I. Samples containing the two perlecan variants can be first passed over an A74 immunoaffinity column, which removes the full-length material that might otherwise contaminate the immunopurified recombinant PG (Fig. 2). This procedure has also been successfully used for recombinant perlecan fragments, which are produced as fusion partners with enhanced green fluorescent protein.

VIII. Quantification and Characterization

Purified proteoglycans can be quantified by various methods. Uronic acid estimation is a useful method to identify chemical quantities of GAGs (Blumenkrantz and Asboe-Hansen, 1973; Vilim, 1985), which usually corresponds well to the yields obtained from tissue extracts. However, more sensitive assays are often required for smaller amounts of material, which is usually the case from tissue culture systems. Bicinchoninic acid (BCA) and Coomassie plus protein assays (both from Pierce), which were developed for protein estimation, have been used for smaller quantities of proteoglycans. These methods are both unsuitable for free GAG chains. However, these assays are prone to interference by the presence of the GAG chains on the proteoglycans which may lead to an under- or overestimation of the amount present (Vogel and Peters, 2001).

ELISAs using antibodies that react with either the GAG chain or the protein core can be used to characterize purified proteoglycans. This is particularly useful in monitoring the presence of PGs through a purification procedure (although the concentration of either salt or urea needs to be lowered to allow proper coating of the

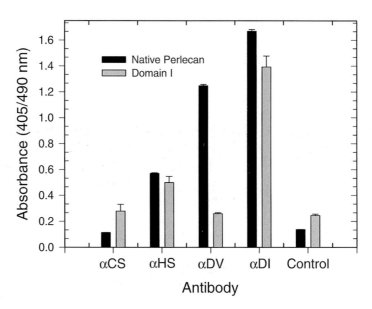

Fig. 3 Characterization of immunopurified full-length perlecan and recombinant Domain I using ELISA. The methodology used is common to all ELISA protocols except that filtered 0.1% casein in PBS is used as both a blocking agent and diluent for the antibodies. The main reason to use casein was high backgrounds obtained for the 10E4 antibody when bovine serum albumin was used. αCS, immunoreactivity obtained using CS56 antibodies (Sigma Cat. No. C8035); αHS, immunoreactivity obtained using 10E4 antibodies (Seikagaku Cat. No. 370255); αDV, immunoreactivity obtained using A74, which is specific for Domain V of perlecan; αDI, immunoreactivity obtained using either A71 or A76, which are specific for Domain I of perlecan. Values are the mean ± S.D. of triplicate determinations.

microtiter plates). ELISAs can also be used to characterize immunopurified preparations of native and recombinant perlecan (Fig. 3). In this case, the antibody that reacts with Domain V will only react with the full-length molecule whereas both the Domain I and V antibodies will react with both species. Also, significant reactivity with both the anti-HS antibody 10E4 (Seikagaku Cat. No. 370255) and the anti-CS antibody CS56 (Sigma Cat. No. C8035) indicates the presence of both types of GAG chains on the recombinant form, but not on the native full-length perlecan (Fig. 3). ELISAs can be useful in characterizing GAG chain differences between preparations of the same proteoglycan. For example, an ELISA has been developed to examine full-length perlecan isolated from various cell sources. It uses an anti-protein core antibody (A76) and an anti-heparan sulfate antibody (10E4) in the following protocol. All of the treatments were 50 μl/well, all of the washes were 200 μl/well, and the substrate was added at 100 μl/well. Obviously this approach can be utilized for other proteoglycans with distinct protein cores, given the increasing availability of monospecific antibodies.

A. General Protocol

1. Serially dilute perlecan across wells of a 96-well microtiter plate. Start the perlecan concentration at 20 μg/ml.

2. Incubate for at least 2 h at RT, wash with PBS, and block the wells with 0.1% casein in PBS for 2 h at RT.

3. Wash the wells twice with PBS and incubate with either the anti-protein core antibody (A76; 5 μg/ml) or 10E4 (anti-HS; 1/1000) for at least 2 h at RT.

4. Wash the wells twice with PBS and incubate with biotinylated anti-mouse antibodies (1/1000 in 0.1% casein in PBS) and incubate for 1 h at RT.

5. Wash the wells twice with PBS and incubate with streptavidin conjugated to horseradish peroxidase (1/500 in 0.1% casein in PBS) for 30 min at RT.

6. Wash the wells four times with PBS and incubate with a solution of 2,2'-azino bis(3-ethylbenzthiazoline-6-sulfonic acid) (ABTS) as substrate (1.1 mg/ml ABTS, with 0.02% (v/v) H_2O_2).

7. Incubate at RT until a suitable color is developed and read in a spectrophotometer at 405 nm.

The results from these ELISAs can be presented by graphing the absorbances obtained with one antibody vs the values obtained for the other antibody. The data are presented as slopes of the lines of best fit of mAb 10E4 vs mAb A76 reactivity. This type of analysis is advantageous since the lines of best fit (obtained with suitable computer packages such as Sigma Plot 2001 or similar programs) are independent of the concentration of perlecan used to coat the ELISA wells. This may reflect either differences in HS substructure or a relative increase in the amount of HS chains that are attached to each protein core.

Western blot analysis is also useful in characterizing the presence and type of PGs present in a purified fraction. This becomes more powerful when you combine this with treatment by specific endoglycosidases. Each of these approaches has advantages and disadvantages, and it is important to utilize more than one of these methods to ensure that the conclusion drawn from the results is reliable. Methods that describe the use of endoglycosidases to identify and characterize proteoglycans are well established (Lindblom and Fransson, 1990). These have been successfully combined with Western blot analyses to identify the type and molecular weight of protein cores of purified proteoglycan fractions (Lories et al., 1992; Couchman and Tapanadechopone, 2001). An example of this is to perform Western blot analyses with the anti-HS stub antibody 3G10 (Seikagaku, Cat. No. 370260) which reacts with HS after heparinase III digestion (David et al., 1992). Another is to use the 10E4 and CS56 antibodies, before and after heparinase III and chondroitinase ABC digestions or a combination of the two. This will enable identification of most of the HS- and some of the CS-containing proteoglycan types present in a pooled sample. It will also indicate if any of the proteoglycans have both HS and CS attached to their protein core (Couchman et al., 1996). This approach can be expanded to include antibodies that react with KS and combine it with keratanase incubations (Midura et al., 1990).

Acknowledgments

The original work in the authors' laboratories was supported by National Institutes of Health grants RO1 CA-39481 and RO1 CA-47282, by Department of the Army grant DAMD17-00-1-0663 (to R.V.I.) and an Australian Government grant through the Cooperative Research Center Scheme (to J.M.W.). We also thank S. Knox, P. Bean, L. Graham, and D. Lock for valuable input and C. C. Clark for critical reading of the manuscript.

References

Aviezer, D., Hecht, D., Safran, M., Eisinger, M., David, G., and Yayon, A. (1994). Perlecan, basal lamina proteoglycan, promotes basic fibroblast growth factor–receptor binding, mitogenesis, and angiogenesis. *Cell* **79,** 1005–1013.

Avnur, Z., and Geiger, B. (1984). Immunocytochemical localization of native chondroitin-sulfate in tissues and cultured cells using specific monoclonal antibody. *Cell* **38,** 811–822.

Benitz, W. E., Kelley, R. T., Anderson, C. M., Lorant, D. E., and Bernfield, M. (1990). Endothelial heparan sulfate proteoglycan. I. Inhibitory effects on smooth muscle cell proliferation. *Am. J. Respir. Cell Mol. Biol.* **2,** 13–24.

Blumenkrantz, N., and Asboe-Hansen, G. (1973). New method for quantitative determination of uronic acids. *Anal. Biochem.* **54,** 484–489.

Castillo, G. M., Cummings, J. A., Ngo, C., Yang, W., and Snow, A. D. (1996). Novel purification and detailed characterization of perlecan isolated from the Engelbreth–Holm–Swarm tumor for use in an animal model of fibrillar A beta amyloid persistence in brain. *J. Biochem. (Tokyo)* **120,** 433–444.

Couchman, J. R., Kapoor, R., Sthanam, M., and Wu, R. R. (1996). Perlecan and basement membrane-chondroitin sulfate proteoglycan (bamacan) are two basement membrane chondroitin/dermatan sulfate proteoglycans in the Engelbreth–Holm–Swarm tumor matrix. *J. Biol. Chem.* **271,** 9595–9602.

Couchman, J. R., and Tapanadechopone, P. (2001). Detection of proteoglycan core proteins with glycosamino-glycan lyases and antibodies. *In* "Proteoglycan Protocols" (R. V. Iozzo, ed.), Vol. 171, pp. 329–334. Humana Press, Totowa, NJ.

David, G., Bai, X. M., Van der Schueren, B., Cassiman, J. J., and Van den Berghe, H. (1992). Developmental changes in heparan sulfate expression: in situ detection with mAbs. *J. Cell Biol.* **119,** 961–975.

Esko, J. D., and Lindahl, U. (2001). Molecular diversity of heparan sulfate. *J. Clin. Invest.* **108,** 169–173.

Fedarko, N. S. (2001). Purification of proteoglycans from mineralized tissues. *In* "Proteoglycan Protocols" (R. V. Iozzo, ed.), Vol. 171, pp. 19–26. Humana Press, Totowa, NJ.

Fischer, D. C., Henning, A., Winkler, M., Rath, W., Haubeck, H. D., and Greiling, H. (1996). Evidence for the presence of a large keratan sulphate proteoglycan in the human uterine cervix. *Biochem. J.* **320,** 393–399.

Funderburgh, J. L. (2000). Keratan sulfate: structure, biosynthesis, and function. *Glycobiology* **10,** 951–958.

Gallagher, J. T. (2001). Heparan sulfate: growth control with a restricted sequence menu. *J. Clin. Invest.* **108,** 357–361.

Graham, L. D., Whitelock, J. M., and Underwood, P. A. (1999). Expression of human perlecan domain I as a recombinant heparan sulfate proteoglycan with 20-kDa glycosaminoglycan chains. *Biochem. Biophys. Res. Commun.* **256,** 542–548.

Hascall, V. C., Calabro, A., Midura, R. J., and Yanagishita, M. (1994). Isolation and characterization of proteoglycans. *Methods Enzymol.* **230,** 390–417.

Iozzo, R. V. (1989). Presence of unsulfated heparan chains on the heparan sulfate proteoglycan of human colon carcinoma cells. Implications for heparan sulfate proteoglycan biosynthesis. *J. Biol. Chem.* **264,** 2690–2699.

Iozzo, R. V. (1998). Matrix proteoglycans: from molecular design to cellular function. *Annu. Rev. Biochem.* **67,** 609–652.

Iozzo, R. V. (2001a). Heparan sulfate proteoglycans: intricate molecules with intriguing functions. *J. Clin. Invest.* **108,** 165–167.

Iozzo, R. V. (2001b). "Proteoglycan Protocols." Humana Press, Totowa, NJ.

Lindblom, A., and Fransson, L. Å. (1990). Endothelial heparan sulphate: compositional analysis and comparison of chains from different proteoglycan populations. *Glycoconjugate J.* **7,** 545–562.

Lories, V., Cassiman, J. J., Van den Berghe, H., and David, G. (1992). Differential expression of cell surface heparan sulfate proteoglycans in human mammary epithelial cells and lung fibroblasts. *J. Biol. Chem.* **267,** 1116–1122.

Maeda, N., Hamanaka, H., Oohira, A., and Noda, M. (1995). Purification, characterization and developmental expression of a brain-specific chondroitin sulfate proteoglycan, 6B4 proteoglycan/phosphacan. *Neuroscience* **67,** 23–35.

Melrose, J., and Ghosh, P. (1993). Determination of the average molecular size of glycosaminoglycans by fast protein liquid chromatography. *J. Chromatogr.* **637,** 91–95.

Merry, C. L., Lyon, M., Deakin, J. A., Hopwood, J. J., and Gallagher, J. T. (1999). Highly sensitive sequencing of the sulfated domains of heparan sulfate. *J. Biol. Chem.* **274,** 18455–18462.

Midura, R. J., Hascall, V. C., MacCallum, D. K., Meyer, R. F., Thonar, E. J., Hassell, J. R., Smith, C. F., and Klintworth, G. K. (1990). Proteoglycan biosynthesis by human corneas from patients with types 1 and 2 macular corneal dystrophy. *J. Biol. Chem.* **265,** 15947–15955.

Oda, O., Shinzato, T., Ohbayashi, K., Takai, I., Kunimatsu, M., Maeda, K., and Yamanaka, N. (1996). Purification and characterization of perlecan fragment in urine of end-stage renal failure patients. *Clin. Chim. Acta* **255,** 119–132.

Poe, M., Stein, R. L., and Wu, J. K. (1992). High pressure gel-permeation assay for the proteolysis of human aggrecan by human stromelysin-1: Kinetic constants for aggrecan hydrolysis. *Arch. Biochem. Biophys.* **298,** 757–759.

Staatz, W. D., Toyoda, H., Kinoshita-Toyoda, A., Chhor, K., and Selleck, S. B. (2001). Analysis of proteoglycans and glycosaminoglycans from *Drosophila. In* "Proteoglycan Protocols" (R. V. Iozzo, ed.), Vol. 171, pp. 41–52. Humana Press, Totowa, NJ.

Turnbull, J. E., Hopwood, J. J., and Gallagher, J. T. (1999). A strategy for rapid sequencing of heparan sulfate and heparin saccharides. *Proc. Natl. Acad. Sci. U.S.A* **96,** 2698–2703.

Van Kuppevelt, T. H., Dennissen, M. A., van Venrooij, W. J., Hoet, R. M., and Veerkamp, J. H. (1998). Generation and application of type-specific anti-heparan sulfate antibodies using phage display technology. Further evidence for heparan sulfate heterogeneity in the kidney. *J. Biol. Chem.* **273,** 12960–12966.

Vilim, V. (1985). Colorimetric estimation of uronic acids using 2-hydroxydiphenyl as a reagent. *Biomed. Biochim. Acta* **44,** 1717–1720.

Vogel, K. G., and Peters, J. A. (2001). Isolation of proteoglycans from tendon. *In* "Proteoglycan Protocols" (R. V. Iozzo, ed.), Vol. 171, pp. 9–18. Humana Press, Totowa, NJ.

Whitelock, J. (2001). Purification of perlecan from endothelial cells. *In* "Proteoglycan Protocols" (R. V. Iozzo, ed.), Vol. 171, pp. 27–34. Humana Press, Totowa, NJ.

Whitelock, J., Mitchell, S., Graham, L., and Underwood, P. A. (1997). The effect of human endothelial cell-derived proteoglycans on human smooth muscle cell growth. *Cell Biol. Int.* **21,** 181–189.

Whitelock, J. M., Murdoch, A. D., Iozzo, R. V., and Underwood, P. A. (1996). The degradation of human endothelial cell-derived perlecan and release of bound basic fibroblast growth factor by stromelysin, collagenase, plasmin, and heparanases. *J. Biol. Chem.* **271,** 10079–10086.

Yamaguchi, Y. (2001). Isolation and characterization of nervous tissue proteoglycans. *In* "Proteoglycan Protocols" (R. V. Iozzo, ed.), Vol. 171, pp. 35–40. Humana Press, Totowa, NJ.

Yanagishita, M. (2001). Isolation of proteoglycans from cell cultures and tissues. *In* "Proteoglycan Protocols" (R. V. Iozzo, ed.), Vol. 171, pp. 3–8. Humana Press, Totowa, NJ.

Yanagishita, M., Midura, R. J., and Hascall, V. C. (1987). Proteoglycans: Isolation and purification from tissue cultures. *Methods Enzymol.* **138,** 279–289.

CHAPTER 3

Expression of Recombinant Matrix Components Using Baculoviruses

Deane F. Mosher, Kristin G. Huwiler, Tina M. Misenheimer, and Douglas S. Annis

Department of Medicine
University of Wisconsin
Madison, Wisconsin 53706

I. Introduction: The Importance of Recombinant Extracellular Matrix Molecules for Protein Studies

Expression of recombinant matrix molecules presents a special challenge because of the complex modular structures of the molecules and the many posttranslational modifications that determine structure and function. It is an important challenge, because an understanding of the structure and function of extracellular matrix requires that adequate proteins folded in their native states be available for study. Studies by Timpl and colleagues of the interaction between the γ1-chain of laminins and nidogen (entactin) are an excellent example of what can be learned with purified matrix proteins. These investigators identified a fragment of the γ1-chain in laminin-1 that interacts with nidogen. They then identified a single laminin-type of epidermal growth factor-like module that is responsible for the interaction. This module was expressed recombinantly either as a single module, or within an array of three modules, and the solution structure of the single module (Baumgartner *et al.,* 1996) and crystal structure of the array (Stetefeld *et al.,* 1996) were determined. The structures validated and gave fresh insight into results based on synthetic peptides that mimicked loops within the module (Pöschl *et al.,* 1994) and *in vitro* mutagenesis of the recombinant module (Pöschl *et al.,* 1996). The information has been exploited to produce transgenic mice expressing a laminin γ1-chain that is unable to interact with nidogen (Mayer *et al.,* 1998).

Studies of laminins and other matrix components by the Timpl group have utilized human embryonic kidney cells stably transfected with plasmids encoding fusions of the signal peptide of human BM-40 (osteonectin) with the desired proteins. The fusion allows processing and secretion of the recombinant product. The recombinant product is purified from medium of transfected cells in mass culture by strategies that are specific for the particular protein product. We have developed an expression system that is complementary to the Timpl system, based on the pAcGP67A.coco (pCOCO) baculoviral transfer plasmid. pCOCO allows baculovirally mediated expression of a fusion protein with an N-terminal signal sequence, the matrix component of interest, a thrombin cleavage site, and a C-terminal polyhistidine tag. The fusion protein is produced in large amount late in the viral infectious cycle, is secreted efficiently, and can be purified by a general strategy, i.e., binding to a Ni^{2+}-chelate resin. The polyhistidine tag then can be removed with thrombin. More than 30 different matrix molecules or modular segments of matrix molecules have been expressed utilizing pCOCO. Yields have ranged from 1 to 80 mg per liter of medium from infected cells. In all cases studied, the recombinant proteins have been shown to undergo the same posttranslational modifications and to adopt the same structures as molecules purified from natural sources.

II. Requirements and Rationale for Development of a General Protein Expression System

A. Requirements

Five features are desirable for facile expression of recombinant modules of extracellular proteins for structural and functional studies. First, constructs should contain a signal sequence to direct the recombinant protein to the variety of mechanisms in the endoplasmic reticulum that facilitate proper folding and pass properly folded proteins into the secretory pathway (Ellgaard *et al.,* 1999). The N-terminal tail between the site of cleavage by signal protease and module(s) of interest should be kept to a minimum. Second, constructs should contain an affinity tag to allow easy purification of protein without denaturation. The affinity tag should preferably be at the C terminus so that full-length protein is selected by the purification. Third, it should be possible to remove the affinity tag by proteases or chemical treatments that will not attack the module(s) of interest. The C-terminal tail between the module(s) of interest and engineered cleavage site should also be kept to a minimum. Fourth, the system should allow multiple recombinant proteins to be expressed in the same cell, thus allowing study of matrix components that are secreted as heteromultimers. Fifth, the secretory apparatus of the expressing cell should contain the molecules necessary to introduce posttranslational modifications that are key for the structure and function of matrix components.

B. Rationale

In late 1996, we decided to develop a standard strategy to express a wide variety of segments of extracellular proteins for biophysical and functional studies. We hoped that a single strategy would allow us to make valid protein-to-protein comparisons and to develop an accumulated experience with which to troubleshoot problems that arose with new proteins. At that time, there were no commercially available vectors that incorporated the first three features listed. We had considerable experience in expression of proteins using recombinant baculoviruses in Sf9 or High-five insect cells. We found that two different baculoviruses could be used to express V^L and V^H segments in the same cell and thus form functional recombinant immunoglobulin V segments (Kunicki *et al.,* 1995), and that baculovirally expressed recombinant Ca^{2+}-binding EGF repeats of blood coagulation IX were correctly β-hydroxylated (Astermark *et al.,* 1994). We therefore constructed the pAcGP67A.coco (pCOCO) baculovirus transfer plasmid in the expectation that it would allow not only the first three but all five requirements to be met. pCOCO is the pAcGP67A baculovirus transfer vector (BD Biosciences Pharmingen, San Diego, CA) modified by addition of DNA encoding for a thrombin cleavage site and a polyhistidine tag (Fig. 1). There are now a number of commercially available vectors that have some of the features of pCOCO. However, none is as "stripped-down" as pCOCO, i.e., yields such short and simple N- and C-terminal tails.

C. Design of pAcGP67A.coco

pAcGP67A has DNA encoding the signal sequence of the GP67 acidic glycoprotein of AcNPV virus 5′ to a multiple cloning site and 3′ to the polyhedrin promoter (Fig. 1). pAcGP67A was designed such that cDNA encoding the protein of interest can be amplified by polymerase chain reaction (PCR) using appropriate primers and placed in the multiple cloning site in frame with the signal sequence. The recombinant plasmid can then be transfected into Sf9 insect cells along with defective linearized AcNPV virus. Surrounding viral sequences (Fig. 1) allow the plasmid to recombine with defective

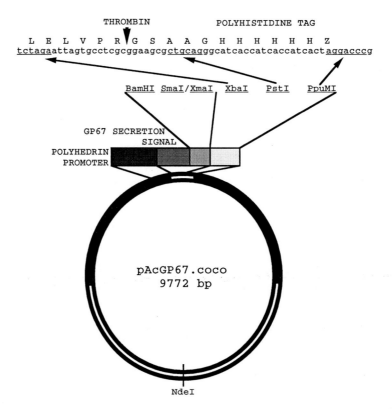

Fig. 1 Diagram of COCO. The plasmid contains viral sequences (blackened) surrounding the polyhedrin promoter (boxed, darkest shading). In place of the coding sequence for the polyhedrin protein, pAcGP67 contains DNA encoding for the signal peptide of viral GP67 (darker shading) and a multiple cloning region. In construction of COCO (pAcGP67.coco), part of the multiple cloning region was removed, leaving only the *Bam*HI and *Sma*I/*Xma*I sites from the original cloning site (lighter shading). The remainder of the multiple cloning site (lightest shading) was replaced with the sequence shown at the top (lowercase letters). This sequence codes (uppercase letters) for a thrombin cleavage site, short linker, and polyhistidine tag.

AcNPV virus and to form a competent recombinant virus that expresses the protein of interest. The fusion protein is directed to the lumen of the endoplasmic reticulum and into the secretory pathway by the N-terminal GP67 signal sequence.

We modified pAcGP67A to create pAcGP67A.coco (pCOCO) by insertion of DNA encoding a sequence susceptible to cleavage by thrombin (LVPRGS) and polyhistidine tag into the multiple cloning region of pAcGP67A at *Xba*I and *Ppu*MI sites (Fig. 1). A *Pst*I restriction site was engineered into the small linker between sequences encoding the thrombin cleavage site and the polyhistidine tag. DNA encoding molecules or modular segments of interest can be amplified by PCR and, depending on restriction sites present in the PCR primers, cloned into the *Bam*HI or *Xma*I/*Sma*I site on the 5′ end and the *Xba*I or *Pst*I site at its 3′ end. After cleavage of the signal peptide and secretion, recombinant proteins expressed by DNA cloned into the *Bam*HI or *Xma*I/*Sma*I site will have the sequence ADP or ADPG, respectively, at the N terminus followed by the introduced protein. The polyhistidine tag allows the secreted protein to be purified readily from the medium of infected cells on a Ni^{2+}-chelate resin. The thrombin cleavage site allows the polyhistidine tag to be removed. Proteins expressed by DNA cloned using the *Xba*I site and cleaved with thrombin will have the sequence ZLELVPR added at the C terminus (where Z is V, A, D, G, F, S, Y, C, L, P, H, R, I, T, or N). If the protein of interest is suspected to be susceptible to cleavage by thrombin, thus making the thrombin cleavage site encoded by pCOCO superfluous, the *Pst*I site can be used. Proteins expressed by DNA cloned using the *Pst*I site will have the polyhistidine tag in a shorter C-terminal tail of XAGHHHHHH, (where X is A, T, P, or S). Thus, the final product of expression using the pCOCO strategy will have, depending on the restriction sites used to clone the PCR-derived cDNA in pCOCO, an N-terminal tail of three or four residues and a C-terminal tail of seven (after cleavage with thrombin) or nine residues (Fig. 2). The tails do not contain a tryptophan residue, which has the potential to confound spectroscopic studies, and the 3' PCR primer can be designed so that the ZLEL . . . sequence lacks cysteine, tyrosine, or phenylalanine at Z.

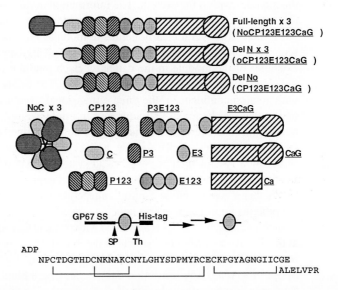

Fig. 2 Schematic representation of the various modular arrays of thrombospondin-1 and thrombospondin-2 that have been expressed using pCOCO. The structural organizations of thrombospondins are described by Adams (2001). Abbreviations are: N, N-terminal module (large dark oval); o, oligomerization sequence (stick); P, properdin or thrombospondin type 1 module (cross-hatched ovals); E, EGF-like or type 2 module (shaded circles); Ca, Ca^{2+}-binding repeats (cross-hatched rectangle); G, C-terminal globule (cross-hatched oval); and Del, deleted. All proteins containing o form trimers (×3). Trimerization is shown for NoC. At the bottom, the fusion protein encoded by the E3 baculovirus is shown before and after cleavage of the signal sequence (SS) by signal protease (SP) and of the polyhistidine tag (His-tag) by thrombin (Th). The sequence of E3 after processing is shown. This construct had been cloned using the *Bam*HI and *Xba*I sites of pCOCO. Note the N-terminal ADP and C-terminal ALELVPR tails at the ends of the module, shown with likely 1-3, 2-4, 5-6 disulfide connectivity.

III. Expression in Insect Cells Using pCOCO

A. Planning and Production of Recombinant Proteins from pCOCO

A key to expression of segments of extracellular matrix molecules is to respect their modularity (Bork *et al.*, 1996). A protein module is a sequence that contains within itself the information to adopt a global fold characteristic of other modules of the same homology type. Thus, the modules in a properly designed recombinant protein are expected to each be able to fold efficiently with the help of the secretory quality control machinery. The limits of a module can be defined by a variety of criteria. As much information as can be found about a specific module and module type should be considered in defining these limits and planning amplification of cDNA for cloning into pCOCO. Several sites devoted to protein modules are available through the ExPASy (Expert Protein Analysis System) server (www.expasy.ch). The matrix protein can be called up by name from the combined Swiss-Prot and TrEMBL databases and is displayed in the NiceProt format. NiceProt provides considerable information about the protein and directs the user to links with InterPro, Pfam, PRINTS, ProDom, SMART, and PROSITE for further analyses of the modules within the protein. The information at the various sites is overlapping, but it is worth looking at each for its unique features. One can be confident about a module when there are several clearly definable modular repeats connected to one another by sequences of variable length, and each has the "signature" residues that define other examples of the module in the database. Each modular repeat may be encoded by an individual exon, but such need not be the case. The limits of a module can be predicted with greatest confidence if the three-dimensional structure of an example of the particular module, or better still structures of multiple examples of the module, has or have been deduced. Nevertheless, it is likely to be some time before all the "exceptions" are found that prove the "rule" for how a given module type folds. One of the advantages of having a system such as the pCOCO system is that after successful expression of a number of proteins, the presence or absence of expression in the system becomes a test of whether a given module can fold independently or require other modules or parts of the proteins for efficient expression.

In most cases, the cDNA is amplified with primers that add a *Bam*HI or *Xma*I/*Sma*I site at the 5′ end and a *Xba*I or *Pst*I site at the 3′ end. Cohesive ends compatible for cloning into pCOCO can also be generated by substituting *Bgl*II for *Bam*HI, *Nsi*I for *Pst*I, or *Spe*I for *Xba*I. Any restriction enzyme leaving a blunt end can substitute for *Sma*I. Such substitutions allow cloning of cDNA containing an internal *Bam*HI, *Pst*I, *Xba*I, or *Xma*I/*Sma*I site. A high-fidelity polymerase such as Deep Vent (New England BioLabs) should be used to minimize the number of mutations.

Generation of recombinant pCOCO followed by sequencing of the insert should take about 2 weeks (Table I). The rate-limiting step is often cloning. In designing primers, we add an additional 5 bases 5′ to the restriction site to ensure efficient digestion of the PCR product. The largest cDNA insert that we have cloned into pCOCO has been human fibronectin, which is 7 kb. For such large inserts, our strategy is to insert the cDNA in two phases. We first insert the 5′ and 3′ ends as two pieces of amplified cDNA

Table I
Time Line for Production of a Recombinant Protein

	Days
Initial planning, primer design, and primer production	3
PCR amplification of desired gene	1
Cloning of PCR product and isolation of recombinant pCOCO cDNA	3
Sequencing of recombinant pCOCO cDNA	3
Generation of recombinant baculoviruses by co-transfection of linearized baculovirus DNA and recombinant pCOCO	5
Plaque purification of recombinant baculoviral clones	5
Generation of pass 1 and 2 baculoviral stocks for recombinant clones and analysis of protein expression	8
Generation of pass 3 baculovirus stock	4
Infection of cells and harvest of conditioned media for purification of recombinant protein	3
Total	35

ligated together but each extending far enough into the full-length cDNA as to contain restriction sites unique to a full-length insert. These restriction sites are used to excise the remaining cDNA from a plasmid containing the full-length insert. In the second phase, pCOCO containing the PCR-amplified ends is treated with the restriction enzymes, and the large fragment is ligated into the plasmid. This strategy minimizes the amount of confirmatory DNA sequencing that needs to be done.

B. Production of Recombinant Baculovirus

Recombinant baculovirus is generated by cotransfection of Baculogold linearized AcNPV viral DNA (BD Biosciences Pharmingen) and recombinant pCOCO transfer vector into monolayer Sf9 cells (InVitrogen) in SF900II SFM serum-free medium (Gibco-BRL). The support of baculoviral expression by commercial suppliers is excellent, and we recommend their technique manuals. The cotransfection is performed using Cell-Fectin (Gibco-BRL) according to manufacturer's instructions. The medium is harvested 4 days after transfection, and recombinant baculoviruses are cloned by plaque purification on monolayers of Sf9 cells overlayed with agarose. A plaque is plucked, vortexed, and added to monolayer Sf9 cells. Medium from these cells is collected as pass 1 virus. Pass 1 virus from several clones is expanded on monolayer Sf9 cells to produce pass 2 virus and also used to infect High five cells. Medium from the High five cells is tested for production of secreted recombinant protein. Medium, 5 ml, is incubated with nickel-nitrilotriacetic acid (NiNTA) resin (Qiagen, Valencia, CA). The resin is gently pelleted; suspended in 1 ml of 10 mM Tris or MOPS, 10 mM imidazole, and 300 mM NaCl, pH 7.5; placed in a microfuge tube; and washed three times with the suspension buffer. Bound protein is eluted in 100 μl of 10 mM Tris or MOPS, 300 mM NaCl, 250–300 mM imidazole, pH 7.5, and analyzed by polyacrylamide gel electrophoresis in sodium dodecyl sulfate (SDS–PAGE) alongside size standards of known amount. The expression level is estimated by the staining intensity of a protein band with the expected size of the product. If the expression level is 1 μg/ml (1 mg/liter), 20 μl of the eluate should contain 1 μg of recombinant protein. Based on these results, a single baculovirus clone is selected for generation of high titer ($1-5 \times 10^8$ pfu/ml) pass 3 virus stock that is used for subsequent recombinant protein expression. The pass 1 and 2 virus stocks from this clone are saved at 4°C in polypropylene tubes for future generation of additional pass 3 virus.

C. Expression and Purification of Recombinant Protein

High Five cells are grown in suspension at 27°C with SF900 II SFM. Cells are infected at a density of 1×10^6 cells/ml. A multiplicity of infection (MOI) of 2–10 pfu/cell is routinely used. After 60–72 h, cells are pelleted, and medium is decanted and prepared for purification on NiNTA by one of two methods.

For conditioned medium with <2 mg/liter recombinant protein, imidazole, pH 6.7, phenylmethylsulfonyl fluoride (PMSF), and NaCl are added to medium to obtain final concentrations of 10–15 mM, 1–2 mM, and 300 mM, respectively. Medium is then filtered through Whatman No. 2 filter paper and concentrated 10- to 20-fold using a CH2S Spiral Wound Cartridge ultrafiltration system (Millipore, Bedford, MA). During the concentration process, 15 mM imidazole, 500 mM NaCl, pH 6.7, is added stepwise to result in a 20-fold buffer exchange. Concentrated medium is clarified by centrifugation prior to affinity purification. Medium with recombinant protein >2 mg/ml is subjected to purification without concentration. Imidazole, pH 7.5, and PMSF are added to final concentrations of 5 mM and 1–2 mM, respectively, followed by dialysis at 4°C against 10 mM phosphate, Tris, or MOPS, 300–500 mM NaCl, 5–10 mM imidazole, pH 7.5–8.0.

Medium treated by either method is incubated with NiNTA resin, approximately 1 ml per 100 ml of concentrated or dialyzed medium, for 2 h at room temperature, followed by overnight incubation at 4°C. The NiNTA resin is placed in a small column; washed with 10 mM Tris or MOPS, 10–15 mM imidazole, 300–500 mM NaCl, pH 7.5; and eluted with 10 mM Tris or MOPS, 300 mM NaCl, 250–300 mM imidazole-Cl, pH 7.5. Fractions are analyzed by SDS–PAGE, and those containing the recombinant protein are pooled and dialyzed at 4°C into buffer specific for the subsequent use of the protein.

Binding to NiNTA requires that histidine be unprotonated, i.e., on the basic side of its pI. Bound protein is specifically eluted by imidazole, which is the chemical grouping of the side chain of histidine. In general, contaminating proteins not containing the polyhistidine tag bind to the NiNTA resin at imidazole concentrations less than 10 mM at pH 7.5, whereas proteins bearing the polyhistidine tag bind at concentrations between 10 and 100 mM at pH more basic than 6.7. In addition to pH and imidazole concentration, variables include concentrations of the recombinant proteins and nonspecifically binding proteins and amount of NiNTA resin. In cases in which the recombinant protein is heavily contaminated, a second purification with less resin may clean up the protein. In cases of a poor yield of recombinant protein, the pH of the unbound medium should be checked to be sure it is more basic than 6.7, and a second attempt made to purify the recombinant protein.

The removal of the His-tag from recombinant proteins can be accomplished with biotinylated thrombin (Novagen, Madison, WI) (Misenheimer *et al.*, 2000). Following cleavage, pefabloc SC (Roche Molecular Biochemicals, Indianapolis, IN) is added to a final concentration of 4 mM, and biotinylated thrombin is removed by incubation with streptavidin–agarose (Novagen). Uncleaved recombinant protein is removed by incubation with NiNTA resin as described above. Cleaved protein is concentrated by ultrafiltration and dialyzed into the appropriate solution for further purification or study.

D. Characterization of Recombinant Proteins

Purity, homogeneity, posttranslational modification(s), and structural features of the purified recombinant protein can be assessed by a variety of methods. Purity and the presence of disulfide-dependent oligomerization are assessed by SDS–PAGE analysis of reduced and nonreduced samples. N-Terminal sequencing can be utilized to verify removal of the signal sequence, although in all proteins produced using pCOCO that we have studied to date, the expected sequence of ADP or ADPG (Figs. 1 and 2) has been found. Mass spectrometry (MS) offers unequivocal information about masses of molecules present in the sample and by inference the presence and heterogeneity of posttranslational modifications such as glycosylation, signal sequence removal, and disulfide bond formation. Coupling HPLC to MS (LC/MS) allows further quantification of the overall purity and/or heterogeneity of the recombinant protein. If the site(s) of glycosylation and/or the disulfide bond pairings are unknown, LC/MS of digests of the recombinant protein can be utilized to solve these questions. We cannot overemphasize the importance of MS in protein characterization. If such is possible, MS is a more humbling instrument to use than a computer. The accuracy is amazing to the novice user. In each of the several cases in which the mass was "wrong" after accounting for posttranslational modification, we found that we had made a simple error in recording the protein sequence and predicting the protein mass.

Far-UV circular dichroism (CD) is a useful method to assess the secondary structure of recombinant proteins. Because many proteins from natural sources have been studied by CD, comparisons of natural and recombinant proteins by CD is a good test of whether the two proteins have the same secondary structure (Misenheimer *et al.*, 2000). Far-UV CD is also useful to monitor alterations in peptide backbone conformation induced by heat or other agents, which may be characteristic of the protein under study (Misenheimer *et al.*, 2000).

Many protein modules have highly conserved tryptophan and/or tyrosine residues that contribute to the global fold of the module. The environment and stereochemistry of these residues can be assessed by near-UV CD. Intrinsic protein fluorescence spectroscopy is also useful to examine the environment of conserved tryptophans and has the advantage of requiring very little protein. In proteins expressed using pCOCO, tryptophans are present only in the protein of interest and not in the tails.

▬▬▬ IV. Results Obtained from the COCO System

A. Variety of Expressed Proteins

We have expressed and purified more than 30 different proteins: full-length human fibronectin, N-terminal 70-kDa region of fibronectin, extracellular domain of vascular cell adhesion molecule-1 (VCAM), Ca^{2+}-binding EGF modules 11-12 and 34-35 of human notch-1, DSL module of human jagged-1 notch ligand, and various modular arrays from human thrombospondin-1 and -2 (Fig. 2). VCAM was expressed with and without the alternatively spliced 4th Ig module and with and without the 7th Ig module. Module types present at least once in these proteins include fibronectin types I, II, and III modules; the N-terminal module of thrombospondins; procollagen modules; thrombospondin type 1 or properdin modules; Ca^{2+}-binding and -nonbinding EGF-like modules; DSL module; C-terminal modules of thrombospondin; and Ig modules. A conservative missense mutation has been introduced into the C-terminal module of thrombospondin-2 without alteration of protein expression. The largest protein produced was fibronectin, a dimer of subunits each 2324 residues in length. The smallest protein was the C-terminal EGF module of human thrombospondin-1, which is 44 residues in length. This experience indicates that most matrix proteins or modular parts of matrix proteins can be expressed using the pCOCO strategy and that the proteins can be used to study structure-function consequences of *in vitro* mutagenesis or nonsynonymous single nucleotide polymorphisms.

B. Yields of Proteins

Yields of proteins, i.e., amount purified from medium of cultures 60–72 h after infection, have varied from 1 to 80 mg per liter. In general, the chief determinant of yield is the particular protein undergoing production. We have not detected a correlation between size of the product and production levels. A typical yield is in the range 10–20 mg/liter.

C. Oligomerization of Expressed Proteins

Human fibronectin efficiently formed a disulfide-bonded dimer via cysteines at the C terminus. Full-length thrombospondin-1 or thrombospondin-2 constructs containing an N-terminal oligomerization sequence efficiently formed disulfide-bonded trimers through cysteines in the oligomerization sequence (Fig. 2). NoC and delN constructs, each of which contains the oligomerization sequence (Fig. 2), also form trimers. We have found that cells dually infected at MOIs of 5 with viruses encoding NoC and delN constructs of thrombospondin-1 (Fig. 2) formed homotrimers [(delN)$_3$, (NoC)$_3$] and heterotrimers [(delN)$_2$NoC, delN(NoC)$_2$] in the expected ratio. These results indicate that baculoviruses can be used routinely to produce extracellular matrix proteins that are obligate homomultimers or heteromultimers.

D. Posttranslational Modifications

We have carried out extensive analyses of posttranslational modification of thrombospondin segments expressed using pCOCO. We found N-glycosylation modifications at predicted sites in the procollagen (Misenheimer *et al.,* 2000) and C-terminal (Misenheimer *et al.,* 2001) modules of thrombospondins. The procollagen module, which contains 10 cysteines and is dependent on disulfides for structural stability, lacked any free cysteine; all were in disulfides (Misenheimer *et al.,* 2000). Similarly, the 18 cysteines of the CaG proteins (Fig. 2) from thrombospondin-2 were all involved in disulfides (Misenheimer *et al.,* 2001). Two novel glycosylations were discovered for the type 1 "properdin" repeats of thrombospondin-1, C-mannosylation of conserved tryptophans and O-fucosylation of conserved serines or threonines (Hofsteenge *et al.,* 2001). Completeness of the C-mannosylation and O-fucoslyation was less than in thrombospondin-1 purified from platelets. These results indicate that mechanisms to introduce various posttranslational modifications into the recombinant protein remain intact in cells infected with baculoviruses. We have found no evidence that these mechanisms are overwhelmed, even though the flux of recombinant protein in the secretory machinery is large and the cells are late in the lytic cycle of the virus.

The major modification of the glycosylation site at Asn710 near the N terminus of the CaG protein (Fig. 2) of thrombospondin-2 was $[(mannose)_2(GlNac)_2(Fucose)_2]$ (Misenheimer *et al.,* 2001). The absence of sialic acid from N-linked oligosaccharide and the presence of a difucosylated product is described in authentic insect glycoproteins (Kubelka *et al.,* 1994). The structural and functional significances of the altered glycans are not known. Thus, the differences in N-glycosylation represent a potential limitation of pCOCO and expression of recombinant proteins in insect cells.

E. Antigenicity of Expressed Proteins

Panels of overlapping recombinant modular arrays such as shown in Fig. 2 are ideal for identification of epitopes for monoclonal antibodies to extracellular matrix proteins. The antibodies can be mapped to specific modules, and the conformational dependence of the epitope can be determined. In favorable cases, the epitope will map to a module in which there are very few differences between the module in the human protein and in the protein of the species in which the antibody was generated. *In vitro* mutagenesis can then be carried out to change these residues one by one and thus define the epitope exactly.

Use of the products of pCOCO expression to produce polyclonal antibodies was disappointing in a preliminary attempt. Antibodies elicited to the procollagen module of thrombospondin-1 reacted equally well with pCOCO-expressed properdin and procollagen modules. We presume that the bulk of the antibodies were directed to epitopes of the N- and C-terminal tails that are common to both constructs. We have been successful, however, in using proteins expressed with pCOCO to produce new monoclonal antibodies. Clones are screened for antibodies that react with the recombinant protein of interest and not with an irrelevant recombinant protein expressed in parallel using pCOCO. Only clones that produce antibodies to the target insert are selected for further analysis.

F. Formation of Protein Aggregates

We have noted variable amounts of purified protein that migrates as a protein aggregate by SDS–PAGE in the absence of reducing agent. We suspect that the aggregates arise from aberrant interactions among recombinant proteins during secretion. The aggregates are polyvalent in respect to polyhistidine tags and bind preferentially to NiNTA resin if the resin is limiting during purification. The high avidity of aggregates for NiNTA can be exploited to purify the desired nonaggregated protein. Conditioned medium is simply "precleared" of aggregates by a preliminary incubation with a small amount of NiNTA resin.

G. Failure of Expression

We have had only one example in which no protein was secreted. The C-terminal globular sequence, G, of thrombospondin-2 (Fig. 2) could not be produced although proteins composed of G and the more N-terminal Ca^{2+}-binding sequences, CaG, were produced in large amounts (Misenheimer *et al.,* 2001). Against the background of many successful protein products, our hypothesis is that failure of secretion of G is due to the requirement for G to fold in the context of the more N-terminal sequence.

References

Adams, J. C. (2001). Thrombospondins: multifunctional regulators of cell interactions. *Annu. Rev. Cell Dev. Biol.* **17,** 25–51.

Astermark, J., Sottile, J., Mosher, D. F., and Stenflo, J. (1994). Baculovirus-mediated expression of the EGF-like modules of human factor IX fused to the factor XIIIa transamidation site in fibronectin. *J. Biol. Chem.* **269,** 3690–3697.

Baumgartner, R., Czisch, M., Mayer, U., Pöschl, E., Huber, R., Timpl, R., and Holak, T. A. (1996). Structure of the nidogen binding LE module of the laminin γ 1 chain in solution. *J. Mol. Biol.* **257,** 658–668.

Bork, P., Downing, A. K., Kiefer, B., and Campbell, I. D. (1996). Structure and distribution of modules in extracellular proteins. *Quart. Rev. Biophys.* **29,** 119–167.

Ellgaard, L., Molinari, M., and Helenius, A. (1999). Setting the standards: quality control in the secretory pathway. *Science* **286,** 1882–1888.

Hofsteenge, J., Huwiler, K. G., Macek, B., Hess, D., Lawler, J., Mosher, D. F., and Peter-Katalinic, J. (2001). C-Mannosylation and O-fucosylation of the thrombospondin type 1 module. *J. Biol. Chem.* **276,** 6485–6498.

Kubelka, V., Altmann, F., Kornfeld, G., and Marz, L. (1994). Structures of the N-linked oligosaccharides of the membrane glycoproteins from three lepidopteran cell lines (Sf-21, IZD-Mb-0503, Bm-N). *Arch. Biochem. Biophys.* **308,** 148–157.

Kunicki, T. J., Ely, K. R., Kunicki, T. C., Tomiyama, Y., and Annis, D. (1995). The exchange of Arg-Gly-Asp (RGD) and Arg-Tyr-Asp (RYD) binding sequences in a recombinant murine fab fragment specific for the integrin $\alpha_{IIb}\beta_3$ does not alter integrin recognition. *J. Biol. Chem.* **270,** 16660–16665.

Mayer, U., Kohfeldt, E., and Timpl, R. (1998). Structural and genetic analysis of laminin-nidogen interaction. *Ann. N.Y. Acad. Sci.* **857,** 130–142.

Misenheimer, T. M., Hahr, A. J., Harms, A. C., Annis, D. S., and Mosher, D. F. (2001). Disulfide connectivity of recombinant C-terminal region of human thrombospondin-2. *J. Biol. Chem.* **276,** 45882–45887.

Misenheimer, T. M., Huwiler, K. G., Annis, D. S., and Mosher, D. F. (2000). Physical characterization of the procollagen module of human thrombospondin 1 expressed in insect cells. *J. Biol. Chem.* **275,** 40938–40945.

Pöschl, E., Fox, J. W., Block, D., Mayer, U., and Timpl, R. (1994). Two non-contiguous regions contribute to nidogen binding to a single EGF-like motif of the laminin $\gamma 1$ chain. *EMBO J.* **13,** 3741–3747.

Pöschl, E., Mayer, U., Stetefeld, J., Baumgartner, R., Holak, T. A., Huber, R., and Timpl, R. (1996). Site-directed mutagenesis and structural interpretation of the nidogen binding site of the laminin $\gamma 1$ chain. *EMBO J.* **15,** 5154–5159.

Stetefeld, J., Mayer, U., Timpl, R., and Huber, R. (1996). Crystal structure of three consecutive laminin-type epidermal growth factor-like (LE) modules of laminin $\gamma 1$ chain harboring the nidogen binding site. *J. Mol. Biol.* **257,** 644–657.

CHAPTER 4

Heparan Sulfate–Growth Factor Interactions

Alan C. Rapraeger

Department of Pathology and Laboratory Medicine
University of Wisconsin-Madison
Madison, Wisconsin 53706

I. Introduction: HS Regulation of Growth Factor Signaling

Polypeptide growth factors evoke cellular responses through the activation of cell surface receptor tyrosine kinases. The recognition of these receptors is often complex, involving interactions at more than one site, the formation of a high-affinity complex, and receptor oligomerization. Numerous growth factors have "heparin-binding" domains, so named because they exhibit an affinity for mast-cell derived heparin and are often

Table I
The FGF Family of Growth Factors

Name	Alternate name	Reference
FGF-1	Acidic FGF (aFGF)	Gimenez-Gallego et al., 1985; Jaye et al., 1986
FGF-2	Basic FGF (bFGF)	Abraham et al., 1986a,b; Esch et al., 1985
FGF-3	int-2	Moore et al., 1986
FGF-4	hst, K-FGF	Huebner et al., 1988; Yoshida et al., 1987
FGF-5	—	Zhan et al., 1988
FGF-6	—	Marics et al., 1989
FGF-7	Keratinocyte growth factor (KGF)	Finch et al., 1989
FGF-8	Androgen-induced growth factor (AIGF)	Tanaka et al., 1992
FGF-9	Glia-activating factor (GAF)	Miyamoto et al., 1993
FGF-10	—	Yamasaki et al., 1996
FGF-11	FGF homology factor-3 (FHF-3)	Smallwood et al., 1996
FGF-12	FGF homology factor-1 (FHF-1)	Smallwood et al., 1996
FGF-13	FGF homology factor-2 (FHF-2)	Greene et al., 1998; Smallwood et al., 1996
FGF-14	FGF homology factor-4 (FHF-4)	Smallwood et al., 1996
FGF-15	—	McWhirter et al., 1997
FGF-16	—	Miyake et al., 1998
FGF-17	—	Hoshikawa et al., 1998
FGF-18	—	Hu et al., 1998; Ohbayashi et al., 1998
FGF-19	—	Nishimura et al., 1999; Xie et al., 1999
FGF-20	—	Jeffers et al., 2001; Kirikoshi et al., 2000; Ohmachi et al., 2000
FGF-21	—	Nishimura et al., 2000
FGF-22	—	Nakatake et al., 2001
FGF-23	—	White et al., 2000; Yamashita et al., 2000

purified on heparin affinity columns; it is now recognized that such domains may be part of a regulatory mechanism in which HS proteoglycans at the cell surface and in the extracellular matrix regulate the binding and signaling of the growth factor.

The prototypes of the heparin-binding growth factors are acidic and basic fibroblast growth factor (FGF), members of a family of FGFs that has now grown to at least 23 related polypeptides—all of them displaying an apparent affinity for HS (Table I). These interactions are not confined to the FGFs, however, as other growth factors including HGF (Sakata et al., 1997), VEGF (Poltorak et al., 2000; Robinson and Stringer, 2001), (Lin and Perrimon, 2000), HB-GAM (Kilpelainen et al., 2000), and members of the EGF family (amphiregulin, HB-EGF)(Piepkorn et al., 1998) also bind to HS and rely on HS for their activity.

The means by which HS participates in signaling by each of these growth factors remains under investigation. The mechanism is becoming more clear, however, for certain members, such as the FGFs; current models indicate that HS interacts with both the growth factor and its receptor, stabilizing their interactions as a high-affinity complex (Fig. 1). Thus, whereas the affinity of FGF for its receptor is relatively low ($K_d \sim 5-10$ nM), and the affinity of FGF for the HS chain is in the same range, the

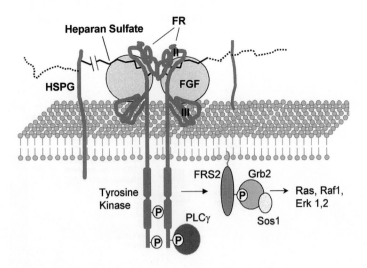

Fig. 1 Model depicting the assembly of a transmembrane HS proteoglycan, FGF, and FGF receptor tyrosine kinase into a ternary signaling complex. Note the simultaneous interaction of each of the HS chains with the FGF and domain II of the receptor.

affinity of FGF in the assembled ternary complex comprising FGF, receptor, and HS glycosaminoglycan is much greater ($K_d \sim 50-100$ pM) (Pantoliano *et al.*, 1994).

Each of the FGFs signals through an FGF receptor (FR), one of a family of four receptor tyrosine kinases (Fig. 2). These kinases contain three extracellular Ig-like domains (I, II, and III), a membrane-spanning domain, and a split tyrosine-kinase intracellular domain (Basilico and Moscatelli, 1992; Dionne *et al.*, 1991). The distal domain I is not required for ligand binding and signaling; receptor splice variants are expressed that lack this region. Domains II and III largely regulate FGF recognition. Importantly for the role of HS in the regulation of FGF signaling, the distal portion of domain II has a basic amino acid motif that binds HS, an interaction that is crucial for activity of the receptor (Kan *et al.*, 1993). In addition, amino acids in this domain make contacts with the FGF ligand and provide specificity for FGF binding (Chellaiah *et al.*, 1999). The major site for direct interaction of the FGF with the FR, however, is in the membrane proximal portion of domain III. This region in FRs 1, 2, and 3 (but not FR4) is also subject to splicing variation, leading to the expression of FR extracellular domains containing either domain IIIa, IIIb, or IIIc (Basilico and Moscatelli, 1992; Dionne *et al.*, 1991; Miki *et al.*, 1991). IIIa encodes a stop codon, leading to expression of a soluble FR extracellular domain. The remaining two (e.g., IIIb and IIIc) encode membrane receptors. Thus, ignoring for the moment the possible roles of the Ig-I domain, the FR family has at least seven receptors with differing FGF binding potential (FR1b, FR1c, FR2b, FR2c, FR3b, FR3c, and FR4). Within each of these receptor isoforms, the ligand binding is determined by the variable domain III, and by the simultaneous binding of HS to domain II and to the FGF.

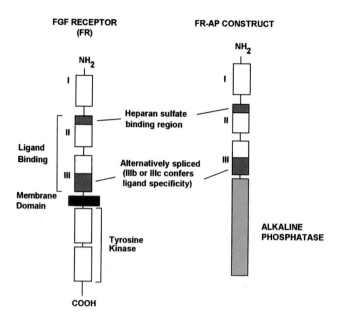

Fig. 2 The domain structure of the FGF receptor (FR) and the FR–alkaline phosphatase (AP) chimera used as an HS probe. The extracellular domain of the FR consists of three Ig-like domains (I, II, and III). Domain II contains the HS binding region. The membrane receptors may be subject to splicing variation in domain III.

Based on current crystal structures (Pellegrini *et al.,* 2000; Plotnikov *et al.,* 2000; Schlessinger *et al.,* 2000; Stauber *et al.,* 2000), a potential model for the assembly of FGF with its receptor is shown (Fig. 1), in which a single HS chain lies in a groove within the FGF/FR complex where sulfate residues in the HS chain make contacts simultaneously with both proteins. As it is likely that these contact sites are not identical, HS with different sulfation patterns may be necessary to promote the interactions of each different FGF with its FR.

Which FGFs bind to which FRs? At the current time, FR specificity for each FGF has been determined using BaF3 cells deficient in HS and FR expression (Ornitz *et al.,* 1996). Expression of individual FRs in these cells allows one to assess which FGFs will bind and signal. This assessment is routinely done in the presence of added heparin to generate the ternary signaling complex. A problem with this assessment, however, is that it does not use endogenous HS. Heparin is a close cousin of HS, synthesized in mast cells with a role in packaging proteins into α-granules (Forsberg *et al.,* 1999; Humphries *et al.,* 1999). It is highly modified and sulfated for this role (Lindahl *et al.,* 1998), a characteristic that apparently allows it to bind all FGFs and function with them in signaling. Thus, while heparin allows for the determination of possible FGF/FR interactions, it does not guarantee that HS at the cell surface and in the extracellular matrix will specify the same interactions.

The potential for specific regulation of growth factor signaling by HS is bolstered by the finding that the pattern of sulfation within the HS chain can be quite variable. HS is synthesized on a core protein, with chain assembly occurring on a linkage tetrasaccharide attached to a serine (usually an SGxG motif) in the core. The chain is elongated through the action of an HS-copolymerase, which adds alternate N-acetyl-D-glucosamine and D-glucuronate residues (Fig. 3). The growing chain is then acted upon by a battery of other modifying enzymes, starting with a family of four N-deacetylase/N-sulfotransferases (NDSTs) (Aikawa and Esko, 1999; Aikawa *et al.*, 2001; Kusche-Gullberg *et al.*, 1998; Pikas *et al.*, 2000) and an epimerase capable of converting glucuronate to iduronate. The chain is then modified by other sulfotransferases, including a single 2-O-sulfotransferase (HS2OST) (Kobayashi *et al.*, 1997), three glucosaminyl-6-O-sulfotransferases (Habuchi *et al.*, 2000), and at least five glucosaminyl-3-O-sulfotransferases (Liu *et al.*, 1999; Shworak *et al.*, 1999). Importantly, these enzymes act in a concerted fashion; thus, most of the enzymes require a prior modification by a preceding enzyme before they will act. Although the process is not fully understood, it appears that the variability in the sulfation pattern of the chain is a function of the enzyme profile expressed by a cell, perhaps subject to regulatory factors that govern the enzyme activity as well. The outcome is the appearance of HS proteoglycans at the cell surface bearing HS chains with specific sulfation patterns that can regulate FGF signaling. Emerging work indicates that these sulfation patterns may be expressed in a tissue-specific manner during embryonic development (Allen and Rapraeger, unpublished).

It is clear that the HS sulfation sequence is a critical factor in FGF recognition and signaling. A comparison using several different FGFs together with a panel of heparin oligosaccharides shows that the FGF activity depends on recognition of different sulfation patterns within the HS chain (Guimond *et al.*, 1993). The model that emerges is one where a particular FGF and FR pair recognizes complementary and overlapping sulfation domains within a sequence of 12–14 sugars that promotes their high-affinity binding. Failure to provide either the FGF or the FR-specific sulfation sites within that domain prevents that pair from assembling and signaling.

Finally, genetic evidence indicates important roles for HS in developmental processes. *Drosophila* mutants that have defects in HS synthesis (e.g., *tout velu*, an HS-copolymerase defect that impairs chain synthesis (Bellaiche *et al.*, 1998); *sugarless*, a defect in sugar precursor synthesis (Binari *et al.*, 1997; Hacker *et al.*, 1997); *sulfateless*, a defect in the N-deacetylase/N-sulfotransferase that is the first step in HS chain sulfation (Lin *et al.*, 1999)) show impaired FGF and wnt signaling. More specific defects are seen with individual proteoglycans, such as the Dally mutant, a GPI-linked cell surface HS proteoglycan that leads to disrupted cell cycle progression during *Drosophila* eye development (Selleck, 2000). Not all of these defects will trace to an FGF, however, as a large number of other regulatory molecules bind to HS as well.

One difficulty in confirming the presence of specific HS on cell and tissue types has been in sequencing the HS chains. Enzyme and chemical analyses can provide general information on overall levels of N-sulfation, 2-O-sulfation, and 6-O-sulfation, as well as sugar identity, but the exact sequence of these modifications is more difficult to ascertain. Nonetheless, it has been shown that certain tissues have HS that differs chemically

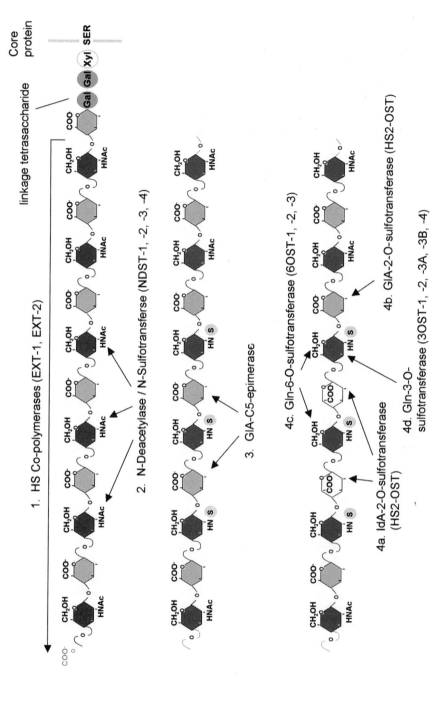

Fig. 3 Synthesis of HS. The HS chain is synthesized on a linkage tetrasaccharide attached to a serine in the core protein. The HS copolymerase extends the chain (step 1), which is acted upon by a series of modifying enzymes (see text). Sites in the chain modified by the N-deacetylase/N-sulfotransferase (step 2) are subject to epimerization (step 3) and sulfation (step 4).

from others. More recently, procedures for sequencing HS chains have been developed (Turnbull *et al.,* 1999; Venkataraman *et al.,* 1999). One procedure relies on sequencing HS fragments using lysosomal enzymes that are now commercially available (Turnbull *et al.,* 1999). These enzymes sequentially remove specific sulfate groups or sugars from the nonreducing end of the fragment. This is particularly useful when combined with a method for precise determination of the mass of the resulting oligosaccharides, such as MALDI mass spectroscopy (Keiser *et al.,* 2001; Rhomberg *et al.,* 1998; Venkataraman *et al.,* 1999). As these methods become more sophisticated, they are being applied to learning the sequences of HS fragments as long as 18 sugars.

This chapter will examine the techniques that are typically used to demonstrate the dependence of receptor–growth factor interactions on HS, focusing on the FGFs. It will describe methods for blocking the binding of growth factor to cell surface HS, and a novel method in which FGF and a soluble form of FGF receptor are used to detect endogenous HS that is capable of mediating their assembly into a signaling complex.

II. Methods for Identifying the Role of HS in Growth Factor Binding and Signaling

A. Blockade of HS–Growth Factor Interactions by Chlorate Treatment: General Considerations

Binding of growth factors to HS is dependent on the sulfation of the HS chain, a process that occurs in the Golgi apparatus through the action of specialized sulfotransferases (Fig. 3). These enzymes utilize the high-energy sulfate donor phosphoadenosine phosphosulfate (PAPS), transferring sulfate from PAPS to the growing HS chain. PAPS is synthesized in the cytosol by enzymes with dual sulfurylase and kinase activity (SK1 and SK2) (Kurima *et al.,* 1999; Li *et al.,* 1995). The ATP sulfurylase (ATP sulfate adenylyltransferase, EC 2.7.7.4) activity transfers a sulfate group to ATP to yield adenosine 5′-phosphosulfate (APS) and pyrophosphate. Subsequently, the APS kinase activity (ATP adenosine-5′-phosphosulfate 3′-phosphotransferase, EC 2.7.1.25) transfers a phosphate group from ATP to APS to yield ADP and adenosine 3′-phosphate 5′-phosphosulfate (PAPS). PAPS is the sole source of sulfate for sulfate esters in mammals. The PAPS in then transported into the lumen of the Golgi by PAPS translocase (Ozeran *et al.,* 1996).

The ATP sulfurylase activity is competitively inhibited by chlorate (Farley *et al.,* 1978). Thus, as cells are treated with increasing concentrations of chlorate, the level of PAPS available for the sulfation of lipids, glycoproteins, and glycosaminoglycans drops. Whereas 30 mM chlorate is typically sufficient to block the sulfation of most glycosaminoglycans, this depends on the K_m of the sulfotransferases for PAPS and the ability of the treated cells to scavenge sulfate from other metabolic pathways. An important control for the chlorate treatment is addition of excess sulfate to the culture medium, which competes with the chlorate and restores sulfation despite the continued inhibitor treatment. The use of chlorate to block FGF signaling was first demonstrated by Rapraeger *et al.* (1991). Treated fibroblasts failed to proliferate in response to FGF2 (basic FGF) and mouse MM14 myoblasts, which are dependent upon FGF to block their differentiation, would exit the cell cycle and fuse into myosin-expressing myotubes in chlorate. An important aspect of this assay is that the chlorate treatment can be reversed by adding soluble heparin or HS to the assay. This demonstrates that the HS is not functioning by trapping or concentrating the FGF at the cell surface, a prevailing notion at the time, as the active FGF/heparin complexes are fully soluble. Secondly, it provides an assay to question what structural features of the HS might be needed for assembly of the signaling complex. Using heparin treated chemically to remove specific sulfate groups demonstrated that whereas a pentasaccharide bearing iduronysyl-2-O-sulfates was sufficient for FGF2 binding to HS, this did not suffice for FGF2 signaling (Guimond *et al.,* 1993; Maccarana *et al.,* 1993). Rather, signaling required a dodecasaccharide in which the FGF2 binding domain was present along with at least one glucosaminyl-6-O-sulfate residue (Guimond *et al.,* 1993; Pye *et al.,* 1998). As signaling requires HS binding to the FR as well as the FGF, this suggested that this additional information was required by the FR, and that distinct information in the HS chain is needed for FGF recognition and for FR recognition. Only if both are satisfied is the ternary signaling complex assembled.

B. Blockade of HS Sulfation Using Chlorate Treatment: Methodology

1. Chlorate Treatment

The blockade to proteoglycan sulfation leads to a reduction in the response of the cells to FGF. Of course, this is dependent on the particular cell type, the receptor tyrosine kinases that it expresses, and the structure of its HS chains. The response of the cell can be measured in several ways, including proliferation as assessed by cell number or ^{3}H-thymidine incorporation, activation of downstream signaling pathways (e.g., phosphorylation of the receptor, activation of MAPK, phosphorylation of PLCγ), or measurements of cell behavior, such as cell migration. The response time for these measurements can vary from a matter of minutes for MAPK activation or receptor phosphorylation, to days for proliferation studies. The chlorate treatment provides a uniform method for blocking HS binding to growth factors that is effective for either short- or long-term assays. The cells to be examined are subjected to a pretreatment in 10–50 mM sodium chlorate for 24–48 h. They are then suspended by trypsinization and replated in chlorate and serum-starved to quiescence. The pretreatment and subsequent trypsinization removes any preexisting proteoglycan at the cell surface or secreted matrix that would bear native sulfation. The concentration of chlorate used can vary with cell type and needs to be established. Some cells, such as Swiss 3T3 fibroblasts, are difficult to competitively inhibit with chlorate as they can utilize cysteine and methionine to generate sulfate. Thus, such cells are cultured in a sulfate-deficient DMEM where the serum is dialyzed against 20 mM Hepes, 0.15 M NaCl (pH 7.4), MgSO$_4$ is substituted with MgCl$_2$, methionine is removed from the formulation, and cysteine levels are reduced (Rapraeger *et al.,* 1994). The chlorate concentration necessary for inhibition can also vary depending on the K_m of the sulfotransferases for PAPS. Thus, 6-O-sulfation of HS appears most sensitive to chlorate treatment, followed by 2-O-sulfation and finally N-sulfation (Safaiyan *et al.*, 1998). The overall reduction can be measured by incubating cells treated with different chlorate concentrations with 50–$100\ \mu Ci/ml\ H_2{}^{35}SO_4$, followed by suspension of the cells with trypsin, and capture of radiolabeled glycosaminoglycan on DEAE-Sephacel (Pharmacia Biotech, Uppsala, Sweden) with or without pretreatment with heparin lyase I and heparin lyase III (see Section II.B.3) to cleave the endogenous HS to disaccharides.

2. Measuring the Cellular Response

Once the cells have been pretreated to remove existing proteoglycans, they are replated in chlorate and serum-starved in medium in which serum has been replaced with 0.1% BSA. Some cells may require serum for adherence following the trypsinization; thus, the serum starvation may commence 24 h after replating. As the chlorate is a competitive inhibitor, the investigator can return sulfate (10 mM) to the culture medium at this point to restore sulfation despite the continued presence of the chlorate; this is an effective control for potential nonspecific effects of the chlorate. The starvation time will vary among cell types and different assays. A 6-h starvation is typically sufficient for a MAPK assay, whereas overnight is customary for proliferation assays. Following this starvation period, the FGF is added for times ranging from 15 min (for MAPK stimulation) to 18 h for ^{3}H-thymidine incorporation) or several days (for cell number assessment). Usually, 1–5 nM FGF is an effective dose. For MAPK activation, the cells are directly solubilized in hot SDS–PAGE sample buffer, electrophoresed on PAGE, transferred to blots, and probed with anti-ACTIVE antibodies that detect only the dual phosphorylated active MAPK in parallel with antibodies that detect the total MAPK pool (Promega Corp., Madison, WI).

3. Recovering the Cellular Response Using Exogenous Heparan Sulfate

An alternative means of confirming the specificity of the chlorate treatment is to include soluble HS or heparin (10–100 μg/ml) with the FGF on chlorate-treated cells to mimic the endogenous HS (Krufka *et al.*, 1996). For FGF signaling, the soluble HS will combine with added FGF to restore the FGF signal. This is a powerful tool, as it not only corroborates the specific effect of the chlorate, but also demonstrates the specificity of the inhibition for HS. In addition, the specificity of the signaling complex for certain sulfate residues can be examined by using HS or heparin that has been depleted for those residues. Thus, chemical procedures can be used to remove 6-O-sulfates, 2-O-sulfates, or N-sulfates from their constituent sugars within the chain (Kariya *et al.*, 2000; Rapraeger *et al.*, 1994). One commercial source of these modified heparins is Neoparin, Inc. (San Leandro, CA; http://www.heparinoids.com). The efficacy of these modified glycosaminoglycans, compared to the native polymer, in restoring FGF signaling can provide useful information on the HS binding requirements of the FGF and/or receptor.

C. Blockade of Glycosaminoglycan–Growth Factor Interactions—Removal of Specific Glycosaminoglycans with Enzyme Treatment: General Considerations

Blockade of sulfation by treatment with chlorate is not specific for HS, or even glycosaminoglycans (e.g., sulfated glycoproteins and sulfated glycolipids are affected). Another useful tool is the use of enzymes to remove specific glycosaminoglycans. Although not as effective as chlorate, as the cell is continually replacing the cell surface glycosaminoglycan through synthetic mechanisms, the availability of enzymes with varying specificity provides the opportunity to confirm the specificity of the glycosaminoglycan activity.

Prokaryotic sources provide enzymes for degradation of glycosaminoglycans (available from Sigma Corp., St. Louis, MO, and Seikagaku America, Falmouth, MA) (Table II). These enzymes are lyases, in contrast to the hydrolases found in eukaryotic systems for glycosaminoglycan degradation. Lyases carry out eliminative cleavage, in which the glycosidic bond is broken in conjunction with the generation of a $\Delta_{4,5}$-unsaturated uronic acid at the newly generated nonreducing ends of the fragments. For the fragment remaining attached to the core protein, the unsaturated uronic acid is exposed as the distal terminus of the chain and, for heparan sulfate, can be recognized by commercial antibodies (3G10 specific for heparan sulfate (Seikagaku America, Falmouth, MA)). This is useful to verify the cleavage event, as well as to visualize the distribution of the core proteins and the sites where the chains were expressed prior to their removal by the enzyme. For chemical amounts of glycosaminoglycan, the unsaturated bond introduced by the cleavage can be monitored by absorbance at 232 nm, with an extinction coefficient of 5500 M^{-1} (Hovingh and Linker, 1974).

Table II
Glycosaminoglycan-Degrading Enzymes

Name	Substrate	Optimum pH/temp	Commercial source
Chondroitin ABC Lyase (ABCase)	Chondroitin 4-sulfate Chondroitin 6-sulfate Dermatan sulfate	8.0/37°C	Sigma Seikagaku America
Chondroitin AC Endolyase (ACase I)	Chondroitin 4-sulfate Chondroitin 6-sulfate	7.5/37°C	Sigma Seikagaku America
Chondroitin AC Exolyase (ACase II)	Chondroitin 4-sulfate Chondroitin 6-sulfate	6.0/37°C	Sigma Seikagaku America
Chondroitin B Lyase	Dermatan sulfate	7.5/25°C	Sigma Seikagaku America
Chondroitin C Lyase	Chondroitin 6-sulfate	8.0/25°C	Sigma
Heparin lyase I (EC 4.2.2.7) ("heparinase")	Heparin (also some sites in heparan sulfate)	7.1/30°C	Sigma Seikagaku America
Heparin lyase II	Heparin and heparan sulfate	7.1/35°C	Sigma
Heparin lyase III (EC 4.2.2.8) ("heparitinase")	Heparan sulfate (also some sites in heparin)	7.6/35°C	Sigma Seikagaku America

1. Chondroitin Sulfate

Chondroitin sulfate is composed of a repeating *N*-acetyl-D-galactosamine and D-glucuronate disaccharide, in which the *N*-acetylgalactosamine may be sulfated in either the 4- or 6-position. In addition to these modifications, the 4-O-sulfated form may be acted upon by an epimerase during chain synthesis that converts the glucuronate residues to iduronate, thus generating dermatan sulfate.

The standard enzyme available for digestion of chondroitin sulfate is chondroitinase ABC (ABCase). This enzyme cleaves chondroitin 4-*O*-sulfate (chondroitin sulfate A), chondroitin 6-*O*-sulfate (chondroitin sulfate C), and dermatan sulfate (chondroitin sulfate B), as well as oversulfated chondroitin sulfates (Sugahara *et al.*, 1996). The commercially available enzyme is actually a mixture of two derived from the bacterium *Proteus vulgaris,* an ABC endolyase and an ABC exolyase that work together to depolymerize the chondroitin sulfate chain (Hamai *et al.*, 1997). Two similar enzymes (chondroitinases AC or ACases) cleave chondroitin sulfate A and C but lack specificity for dermatan sulfate. Chondroitinase AC-I is an endolyase derived from *Flavobacterium heparinum* and chondroitinase AC-II is an exolyase derived from *Arthrobacter aurescens* (Gu *et al.*, 1993). Despite the lack of activity toward dermatan sulfate, however, these enzymes are unlikely to be useful for removing chondroitin sulfate alone from cells as dermatan sulfate is present as modified stretches within the chondroitin sulfate chains; thus, ACase-I is likely to remove dermatan sulfate from cells along with chondroitin sulfate, although the regions of dermatan sulfate within the released chains would be resistant to digestion. Chondroitinase B, from *Flavobacterium heparinum,* is also available from Seikagaku America for cleavage of dermatan sulfate at cell surfaces (Michelacci and Dietrich, 1975).

2. Heparan Sulfate and Heparin

HS is synthesized as disaccharides of *N*-acetyl-D-glucosamine and D-glucuronic acid (Fig. 3), which are variably modified. Heparin, which is synthesized in mast cells, undergoes these modifications to a high degree for its role in packaging proteins into mast cell α-granules (Forsberg *et al.*, 1999; Humphries *et al.*, 1999). HS is modified to a lesser extent, ostensibly to endow the chain with the ability to discern discrete ligands.

There are three enzymes with differing specificity for HS and heparin. Heparin lyase I (also known as heparinase or heparinase I, E.C. 4.2.2.7) recognizes a highly modified disaccharide common in heparin in which the *N*-acetyl-D-glucosamine is deacetylated and sulfated and the glucuronate has been epimerized to iduronate and sulfated in the 2-*O* position; this enzyme degrades heparin to di- and oligosaccharides (Ernst *et al.*, 1995; Linhardt *et al.*, 1990). Although more rare, the same cleavage site also exists in HS. Thus, treatment of intact cells with heparin lyase I will partially, but incompletely, trim HS chains from cell surface proteoglycans. Heparin lyase III (heparitinase or heparinase III, E.C. 4.2.2.8) recognizes sites that are common in HS, but rare in heparin; these sites contain disaccharides in which the glucuronate residue remains unmodified (Ernst *et al.*, 1995; Linhardt *et al.*, 1990). Thus, this enzyme cleaves HS into block fragments and removes almost all HS from cell surfaces. Heparin is relatively resistant to this enzyme. Of course, the actual degree of cleavage of either HS or heparin depends on the modifications introduced into the glycosaminoglycan during synthesis, and hence the source of the glycosaminoglycan. Treatment with both heparin lyases I and III will cleave heparin and HS to short oligosaccharides and disaccharides. A third enzyme, heparin lyase II, is less stringent and will cleave both heparin and heparan sulfate. As described for the chondroitin sulfate degrading enzymes, the heparinases are lyases and generate fragments bearing $\Delta_{4,5}$-glucuronate (or iduronate) residues at their nonreducing termini. Commercial antibodies are available (Seikagaku America) that recognize this novel unsaturated sugar and can be used to detect HS fragments, including the fragment remaining attached to the core protein.

D. Enzyme Treatment Protocols

The HS and chondroitin sulfate degrading enzymes are active at neutral pH, although their pH optima range from 6.0 to 8.0 (Table II). This allows the investigator to remove the glycosaminoglycan chains directly from cultured cells without extreme treatments. Our procedure is to add the enzyme directly to normal culture media containing serum. As the pH of bicarbonate-buffered media can change rapidly when cells are removed from the CO_2 incubator, we often use medium buffered to pH 7.4 with 20–50 mM Hepes. Hepes is also reported to stabilize the heparin lyases used for heparan sulfate removal (Lohse and Linhardt, 1992).

To remove chondroitin sulfate, we use chondroitin ABC lyase at 0.5 U/ml for 2 h. For HS removal, the cells are treated with 0.01 IU of heparin lyase/ml at 37°C for 2 h (note that one international unit (IU) is equal to 50 conventional units (U)). Addition of fresh enzyme for an additional 2 h ensures complete removal. Although heparan lyase III is usually sufficient for removal of HS from the cell, equal amounts of heparin lyase I and III are used if the HS is to be destroyed completely. Unlike chlorate treatment, which can be used over a period of days, the enzyme treatments are done over shorter time periods because of the relative instability of heparin lyase III in physiological culture media. Although the pH optimum of heparin lyase III is 7.6 when measured in brief assays, it loses 80% of its activity over 3.5 h at 35°C at this pH (Lohse and Linhardt, 1992).

To test the involvement of HS (or its removal) on a growth factor mediated signaling pathway, the MAPK kinase pathway is used because it is a rapid assay that can be done in the time frame of the enzyme treatment (Section II.A.2).

E. Alternatives for Examining HS Dependence of Growth Factor Signaling: HS-Deficient Cells

The procedures just described allow the investigator to examine the role of HS in the regulation of growth factor signaling on cells that are expressing endogenous HS proteoglycans. Another method to confirm this dependence is to use cells that are deficient in HS synthesis. HS-deficient cells are rare, as most or all adherent cells express HS proteoglycans. However, some lymphoid cells do not express HS (e.g., BaF3 pro-B cell line). These cells prove useful because they are also devoid of growth factor receptors and are dependent on IL-3 for their survival and proliferation (Daley and Baltimore, 1988; Palacios and Steinmetz, 1985); cells expressing growth factor receptors following transformation with cDNA will often respond to the activated receptors and survive in the absence of IL-3 (Daley and Baltimore, 1988). BaF3 cells transfected to express FGF receptors do not survive and proliferate in response to FGF (in the absence of IL-3), but do respond if an exogenous source of heparin or HS is supplied (Filla *et al.*, 1998; Ornitz *et al.*, 1996). Other cells that have proven useful are CHO cells that contain mutations in the HS synthetic enzymes, which have been isolated following mutagenesis screening. These include defects in synthesis of the linkage tetrasaccharide necessary for either chondroitin sulfate or HS synthesis (pgsA (Esko *et al.*, 1985), pgsB (Esko *et al.*, 1987), and pgsG (Bai *et al.*, 1999)) leading to glycosaminoglycan-deficient cells, or loss of the HS copolymerase EXT-1 activity necessary for extension of HS chains (pgsD CHO mutant; Lidholt *et al.*, 1992), which leads to HS-deficient cells, among others.

III. Identification of Regulatory HS

A. General Considerations

A useful method for examining the binding specificity of a growth factor–receptor pair is to use the isolated proteins for direct binding studies *in vitro*. For the FGFs, this has been used extensively to determine the general sulfation requirements for binding. In addition, as the assembly of the FGF with the receptor is also dependent on HS, this assembly can be examined *in vitro* using FGF and soluble FR constructs as probes. FR extracellular domains expressed as chimeras with human placental alkaline phosphatase (AP) have been used effectively to examine binding specificity (Chang *et al.*, 2000; Chellaiah *et al.*, 1999; Friedl *et al.*, 1997). The AP tag is useful for facile purification of the chimera from conditioned medium of mammalian cells, as well as for detection of the protein during binding studies.

B. Protocols for Preparation of Binding Probes

1. Production of FGF

Methods for purification of FGF directly from tissues, such as bovine brain, are available (Rapraeger *et al.*, 1994), but are tedious compared to the purification of bacterially expressed protein. A method for purification of mammalian FGF2 expressed in yeast can be found in Rapraeger *et al.* (1994). These methods rely on several purification steps, including differential binding and elution from heparin–agarose, which utilizes the heparin affinity shared by all FGFs. Many of the FGFs are now available from multiple commercial sources (R&D Systems, Minneapolis, MN; Peprotech, Rocky Hill, NJ; Intergen Co., Purchase, NY). These FGFs are typically produced in bacteria and appear to retain full biological activity. The following is a procedure for isolation of FGF expressed in bacteria with a polyhistidine tag, using the constructs and procedure published by MacArthur *et al.* (1995) for the expression and purification of FGF8:

(a) The *Escherichia coli* host strain SG13009, transformed with a version of the QE16 bacterial expression plasmid (Qiagen) containing a 4.8-kb FGF8 cDNA, are grown in 500 ml standard LB broth to an A_{600} of 0.6–0.8. IPTG (1 ml of 1 M stock) is added to induce protein expression over the following 5 h. The cells are harvested by centrifugation at $7000 \times g$. The bacterial pellets can be frozen at this point. The FGF8 is present in inclusion bodies as insoluble protein, but is solubilized under denaturation conditions. The cells are thawed and protein purified using the Qiagen procedure for isolating $6\times$ His-tagged proteins on nickel-resin affinity columns. Details on this procedure are available from Qiagen Corporation (the QIAexpressionist, third edition) and are summarized here for $6\times$ His-tagged FGF8. Cell pellets (approx. 1.7 g/500 ml culture) are solubilized in Qiagen buffer A (6 M guanidine-HCl, 0.1 M NaPO$_4$, 10 mM Tris-HCl (pH 8.0)) at 5 ml per gram wet weight and stirred for 1 h at room temp. The supernatant containing the proteins is isolated by centrifugation at $7000 \times g$.

(b) The affinity column is prepared by loading 4 ml packed resin volume of Ni-NTA agarose resin (Qiagen) onto a column at room temperature and equilibrating the resin in 40 ml of Qiagen buffer A at 25 ml/h. Load the lysate supernatant onto the nickel column at 15 ml/h at room temperature. Wash the column with 40 ml of Qiagen buffer A (or until the A_{280} is <0.01) at 15 ml/h at room temperature. Wash the column with 20 ml of Qiagen buffer B (8 M urea, 0.1 M NaPO$_4$, 10 mM Tris-HCl (pH 8.0)), followed by 20 ml of Qiagen buffer C (8 M urea, 0.1 M NaPO$_4$, 10 mM Tris-HCl (pH 6.3). This is followed by 20 ml of Qiagen buffer D (8 M urea, 0.1 M NaPO$_4$, 10 mM Tris-HCl (pH 5.9)). Although some His-tagged protein may elute in Qiagen buffer D, FGF elution occurs primarily in 20 ml Qiagen buffer E (8 M urea, 0.1 M NaPO$_4$, 10 mM Tris-HCl (pH 4.5)). The column is then washed in 6 M guanidine-HCl in 0.2 M acetic acid (Qiagen buffer F) to ensure complete protein removal. Elution of protein is monitored at A_{280} and verified by SDS–PAGE of TCA-precipitated samples.

(c) The denatured FGF is refolded by dialysis against 30 volumes of 1 M urea, 50 mM Tris (pH 8.0), 30 mM HCl, 0.05% Tween-20, and 2.2 μM reduced gluathione for 24 h at 4°C, followed by 2-fold dialysis against 30 volumes of PBS (pH 7.4) containing 0.05% Tween-20, 2.2 μM reduced glutathione, 1 mM PMSF, 1 μg/ml leupeptin, 5 μg/ml aprotinin, and 1 μg/ml pepstatin A for 24 h at 4°C.

(d) Following the final renaturation dialysis step, the FGF is passed over a heparin–agarose affinity column. Two ml heparin–agarose (Sigma) is equilibrated by pre-washing in PBS containg 0.15 M NaCl (pH 8.0). The FGF preparation is mixed with an equal volume of PBS/0.15 M NaCl and passed over the heparin–agarose, followed by a wash with 20 ml PBS/0.15 M NaCl. Bound FGF is then eluted in PBS with a 40 ml gradient from 0.15 to 2.00 M NaCl, with elution of the protein expected around 1.2 M NaCl. Eluted FGF is dialyzed against PBS/0.15 M NaCl. Protein concentration is determined using the BioRad BCA assay.

(e) The final step is to confirm the activity of the growth factor. A useful method is to measure ^{3}H-thymidine incorporation in serum-starved cells (Krufka *et al.,*

1996). The cell of choice, of course, must be responsive to the growth factor in question. Most cells are typically quiescent in 0.1% serum, which still provides factors necessary for cell attachment, but does not provide sufficient mitogenic factors. Cells are plated in 24-well plates in serum-containing culture medium for 24 h, then starved by transfer to medium containing 0.1% serum (or 0.1% BSA if serum is not required) for 24 h. Following starvation, the FGF to be tested is added to the medium at concentrations ranging from 0.1 to 100 ng/ml. It is useful if a known standard of the FGF is available for comparison. Addition of serum is a useful positive control. ^{3}H-Thymidine (1 μCi/ml) is added to the culture 18 h after FGF stimulation for a 6-h pulse. Incorporation into DNA is measured following fixation of the cells with 5% TCA, 3× washes with PBS, and solubilization of the monolayer in 0.1 N NaOH and scintillation counting. Although the cellular response can vary with cell type, a typical half-maximal response would be seen in the 100 pM range of added FGF.

2. Purification of Chimeric Proteins Comprising Receptor Extracellular Domain and Alkaline Phosphatase

(a) The cDNA for the FR-AP chimera in pcDNA3 is expressed in COS-7 cells (Chang *et al.,* 2000). These cells are easily transfectable using calcium-phosphate trans-fection techniques, produce high amounts of protein (10 μg/ml) in response to the CMV promoter, and secrete a properly folded and disulfide-bonded receptor into the conditioned medium. Transfected cells are initially grown in T175 flasks to confluence, then each flask is passaged by routine trypsinization into an 850 cm^2 roller bottle. The initial culture in T175 flasks in conducted in 500 μg/ml Ge-neticin (G418) for plasmid selection; this is not added to the medium once the cells are placed in the roller bottles. The roller bottles are useful as a small amount of medium can be used to bathe a relatively large number of cells. The cells are grown in 100 ml of 10% calf serum-containing complete DMEM buffered with 25 mM Hepes (pH 7.4) and gassed with a 5% CO$_2$/95% air mixture; this is done by blowing the gas through a sterile pipette into the flask for several minutes in a sterile hood. The flask is tightly capped and placed in a 37°C roller bottle apparatus. Cell con-fluence can be monitored by observing the monolayers directly on an inverted tissue culture microscope, and once cells are 60% confluent the medium in the roller bottle is changed daily. Harvested conditioned medium is quickly frozen in liquid nitrogen and stored at −70°C.

(b) The FR-AP construct is isolated from the conditioned medium on an anti-AP affinity column. The agarose affinity beads are available from Sigma Corporation (St. Louis, MO; Cat. #A-2080) and can be reutilized, with the column stored in 0.02% NaN$_3$ at 4°C between uses. A 1.0 ml column is washed with 5 column volumes of PBS prior to use, then 300 ml of filtered conditioned medium is applied. The column is washed in 5 column volumes of PBS followed by 10 ml distilled water, then eluted in 3.0 ml 0.1 M glycine (pH 2.5). As the construct is not stable in this low pH, care is taken to collect 0.3 ml fractions directly into an equal volume of 1.0 M Tris (pH 8.0) to neutralize the pH. The elution profile is determined by mixing 50 μl of a 1:10 dilution (in PBS, pH 8.0) of each eluted fraction with an equal volume of assay solution containing p-nitrophenylphosphate (24 mM; Sigma 104, Cat. #1040-G), 0.4 M diethanolamine, 1 mM MgCl$_2$ (from a stock 10 mM solution in water), and 20 mM L-homoarginine. Absorbance at 405 nm generated by released p-nitrophenol is recorded as a measure of AP activity. Pooled fractions are dialyzed against PBS. The relative amount of FR-AP harvested can be determined by comparison with a known human placental alkaline phosphatase standard obtained from Sigma (Cat. #P1391). The FR-AP chimera is typically pure following this procedure. This can be checked by SDS–PAGE followed by Coomassie blue staining. The position of the chimera can be verified using the anti-AP antibody; however, some antibodies only detect native protein, which is present as a dimer even in SDS. For this reason, boiling of the sample prior to electrophoresis should be avoided. After confirmation of purity, final protein concentration can be determined using a BCA protein assay (Pierce).

C. Using FGF and FR-AP as Probes to Measure Binding Specificity of HS

The activity of the FR-AP constructs is tested via binding to FGFs immobilized on heparin. In addition, similar assays are used to explore the binding requirements of HS for these FGF and FR-AP pairs, as the binding specificity of the HS may vary depending on its source.

1. Binding to Heparin or HS Immobilized in 96-Well Plates

We use an assay that allows us to quantitatively measure FR binding to FGF immobilized on HS. The assay uses 96-well plates in which the HS (or heparin) is attached via its reducing end. This is important because the entire chain is available for interaction with the ligands (as it would be if attached to a core protein) and the chain has not been chemically modified by the attachment procedure. In this procedure (Satoh *et al.,* 1998), the wells are initially coated with methylvinyl ether/maleic anhydride copolymer (Fisher Scientific-Acros Organics). Adipic acid dihydrazide (Sigma) is added to form hydrazine groups, followed by heparin or HS. The hydrazine groups react with the reducing end of the heparin or HS, forming Schiff bases. This is followed by sodium cyanoborohydride (Sigma) to reduce the Schiffs base to stable alkylamine bonds, covalently linking the HS to the dish. The well is blocked with BSA. Heparin is commercially available, as is HS (Sigma Corp.; Seikagaku America; Neoparin, San Leandro, CA). The source and size of the glycosaminoglycan is important. Low-MW heparin (e.g., less than 3000 Dal) is on the order of 12 saccharides in length. This is estimated to be the minimal size necessary for an FGF–FR complex to form. Thus, heparin or HS that is 10,000 Dal or greater is recommended. The glycosaminoglycans are typically from adult porcine or bovine organs, such as lung, or intestinal mucosa, and the structures of the heparin and almost certainly the HS from these or different sites may vary. In addition, the HS may be contaminated to varying degrees with heparin. This can be checked by treatment with heparinases; heparin should be degraded almost to completion by heparin lyase I, whereas HS should be largely resistant, and vice versa if heparin lyase III is used. The digestion can be monitored at A_{232} or by electrophoresis on PAGE followed by Azure A and ammoniacal silver staining (Lyon and Gallagher, 1990).

Procedure for binding of FGF and FR-AP probes to heparin (or HS) on microtiter plates:

(a) Coat wells of 96-well microtiter plates with 150 μl of MMAC (methyl vinyl ether/maleic anhydride copolymer—low MW) dissolved (5 mg/ml) in DMSO. Aspirate after 30 min. Add 200 μl ADHZ (adipic acid dihydrazide) dissolved in water at 10 mg/ml. Aspirate after 2.5 h incubation, wash 3-fold with water. Add 50 μl of heparin dissolved at 1 mg/ml in 25 mM citrate–phosphate buffer (pH 5.0); incubate overnight at room temp. Add 10 μl of 1% sodium cyanoborohydride and incubate for 1 h. Wash wells 3-fold with Tris-buffered saline (TBS; pH 7.4). Block wells with TBS containing 3% powdered milk.

(b) Binding of FGF and FR-AP to prepared plates is assessed by incubation of 40 μl 10–30 nM FGF in TBS + 0.3% Tween-20 with each well for 60 min at room temp. After washing, FGF binding is detected by FGF-specific antibodies. Bound antibodies are detected by AP-conjugated secondary antibodies and AP activity is assessed using p-nitrophenylphosphate cleavage and a 96-well ELISA plate reader. If antibodies are not available, the FGF can be labeled via iodination or biotinylation (Friedl et al., 1997; Rapraeger et al., 1994). To test the activity of the FR-AP, the FGF/heparin or FGF/HS complexes on the plate are incubated with 40 μl of 10–100 nM FR-AP in TBS + 0.3% Tween-20 for 60 min, followed by 5-fold washes in TBS. The bound ligand is assessed by incubating the wells with p-nitrophenylphosphate (12 mM; Sigma 104, Cat. #1040-G), 0.2 M diethanolamine, 0.5 mM MgCl$_2$ (from a stock 10 mM solution in water), and 10 mM L-homoarginine. Absorbance at 405 nm generated by released p-nitrophenol is recorded as a measure of AP activity.

2. Binding to Fixed Cell Monolayers

Similar to using 96-well plates coated with heparin or HS, confluent cells displaying cell surface HS can be used to assess FGF and FR-AP binding. This is an extremely useful technique to predict the binding specificity of the cell-type-specific HS. Cells to be examined are suspended in EDTA, then plated at confluent densities (ca. 5×10^5 cells per cm^2) in 96-well plates. As many cell surface HSPGs are extremely sensitive to trypsin (e.g., the syndecans), suspension of the cells using proteases should be avoided. If this is not possible, then the cells should be allowed to recover for 6–12 h under standard culture conditions in the wells. The cells are then washed 5-fold in PBS and fixed in 4% paraformaldehyde in PBS for 1 h. The glutaraldehyde is then removed by 5× washes in PBS. Following blocking of the wells for 60 min in TBS containing 5% (w/v) BSA, the fixed monolayers can now be used for measuring FGF and FR-AP binding, using the procedure described for heparin-coated wells.

3. Confirmation of Binding Specificity by Activity Measurement

The binding of FGF and FR-AP to fixed cells provides an accurate measure of which FGFs and receptors would bind and signal using the endogenous HS. An additional means for assessing this signaling role is to use BaF3 cells expressing individual FGF receptors in lieu of the FR-AP and observing the response of the cells to FGF. BaF3 lymphoid cells are devoid of endogenous FGF receptors and devoid of HS proteoglycan. However, addition of FGF together with heparin will stimulate the survival and proliferation of cells that have been engineered to express an FGF receptor specific for that FGF family member (von Gise *et al.,* 2001). Thus, whereas the cells are typically dependent upon Il-3 for prevention of apoptosis (von Gise *et al.,* 2001), an active FGF signal will suffice (von Gise *et al.,* 2001). In addition, fixed cells bearing endogenous HS proteoglycans will suffice in lieu of the added heparin as long as the HS that the fixed cells have produced is sulfated appropriately for the FGF–receptor pair being studied (Filla *et al.,* 1998).

Procedure:

(a) The cells to be tested are plated at confluent densities in multiwell plates (Filla *et al.,* 1998). The assay works well in 96-well flat bottom plates (Fisher Scientific), or standard 24-well or 6-well culture dishes. Approximately 5×10^5 cells in complete, serum-containing culture medium are plated per cm^2. Once the cells are attached and confluent, which may range from 1 h to overnight culture, they are washed 3-fold in PBS (pH 7.4) and fixed in 0.5% glutaraldehyde for 1 h. The glutaraldehyde is removed by 5-fold washes with PBS containing 0.2 M glycine, following which the cells are incubated overnight in RPMI-1640 containing 10% calf serum. It is essential that all of the fixative be removed or it will affect the survival of the BaF3 cells in subsequent steps.

(b) Fixed, washed monolayers are incubated with BaF3 cells expressing a specific FGF receptor. The BaF3 cells respond to activated FR1, and chimeras in which the extracellular domains of other receptors are expressed with the FR1 cytoplasmic domain are useful to ensure that the signaling response is uniform for different receptors. The FR4 signal, for example, does not promote survival and proliferation in the BaF3 cells and is useful only as an $FR4_{ecto}/FR1_{cyto}$ chimera (Wang *et al.,* 1994). The cells (10^5/ml in complete RPMI-1640 containing 10% calf serum, 4 mM L-glutamine, and 1% penicillin/streptomycin) are added to the wells in the presence or absence of 1–10 nM FGF and with or without 100 ng/ml heparin as a positive control. An additional positive control is inclusion of 10% conditioned medium from WEHI-3 cells, which provides Il-3 (Filla *et al.,* 1998), to prevent apoptosis of the BaF3 cells. The cultures are incubated at 37°C in a CO_2 incubator for 72–96 h. During this time period, the cells that fail to be stimulated either by Il-3 or FGF will undergo apoptosis, whereas surviving cells will proliferate. An additional control is to treat the fixed monolayer with heparin lyase I and III (0.01 IU each for 2 h followed by enzyme renewal for an additional 2 h) to destroy the HS provided by the cells being tested.

(c) Survival and proliferation of the BaF3 cells can be monitored in several ways, including measuring incorporation of ^{3}H-thymidine for 6 h (Filla *et al.,* 1998),

by performing cell counts, or by using commercially available cell quantification kits, such as the Celltiter 96 Aqueous One Solution reagent (Promega Corporation, Fitchburg, WI). The last is especially useful as it does not require any additional washing of the cells or the wells and can be read directly on a spectrophotometic plate reader.

4. Binding of Probes to HS *in Vivo:* Use of Frozen or Paraffin-Embedded Tissue Sections

A major use of the FGF and FR-AP probes is to detect specificity that may reside in HS proteoglycans expressed in different tissues. Our current work indicates that there is substantial variability in HS that governs which FGF and receptor pairs can assemble (Chang *et al.,* 2000; Friedl *et al.,* 1997; Allen and Rapraeger, unpublished). As discussed for heparin or HS-coated wells, or fixed cells, the FGF and FR-AP probes can be used to examine the prevalent HS present around cells and in the extracellular matrix in 4–5 μm sections. The distribution of total HS can be demonstrated by using specific heparinases to cleave the HS chains, leaving a glycosaminoglycan fragment attached to the endogenous core protein bearing a terminal $\Delta_{4,5}$-uronic acid that can be detected with antibodies. In addition, competition with modified heparin during the binding procedure can provide information on the binding specificity of the probes.

Procedure:

(a) For frozen sections, tissue (embryonic, adult, pathological) is embedded in OCT without prior fixation. Paraffin-embedded tissues are fixed in formalin. Sections (4–5 μm) are cut and mounted on charged slides (Fisher Superfrost Plus slides). The paraffin sections are rehydrated according to standard procedures utilizing xylene and a series of graded alcohols. The frozen sections are allowed to dry for 1 min, then are fixed for 5–10 min in 4% formaldehyde in PBS at room temp. The sections are treated with sodium borohydride (mixed at 0.5 mg/ml in PBS at 4°C, then allowed to warm to room temp.) for 10 min, then are washed in PBS and incubated with 0.1 M glycine overnight for 30 min at 4°C. This is followed by a blocking solution (2% BSA in TBS, pH 7.4) overnight at 4°C.

(b) Select sections are treated with heparin lyase III either as the initial step in HS localization or as a negative control for probe binding to HS. The enzyme (0.01 IU/ml in 50 mM Hepes, 50 mM NaOAc, 150 mM NaCl, 9 mM CaCl$_2$, 0.1% BSA, pH 6.5) is incubated for 2 h, replenishing the enzyme after 2 h if necessary. To detect the remaining HS stubs bearing $\Delta_{4,5}$-uronate, the sections are incubated with 2 μg/ml mAb 3G10 (Seikagaku America, Falmouth, MA) for 60 min in PBS. The bound antibody is detected with a fluorescent (frozen sections) or HRP-conjugated (paraffin sections) secondary antibody. Sections that are not treated with the heparin lyase should be negative for 3G10 staining.

(c) For FGF binding, the growth factor is incubated at 3–30 nM concentrations in TBS containing 10% BSA for 60 min at room temp. Unbound FGF is removed by 3–5× washes, followed by FGF-specific antibody if the FGF is to be localized. Alternatively, the section is incubated with 10–100 nM FR-AP chimera in TBS containing 10% BSA for 60 min at room temp. Alternatively, the staining with FGF and FR-AP is performed in RPMI-1640 buffered with 20 mM Hepes and containing 10% calf serum to ensure physiological calcium/magnesium concentrations. Unbound FR-AP is removed by 3–5 washes with TBS, followed by detection with anti-placental alkaline phosphatase antibody (Biomeda, Hayward, CA, Cat. #A67). Although the alkaline phosphatase activity itself might be used for detection, its signal is typically not sufficiently strong.

IV. Analyzing the Structure of HS: General Considerations and Methodologies

Once it is clear that HS is acting as a regulatory agent in the signaling process, it becomes important to understand the actual HS structure that confers the specificity. This has been a difficult area of investigation because of the inherent difficulty in determining an exact sequence and sulfation pattern of the chain. However, new methods are now becoming available that will make this a more realistic endeavor. Traditionally, information on the HS structure has relied on two approaches: (1) the use of modified heparin as an activator or inhibitor of the signaling process, and (2) isolation of an HS fragment, using the growth factor as an affinity ligand, and determination of the oligosaccharide sequence.

Chemical treatments can be used to selectively remove specific sulfate groups from heparin or heparan sulfate and the failure of these modified forms to display activity suggest that the removal of sulfate moieties were important for recognition by the growth factor or its receptor, or both (Rapraeger *et al.*, 1994). This information is highly useful in comparing the relative HS binding requirements of different growth factors and receptors. It does not, however, provide information on how such sulfate groups may be arranged into more discrete domains within the HS chain.

To derive more information, ligand affinity columns can be used to isolate specific HS oligosaccharides from a library of HS fragmented by partial heparitinase digestion (Pye *et al.*, 1998; Turnbull *et al.*, 1992). This is done most easily using HS obtained from commercial sources; cultured cells do not provide sufficient amounts unless cultured in very high amounts. For FGF, the sequences that are sufficient for ligand binding are on the order of 6–8 saccharides (Maccarana *et al.*, 1993), whereas active oligosaccharides are on the order of 12–14 sugars (Guimond *et al.*, 1993). The difference in size likely represents the difference between binding to ligand and binding in a complex with the ligand and the receptor tyrosine kinase. Oligosaccharides that bind to the ligand affinity column can be separated based on size, then screened for signaling activity, for example, on chlorate-treated cells to which the growth factor has been added. The smallest oligosaccharide that can promote signaling would be a candidate for sequencing. The methods for sequencing the HS oligosaccharides will be summarized here; more details on the procedures are available in several excellent papers (Merry *et al.*, 1999; Rhomberg *et al.*, 1998; Turnbull *et al.*, 1999; Venkataraman *et al.*, 1999; Vives *et al.*, 1999). A highly useful procedure relies on MALDI-mass spectroscopy to identify the exact molecular mass of the oligosaccharide. A current procedure devised by the Sasisekharan group (Keiser *et al.,* 2001) uses polyarginine as a basic polypeptide $(RG)_{19}R$ to combine with the HS polyanion. The polyarginine has a known protonated $(M + H^+)$ molecular mass of 4226.8. The HS fragment is lyophilized and solubilized in 5 μl of matrix solution (12 mg/ml caffeic acid in 30% acetonitrile) containing a 2-fold molar excess of the basic peptide. The mass spectrum of the sample comprises the $(M + H)^+$ ion of the peptide and the $(M + H)^+$ ion of the 1:1 oligosaccharide/peptide complex. The mass of the peptide is subtracted to obtain the oligosaccharide mass to <1 Dal. This mass allows the calculation of the number of sugars and sulfate or acetyl groups in the oligosaccharide for any oligosaccharide that is 14 sugars in length or less. Longer oligosaccharides can be subjected to limited cleavage with heparinase I or III, which will cleave at known sites. This will reduce the oligosaccharide to two or more fragments, depending on the extent of enzyme cleavage. The size of these fragments can then be determined and added together to provide the size of the whole.

Next, a compositional analysis of the fragment is carried out by degrading the fragment to disaccharides using a mixture of heparin lyases I and III. The disaccharides and their relative stoichiometry are identified by capillary electrophoresis using known standards (Rhomberg *et al.,* 1998).

Given the known size of the fragment, its constituent disaccharides, and their stoichiometry, a master list of possible structures is compiled. If the fragment comprised only one disaccharide, the sequence would obviously now be clear. As it is likely to be more complex, the next approach is to carry out further fragmentation using enzymatic and chemical digestion. Heparin lyase I or III cleavage sites are discerned in the library of possible oligosaccharide sequences. Limited or exhaustive cleavage of the unknown fragment with one or more of the enzymes will then produce products, the sizes of which can rule candidate sequences in or out. Since the enzymes introduce a $\Delta_{4,5}$ unsaturated bond in the uronic acid at the nonreducing end of the cleavage site, this alteration in size defines which of the resulting fragments was derived from the reducing vs nonreducing side of the cut. In addition, the specificity of the enzyme defines the structure of the oligosaccharide at that site. Further information is provided by nitrous acid cleavage, of which there are procedures that cleave the oligosaccharide depending on the N-substitution group on the glucosamine. After each digestion, the molecular mass of the resulting fragments can be compared to the table of possible oligosaccharide sequences and will quickly rule out false candidates.

At this point, the possible structures for the oligosaccharide will be known or narrowed substantially. Depending on what part of the fragment remains uncertain, other procedures are employed. The reducing end of the fragment will be labeled by treatment with semicarbozole, followed by enzymatic digestion to identify the enzyme-generated fragment that contains the reducing end (Keiser *et al.,* 2001). In addition, the lysosomal exoenzymes (available from Glycoscience) can now be employed to remove specific sulfate groups or terminal iduronate, glucuronate, or glucosaminyl residues (Keiser *et al.,* 2001; Turnbull *et al.,* 1999), thus completing the identification of the fragment.

Acknowledgments

The author thanks Ben Allen, Mark Filla, Andreas Friedl, and Jennifer Peterson for their contributions to these methods. Work in the author's laboratory is supported by funding from the National Institutes of Health (R01-GM48850 and R01-HD21881).

References

Abraham, J. A., Mergia, A., Whang, J. L., Tumolo, A., Friedman, J., Hjerrild, K. A., Gospodarowicz, D., and Fiddes, J. C. (1986a). Nucleotide sequence of a bovine clone encoding the angiogenic protein, basic fibroblast growth factor. *Science* **233,** 545–548.

Abraham, J. A., Whang, J. L., Tumolo, A., Mergia, A., and Fiddes, J. C. (1986b). Human basic fibroblast growth factor: Nucleotide sequence, genomic organization, and expression in mammalian cells. *Cold Spring Harbor Symp. Quant. Biol.* **51 Pt 1,** 657–668.

Aikawa, J., and Esko, J. D. (1999). Molecular cloning and expression of a third member of the heparan sulfate/heparin GlcNAc N-deacetylase/N-sulfotransferase family. *J. Biol. Chem.* **274,** 2690–2695.

Aikawa, J., Grobe, K., Tsujimoto, M., and Esko, J. D. (2001). Multiple isozymes of heparan sulfate/heparin GlcNAc N-deacetylase/GlcN N-sulfotransferase. Structure and activity of the fourth member, NDST4. *J. Biol. Chem.* **276,** 5876–5882.

Bai, X., Wei, G., Sinha, A., and Esko, J. D. (1999). Chinese hamster ovary cell mutants defective in gly-cosaminoglycan assembly and glucuronosyltransferase I. *J. Biol. Chem.* **274,** 13017–13024.

Basilico, C., and Moscatelli, D. (1992). The FGF family of growth factors and oncogenes. *Adv. Cancer Res.* **59,** 115–165.

Bellaiche, Y., The, I., and Perrimon, N. (1998). Tout-velu is a *Drosophila* homologue of the putative tumour suppressor EXT-1 and is needed for Hh diffusion. *Nature* **394,** 85–88.

Binari, R. C., Staveley, B. E., Johnson, W. A., Godavarti, R., Sasisekharan, R., and Manoukian, A. S. (1997). Genetic evidence that heparin-like glycosaminoglycans are involved in wingless signaling. *Development* **124,** 2623–2632.

Chang, Z., Meyer, K., Rapraeger, A. C., and Friedl, A. (2000). Differential ability of heparan sulfate proteo-glycans to assemble the fibroblast growth factor receptor complex in situ. *FASEB J.* **14,** 137–144.

Chellaiah, A., Yuan, W., Chellaiah, M., and Ornitz, D. M. (1999). Mapping ligand binding domains in chimeric fibroblast growth factor receptor molecules. Multiple regions determine ligand binding specificity. *J. Biol. Chem.* **274,** 34785–34794.

Daley, G. Q., and Baltimore, D. (1988). Transformation of an interleukin 3-dependent hematopoietic cell line by the chronic myelogenous leukemia-specific P210bcr/abl protein. *Proc. Natl. Acad. Sci. USA* **85,** 9312–9316.

Dionne, C. A., Jaye, M., and Schlessinger, J. (1991). Structural diversity and binding of FGF receptors. *Ann. N.Y. Acad. Sci.* **638,** 161–166.

Ernst, S., Langer, R., Cooney, C. L., and Sasisekharan, R. (1995). Enzymatic degradation of glycosaminogly-cans. *Criti. Rev. Biochem. Mol. Biol.* **30,** 387–444.

Esch, F., Baird, A., Ling, N., Ueno, N., Hill, F., Denoroy, L., Klepper, R., Gospodarowicz, D., Bohlen, P., and Guillemin, R. (1985). Primary structure of bovine pituitary basic fibroblast growth factor (FGF) and comparison with the amino-terminal sequence of bovine brain acidic FGF. *Proc. Natl. Acad. Sci. USA* **82,** 6507–6511.

Esko, J. D., Stewart, T. E., and Taylor, W. H. (1985). Animal cell mutants defective in glycosaminoglycan biosynthesis. *Proc. Natl. Acad. Sci. USA* **82,** 3197–3201.

Esko, J. D., Weinke, J. L., Taylor, W. H., Ekborg, G., Roden, L., Anantharamaiah, G., and Gawish, A. (1987). Inhibition of chondroitin and heparan sulfate biosynthesis in Chinese hamster ovary cell mutants defective in galactosyltransferase I. *J. Biol. Chem.* **262,** 12189–12195.

Farley, J. R., Nakayama, G., Cryns, D., and Segel, I. H. (1978). Adenosine triphosphate sulfurylase from *Penicillium chrysogenum* equilibrium binding, substrate hydrolysis, and isotope exchange studies. *Arch. Biochem. Biophys.* **185,** 376–390.

Filla, M. S., Dam, P., and Rapraeger, A. C. (1998). The cell surface proteoglycan syndecan-1 mediates fibroblast growth factor-2 binding and activity. *J. Cell. Physiol.* **174,** 310–321.

Finch, P. W., Rubin, J. S., Miki, T., Ron, D., and Aaronson, S. A. (1989). Human KGF is FGF-related with properties of a paracrine effector of epithelial cell growth. *Science* **245,** 752–755.

Forsberg, E., Pejler, G., Ringvall, M., Lunderius, C., Tomasini-Johansson, B., Kusche-Gullberg, M., Eriksson, I., Ledin, J., Hellman, L., and Kjellen, L. (1999). Abnormal mast cells in mice deficient in a heparin-synthesizing enzyme. [see comments]. *Nature* **400,** 773–776.

Friedl, A., Chang, Z., Tierney, A., and Rapraeger, A. C. (1997). Differential binding of fibroblast growth factor-2 and -7 to basement membrane heparan sulfate: Comparison of normal and abnormal human tissues. *Ame. J. Pathol.* **150,** 1443–1455.

Gimenez-Gallego, G., Rodkey, J., Bennett, C., Rios-Candelore, M., DiSalvo, J., and Thomas, K. (1985). Brain-derived acidic fibroblast growth factor: Complete amino acid sequence and homologies. *Science* **230,** 1385–1388.

Greene, J. M., Li, Y. L., Yourey, P. A., Gruber, J., Carter, K. C., Shell, B. K., Dillon, P. A., Florence, C., Duan, D. R., Blunt, A., Ornitz, D. M., Ruben, S. M., and Alderson, R. F. (1998). Identification and characterization of a novel member of the fibroblast growth factor family. *Eur. J. Neurosci.* **10,** 1911–1925.

Gu, K., Liu, J., Pervin, A., and Linhardt, R. J. (1993). Comparison of the activity of two chondroitin AC lyases on dermatan sulfate. *Carbohydr. Res.* **244,** 369–377.

Guimond, S., Maccarana, M., Olwin, B. B., Lindahl, U., and Rapraeger, A. C. (1993). Activating and inhibitory heparin sequences for FGF-2 (basic FGF). Distinct requirements for FGF-1, FGF-2, and FGF-4. *J. Biol. Chem.* **268,** 23906–23914.

Habuchi, H., Tanaka, M., Habuchi, O., Yoshida, K., Suzuki, H., Ban, K., and Kimata, K. (2000). The occurrence of three isoforms of heparan sulfate 6-O-sulfotransferase having different specificities for hexuronic acid adjacent to the targeted N-sulfoglucosamine. *J. Biol. Chem.* **275,** 2859–2868.

Hacker, U., Lin, X., and Perrimon, N. (1997). The Drosophila sugarless gene modulates Wingless signaling and encodes an enzyme involved in polysaccharide biosynthesis. *Development* **124,** 3565–3573.

Hamai, A., Hashimoto, N., Mochizuki, H., Kato, F., Makiguchi, Y., Horie, K., and Suzuki, S. (1997). Two distinct chondroitin sulfate ABC lyases. An endoeliminase yielding tetrasaccharides and an exoeliminase preferentially acting on oligosaccharides. *J. Biol. Chem.* **272,** 9123–9130.

Hoshikawa, M., Ohbayashi, N., Yonamine, A., Konishi, M., Ozaki, K., Fukui, S., and Itoh, N. (1998). Structure and expression of a novel fibroblast growth factor, FGF-17, preferentially expressed in the embryonic brain. *Biochem. Biophys. Res. Commun.* **244,** 187–191.

Hovingh, P., and Linker, A. (1974). The disaccharide repeating-units of heparan sulfate. *Carbohydr. Res.* **37,** 181–192.

Hu, M. C., Qiu, W. R., Wang, Y. P., Hill, D., Ring, B. D., Scully, S., Bolon, B., DeRose, M., Luethy, R., Simonet, W. S., Arakawa, T., and Danilenko, D. M. (1998). FGF-18, a novel member of the fibroblast growth factor family, stimulates hepatic and intestinal proliferation. *Mol. Cell. Biol.* **18,** 6063–6074.

Huebner, K., Ferrari, A. C., Delli Bovi, P., Croce, C. M., and Basilico, C. (1988). The FGF-related oncogene, K-FGF, maps to human chromosome region 11q13, possibly near int-2. *Oncogene Res.* **3,** 263–270.

Humphries, D. E., Wong, G. W., Friend, D. S., Gurish, M. F., Qiu, W. T., Huang, C., Sharpe, A. H., and Stevens, R. L. (1999). Heparin is essential for the storage of specific granule proteases in mast cells. [see comments]. *Nature* **400,** 769–772.

Jaye, M., Howk, R., Burgess, W., Ricca, G. A., Chiu, I. M., Ravera, M. W., O'Brien, S. J., Modi, W. S., Maciag, T., and Drohan, W. N. (1986). Human endothelial cell growth factor: Cloning, nucleotide sequence, and chromosome localization. *Science* **233,** 541–545.

Jeffers, M., Shimkets, R., Prayaga, S., Boldog, F., Yang, M., Burgess, C., Fernandes, E., Rittman, B., Shimkets, J., LaRochelle, W. J., and Lichenstein, H. S. (2001). Identification of a novel human fibroblast growth factor and characterization of its role in oncogenesis. *Cancer Res.* **61,** 3131–3138.

Kan, M., Wang, F., Xu, J., Crabb, J. W., Hou, J., and McKeehan, W. L. (1993). An essential heparin-binding domain in the fibroblast growth factor receptor kinase. *Science* **259,** 1918–1921.

Kariya, Y., Kyogashima, M., Suzuki, K., Isomura, T., Sakamoto, T., Horie, K., Ishihara, M., Takano, R., Kamei, K., and Hara, S. (2000). Preparation of completely 6-O-desulfated heparin and its ability to enhance activity of basic fibroblast growth factor. *J. Biol. Chem.* **275,** 25949–25958.

Keiser, N., Venkataraman, G., Shriver, Z., and Sasisekharan, R. (2001). Direct isolation and sequencing of specific protein-binding glycosaminoglycans. *Natl. Med.* **7,** 123–128.

Kilpelainen, I., Kaksonen, M., Avikainen, H., Fath, M., Linhardt, R. J., Raulo, E., and Rauvala, H. (2000). Heparin-binding growth-associated molecule contains two heparin-binding beta-sheet domains that are homologous to the thrombospondin type I repeat. *J. Biol. Chem.* **275,** 13564–13570.

Kirikoshi, H., Sagara, N., Saitoh, T., Tanaka, K., Sekihara, H., Shiokawa, K., and Katoh, M. (2000). Molecular cloning and characterization of human FGF-20 on chromosome 8p21.3–p22. *Biochem. Biophys. Res. Commun.* **274,** 337–343.

Kobayashi, M., Habuchi, H., Yoneda, M., Habuchi, O., and Kimata, K. (1997). Molecular cloning and expression of Chinese hamster ovary cell heparan-sulfate 2-sulfotransferase. *J. Biol. Chem.* **272,** 13980–13985.

Krufka, A., Guimond, S., and Rapraeger, A. C. (1996). Two hierarchies of FGF-2 signaling in heparin: Mitogenic stimulation and high-affinity binding/receptor transphosphorylation. *Biochemistry* **35,** 11131–11141.

Kurima, K., Singh, B., and Schwartz, N. B. (1999). Genomic organization of the mouse and human genes encoding the ATP sulfurylase/adenosine 5′-phosphosulfate kinase isoform SK2. *J. Biol. Chem.* **274,** 33306–33312.

Kusche-Gullberg, M., Eriksson, I., Pikas, D. S., and Kjellen, L. (1998). Identification and expression in mouse of two heparan sulfate glucosaminyl N-deacetylase/N-sulfotransferase genes. *J. Biol. Chem.* **273,** 11902–11907.

Li, H., Deyrup, A., Mensch, J. R. Jr., Domowicz, M., Konstantinidis, A. K., and Schwartz, N. B. (1995). The isolation and characterization of cDNA encoding the mouse bifunctional ATP sulfurylase-adenosine 5′-phosphosulfate kinase. *J. Biol. Chem.* **270,** 29453–29459.

Lidholt, K., Weinke, J. L., Kiser, C. S., Lugemwa, F. N., Bame, K. J., Cheifetz, S., Massague, J., Lindahl, U., and Esko, J. D. (1992). A single mutation affects both N-acetylglucosaminyltransferase and glucuronosyltransferase activities in a Chinese hamster ovary cell mutant defective in heparan sulfate biosynthesis. *Proc. Natl. Acad. Sci. USA* **89,** 2267–2271.

Lin, X., and Perrimon, N. (2000). Role of heparan sulfate proteoglycans in cell-cell signaling in Drosophila. *Matrix Biol.* **19,** 303–307.

Lin, X., Buff, E. M., Perrimon, N., and Michelson, A. M. (1999). Heparan sulfate proteoglycans are essential for FGF receptor signaling during *Drosophila* embryonic development. *Development* **126,** 3715–3723.

Lindahl, U., Kusche-Gullberg, M., and Kjellen, L. (1998). Regulated diversity of heparan sulfate. *J. Biol. Chem.* **273,** 24979–24982.

Linhardt, R. J., Turnbull, J. E., Wang, H. M., Loganathan, D., and Gallagher, J. T. (1990). Examination of the substrate specificity of heparin and heparan sulfate lyases. *Biochemistry* **29,** 2611–2617.

Liu, J., Shworak, N. W., Sinay, P., Schwartz, J. J., Zhang, L., Fritze, L. M., and Rosenberg, R. D. (1999). Expression of heparan sulfate D-glucosaminyl 3-O-sulfotransferase isoforms reveals novel substrate specificities. *J. Biol. Chem.* **274,** 5185–5192.

Lohse, D. L., and Linhardt, R. J. (1992). Purification and characterization of heparin lyases from Flavobacterium heparinum. *J. Biol. Chem.* **267,** 24347–24355.

Lyon, M., and Gallagher, J. T. (1990). A general method for the detection and mapping of submicrogram quantities of glycosaminoglycan oligosaccharides on polyacrylamide gels by sequential staining with azure A and ammoniacal silver. *Anal. Biochem.* **185,** 63–70.

MacArthur, C. A., Lawshe, A., Xu, J., Santos-Ocampo, S., Heikinheimo, M., Chellaiah, A. T., and Ornitz, D. M. (1995). FGF-8 isoforms activate receptor splice forms that are expressed in mesenchymal regions of mouse development. *Development* **121,** 3603–3613.

Maccarana, M., Casu, B., and Lindahl, U. (1993). Minimal sequence in heparin/heparan sulfate required for binding of basic fibroblast growth factor. [erratum appears in *J. Biol. Chem.* (1994) **269,** 3903]. *J. Biol. Chem.* **268,** 23898–23905.

Marics, I., Adelaide, J., Raybaud, F., Mattei, M. G., Coulier, F., Planche, J., de Lapeyriere, O., and Birnbaum, D. (1989). Characterization of the HST-related FGF.6 gene, a new member of the fibroblast growth factor gene family. *Oncogene* **4,** 335–340.

McWhirter, J. R., Goulding, M., Weiner, J. A., Chun, J., and Murre, C. (1997). A novel fibroblast growth factor gene expressed in the developing nervous system is a downstream target of the chimeric homeodomain oncoprotein E2A-Pbx1. *Development* **124,** 3221–3232.

Merry, C. L., Lyon, M., Deakin, J. A., Hopwood, J. J., and Gallagher, J. T. (1999). Highly sensitive sequencing of the sulfated domains of heparan sulfate. *J. Biol. Chem.* **274,** 18455–18462.

Michelacci, Y. M., and Dietrich, C. P. (1975). A comparative study between a chondroitinase B and a chondroitinase AC from Flavobacterium heparinum: Isolation of a chondroitinase AC-susceptible dodecasaccharide from chondroitin sulphate B. *Biochem. J.* **151,** 121–129.

Miki, T., Fleming, T. P., Bottaro, D. P., Rubin, J. S., Ron, D., and Aaronson, S. A. (1991). Expression cDNA cloning of the KGF receptor by creation of a transforming autocrine loop. *Science* **251,** 72–75.

Miyake, A., Konishi, M., Martin, F. H., Hernday, N. A., Ozaki, K., Yamamoto, S., Mikami, T., Arakawa, T., and Itoh, N. (1998). Structure and expression of a novel member, FGF-16, on the fibroblast growth factor family. *Biochem. Biophys. Res. Commun.* **243,** 148–152.

Miyamoto, M., Naruo, K., Seko, C., Matsumoto, S., Kondo, T., and Kurokawa, T. (1993). Molecular cloning of a novel cytokine cDNA encoding the ninth member of the fibroblast growth factor family, which has a unique secretion property. *Mol. Cell. Biol.* **13,** 4251–4259.

Moore, R., Casey, G., Brookes, S., Dixon, M., Peters, G., and Dickson, C. (1986). Sequence, topography and protein coding potential of mouse int-2: a putative oncogene activated by mouse mammary tumour virus. *EMBO J.* **5,** 919–924.

Nakatake, Y., Hoshikawa, M., Asaki, T., Kassai, Y., and Itoh, N. (2001). Identification of a novel fibroblast growth factor, FGF-22, preferentially expressed in the inner root sheath of the hair follicle. *Biochim. Biophys. Acta* **1517,** 460–463.

Nishimura, T., Nakatake, Y., Konishi, M., and Itoh, N. (2000). Identification of a novel FGF, FGF-21, preferentially expressed in the liver. *Biochim. Biophys. Acta* **1492,** 203–206.

Nishimura, T., Utsunomiya, Y., Hoshikawa, M., Ohuchi, H., and Itoh, N. (1999). Structure and expression of a novel human FGF, FGF-19, expressed in the fetal brain. *Biochim. Biophys. Acta* **1444,** 148–151.

Ohbayashi, N., Hoshikawa, M., Kimura, S., Yamasaki, M., Fukui, S., and Itoh, N. (1998). Structure and expression of the mRNA encoding a novel fibroblast growth factor, FGF-18. *J. Biol. Chem.* **273,** 18161–18164.

Ohmachi, S., Watanabe, Y., Mikami, T., Kusu, N., Ibi, T., Akaike, A., and Itoh, N. (2000). FGF-20, a novel neurotrophic factor, preferentially expressed in the substantia nigra pars compacta of rat brain. *Biochem. Biophys. Res. Commun.* **277,** 355–360.

Ornitz, D. M., Xu, J., Colvin, J. S., McEwen, D. G., MacArthur, C. A., Coulier, F., Gao, G., and Goldfarb, M. (1996). Receptor specificity of the fibroblast growth factor family. *J. Biol. Chem.* **271,** 15292–15297.

Ozeran, J. D., Westley, J., and Schwartz, N. B. (1996). Identification and partial purification of PAPS translocase. *Biochemistry* **35,** 3695–3703.

Palacios, R., and Steinmetz, M. (1985). Il-3-dependent mouse clones that express B-220 surface antigen, contain Ig genes in germ-line configuration, and generate B lymphocytes in vivo. *Cell* **41,** 727–734.

Pantoliano, M. W., Horlick, R. A., Springer, B. A., Van Dyk, D. E., Tobery, T., Wetmore, D. R., Lear, J. D., Nahapetian, A. T., Bradley, J. D., and Sisk, W. P. (1994). Multivalent ligand-receptor binding interactions in the fibroblast growth factor system produce a cooperative growth factor and heparin mechanism for receptor dimerization. *Biochemistry* **33,** 10229–10248.

Pellegrini, L., Burke, D. F., von Delft, F., Mulloy, B., and Blundell, T. L. (2000). Crystal structure of fibroblast growth factor receptor ectodomain bound to ligand and heparin. *Nature* **407,** 1029–1034.

Piepkorn, M., Pittelkow, M. R., and Cook, P. W. (1998). Autocrine regulation of keratinocytes: The emerging role of heparin-binding, epidermal growth factor-related growth factors. *J. Invest. Dermatol.* **111,** 715–721.

Pikas, D. S., Eriksson, I., and Kjellen, L. (2000). Overexpression of different isoforms of glucosaminyl N-deacetylase/N-sulfotransferase results in distinct heparan sulfate N-sulfation patterns. *Biochemistry* **39,** 4552–4558.

Plotnikov, A. N., Hubbard, S. R., Schlessinger, J., and Mohammadi, M. (2000). Crystal structures of two FGF-FGFR complexes reveal the determinants of ligand-receptor specificity. *Cell* **101,** 413–424.

Poltorak, Z., Cohen, T., and Neufeld, G. (2000). The VEGF splice variants: Properties, receptors, and usage for the treatment of ischemic diseases. *Herz* **25,** 126–129.

Pye, D. A., Vives, R. R., Turnbull, J. E., Hyde, P., and Gallagher, J. T. (1998). Heparan sulfate oligosaccharides require 6-O-sulfation for promotion of basic fibroblast growth factor mitogenic activity. *J. Biol. Chem.* **273,** 22936–22942.

Rapraeger, A. C., Guimond, S., Krufka, A., and Olwin, B. B. (1994). Regulation by heparan sulfate in fibroblast growth factor signaling. *Methods Enzymol.* **245,** 219–240.

Rapraeger, A. C., Krufka, A., and Olwin, B. B. (1991). Requirement of heparan sulfate for bFGF-mediated fibroblast growth and myoblast differentiation. *Science* **252,** 1705–1708.

Rhomberg, A. J., Ernst, S., Sasisekharan, R., and Biemann, K. (1998). Mass spectrometric and capillary electrophoretic investigation of the enzymatic degradation of heparin-like glycosaminoglycans. *Proc. Natl. Acad. Sci. USA* **95,** 4176–4181.

Robinson, C. J., and Stringer, S. E. (2001). The splice variants of vascular endothelial growth factor (VEGF) and their receptors. *J. Cell. Sci.* **114,** 853–865.

Safaiyan, F., Lindahl, U., and Salmivirta, M. (1998). Selective reduction of 6-O-sulfation in heparan sulfate from transformed mammary epithelial cells. *Eur. J. Biochem.* **252,** 576–582.

Sakata, H., Stahl, S. J., Taylor, W. G., Rosenberg, J. M., Sakaguchi, K., Wingfield, P. T., and Rubin, J. S. (1997). Heparin binding and oligomerization of hepatocyte growth factor/scatter factor isoforms. Heparan sulfate glycosaminoglycan requirement for Met binding and signaling. *J. Biol. Chem.* **272,** 9457–9463.

Satoh, A., Kojima, K., Koyama, T., Ogawa, H., and Matsumoto, I. (1998). Immobilization of saccharides and peptides on 96-well microtiter plates coated with methyl vinyl ether–maleic anhydride copolymer. *Anal. Biochem.* **260,** 96–102.

Schlessinger, J., Plotnikov, A. N., Ibrahimi, O. A., Eliseenkova, A. V., Yeh, B. K., Yayon, A., Linhardt, R. J., and Mohammadi, M. (2000). Crystal structure of a ternary FGF–FGFR–heparin complex reveals a dual role for heparin in FGFR binding and dimerization. *Mol. Cell* **6,** 743–750.

Selleck, S. B. (2000). Proteoglycans and pattern formation: Sugar biochemistry meets developmental genetics. *Trends Genet.* **16,** 206–212.

Shworak, N. W., Liu, J., Petros, L. M., Zhang, L., Kobayashi, M., Copeland, N. G., Jenkins, N. A., and Rosenberg, R. D. (1999). Multiple isoforms of heparan sulfate D-glucosaminyl 3-O-sulfotransferase. Isolation, characterization, and expression of human cdnas and identification of distinct genomic loci. *J. Biol. Chem.* **274,** 5170–5184.

Smallwood, P. M., Munoz-Sanjuan, I., Tong, P., Macke, J. P., Hendry, S. H., Gilbert, D. J., Copeland, N. G., Jenkins, N. A., and Nathans, J. (1996). Fibroblast growth factor (FGF) homologous factors: new members of the FGF family implicated in nervous system development. *Proc. Natl. Acad. Sci. USA* **93,** 9850–9857.

Stauber, D. J., DiGabriele, A. D., and Hendrickson, W. A. (2000). Structural interactions of fibroblast growth factor receptor with its ligands. *Proc. Natl. Acad. Sci. USA* **97,** 49–54.

Sugahara, K., Nadanaka, S., Takeda, K., and Kojima, T. (1996). Structural analysis of unsaturated hexasaccharides isolated from shark cartilage chondroitin sulfate D that are substrates for the exolytic action of chondroitin ABC lyase. *Eur. J. Biochem.* **239,** 871–880.

Tanaka, A., Miyamoto, K., Minamino, N., Takeda, M., Sato, B., Matsuo, H., and Matsumoto, K. (1992). Cloning and characterization of an androgen-induced growth factor essential for the androgen-dependent growth of mouse mammary carcinoma cells. *Proc. Natl. Acad. Sci. USA* **89,** 8928–8932.

Turnbull, J. E., Fernig, D. G., Ke, Y., Wilkinson, M. C., and Gallagher, J. T. (1992). Identification of the basic fibroblast growth factor binding sequence in fibroblast heparan sulfate. *J. Biol. Chem.* **267,** 10337–10341.

Turnbull, J. E., Hopwood, J. J., and Gallagher, J. T. (1999). A strategy for rapid sequencing of heparan sulfate and heparin saccharides. *Proc. Natl. Acad. Sci. USA* **96,** 2698–2703.

Venkataraman, G., Shriver, Z., Raman, R., and Sasisekharan, R. (1999). Sequencing complex polysaccharides. *Science* **286,** 537–542.

Vives, R. R., Pye, D. A., Salmivirta, M., Hopwood, J. J., Lindahl, U., and Gallagher, J. T. (1999). Sequence analysis of heparan sulphate and heparin oligosaccharides. *Biochem. J.* **339,** 767–773.

von Gise, A., Lorenz, P., Wellbrock, C., Hemmings, B., Berberich-Siebelt, F., Rapp, U. R., and Troppmair, J. (2001). Apoptosis suppression by Raf-1 and MEK1 requires MEK- and phosphatidylinositol 3-kinase-dependent signals. *Mol. Cell. Biol.* **21,** 2324–2336.

Wang, J. K., Gao, G., and Goldfarb, M. (1994). Fibroblast growth factor receptors have different signaling and mitogenic potentials. *Mol. Cell. Biol.* **14,** 181–188.

White, K., Evans, W., O'Riordan, L., Speer, M., Econs, M., Lorenz-Depiereux, B., Grabowski, M., Meitinger, T., and Strom, T. (2000). Autosomal dominant hypophosphataemic rickets is associated with mutations in FGF23. *Nat. Genet.* **26,** 345–348.

Xie, M. H., Holcomb, I., Deuel, B., Dowd, P., Huang, A., Vagts, A., Foster, J., Liang, J., Brush, J., Gu, Q., Hillan, K., Goddard, A., and Gurney, A. L. (1999). FGF-19, a novel fibroblast growth factor with unique specificity for FGFR4. *Cytokine* **11,** 729–735.

Yamasaki, M., Miyake, A., Tagashira, S., and Itoh, N. (1996). Structure and expression of the rat mRNA encoding a novel member of the fibroblast growth factor family. *J. Biol. Chem.* **271,** 15918–15921.

Yamashita, T., Yoshioka, M., and Itoh, N. (2000). Identification of a novel fibroblast growth factor, FGF-23, preferentially expressed in the ventrolateral thalamic nucleus of the brain. *Biochem. Biophys. Res. Commun.* **277,** 494–498.

Yoshida, T., Miyagawa, K., Odagiri, H., Sakamoto, H., Little, P. F., Terada, M., and Sugimura, T. (1987). Genomic sequence of hst, a transforming gene encoding a protein homologous to fibroblast growth factors and the int-2-encoded protein. [erratum appears in *Proc. Natl. Acad. Sci. USA* (1988) **85,** 1967]. *Proc. Natl. Acad. Sci. USA* **84,** 7305–7309

Zhan, X., Bates, B., Hu, X. G., and Goldfarb, M. (1988). The human FGF-5 oncogene encodes a novel protein related to fibroblast growth factors. *Mol. Cell. Biol.* **8,** 3487–3495.

CHAPTER 5

Analysis of Basement Membrane Self–Assembly and Cellular Interactions with Native and Recombinant Glycoproteins

Peter D. Yurchenco, Sergei Smirnov, and Todd Mathus

Department of Pathology and Laboratory Medicine
Robert Wood Johnson Medical School
Piscataway, New Jersey 08854

I. Introduction

Self-assembly, following mass action and resulting from macromolecular structure, is an established mechanism in the formation of extracellular matrices (ECMs). The binding interactions range from relatively simple dimerization, to cooperative multi-step interactions characteristic of biological polymerizations. A paradigm for the latter in ECM is the self-assembly of type I collagen (Prockop and Hulmes, 1994; Veis and

George, 1994). Individual triple helical collagen molecules associate with each other by nucleation and propagation to form fibrils, each consisting of a staggered periodic array of overlapping monomers. The fibrils, whose diameters can be regulated by type V collagen intercalation (Birk and Linsenmayer, 1994), become grouped into larger fibers. While there are cellular contributions beyond secretion, in which fibrils propagate out of fibroblast recesses in an orderly fashion (Birk and Linsenmayer, 1994), the fibril assembly process can largely be recapitulated *in vitro*. A different assembly model has been envisioned for fibronectin, a noncollagenous glycoprotein found ubiquitously in ECMs (Pesciotta Peters and Mosher, 1994; Schwarzbauer and Sechler, 1999). Fibronectin polymerization is not observed *in vitro*. However, upon contact with living cell surfaces, fibronectin accumulates into insoluble fibrillar-like arrays. The process requires receptor engagement and is mediated in substantial part by cytoskeletal events. It has been proposed that the fibronectin conformation is altered upon cell contact in a Rho-dependent fashion, enabling cell surface self-assembly (Zhong *et al.*, 1998). Here, complex cellular activity requiring energy expenditure appears to represent a driving force in fibronectin matrix assembly.

Basement membrane formation, to a significant degree, is one of self-assembly in which both laminins and type IV collagens polymerize and bind to nidogen, perlecan, and, where present, agrin. These ECMs can be many microns thick, and thus only a small fraction of components may be close enough to the cell surface to be able to interact with a receptor. Nonetheless, unlike interstitial collagens and their associated complexes, basement membranes are cell surface-associated entities, present on a subset of plasma membranes, in which interaction with receptors or receptor-like molecules is required for the ECM to assemble *in situ*. Since the architecture observed in basement membranes is quite similar to that observed in polymers reconstituted *in vitro*, it is unlikely that receptor interactions drive the creation of supramolecular structure and order. What then are the roles of cell surface anchors (used here to describe cell surface-binding molecules that may not in themselves mediate signal transduction), and/or receptors (used here to describe a cell surface protein mediating signal transduction such as integrins and dystroglycan), in basement membrane assembly?

Major macromolecular components of basement membranes are characterized by two different classes of activities. On the one hand, they can participate in intercomponent binding interactions ("matrix binding" that occurs through self-assembly) that contribute to supramolecular architecture. On the other hand, they can interact with cell surface receptors or anchors. Macromolecules that exhibit these properties are the laminins and type IV collagens. Furthermore, although nidogens, perlecan and agrin may not be essential for supramolecular architecture per se in most basement membranes, they exhibit both cell- and matrix-binding activities, accumulate within basement membranes, and mediate biological functions. Each component is a multidomain, and in some cases, multisubunit, entity in which matrix-binding activities and receptor-interacting activities map to different domains. These two types of interactions have been proposed to act in concert to generate a basement membrane on a selected cell surface (Colognato and Yurchenco, 2000).

Laminin is required for basement membrane formation and may initiate the accumulation of other structural components (De Arcangelis *et al.*, 1996; Smyth *et al.*, 1999).

In the nematode, laminin deposition precedes that of type IV by several developmental stages (Graham *et al.,* 1997; Huang, 2001). A working hypothesis, in which self-assembly is receptor/anchor facilitated and conveys information to the cell, is as follows (Fig. 1). Laminin, secreted by a targeting cell or by a target cell, binds to surface anchors and/or receptors and becomes concentrated on the cell surface. This favors laminin self-assembly into a polymer on the interacting membrane surface, ("anchorage facilitated self-assembly"), creating a laminin ECM that initially lacks type IV collagen (a "nascent" matrix). As receptors become engaged, signal transduction is enabled. In

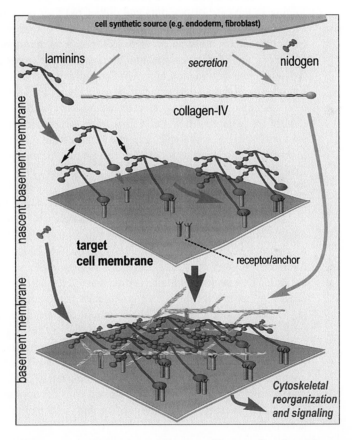

Fig. 1 A working hypothesis of basement membrane assembly on a cell surface. A basement membrane forms between two cell layers, between a cell layer and a fibrous stroma, or on otherwise exposed cell surfaces (e.g., myotubes and fat cells). The site of assembly need not be the cell of synthesis. The diagram illustrates the anchorage of laminin to a target cell surface through its LG modules and its engagement by integrin and/or dystroglycan receptors, possibly requiring an additional anchor. Laminin, now at a high local concentration on a plasma membrane, polymerizes through its short arms, co-aggregates ligated receptors, and reorganizes the cytoskeletal elements that interact with the cytoplasmic domains of the receptors. The other basement membrane components accumulate into the nascent basement membrane and self-assemble either on a laminin scaffolding or on receptors that can engage the components.

the case of integrins, this may be due to a combination of active-site ligation and receptor "clustering" that is transmitted to the cytoskeleton (Miyamoto *et al.,* 1995). The triggering event during basement membrane assembly may be a direct consequence of receptor reorganization or, possibly, a consequence of the increase in force required to displace engaged receptors (from increased topographical receptor immobility), resulting from the formation of an overlying and connected laminin polymer. As laminin accumulates, the binding of other basement membrane components to laminin and/or other receptors and their assembly into a matrix is initiated.

Basement membrane formation, as it occurs *in vivo,* can be separated into several steps. These are (a) the synthesis of structural components and their folding (and subunit joining in the case of laminins and collagens) to form active molecules; (b) secretion into the extracellular compartment; (c) adhesion of components to target cell surfaces; (d) self-assembly through polymerization and intercomponent bond formation to create an extracellular supramolecular complex; and (e) turnover, reflecting the balance of formation and degradation of assembled components that is an important part of tissue remodeling. A defect in any one of these steps may lead to the failure to develop or detect a basement membrane in a correct tissue location. When studying the effect of mutations that affect basement membrane formation, it becomes important to determine which step is affected. For example, a failure of basement membrane formation in embryonic cells expressing nonfunctional FGF receptors appears to be due to a loss of endoderm that synthesizes laminin-1 and type IV collagen (Li *et al.,* 2001). Defects in the ability of laminin to polymerize, on the other hand, may be the basis for basement membrane failure in a mouse model of muscular dystrophy (Colognato and Yurchenco, 1999).

The evaluation of basement membrane models which involve both self-assembly and complex receptor-dependent processes can be facilitated by the application of a multi-disciplined analytical approach that can include biophysics, biochemistry, electron microscopy, X-ray crystallography, cell biology, molecular biology, and genetics. Evaluation of basement membrane components and their functions represents a particular challenge because of their size, complexity, and sensitivity to denaturation and post-translational modifications. In the course of this article which we hope will serve as a guide, we will discuss some of the methods that our colleagues and we have applied, and are beginning to apply, to gain information on basement membrane assembly and the relationship of assembly to cell-surface interactions. This will include some of the biochemical and structural methods used to characterize binding interactions in both native and recombinant glycoproteins, and we will consider approaches that can be used to study cell-surface interactions and the relationships between matrix binding and cell-surface accumulation.

II. Native and Recombinant Basement Membrane Proteins

Early molecular insights into basement membrane self-assembly were gained from the analysis of extracts of basement membrane-enriched tissue, by isolation of components from a limited number of cell lines, and from the study of how different purified components polymerized and interacted with each other. The tissues analyzed included lens capsule, renal glomerulus, placenta, and murine Englebreth–Holm–Swarm (EHS) tumor. Cultured cells used to obtain laminin-1 included the M1536B3 embryonal carcinoma cell line and the PYS (parietal yolk sac) tumor (Amenta *et al.,* 1983; Chung *et al.,* 1979). Cells used as a source for type IV collagen were the mouse PF-HR9 and the human HT1080 cell (ATCC) lines (Bachinger *et al.,* 1982; Brazel *et al.,* 1988). More recently, the repertoire of available basement membrane components has been increased both by the discovery of cell lines that secrete one or more laminin isoforms and by the application of recombinant technology. The latter creates the potential to engineer specific modifications, either for the purpose of purification and identification or for functional analysis.

A. Native Basement Membrane Proteins and Their Fragments

1. EHS Components

The discovery that basement membrane components were present in abundance in the EHS tumor provided the first opportunity to purify substantial biochemical quantities of intact basement membrane laminin-1, $\alpha 1_2\alpha 2$(IV)-type IV collagen, nidogen-1 (entactin-1), and perlecan, a low-density heparan sulfate proteoglycan (Kleinman *et al.,* 1982; Paulsson *et al.,* 1987; Timpl *et al.,* 1978, 1979, 1981). For example, hundreds of milligrams of laminin-1 can be purified from a quarter of a kilogram of tumor. With these components in hand, it became possible to conduct detailed analyses of self-assembly. The very thickness of the EHS basement membrane, it can be argued, led to the early recognition of the importance of binding interactions among basement membrane components. Contributions from cellular receptors, representing only a small fraction of the total possible ECM component bonds in the EHS tumor, were recognized from other types of studies. Since the EHS tumor possessed only a single laminin and single type IV collagen chain composition, the discovery of variant laminins and collagens awaited the analysis of other tissue sources or gene analysis.

2. Components from Placenta and Conditioned Media

The human placenta is another substantial source of extracellular matrix proteins, providing milligram quantities. In the case of laminins, several isoforms can be extracted using conventional biochemical purification techniques (discussed later). Antibody affinity chromatography is an alternative method for isoforms such as those bearing the $\alpha 5$ subunit; however, the method harbors potential pitfalls of cost/availability of antibody reagents and possibly even harshness of elution conditions. Type IV collagens, along with other collagens, are also present in placenta and can be extracted and purified. Since the collagens contain nonreducible cross-links (prevented from forming in other tissues with lathyrogens), proteolytic fragments and complexes rather than intact monomers are obtained from this source.

Laminin-2 ($\alpha 2\beta 1\gamma 1$) and -4 ($\alpha 2\beta 1\gamma 1$) can be extracted and purified from the placental chorionic tissue. Yields are substantially increased if the tissue is first treated with bacterial collagenase (50 μg/ml, 37°C, 20 h). One method of purification utilizes EDTA extraction of a collagenase-treated tissue slurry, followed by gel filtration, DEAE ion exchange, and HPLC high-resolution DEAE-5PW chromatography in sequential steps with low milligram yields per placenta (Cheng *et al.*, 1997). A laminin-4 fraction separates from a laminin-2/4 mixture in the DEAE-5PW ion exchange step. Oddly, it is less easy to separate laminin-2 from the laminin-2/4 mixture (phenyl-Superose hydrophobic interaction chromatography can provide partial separation). Low levels of other laminins can be detected through most steps (Cheng *et al.*, 1997). Laminin-10 and -11 ($\alpha5\beta1\gamma1$ and $\alpha5\beta2\gamma1$) is also present in the initial placental extract and can be purified by antibody affinity chromatography using the general approaches outlined in Champliaud *et al.* (2000) with the $\alpha5$-specific monoclonal antibody 4C7 (Engvall *et al.*, 1986), which is commercially available (Life Technologies).

As new laminin isoforms have been identified, a number of cell lines have been found to secrete them into conditioned medium. The laminins have been isolated by standard biochemical techniques and/or antibody affinity chromatography. Human keratinocytes, epithelial cell lines, and placenta are a source of laminin-5, laminin-6, and laminin-7 (Champliaud *et al.*, 1996; Marinkovich *et al.*, 1992; Rousselle and Aumailley, 1994); A549 cells are a source of laminin-10/-11 while T98G glioblastoma cells are a source of laminin-8 (Fujiwara *et al.*, 2001; Kikkawa *et al.*, 1998). The conditioned medium yields, however, are limited ($\sim$0.1 to <1 μg/ml).

3. Native Fragments

A variety of fragments prepared from native protein exist for laminin-1 (Fig. 2). These have proved to be valuable reagents, important for characterization and low-resolution mapping of laminin functions. Defined fragments, bearing a subset of binding activities, can then be prepared from the purified proteins following limited proteolysis. The laminin-1 fragments include the three short-arm complex C1-4, the $\alpha1/\gamma1$ short arm complexes E1′ E1, E1X, and P1′, the $\beta1$ short arm-derived E4, and the long arm-derived C8-9, E8, and E3 (Bruch *et al.*, 1989; Ott *et al.*, 1982; Paulsson, 1988; Rohde *et al.*, 1980; Timpl *et al.*, 1983). The letters refer to the enzymes cathepsin-G, elastase, and pepsin used to prepare the fragments. These reagents have been found to polymerize, inhibit polymerization, bind heparin, bind dystroglycan, bind nidogen, and/or support integrin-mediated cell adhesion and thus assign these activities to different domains of laminin (Table I). Their purification is well described elsewhere (see Cheng *et al.*, 1997; Yurchenco and O'Rear, 1994). While the short arm complexes E1′, E1, and P1′ are relatively large, differences among these complexes have permitted further identification of active sites within the short arms. For example, E4 and calcium bind to E1′, but not E1, which lacks γ1-VI (Yurchenco and Cheng, 1993), implicating γ1-VI as a critical binding domain. Fragment E8, which extends from a middle portion of the long coiled coil to the middle of G domain, supports $\alpha6\beta1$ integrin ligation. Additional digestion with trypsin generated specific subfragments that were used to limit the binding site to a region near the junction of the coiled coil and proximal LG modules (Deutzmann *et al.*, 1990). More precise activity assignments have required, or are requiring, recombinant approaches.

A relatively small repertoire of proteolytic fragments exist for nidogen, perlecan, and type IV collagen (Mann *et al.*, 1988; Yurchenco and Furthmayr, 1984; Yurchenco

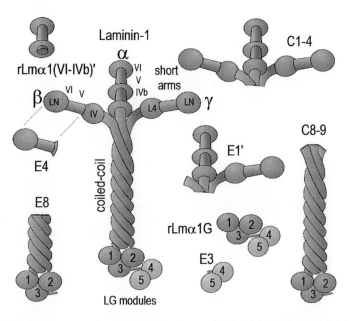

Fig. 2 Laminin-1 and fragments. Laminin-1 is a large heterotrimeric ($\alpha 1\beta 1\gamma 1$) glycoprotein joined through a three-chain parallel coiled coil that spans about 550 amino acid residues. The short arms constitute the N-terminal moieties of the three chains and contain polymerization, nidogen-binding, heparin-binding, and integrin-binding activities. The C-terminal moiety G-domain, with its five LG modules, is a major site of cell-surface interaction and can be ligated by integrins $\alpha 6\beta 1$ and $\alpha 7\beta 1$. Dystroglycan, a component of the dystrophin-associated glycoprotein complex, and heparin bind to related sites in LG-4. A number of defined fragments can be prepared from laminin following elastase (E), cathepsin G (C), and pepsin (P) digestion.

and Ruben, 1988). Methods of purification of type IV collagen, nidogen/entactin, and perlecan from the EHS tumor and of placental laminins have been previously described in detail (Yurchenco and O'Rear, 1994). The collagen fragments include the C-terminal NC1 domain (a fragment derived from the site of collagen dimerization which can self-assemble into "hexamers"), the short form of the 7S domain (proteolytically extracted from placenta and other tissues), and a triple helical fragment that includes the 7S binding domain (designated Col-IV$_{1P}$) which can self-assemble into tetramers (Yurchenco and Furthmayr, 1986; Yurchenco and Furthmayr, 1984).

B. Recombinant Basement Membrane Laminins and Fragments

Increasingly, both recombinant fragments and recombinant proteins bearing deletions, domain "swaps," and point mutations have been, and are being, used to obtain information on site-specific functions. The use of whole laminins bearing site-specific modifications is particularly valuable, since it allows study of loss of a single function in a multifunctional molecule.

Table I
Basement Membrane Component Fragments

Protein	Fragment	Description	Activities
Laminin-1[1]	C1-4	Complex of all short arms	Polymerizes, binds to heparin, supports $\alpha1\beta1$ and $\alpha2\beta1$ integrin-mediated cell adhesion[4]
	E1'	Complex of α- and γ-short arms with part of β-short arm[5]	Inhibits polymerization, binds to E4, binds to heparin less strongly than E3, supports $\alpha1\beta1$ and $\alpha2\beta1$ integrin-mediated cell adhesion
	P1', P1, P1X	P1' is similar to E1', but lacks globular domains	Little or no polymerization-inhibition activity, may expose cryptic cell adhesive site
	E4	β domains VI-V[6]	Inhibits polymerization
	C8-9	Entire coiled-coil domain + LG1-3	Binds to agrin, supports $\alpha6\beta1$, $\alpha6\beta4$, and $\alpha7\beta1$ integrin-mediated cell adhesion, weak heparin binding[7]
	E8	Lower half of coiled-coil domain + LG1-3	Supports $\alpha6\beta1$, $\alpha6\beta4$, and $\alpha7\beta1$ integrin-mediated cell adhesion, weak heparin binding
	E3	LG modules 4-5[8]	Dystroglycan and heparin/sulfatide binding
Nidogen-1[2]	E-50	G3 and part of rod	Laminin γ1-III binding
Collagen-IV[3]	NC1	C-terminal NC1 fragment	Derived from different tissues to give NC1 domains of type IV collagen isoforms; oligomerizes to form "hexamers"
	7S	7S complex, short form	Tetrameric complex (containing 12 N-terminal peptide fragments) does not dissociate[9]
	Col-IV[1P]	Collagenous and 7S binding regions	Self-assembles into 7S-based tetramers that become disulfide-stabilized but does not polymerize[10]

References and Notes: 1, Bruch *et al.,* 1989; Mann *et al.,* 1988; Ott *et al.,* 1982; Paulsson, 1988; Rohde *et al.,* 1980; Timpl *et al.,* 1983; Yurchenco and Cheng, 1993; Yurchenco and O'Rear, 1994. 2, Mann *et al.,* 1988. 3, Langeveld *et al.,* 1988; Risteli *et al.,* 1980; Weber *et al.,* 1984; Yurchenco and Furthmayr, 1984. 4, Activity deduced from behavior of included E1'. 5, E1' is similar to E1XNd and contains nidogen fragment. 6, E4 includes part of domain IV. 7, Adhesive activity deduced from included E8. 8, Start of E3 in "hinge" region of G-domain. 9, Risteli *et al.,* 1980; Siebold *et al.,* 1987. 10, Yurchenco, 1986; Yurchenco and Furthmayr, 1984.

1. Recombinant Fragments and Single–Subunit Glycoproteins

Basement membrane proteins are glycosylated and bear intra- and intermolecular disulfide links; however, such posttranslational modifications are not provided by standard prokaryotic expression systems. Carbohydrate and/or correct disulfide bonding can be important for folding, secretion, and function (Beck *et al.,* 1990; Dean *et al.,* 1990; Deutzmann *et al.,* 1990; Howe, 1984; Stetefeld *et al.,* 1996). Functionally active recombinant fragments have been produced by Baculovirus insect expression (Sung *et al.,* 1993; Yurchenco *et al.,* 1993). Insect cells such as Sf9, used to express the recombinant glycoprotein, possess disulfide isomerase and add N-linked oligosaccharide of the simple

Table II

Recombinant Basement Membrane Single-Subunit Glycoproteins and Fragments

Protein	Composition	Production	Activity
Laminin-α1 long arm fragments	(a) LG1-5 (globular domain) and (b) lower portion of coiled-coil + LG1-5[1]	Baculovirus-infected Sf9 insect cells	(a) Heparin binding, dystroglycan binding[2], myoblast adhesion, (b) heparin binding, dystroglycan binding[2], and, when combined with β1/γ1 coiled-coil fragments, assembles into structure that supports α6β1 integrin-mediated cell adhesion
Laminin-α1 G-domain	(a) LG4, (b) LG5[3]	Mammalian expression	(a) Heparin/sulfatide and dystroglycan binding, (b) neither activity.
Laminin-α2 G-domain	α2-LG1-5[4]	Baculovirus	Heparin binding, binding to *Mycobacterium leprae*
Laminin-α2 G-domain fragments	(a)LG1-3, (b) LG4-5[5]	Mammalian	(a) dystroglycan binding, (b) dystroglycan and heparin binding
Laminin-α1 short arm fragment	α1(VI-IVb)$'$[6]	Single transfected 293 cells	Heparin binding, supports α1β1 and α2β1 integrin-mediated cell adhesion
Laminin-α1/β1 short arm	β1(VI)α1(V-IVb)$'$[7] chimeric fragment	Same	Lacks heparin binding and cell adhesion
Laminin-α2 fragment	α2(VI-IVb)$'$[7]	Same	Similar to α1(VI-IVb)$'$
Nidogen-1	(a) intact protein, (b) Nd-I (1-905), (c) Nd-II (906-1217)[8]	Mammalian	(a) Binds to laminin, collagen-IV, perlecan, fibulin-1, (b) binds to collagen-IV, (c) binds to laminin
Nidogen-2	intact nidogen-2[9]	Same	Binds collagens I and IV, and perlecan; binds weakly to laminin; does not bind to fibulin-1
Perlecan	various domain fragments[10]	Same	Various, including beta1 integrin-mediated cell adhesion, binds heparin, nidogen and fibulin-2

References and Notes: 1, Sung *et al.,* 1993; Yurchenco *et al.,* 1993. 2, Predicted activity. 3, Talts *et al.,* 1999. 4, Rambukkana *et al.,* 1997. 5, Talts *et al.,* 1999; Talts and Timpl, 1999. 6, Colognato-Pyke *et al.,* 1995. 7, Colognato *et al.,* 1997. 8, Fox *et al.,* 1991. 9, Kohfeldt *et al.,* 1998. 10, Brown *et al.,* 1997; Friedrich *et al.,* 1999; Hopf *et al.,* 1999.

mannose type. Recombinant laminin α1-LG1-5 was found to support cell adhesion and heparin binding. However, α6-integrin recognition was not observed unless the modules were expressed with a portion of the coiled-coil domain that was allowed to join with corresponding β1 and γ1 domain segments (Sung *et al.,* 1993).

Expression of recombinant basement membrane glycoproteins by cultured mammalian cells provides, in our opinion, the most promising approach to map functional activities to specific sites within basement membrane components and to characterize the effect of these activities on cells and tissues. A fair number of such fragments or intact single chain components have been reported (Table II). While hampered by low expression levels in the past, improved vectors, many employing the CMV promoter aided by enhancing sequences and offering several choices of antibiotic selection, have made it possible to generate biochemical quantities of glycoprotein for analysis. Yield can be dependent on the particular construct and has, in our hands, ranged from 0.5 to 20 μg/ml of conditioned medium (more typically one to several μg/ml). Such levels, though below those obtainable from the EHS tumor, nonetheless enable investigations not possible in the past and hold great promise for the analysis and mapping of complex functions and for the study of rarer forms of basement membrane components for which no adequate natural source is available. Purification can be greatly simplified by the engineering of an epitope tag used also for identification. Tags for which there are available antibodies and peptides to bind and elute protein include FLAG (DYKDDDDK), hemagglutinin (YPYDVPDYA), and VSV-G protein (YTDIEMNRLGK) epitopes. We have purified recombinant laminins containing the first two tags fused either to the N termini or the C termini.

2. Recombinant Heterotrimeric Laminins

Each laminin is a heterotrimer containing an α-, β-, and γ-chain joined in parallel in a long coiled-coil domain extending some 70–75 nm in length and containing between 500 and 600 amino acid residues per chain (Table III). In the case of polymerization and α6-integrin recognition, activity is dependent on coordinate chain contributions. For example, $\alpha6\beta1$ integrin binding activity requires both LG modules and coiled-coil in laminin-1 and polymerization requires all three chains (Colognato *et al.*, 1999; Sung *et al.*, 1993; Yurchenco and Cheng, 1993; Li *et al.*, 2001). Therefore, to study the full biological effect of laminins with respect to assembly and cellular interactions, full heterotrimeric laminins are required.

The generation of recombinant laminins has presented a special set of technical challenges. In addition to the requirement for disulfide bonding and carbohydrate, the individual chains are much larger than most proteins and recombinant expression may not reach the levels seen with small proteins. Since each laminin is a heterotrimer, it may be necessary to transfect a cell with three different expression vectors. Furthermore, since chain assembly dictates that each transfected cell express all three chains, clonal stability can be a greater problem compared to cells that need to be transfected with a single vector. One way to bypass the need for multiple transfections is to transfect cells that already produce the missing chains. Although there are transfectable cell lines that produce low levels of endogenous laminins (e.g., HT1080, A549) that can provide the complementing subunits for a single recombinant subunit, high expression of these laminins may not be present, limiting expression levels of the laminin bearing a single recombinant chain. In addition, transfection of laminin-producing cells with an expression vector coding for a desired laminin chain will result in the secretion of a mixture of laminins, only a fraction of which include the recombinant chain. The addition of an epitope tag to the recombinant chain may be the only way to effectively permit separation of it from endogenous contaminants. Despite the caveats listed above, the general approach has been fruitful, with the repertoire of reported functional recombinants increasing each year.

Table III
Laminin Subunits

Laminin subunit	Subunit partners	Tissue locations	Null phenotype
$\alpha 1$[1]	$\beta 1, \beta 2, \gamma 1$	Early embryo (blastocyst); fetal kidney, neuroretina and brain, kidney; newborn kidney	ND[2]
$\alpha 2$[3]	$\beta 1, \beta 2, \gamma 1, \gamma 3$	Skeletal and cardiac muscle, peripheral nerve, capillaries, placenta, brain, and other tissues	Postnatal lethal muscular dystrophy and peripheral nerve defect. CNS defects noted in humans bearing $\alpha 2$ mutations
$\alpha 3$[4]	$\beta 1, \beta 2, \gamma 1, \gamma 2$	Skin and other epithelia	Junctional epidermolysis bullosa, ameloblast differentiation
$\alpha 4$[5]	$\beta 1, \beta 2, \gamma 1, \gamma 3$	Primarily mesenchymal cells (adult muscle, lung, nerve, blood vessels, and other tissues)	ND
$\alpha 5$[6]	$\beta 1, \beta 2, \gamma 1$	Diverse epithelia, kidney, developing muscle, synaptic bm	Late embryonic lethal with defects in placental vessels, anterior neural tube closure, and limb development
$\beta 1$[7]	$\alpha 1, \alpha 2, \alpha 3, \alpha 4, \alpha 5, \gamma 1, \gamma 3$	Most tissues	ND
$\beta 2$[8]	$\alpha 1, \alpha 2, \alpha 3, \alpha 4, \alpha 5, \gamma 1$	Neuromuscular junction, glomerulus	Functional defects in neuromuscular junction and renal glomerulus
$\beta 3$[9]	$\alpha 3, \gamma 2$	Skin and other epithelia	Junctional epidermolysis bullosa
$\gamma 1$[10]	$\alpha 1, \alpha 2, \alpha 3, \alpha 4, \alpha 5, \beta 1, \beta 2$	Most tissues	Peri-implantation lethal with failure of blastocyst differentiation
$\gamma 2$[11]	$\alpha 3, \beta 3$	Skin and other epithelia	Junctional epidermolysis bullosa
$\gamma 3$[12]	$\alpha 2, \beta 1, \beta 2, \alpha 4, \alpha 5$	Largely non-basement membrane distributions in nerve, epithelia, brain	ND

References and Notes: 1, Sasaki *et al.,* 1988; Smyth *et al.,* 1999; Vuolteenaho *et al.,* 1994. 2, Not determined. 3, Ehrig *et al.,* 1990; Sunada *et al.,* 1995a,b; Vuolteenaho *et al.,* 1994; Xu *et al.,* 1994. 4, Ryan *et al.,* 1994; McGrath *et al.,* 1995, Ryan, 1999. 5, Iivanainen *et al.,* 1997,1995; Richards *et al.,* 1996, 1994; Champliaud *et al.,* 1996. 6, Miner *et al.,* 1998, 1995, 1997; Patton *et al.,* 1997. 7, Sasaki *et al.,* 1987. 8, Hunter *et al.,* 1989; Noakes *et al.,* 1995a,b. 9, Pulkkinen *et al.,* 1995. 10, Sasaki and Yamada, 1987. 11, Pulkkinen *et al.,* 1994. 12, Koch *et al.,* 1999; Libby *et al.,* 2000.

Three approaches have been used (Table IV). One has been to transfect a cell line with three different expression vectors, one for each of the three different chains. We and our colleagues have been most successful, in the case of laminins-1, -2, and -8 (Fig. 3), in using two consecutive transfection steps and two antibiotic selections, with two chain cDNAs co-transfected (Kortesmaa *et al.,* 2000; Yurchenco *et al.,* 1997; S. Smirnov and P. D. Yurchenco, manuscript in preparation). Ideally, it is preferable to carry out consecutive transfections using three expression vectors, each coding for

Table IV
Recombinant Heterotrimeric Laminins

Protein	Composition	Production	Activity
Laminin-1	Intact and fully recombinant heterotrimeric $\alpha1\beta1\gamma1$[1]	Stable triple-transfected human 293 cells	Wild-type activity (polymerizes, supports $\alpha1\beta1$ and $\alpha6\beta1$ integrin-mediated adhesion)
Laminin-2	(a) Intact heterotrimeric and fully recombinant $\alpha2\beta1\gamma1$, $\alpha2\beta1\gamma1$ with deletions of (b) LG4-5, (c) LG1-3, and (d) LG1-5[2]	Same	(a) Wild-type activity (polymerizes, binds dystroglycan, supports myoblast adhesion, (b) loss of distal dystroglycan binding site, (c) loss of integrin binding and proximal dystroglycan binding site, (d) loss of integrin and dystroglycan binding sites
Laminin-5	(a) Intact heterotrimeric $\alpha3A\beta3\gamma2$, $\alpha3A\beta3\gamma2$ with deletion of (b) LG2-5, (c) LG3-5, (d) LG4-5[3]	$\gamma2$-transfected keratinocytes that express endogenous $\beta3$ and $\alpha3$	(a) Supports $\alpha3$ and $\alpha6$ integrin-mediated cell adhesion, (b-d) LG2-5 and LG3-5 deletions have loss of integrin-mediated cell adhesion
Laminin-5	(a) Heterotrimeric intact $\alpha3A\beta3\gamma2$, $\alpha3A\beta3\gamma2$ with deletion of (b) 358-428 (pyM) and (c) 1-434 (pyC)[4]	$\gamma2$-transfection of $\gamma2$-null immortalized keratinocytes	(a) Wild-type activity, (b-c) loss of ECM deposition (pyM) or deposition and cell adhesion (pyC)
Laminin-8	Intact heterotrimeric $\alpha4\beta1\gamma1$ laminin[5]	Stable triple-transfected human 293 cells	Supports $\alpha6\beta1$ integrin-mediated cell adhesion

References and Notes: 1, Yurchenco *et al.,* 1997. 2, S. Smirnov and P. D. Yurchenco, in preparation. 3, Hirosaki *et al.,* 2000. 4, Gagnoux-Palacios *et al.,* 2001. 5, Kortesmaa *et al.,* 2000.

one of the three chains and utilizing a different antibiotic selection. Furthermore, the most desirable epitope selection approach may be to add a unique N-terminal cleavable tag for each chain, ensuring the greatest degree of control over protein purification and permitting the analysis of multiunit mutations and/or deletions, yet also enabling the removal of all tags should they create functional interference. These approaches are under evaluation.

Another method has been to transfect HT1080, or other cells that express endogenous levels of one or more laminins, using only a single laminin-chain expression vector, to isolate low amounts of heterotrimer that assemble with the endogenous chains. This has been done with recombinant $\gamma2$ in laminin-5 (Hirosaki *et al.,* 2000). The yields were only 0.1 μg/ml of conditioned medium, possibly because the endogenous chains were expressed at low levels. Nonetheless, the approach enabled the mapping of two integrin binding activities ($\alpha6\beta1$ and $\alpha3\beta1$) to LG module-3. We have found that HEK-293 cells can generate quite low levels of endogenous $\beta1$ chain, and clones

rLm-1

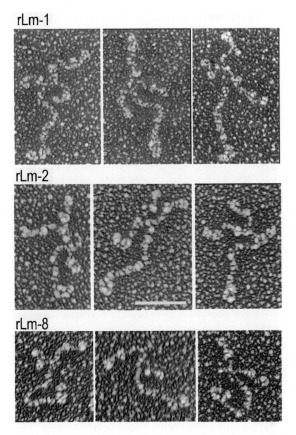

rLm-2

rLm-8

Fig. 3 Recombinant heterotrimeric laminins. Electron micrographs of rotary shadowed Pt/C replicas of recombinant laminin heterotrimers. Recombinant laminin-1 ($\alpha 1 \beta 1 \gamma 1$) was generated by transfecting HEK-293 cells stably expressing laminin-$\gamma 1$ with expression vectors coding for $\alpha 1_{FLAG}$ and $\gamma 1$ (Yurchenco *et al.,* 1997). The laminin was purified by FLAG affinity chromatography. Recombinant laminin-2 ($\alpha 2 \beta 1 \gamma 1$) and laminin-8 ($\alpha 4 \beta 1 \gamma 1$) was generated and purified from transfected cells using the same basic approach, but with epitope tags placed located on the $\gamma 1$ and $\alpha 4$ subunits respectively (S. Smirnov and P. D. Yurchenco, in preparation, and Kortesmaa *et al.,* 2000). While laminin-1 and laminin-2 are seen each to possess three short arms, laminin-8 possesses only two short arms derived from the $\beta 1$ and $\gamma 1$ chains. A small projecting "stub," corresponding to a very short rodlike domain, is present as well. Scale bar $= 50$ nm.

expressing recombinant $\alpha 1$ and $\gamma 1$ chains (only the first was secreted free) were found to secrete about one heterotrimer for every 9 or 10 intact free α-chains as judged by rotary shadowing (Yurchenco *et al.,* 1997). However, yields were considered unnecessarily low to rely on endogenous expression to provide missing trimeric chains in 293 cells. The triple-transfection approach was used instead.

A third method used only to prepare recombinant laminin-5 has depended upon the existence of immortalized keratinocytes (wild-type cells express endogenous laminin-5) derived from a patient with an absolute $\gamma 2$-chain deficiency. These cells could be stably transfected with vectors coding for either wild-type laminin-$\gamma 2$ chain or $\gamma 2$ engineered to contain mutations and deletions (Gagnoux-Palacios *et al.,* 2001). With this approach, evidence was presented to support the concept that a region of the γ2-short arm is critical for deposition of laminin-5 within the extracellular matrix.

III. Analysis of Self-Assembly *in Vitro*

A variety of techniques have been employed to analyze basement membrane interactions. These can be applied to intact native macromolecules, fragments, and recombinant glycoproteins. Affinity is not necessarily a gauge of functional significance, with both low- and high-affinity interactions contributing to architecture (Schittny and Schittny, 1993; Yurchenco and Cheng, 1993). Conventional analysis of low-affinity interactions can require substantial (up to milligram) quantities of component for analysis. It was in part for this reason that the discovery of the EHS tumor and the development of reliable purification techniques provided one of the early breaks in the field.

A. Direct Binding Assays

One of the first activities identified in laminin was that of heparin and heparan sulfate binding (Sakashita *et al.,* 1980). It is thought that heparin-type binding may be important for the attachment of various cell surface and ECM heparan sulfate chains and cell membrane sulfatides. Interacting heparan sulfates can, in turn, bind and activate growth factors such as bFGF, serving to localize growth factor interactions to the basement membrane zone. Heparin binding site may also serve as to anchor laminin to the cell surface during assembly. Methods to detect heparin binding to laminin include those of affinity chromatography, solution-based and solid-phase binding assays, and affinity co-electrophoresis. These can have general applicability to the study of basement membrane interactions and are discussed below.

> *Affinity chromatography:* Heparin affinity is highly dependent upon electrostatic interactions with basic amino acid residues and is therefore quite sensitive to salt concentration. Relevant heparin-binding interactions are likely to occur under conditions of physiological ionic strength (e.g., 140 m*M* NaCl at neutral pH). In the case of laminin, the interaction is sensitive to the presence of calcium. The ionic strength (salt concentration) required to elute a laminin or laminin fragment from a column of immobilized heparin is a relative measure of heparin affinity (Talts *et al.,* 1999). The column can be prepared from commercially available heparin–Sepharose or heparin–agarose (e.g., Sigma-Aldrich) or prepared by coupling heparin to beads following cyanogen bromide activation. This affinity chromatography technique, when scaled up, is useful for the purification of heparin-binding proteins and fragments. High-performance liquid chromatography (HPLC) used with heparin columns with many theoretical plates provide the highest degree of sensitivity. A heparin-5PW column (Toso-Haas) attached in line to a two-pump HPLC system

provides a high-resolution tool with which to compare binding affinities and has been used to identify a laminin short-arm heparin-binding site (Fig. 4).

Affinity retardation chromatography: This method is related to affinity chromatography and is used to detect weak binding interactions (dissociation constants ranging from 3 to 35 μM have been calculated for various laminin interactions; Schittny, 1994). It differs from the conventional analytical and preparative tool by virtue of the behavior of weakly interactive species passing through multiple plates of the immobilized component, characterized by a delay in elution of the mobile component. Elution from the interacting column is compared to a column of identical dimensions and containing an otherwise identical gel coupled to a nonbinding macromolecular component. Elution delay becomes greater as one increases the "concentration" (actually the activity) of immobilized binding component. The analysis is favored by using long columns with high concentrations of immobilized species. The approach has been used to detect interactions among laminin fragments (Schittny and Schittny, 1993; Schittny and Yurchenco, 1990).

Solution binding analysis: If the affinity of interaction is sufficiently high, soluble components that may bind to each other can be incubated with the bound species, separated, and quantitated by a variety of techniques such as immunoprecipitation or selected adhesion. The approach is illustrated with the interaction between heparin and laminin. The highest density of anionic charges in laminin is to be found in the LG modules, the region of highest affinity of heparin and sulfatide binding. Furthermore, the main heparin binding module within the G-domain is LG-4 (Sung *et al.,* 1997; Talts *et al.,* 1999). This interaction has been evaluated by incubating ^{3}H-heparin (New England Nuclear) with laminin fragment E3 in neutral phosphate buffer and capturing the protein with its bound heparin on a nitrocellulose membrane fitted into a 96-well vacuum manifold (Yurchenco *et al.,* 1990). The analysis revealed that 2 mol of E3 bound to 1 mol of heparin under saturating heparin concentrations. For unclear reasons, E1′ and E8 interactions, both measured by affinity chromatography and affinity coelectrophoresis, were not detected by nitrocellulose capture. Using nitrocellulose capture, the binding curve for E3, as well as for intact laminin, was noted to shift in the presence of calcium, reflecting a severalfold increase in affinity (Fig. 5). Evaluation of crystal structure has revealed that a calcium ion binds to each of the terminal LG-modules of the laminin $\alpha2$-G domain, and by analogy, $\alpha1$-G domain (Hohenester *et al.,* 1999; Tisi *et al.,* 2000). A mutagenesis analysis of contiguous groups of basic amino acid residues using recombinant $\alpha1$-LG-4 has placed the heparin-interacting residues on a protein face in the LG4-5 cleft. It is thought that the LG-4/5 calciums contribute to a segregated charged zone, in and near the cleft, increasing heparin affinity (Talts *et al.,* 1999; Tisi *et al.,* 2000). The heparin-binding site is closely linked to the dystroglycan-binding site, with most mutations affecting both activities. Interestingly, LG-4, because of disulfide linkage to the hinge region, is the most topographically distal module.

Affinity coelectrophoresis: Weaker interactions lead to dissociation of bound species during the separation of bound from free species and generally require the

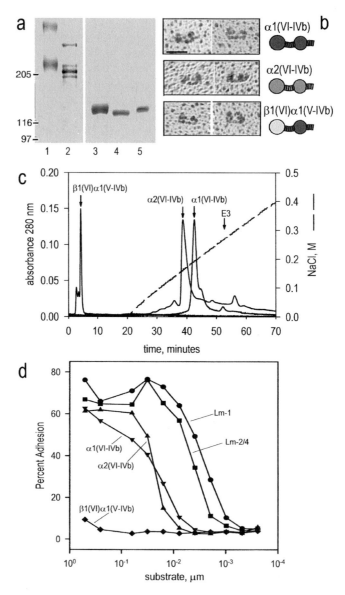

Fig. 4 Recombinant laminin-1 short arm fragments and interactions. Recombinant (r) laminin fragments consisting of short arm N-terminal moieties from the $\alpha1$, $\alpha2$, and $\beta1$ subunits were expressed in 293 cells as follows: r$\alpha1$(VI-IVb)$'$ (prime indicates protein includes 1/3 of domain IIIb), r$\alpha2$(VI-IVb)$'$, and a chimeric protein in which the $\alpha1$ domain-VI globule was replaced by the $\beta1$ domain-VI globule, r$\beta1$(VI)$\alpha1$(V-IVb)$'$. (a) Coomassie blue-stained gels (SDS–PAGE, reducing conditions) of EHS-laminin-1, lane 1; placental laminin-2/-4, lane 2; r$\alpha1$(VI-IVb)$'$, lane 3; r$\alpha2$(VI-IVb)$'$, lane 4; chimeric r$\beta1$(VI)$\alpha1$(V-IVb)$'$, lane 5. (b) Rotary shadow electron micrographs of the three recombinant fragments reveal a dumbbell-like structure for each, the predicted morphology. (c) Heparin 5PW-HPLC salt-elution behavior of the recombinant fragments. Substitution of $\alpha1$ domain VI by $\beta1$ domain-VI removed the heparin binding activity, consistent

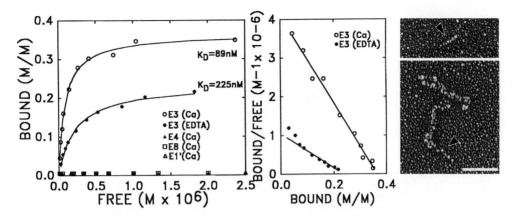

Fig. 5 Heparin binding to the LG modules of laminin-1. (*Left*) Laminin fragments (E3 at 1.7 μM) were incubated in 100 μl aliquots with the indicated concentrations of ³H-heparin (0.43 mCi/mg; average MW 15 kDa) at 35°C for 0.5 h (E3, heparin binding reached apparent equilibrium within 5 min) to 1 h (other fragments) in 50 mM Tris-HCl, 100 mM NaCl, pH 7.4, containing either 1 mM CaCl₂ or 2 mM EDTA. The protein and bound heparin were captured on a 0.45 μm nitrocellulose membrane fitted in a 96-well vacuum manifold and washed briefly (<2 s) with buffer. Radioactivity was determined by scintillation counting. Dissociation constants were calculated by curve fitting with a single-class binding algorithm. (*Middle*) Scatchard transforms of data shown on left. (*Right*) Electron micrographs of rotary shadowed Pt/C replicas, contrast reversed, of heparin (top, arrowhead) and a laminin–heparin complex (bottom). Scale bar = 50 nm. (Reproduced from Yurchenco *et al.*, 1990. *J. Biol. Chem.* **265**, 3981–3991, with permission of The American Society for Biochemistry and Molecular Biology.)

development of special conditions to evaluate affinity. A method that maintains constant protein concentration during analysis is that of affinity coelectrophoresis (Lee and Lander, 1991). ECM protein (e.g., laminin) is mixed with agarose and cast into gel segments at different concentrations, each extending from a perpendicular slot holding radioiodinated tyramine–heparin. A voltage potential is applied between the ends of the gel and one measures the rate of migration of the most highly charged macromolecule (heparin) through the protein field. The higher the protein concentration, the slower the migration of the glycosaminoglycan (Lee and Lander, 1991; San Antonio *et al.*, 1994). The analysis has been employed to measure heparin-binding interactions with a variety of extracellular matrix proteins (Fig. 6). With heparin in the electrophoretic mobile phase, subsets of heparin with different affinities can be separated from each other and extracted for further analysis.

---

with earlier observations that an α1-VI binds to heparin in a blot assay. (d) HT1080 cell adhesion to wells coated at increasing dilutions of recombinant fragments, laminin-1, and laminin-2/-4. Two of the three recombinant fragments support cell adhesion (and spreading). However, substitution of α1-domain VI with β1-domain VI resulted in a loss of adhesion. Both heparin-binding and cell-adhesion activities mapped to this N-terminal α-domain. Evaluation with blocking antibodies revealed that the domain was recognized by the α1β1 and α2β1 integrins. (Reproduced from Colognato *et al.*, 1997. *J. Biol. Chem.* **272**, 29330–29336, with permission of The American Society for Biochemistry and Molecular Biology.)

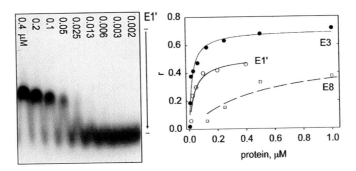

Fig. 6. Affinity coelectrophoresis. Laminin fragments were caste in agarose in vertical slots at two-fold dilutions in the presence of calcium. Radioiodinated heparin (~15 kDa, 10 ng/ml) was placed in a narrow horizontal slot and a voltage potential was established along the long axis of the gel. (*Left*) Autoradiogram of heparin interaction with laminin fragment E1′. Arrow indicates direction of heparin migration from the slot to the nonbinding position. (*Right*) Binding plots of heparin to E3, E1′, and E8 determined from the ratio of bound/unbound migration distances (*r*) measured from the autoradiograms (fitted K_D values of 22, 38, and 450 n*M*, respectively).

Equilibrium gel filtration: Another equilibrium method of binding analysis relies on a separation of species based on Stokes radius rather than net charge differences (Yurchenco and Cheng, 1993). The larger of the two protein components is equilibrated throughout a gel filtration column capable of separating the two interacting components. The second component is radioiodinated and applied to the top of the column as a narrow band. If the two components interact in the column, the labeled species becomes larger and will elute closer to the void volume. By varying the equilibrated protein concentration one obtains a series of elution K_{av} values that in turn can be used to derive an affinity constant. Instead of binding capacity, the saturating K_{av} provides a measure of the Stokes radius for the binding complex that can be used to estimate the mass of the complex. Dissociation constants as high as 20 μM can be determined by this method. Sensitivity of the assay increases as the difference in Stokes radii between the two species increases. This method was used to detect the binding of the larger laminin-1 fragment E1′ (450 kDa) with E4 (75 kDa). The binding data, combined with electron microscopic visualization of the complexes, helped provide the basis of the three-arm hypothesis of laminin polymerization.

Solid-phase binding: For reasons of simplicity, solid-phase binding assays have often been used in the analysis of basement membrane interactions. One protein component is adsorbed onto a plastic surface and the second species is applied in solution over a concentration range, incubated, and then washed to remove the unbound fraction. Detection of the amount of bound species is accomplished by isotopic labeling of the species, or with a noncompeting antibody. Among the most commonly utilized assays, the binding constants one derives need not accurately reflect the true solution-based values. Even when the coat density is known, the binding capacity calculated is likely to provide an underestimate. The assay

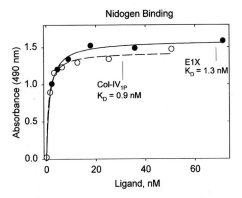

Fig. 7 Recombinant nidogen binding. Recombinant mouse nidogen, purified from the conditioned medium of 293 cells transfected with the pCIS vector containing the mouse nidogen cDNA driven by the CMV promoter, was used to coat 96-well plates at 0.5 μg/well. Laminin-1 fragment E1X and type IV collagen lacking NC1 domain (collagen-IV$_{1P}$) were incubated in 50 mM Tris-HCl, 90 mM NaCl, pH 7.4, in serial twofold dilutions. After washing to remove unbound fragment, specific rabbit antibody (laminin-1 and type IV collagen) were added and washed, and binding was determined colorimetrically following incubation with protein A conjugated to horseradish peroxidase. The data obtained from the plots were fitted for single single-class binding.

can have greater value for comparison among binding species than for the determination of true solution-defined binding constants. Such assays have been used to demonstrate heparin–laminin, nidogen–laminin, nidogen–type IV collagen, and other interactions (Battaglia *et al.,* 1992; Fox *et al.,* 1991). In the case of nidogen, the affinities calculated are not far from those derived from solution-based pull-down assays (Fig. 7).

B. Polymerization Assays

Both laminins and type IV collagen polymerize in buffers of physiologic pH and ionic strength (Yurchenco and Cheng, 1993; Yurchenco and Furthmayr, 1984; Yurchenco *et al.,* 1985). Laminin-1 polymerization has been quantitated by turbidity measurements (useful for following the time-course) and by sedimentation (useful for quantitating polymer; see Yurchenco, 1994; Yurchenco and Cheng, 1993; Yurchenco *et al.,* 1992; Yurchenco and O'Rear, 1994, for experimental details). Assembly is temperature-dependent (thermal gelation), calcium-dependent, and reversible. Reactions have been generally carried out in 50 mM Tris-HCl, pH 7.4, 90 mM NaCl, 1 mM CaCl$_2$ or phosphate-buffered saline. The critical concentration of polymerization is 70–140 nM (in different EHS-laminin preparations) and the gel transition point is 2 μM (Yurchenco *et al.,* 1990). The polymer aggregates that form have been detected by Pt/C rotary shadow electron microscopy (discussed later), and the polymer architecture has been visualized in freeze-etched Pt/C replicas shadowed at high angle (Yurchenco and Cheng, 1993; Yurchenco *et al.,* 1992, 1985).

As more laminin isoforms were discovered, the question arose whether all laminins are capable of polymerization. The model for laminin-1, in which the N-terminal regions of all three short arms participate in polymer bond formation, argued that those laminins which lack these regions in one or more arms should either polymerize less efficiently, or not polymerize at all. A technical problem in addressing this was that the isoforms were sometimes not available in the quantities to conduct a direct self-assembly assay, as was used to analyze laminin-1. Because polymerization is a nucleation–propagation assembly process with a critical concentration, one must be able to exceed that concentration to detect assembly. This limitation could be circumvented by the use of a laminin copolymerization assay in which laminin-1 is used as the "carrier" laminin to drive the polymerization of test components that are maintained at low concentration and detected with chain-specific reagents (or following labeling). In copolymerization, the short arms of different laminins must be able to interact with each other. A "perfect" copolymerization, in which the fraction polymer/total protein concentration curve can be superimposed on that of the carrier laminin, would be seen when both laminins are identical. Partial copolymerization occurs when the two sets of domains can react with each other, but isoform-specific interactions are favored over interisoform binding. A caveat is that if two components appear to copolymerize, one may need to rule out a direct binding of one laminin to the other through a different type of bond, i.e., the two laminins should not form a complex under nonpolymerizing conditions, such as cold, or in the presence of chelating agent. Placental laminins-2 and -4, two laminins with a full complement of short arm domains which co-purify, were found to polymerize in a direct sedimentation assay with a twofold higher critical concentration compared to laminin-1. These placental laminins, which were found not bind to laminin-1 under non-self-assembly conditions, copolymerized with laminin-1 (Fig. 8). In contrast, laminins bearing major domain truncations in the α-short arm (laminin-6) and truncations of nearly all three short arms (laminin-5) did not copolymerize with laminin-1, supporting the self-assembly prediction (Cheng *et al.,* 1997). The assay was also used to show that recombinant laminin-1 has polymerization activity (Fig. 9). Polymerization activity was found to be a property of α2-laminin extracted from wild-type muscle, but not of the laminin extracted from dystrophic muscle bearing a deletion in the α2-LN domain (Colognato and Yurchenco, 1999). These data suggested that a muscular dystrophy can result from a defect in α2-laminin polymerization.

Type IV collagen also self-assembles into a polymer. The process is complex in that it involves three types of interactions, i.e., NC1 domain self-binding to form antiparallel collagenous dimers, 7S bonding to form N-terminal segment end-overlap tetramers, and lateral associations that add great architectural complexity to the network polymer (Timpl *et al.,* 1981; Yurchenco and Furthmayr, 1984; Yurchenco *et al.,* 1993). The network becomes covalently stabilized through reducible and nonreducible bonds at both ends of the molecule (Siebold *et al.,* 1987; Weber *et al.,* 1988). The approaches to analyze this polymerization have been described in detail in previous reviews (Yurchenco, 1994; Yurchenco and O'Rear, 1994).

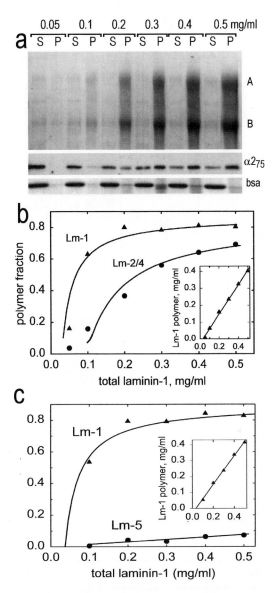

Fig. 8 Laminin coassembly of laminin-2/4 with laminin-1. (a, b) Laminin-2/4 (15 μg/ml) was incubated at 37°C for 3 h in duplicate 50 μl aliquots in 50 m*M* Tris-HCl, 90 m*M* NaCl, 1 m*M* CaCl$_2$ (TBS/Ca) containing increasing concentrations of EHS laminin-1. Both sets of aliquots were analyzed by SDS–PAGE. One was stained with Coomassie blue with laminin-1 nonpolymer (supernatant, S) and polymer (pellet, P) fractions determined by gel densitometry. The other gel was immunoblotted onto nitrocellulose, incubated with polyclonal laminin-α2G domain-specific antibody, and detected with radioiodinated protein A to determine the fraction of α2-75 kDa band in the polymer fraction. The stained gels and blot are shown in (a) and the calculated data are shown in (b). (c) Calculated data from a similar experiment evaluating laminin-5 instead of laminin-2/4. Note copolymerization of laminin-1 with laminin-2/-4 but not with laminin-5. (Reproduced from Cheng *et al.*, 1997. *J. Biol. Chem.* **272,** 31525–31532, with permission of The American Society for Biochemistry and Molecular Biology.)

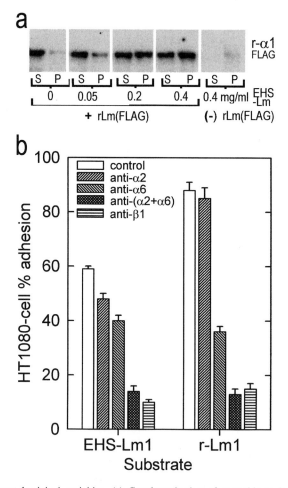

Fig. 9 Recombinant laminin-1 activities. (a) Copolymerization of recombinant laminin-1 with EHS laminin-1. Recombinant mouse laminin-1, purified by FLAG antibody-affinity chromatography, was incubated in small duplicate aliquots with increasing concentrations of EHS laminin-1 in TBS/Ca under the conditions described in the legend to Fig. 8. Recombinant laminin was precipitated with anti-FLAG monoclonal antibody immobilized on agarose beads in supernatant and pellet fraction resolubilized on ice following addition of excess EDTA. Western immunoblots of the immunoprecipitated fractions were probed with FLAG-specific antibody and quantitated by phosphoimaging to determine the polymerized fraction. (b) HT1080 cell adhesion to recombinant laminin-1. Laminin-coated wells (50 μg/ml, 96-well tissue culture plates) were incubated with HT1080 cells, previously radiolabeled with ^{3}H-thymidine, for 70 min at 37°C in culture medium. Where indicated, the cells were incubated in the presence of anti-α6 (GoH3), anti-α2 (IIE10), and/or anti-β1 (AIIB2) integrin-blocking antibody (10 μg/ml each). After 1 h, the cells were washed free of nonadherent cells and the remaining bound radioactivity determined by scintillation counting. Average and s.e.m. of four replicates are shown for each condition. Cell adhesion was blocked by a combination of antibodies to integrin α6 (long arm activity) + α2 (short arm activity), or β1 alone.

C. Analysis of Complexes by Electron Microscopy

Metal shadow casting techniques, coupled with electron microscopy, have served a valuable role in the analysis of basement membrane components and their interactions. The basement membrane glycoproteins and proteoglycans are sufficiently large to be resolved by rotary shadowing, revealing the complexity of globular and rod domains. The most commonly used technique is that of glycerol rotary shadowing, in which the protein structure is protected with 50–70% glycerol in a volatile buffer (0.15 M acetic acid, ammonium acetate, or ammonium bicarbonate), during evacuation and drying, prior to application of a thin (~1 nm) coat of platinum/carbon, applied with sample rotation at an angle of 6–8°. We have carried out evacuation and shadowing in either a Balzers-300 or Balzers-BAF500K freeze-etch/shadowing unit, the latter providing somewhat better resolution (several nm) because of the higher vacuum and lower stage temperature afforded by cryo-pumping and cooling of the sample stage. Following backing with a

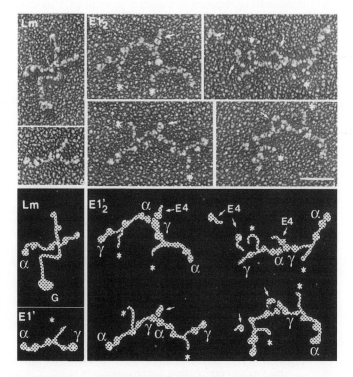

Fig. 10 Laminin fragment complexes. Gallery of electron micrographs of rotary shadowed Pt/C replicas of intact laminin, E1′, and E1′-E4 complexes that form when the two fragments were incubated in a mixture (4.4 μM each, TBS/Ca, 37°C, 30–60 min) and immediately prepared for metal replication. Panels show monomeric laminin (α1 short arm globules indicated with small arrows), E1′ monomer (below intact laminin), and four E1′-dimer/E4 complexes (arrows indicate position of E4, asterisks indicate location of remaining "stub" of β1 short arm in E1′). Interpretive drawings of the laminin, fragments, and complexes are shown at bottom. (Reproduced from Yurchenco and Cheng, 1993. *J. Biol. Chem.* **268,** 17286–17299, with permission of The American Society for Biochemistry and Molecular Biology.)

thicker (8 nm) carbon coat, the replica is lifted off onto the surface of water, picked up with a copper grid, and examined in an electron microscope at 80 kV with a 20–30 μm objective aperture (Yurchenco and O'Rear, 1994). This basic method has been used to visualize most purified ECM components, including laminins, type IV collagen, nidogen, perlecan, fibronectin, and other collagens, sometimes providing the first evidence of overall structure. It has also been employed to examine recombinant fragments and recombinant glycoproteins, providing information on folding at a relatively low, but nonetheless important, level of resolution. The method has also been valuable in providing information on binding interactions. Lower affinity interactions may be detected if full dissociation has not occurred following sample dilution and application to an adherent mica surface. Complexes consisting of laminin-1 fragments E1′ and E4, visualized by this method, helped the development of a three-interaction model of self-assembly (Fig. 10). Nidogen binding to the γ1-short arm of laminin also has been analyzed by this method; however, it was difficult to discriminate between nidogen binding to γ1 domain III (composed of a tandem array of EGF modules), domain IV (inner short-arm globule), or both. A complementing biochemical and recombinant analysis was used to definitively establish that the primary site resided within an EGF module of domain III (Mayer *et al.*, 1993; Poschl *et al.*, 1994, 1996).

The intact laminin and type IV collagen polymer structures have been visualized by a metal replication technique that depends upon exposure of the network (Fig. 11). Freeze-etching exposes the protein structure and high-angle shadowing improves the resolution. The laminin network replica from the basement membrane was nearly identical to that observed with a purified polymer formed *in vitro* (Yurchenco and Ruben, 1987). Similarly, the type IV collagen polymers detected in the EHS basement membrane (shown), detected in the placental basement membrane and formed *in vitro* were nearly identical.

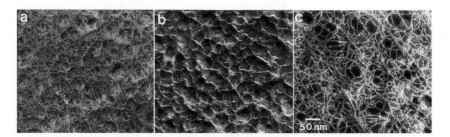

Fig. 11 Basement membrane networks. Extracellular matrix from the EHS basement membrane was vitreous-frozen in a propane-jet freezer, fractured, deep etched, and rotary replicated in a Balzers BAF500K unit with 1.2 nm of Pt/C at a 60° angle (a) untreated, (b) following removal of type IV collagen by incubation overnight with bacterial collagenase to reveal the laminin polymer, or (c) following removal of laminin and soluble components by extraction with 2 *M* guanidine-HCl in 50 m*M* Tris-HCl, 90 m*M* NaCl, pH 7.4, to expose the type IV collagen network. The electron micrographs are shown contrast-reversed. The architecture in (a) reveals the combined contributions of the laminin and type IV collagen polymers. The laminin network in (b) is similar to the network of a purified laminin polymer gel with a ∼35 nm strut dimension. It is quite different from the distinct type IV collagen network shown which shows lateral associations first detected in reconstituted collagen polymers (c). (Reproduced from **The Journal of Cell Biology**, 1992, vol. 117, pp. 1119–1133, by copyright permission of The Rockefeller University Press.)

These similarities of structure were considered evidence that the processes are fundamentally ones of self-assembly (Yurchenco *et al.,* 1992; Yurchenco and Ruben, 1987, 1988).

IV. Analysis of Cell–Surface Contributions to Basement Membrane Assembly

Cell adhesion to plastic coated with ECM ligands, often accompanied by cell spreading, has been used as a general approach to identify interacting receptors. The interaction, in which focal adhesions may form, is seen with the ligands fibronectin, vitronectin, and fibrinogen. The receptors implicated in these interactions are members of the integrin family. Most cells also adhere to members of the laminin family. In the case of laminin-1 coated surfaces, inhibition of $\beta 1$-integrin with specific antibodies frequently prevents adhesion for a variety of cells, implying that the primary anchor corresponds to one or more members of the $\beta 1$-integrin family. The specific $\beta 1$-associated α-chain subunits are $\alpha 6$ and $\alpha 7$. The $\alpha 6$ integrin interaction has been a defining characteristic of members of the laminin family, with the laminins tested all capable of interaction with this integrin subunit (Delwel *et al.,* 1994; Fujiwara *et al.,* 2001; Kikkawa *et al.,* 2000; Sorokin *et al.,* 1990). With the exception of $\alpha 1$-laminin, all laminins have also been found to interact with $\alpha 3 \beta 1$, a more promiscuous integrin that can interact with non-basement membrane matrix as well. For laminin-5 this activity, like that of the $\alpha 6$ integrin, was found to reside within LG-3, using deletions of the LG modules (Hirosaki *et al.,* 2000). Integrins $\alpha 1 \beta 1$ and $\alpha 2 \beta 1$ can ligate both laminins and collagens (Colognato *et al.,* 1997; Colognato-Pyke *et al.,* 1995; Eble *et al.,* 1993, 1996). The $\beta 4$ integrin subunit pairs only with $\alpha 6$, interacting specifically with laminin matrices, most importantly with laminin-5. The $\alpha 6 \beta 4$ integrin, with its exceptionally long $\beta 4$ cytoplasmic tail, is a major component of hemidesmosomes (reviewed in Jones *et al.,* 1998).

Integrins are not the only receptors that interact with basement membranes. However, non-integrin interactions may not be detected in standard cell adhesion and spreading assays. When laminins and other basement membrane components are added to the media of adherent cells, different types of interactions may be observed. Schwann cell adhesion can be completely inhibited by anti-$\beta 1$ antibody under standard adhesion conditions. In contrast, the same antibody only partially blocks laminin adhesion to the cell surface.

A. Receptor–Cytoskeletal Responses to Basement Membrane Components

If laminin-1 is added to the conditioned medium of a CaCo2 epithelial layer cultured on a collagen-I gel with embedded fibroblasts, the laminin will accumulate on the basal side of the epithelial cells, only weakly accumulating on the fibroblast surface (Q. Zhang and P. D. Yurchenco, unpublished observations). This will occur even below its critical concentration. Similarly, exogenous laminin-1 or laminin-2, added to the medium of C2C12 myotubes or to Schwann cells (≤ 10 nM) accumulates on the cell surface and assembles respectively into honeycomb-like or reticular patterns (Colognato *et al.,* 1999; M. Tsiper and P. D. Yurchenco, submitted). The patterns and levels of binding, for example in the case of myotubes (Fig. 12), were found to be dependent on laminin

polymerizing laminin

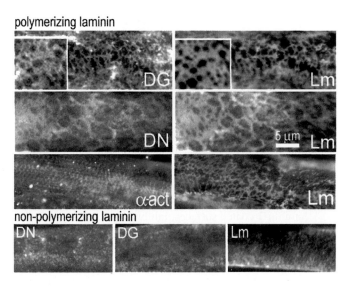

non-polymerizing laminin

Fig. 12 Laminin sarcolemmal membrane-associated complexes. When laminin-1 (or laminin-2) was incubated with C2C12 cultured myotubes, the laminin initially accumulated on the sarcolemmal surface in a diffuse pattern. By several hours the laminin reorganized into a honeycomb-like structure. Polymerization was required as a trigger and reorganization was found to be dependent upon actin-polymerization and tyrosine phosphorylation. The structure that forms was accompanied by a mirrored reorganization of $\alpha7\beta1$ integrin, dystroglycan, and cytoskeletal vinculin and dystrophin. The polygonal structure is much larger than the laminin polymer network as seen by electron microscopy and represents a higher order of organization. Paired immunofluorescence images are shown at 4 h. (Reproduced from *The Journal of Cell Biology,* 1999, vol. 145, pp. 619–631, by copyright permission of The Rockefeller University Press.)

polymerization, with cell anchorage mediated by the LG-modules, and with no detectable binding of short arm to the cell surface. A related observation was that the structural reorganization of laminin was accompanied by a corresponding reorganization of integrin and dystroglycan receptors and of cytoskeletal vinculin and dystrophin. This organization, although triggered by polymerization, required intracellular phosphorylation and actin polymerization. The data support the hypothesis that basement membrane laminin preferentially accumulates on cell surfaces by binding to cognate receptors and can trigger intracellular events through self-assembly.

V. Approaches to Evaluating ECM Molecules for Assembly Interactions

How might one go about determining if and how a new glycoprotein or collagen participates in the structure of an ECM? More likely than not, this protein would have been discovered from its DNA sequence, and not by tissue extraction and protein sequencing. An initial approach could be to express part or all of the glycoprotein/collagen as a recombinant protein followed by the generation of specific antibody. The antibody would be used to determine the tissue location and distribution, as well as to help characterize molecular partners in extracts. Multisubunit glycoproteins such as laminins and collagens present more of a challenge to express in their intact assembled state compared to single polypeptide chains, but nonetheless the expression can be done. At this point there are two obvious experimental routes that can be pursued. One route is the analysis of the component in tissue. Determining whether an ECM component is extractable with acetic acid (collagens), neutral isotonic buffer, or chaotropic agents such as urea, and whether reduction and/or detergent are required, provides information of the nature of the bonds that likely retain the component within the ECM. Further analysis (e.g., rotary shadow EM, gel filtration, sedimentation analysis, light scattering, SDS–PAGE) can then be used to determine whether the component extracts as a monomer, dimer, or oligomer. If the component is covalently bound within the ECM and proves nonextractable with these agents, proteolytic treatment may be used to liberate fragment complexes that can then be analyzed by the foregoing methods.

The other approach is to determine whether the recombinant molecule self-assembles into a supramolecular complex under physiologically relevant buffer conditions and to determine whether the molecule can bind to a repertoire of candidate ECM molecules. As previously discussed, evaluating high-affinity interactions is simpler than evaluating low-affinity interactions because of the greater dissociation that occurs with the latter. Furthermore, whatever screening method of analysis is used, it is important to keep in mind that the folded state of the protein may be critical for binding. With these caveats, solid-phase binding assays and affinity chromatography probably provide the most efficient ways to perform an initial screen.

VI. Integration of Genetic and Molecular Approaches

A decade ago it was found that sequence-altering mutations within the open reading frame of the type IV collagen gene are lethal in the nematode *Caenorhabditis elegans* (Guo *et al.,* 1991). This provided early evidence for an essential biological role for basement membrane and the discovery pointed the way to the application of classical genetics to analyze basement membrane function. For example, it was discovered in the nematode, and confirmed in mice, that nidogen is not essential for basement membrane formation (Kang and Kramer, 2000; Murshed *et al.,* 2000). The glycoprotein was, however, observed to play a role in axonal pathfinding (Kim and Wadsworth, 2000). In the nematode, type IV collagen was found to be assembled on tissues that do not express it, revealing an asymmetrical relationship between synthetic cell and target cell (Graham *et al.,* 1997). The characterization of laminin, nidogen type IV collagen, and perlecan gene expression in the nematode is creating a "roadmap" of basement membrane components, of their expression patterns, and of their roles in developmental steps, all important for the interpretation of genetic screens to detect mutations in basement membrane components, receptors, and other interacting molecules that are involved in matrix assembly. The nematode expresses two laminins that differ only in their α-subunit, i.e., $\alpha_A \beta \gamma$ and $\alpha_B \beta \gamma$, simplifying the analysis compared to mammals. The former subunit is most similar to $\alpha 1/\alpha 2$-laminins while the latter is most similar to $\alpha 5$-laminins. Recently, it was found that a mutation null for α_A results in a pharyngeal defect that can be detected by light microscopy (Huang, 2001).

Mouse genetics have also made significant inroads into the analysis of basement membrane functions. The laminin-$\gamma 1$ knockout revealed that $\gamma 1$-laminin (in particular laminin-1) is essential for early basement membrane formation and gastrulation, while the perlecan knockout revealed that this proteoglycan plays an important role in cartilage development. Lethality first occurred at E10 with a defect developing in a cardiac basement membrane (the brain was also affected). Other basement membranes appeared to be normal, demonstrating that perlecan is not invariably essential for basement membrane assembly (Arikawa-Hirasawa *et al.,* 1999; Costell *et al.,* 1999; Smyth *et al.,* 1999). Collectively, these approaches have revealed important biological roles of basement membrane components, setting limits on, or even eliminating, mechanistic models.

Embryoid bodies can be derived from embryonic stem cells. Null states are created by inactivating both alleles of a gene, providing models for the role of relevant genes in early gastrulation. Laminin is synthesized primarily by the endodermal layer, accumulating into a basement membrane that forms between endoderm and inner cell mass (ICM), and is important for conversion of ICM to epiblast (Murray and Edgar, 2001). The endoderm is not required for basement membrane assembly per se, since embryoid bodies that no longer possess this layer, as seen in a fibroblast growth factor dominant-negative mutant, still form a nascent (lacking type IV collagen), and possibly a complete, basement membrane if exogenous laminin is provided (Li *et al.,* 2001). This separation of cell of origin and site of assembly, seen in other circumstances and with other components, may be a general principle and suggests that basement membrane components behave in a manner reminiscent of developmental induction factors, in which one cell secretes a molecule to instruct other cells. The difference from other induction factors is, of course, that a persisting ECM architecture is created on the cell surface. The embryoid body differentiation system provides several "readouts." One is structurally relevant, i.e., formation of a basement membrane between endoderm and inner cell mass. Another is that of tissue differentiation, i.e., epiblast differentiation and cavitation, which requires formation of basement membrane. The analysis of the effects of recombinant laminins bearing engineered modifications on genetically modified embryonic stem cells has interesting potential for the elucidation of mechanism.

Acknowledgments

Supported by NIH grants R01-DK36425 and R01-NS38469. We thank Dr. Qihang Zhang and Yi-Shan Cheng for their technical assistance.

References

Amenta, P. S., Clark, C. C., and Martinez-Hernandez, A. (1983). Deposition of fibronectin and laminin in the basement membrane of the rat parietal yolk sac: immunohistochemical and biosynthetic studies. *J. Cell Biol.* **96,** 104–111.

Arikawa-Hirasawa, E., Watanabe, H., Takami, H., Hassell, J. R., and Yamada, Y. (1999). Perlecan is essential for cartilage and cephalic development. *Nat. Genet.* **23,** 354–358.

Bachinger, H. P., Fessler, L. I., and Fessler, J. H. (1982). Mouse procollagen IV. Characterization and supramolecular association. *J. Biol. Chem.* **257,** 9796–9803.

Battaglia, C., Mayer, U., Aumailley, M., and Timpl, R. (1992). Basement-membrane heparan sulfate proteoglycan binds to laminin by its heparan sulfate chains and to nidogen by sites in the protein core. *Eur. J. Biochem.* **208,** 359–366.

Beck, K., Hunter, I., and Engel, J. (1990). Structure and function of laminin: anatomy of a multidomain glycoprotein. *FASEB. J.* **4,** 148–160.

Birk, D. E., and Linsenmayer, T. F. (1994). Collagen fibril assembly, deposition, and organization into tissue-specific matrices. *In* "Extracellular Matrix Assembly and Structure," Biology of Extracellular Matrix (P. D. Yurchenco, D. E. Birk, and R. P. Mecham, eds.), pp. 91–128. Academic Press, San Diego.

Brazel, D., Pollner, R., Oberbaumer, I., and Kuhn, K. (1988). Human basement membrane collagen (type IV). The amino acid sequence of the $\alpha2$(IV) chain and its comparison with the $\alpha1$(IV) chain reveals deletions in the $\alpha1$(IV) chain. *Eur. J. Biochem.* **172,** 35–42.

Brown, J. C., Sasaki, T., Gohring, W., Yamada, Y., and Timpl, R. (1997). The C-terminal domain V of perlecan promotes $\beta1$ integrin-mediated cell adhesion, binds heparin, nidogen and fibulin-2 and can be modified by glycosaminoglycans. *Eur. J. Biochem.* **250,** 39–46.

Bruch, M., Landwehr, R., and Engel, J. (1989). Dissection of laminin by cathepsin G into its long-arm and short-arm structures and localization of regions involved in calcium dependent stabilization and self-association. *Eur. J. Biochem.* **185,** 271–279.

Champliaud, M. F., Lunstrum, G. P., Rousselle, P., Nishiyama, T., Keene, D. R., and Burgeson, R. E. (1996). Human amnion contains a novel laminin variant, laminin 7, which like laminin 6, covalently associates with laminin 5 to promote stable epithelial–stromal attachment. *J. Cell Biol.* **132,** 1189–1198.

Champliaud, M. F., Virtanen, I., Tiger, C. F., Korhonen, M., Burgeson, R., and Gullberg, D. (2000). Posttranslational modifications and β/γ chain associations of human laminin $\alpha1$ and laminin $\alpha5$ chains: Purification of laminin-3 from placenta. *Exp. Cell Res.* **259,** 326–335.

Cheng, Y. S., Champliaud, M. F., Burgeson, R. E., Marinkovich, M. P., and Yurchenco, P. D. (1997). Self-assembly of laminin isoforms. *J. Biol. Chem.* **272,** 31525–31532.

Chung, A. E., Jaffe, R., Freeman, I. L., Vergnes, J. P., Braginski, J. E., and Carlin, B. (1979). Properties of a basement membrane-related glycoprotein synthesized in culture by a mouse embryonal carcinoma-derived cell line. *Cell* **16,** 277–287.

Colognato, H., and Yurchenco, P. D. (1999). The laminin $\alpha2$ expressed by dystrophic dy(2J) mice is defective in its ability to form polymers. *Curr. Biol.* **9,** 1327–1330.

Colognato, H., and Yurchenco, P. D. (2000). Form and function: the laminin family of heterotrimers. *Dev. Dyn.* **218,** 213–234.

Colognato, H., MacCarrick, M., O'Rear, J. J., and Yurchenco, P. D. (1997). The laminin $\alpha2$-chain short arm mediates cell adhesion through both the $\alpha1\beta1$ and $\alpha2\beta1$ integrins. *J. Biol. Chem.* **272,** 29330–29336.

Colognato, H., Winkelmann, D. A., and Yurchenco, P. D. (1999). Laminin polymerization induces a receptor–cytoskeleton network. *J. Cell Biol.* **145,** 619–631.

Colognato-Pyke, H., O'Rear, J. J., Yamada, Y., Carbonetto, S., Cheng, Y. S., and Yurchenco, P. D. (1995). Mapping of network-forming, heparin-binding, and $\alpha1\beta1$ integrin-recognition sites within the α-chain short arm of laminin-1. *J. Biol. Chem.* **270,** 9398–9406.

Costell, M., Gustafsson, E., Aszodi, A., Morgelin, M., Bloch, W., Hunziker, E., Addicks, K., Timpl, R., and Fassler, R. (1999). Perlecan maintains the integrity of cartilage and some basement membranes. *J. Cell Biol.* **147,** 1109–1122.

De Arcangelis, A., Neuville, P., Boukamel, R., Lefebvre, O., Kedinger, M., and Simon-Assmann, P. (1996). Inhibition of laminin $\alpha1$-chain expression leads to alteration of basement membrane assembly and cell differentiation. *J. Cell. Biol.* **133,** 417–430.

Dean, J. W., Chandrasekaran, S., and Tanzer, M. L. (1990). A biological role of the carbohydrate moieties of laminin. *J. Biol. Chem.* **265,** 12553–12562.

Delwel, G. O., de Melker, A. A., Hogervorst, F., Jaspars, L. H., Fles, D. L., Kuikman, I., Lindblom, A., Paulsson, M., Timpl, R., and Sonnenberg, A. (1994). Distinct and overlapping ligand specificities of the $\alpha 3A\beta 1$ and $\alpha 6A\beta 1$ integrins: recognition of laminin isoforms. *Mol. Biol. Cell.* **5,** 203–215.

Deutzmann, R., Aumailley, M., Wiedemann, H., Pysny, W., Timpl, R., and Edgar, D. (1990). Cell adhesion, spreading and neurite stimulation by laminin fragment E8 depends on maintenance of secondary and tertiary structure in its rod and globular domain. *Eur. J. Biochem.* **191,** 513–522.

Eble, J. A., Golbik, R., Mann, K., and Kuhn, K. (1993). The alpha 1 beta 1 integrin recognition site of the basement membrane collagen molecule [$\alpha 1(IV)$]2 $\alpha 2(IV)$. *Embo J.* **12,** 4795–4802.

Eble, J. A., Ries, A., Lichy, A., Mann, K., Stanton, H., Gavrilovic, J., Murphy, G., and Kuhn, K. (1996). The recognition sites of the integrins $\alpha 1\beta 1$ and $\alpha 2\beta 1$ within collagen IV are protected against gelatinase A attack in the native protein. *J. Biol. Chem.* **271,** 30964–30970.

Ehrig, K., Leivo, I., Argraves, W. S., Ruoslahti, E., and Engvall, E. (1990). Merosin, a tissue-specific basement membrane protein, is a laminin-like protein. *Proc. Natl. Acad. Sci. USA* **87,** 3264–3268.

Engvall, E., Davis, G. E., Dickerson, K., Ruoslahti, E., Varon, S., and Manthorpe, M. (1986). Mapping of domains in human laminin using monoclonal antibodies: localization of the neurite-promoting site. *J. Cell. Biol.* **103,** 2457–2465.

Fox, J. W., Mayer, U., Nischt, R., Aumailley, M., Reinhardt, D., Wiedemann, H., Mann, K., Timpl, R., Krieg, T., Engel, J., and Chu, M. L. (1991). Recombinant nidogen consists of three globular domains and mediates binding of laminin to collagen type IV. *EMBO J.* **10,** 3137–3146.

Friedrich, M. V., Gohring, W., Morgelin, M., Brancaccio, A., David, G., and Timpl, R. (1999). Structural basis of glycosaminoglycan modification and of heterotypic interactions of perlecan domain V. *J. Mol. Biol.* **294,** 259–270.

Fujiwara, H., Kikkawa, Y., Sanzen, N., and Sekiguchi, K. (2001). Purification and characterization of human laminin-8:laminin-8 stimulates cell adhesion and migration through $\alpha 3\beta 1$ and $\alpha 6\beta 1$ integrins. *J. Biol. Chem.* **276,** 17550–17558.

Gagnoux-Palacios, L., Allegra, M., Spirito, F., Pommeret, O., Romero, C., Ortonne, J. P., and Meneguzzi, G. (2001). The short arm of the laminin $\gamma 2$ chain plays a pivotal role in the incorporation of laminin 5 into the extracellular matrix and in cell adhesion. *J. Cell Biol.* **153,** 835–850.

Graham, P. L., Johnson, J. J., Wang, S., Sibley, M. H., Gupta, M. C., and Kramer, J. M. (1997). Type IV collagen is detectable in most, but not all, basement membranes of *Caenorhabditis elegans* and assembles on tissues that do not express it. *J. Cell Biol.* **137,** 1171–1183.

Guo, X. D., Johnson, J. J., and Kramer, J. M. (1991). Embryonic lethality caused by mutations in basement membrane collagen of *C. elegans. Nature* **349,** 707–709.

Hirosaki, T., Mizushima, H., Tsubota, Y., Moriyama, K., and Miyazaki, K. (2000). Structural requirement of carboxyl-terminal globular domains of laminin $\alpha 3$ chain for promotion of rapid cell adhesion and migration by laminin-5. *J. Biol. Chem.* **275,** 22495–22502.

Hohenester, E., Tisi, D., Talts, J. F., and Timpl, R. (1999). The crystal structure of a laminin G-like module reveals the molecular basis of α-dystroglycan binding to laminins, perlecan, and agrin. *Mol. Cell* **4,** 783–792.

Hopf, M., Gohring, W., Kohfeldt, E., Yamada, Y., and Timpl, R. (1999). Recombinant domain IV of perlecan binds to nidogens, laminin-nidogen complex, fibronectin, fibulin-2 and heparin. *Eur. J. Biochem.* **259,** 917–925.

Howe, C. C. (1984). Functional role of laminin carbohydrate. *Mol. Cell. Biol.* **4,** 1–7.

Huang, C.-C. (2001). *Caenorhabditis elegans* laminin alpha subunits and their distinct roles in organogenesis. Doctoral (Ph.D.) thesis, Robert Wood Johnson Medical School.

Hunter, D. D., Porter, B. E., Bulock, J. W., Adams, S. P., Merlie, J. P., and Sanes, J. R. (1989). Primary sequence of a motor neuron-selective adhesive site in the synaptic basal lamina protein S-laminin. *Cell* **59,** 905–913.

Iivanainen, A., Kortesmaa, J., Sahlberg, C., Morita, T., Bergmann, U., Thesleff, I., and Tryggvason, K. (1997). Primary structure, developmental expression, and immunolocalization of the murine laminin $\alpha 4$ chain. *J. Biol. Chem.* **272,** 27862–27868.

Iivanainen, A., Sainio, K., Sariola, H., and Tryggvason, K. (1995). Primary structure and expression of a novel human laminin $\alpha 4$ chain. *FEBS Lett.* **365,** 183–188.

Jones, J. C., Hopkinson, S. B., and Goldfinger, L. E. (1998). Structure and assembly of hemidesmosomes. *Bioessays* **20,** 488–494.

Kang, S. H., and Kramer, J. M. (2000). Nidogen is nonessential and not required for normal type IV collagen localization in *Caenorhabditis elegans*. *Mol. Biol. Cell* **11,** 3911–3923.

Kikkawa, Y., Sanzen, N., Fujiwara, H., Sonnenberg, A., and Sekiguchi, K. (2000). Integrin binding specificity of laminin-10/11: laminin-10/11 are recognized by $\alpha3\beta1$, $\alpha6\beta1$ and $\alpha6\beta4$ integrins. *J. Cell Sci.* **113,** 869–876.

Kikkawa, Y., Sanzen, N., and Sekiguchi, K. (1998). Isolation and characterization of laminin-10/11 secreted by human lung carcinoma cells. Laminin-10/11 mediates cell adhesion through integrin $\alpha3\beta1$. *J. Biol. Chem.* **273,** 15854–15859.

Kim, S., and Wadsworth, W. G. (2000). Positioning of longitudinal nerves in *C. elegans* by nidogen. *Science* **288,** 150–154.

Kleinman, H. K., McGarvey, M. L., Liotta, L. A., Robey, P. G., Tryggvason, K., and Martin, G. R. (1982). Isolation and characterization of type IV procollagen, laminin, and heparan sulfate proteoglycan from the EHS sarcoma. *Biochemistry* **21,** 6188–6193.

Koch, M., Olson, P. F., Albus, A., Jin, W., Hunter, D. D., Brunken, W. J., Burgeson, R. E., and Champliaud, M. F. (1999). Characterization and expression of the laminin $\gamma3$ chain: A novel, non-basement membrane-associated, laminin chain. *J. Cell Biol.* **145,** 605–618.

Kohfeldt, E., Sasaki, T., Göhring, W., and Timpl, R. (1998). Nidogen-2: A new basement membrane protein with diverse binding properties. *J. Mol. Biol.* **282,** 99–109.

Kortesmaa, J., Yurchenco, P., and Tryggvason, K. (2000). Recombinant laminin-8 ($\alpha4\beta1\gamma1$). Production, purification, and interactions with integrins. *J. Biol. Chem.* **275,** 14853–14859.

Langeveld, J. P., Wieslander, J., Timoneda, J., McKinney, P., Butkowski, R. J., Wisdom, B. J., Jr., and Hudson, B. G. (1988). Structural heterogeneity of the noncollagenous domain of basement membrane collagen. *J. Biol. Chem.* **263,** 10481–10488.

Lee, M. K., and Lander, A. D. (1991). Analysis of affinity and structural selectivity in the binding of proteins to glycosaminoglycans: development of a sensitive electrophoretic approach. *Proc. Natl. Acad. Sci. USA* **88,** 2768–2772.

Li, X., Chen, Y., Scheele, S., Arman, E., Haffner-Krausz, R., Ekblom, P., and Lonai, P. (2001). Fibroblast growth factor signaling and basement membrane assembly are connected during epithelial morphogenesis of the embryoid body. *J. Cell Biol.* **153,** 811–822.

Libby, R. T., Champliaud, M. F., Claudepierre, T., Xu, Y., Gibbons, E. P., Koch, M., Burgeson, R. E., Hunter, D. D., and Brunken, W. J. (2000). Laminin expression in adult and developing retinae: Evidence of two novel CNS laminins. *J. Neurosci.* **20,** 6517–6528.

Mann, K., Deutzmann, R., and Timpl, R. (1988). Characterization of proteolytic fragments of the laminin–nidogen complex and their activity in ligand-binding assays. *Eur. J. Biochem.* **178,** 71–80.

Marinkovich, M. P., Lunstrum, G. P., Keene, D. R., and Burgeson, R. E. (1992). The dermal–epidermal junction of human skin contains a novel laminin variant. *J. Cell. Biol.* **119,** 695–703.

Mayer, U., Nischt, R., Poschl, E., Mann, K., Fukuda, K., Gerl, M., Yamada, Y., and Timpl, R. (1993). A single EGF-like motif of laminin is responsible for high affinity nidogen binding. *EMBO J.* **12,** 1879–1885.

McGrath, J. A., Kivirikko, S., Ciatti, S., Moss, C., Dunnill, G. S., Eady, R. A., Rodeck, C. H., Christiano, A. M., and Vitto, J. (1995). A homozygous nonsense mutation in the $\alpha3$ chain gene of laminin 5 (LAMA3) in Herlitz junctional epidermolysis bullosa: Prenatal exclusion in a fetus at risk. *Genomics* **29,** 282–284.

Miner, J. H., Cunningham, J., and Sanes, J. R. (1998). Roles for laminin in embryogenesis: Exencephaly, syndactyly, and placentopathy in mice lacking the laminin $\alpha5$ chain. *J. Cell Biol.* **143,** 1713–1723.

Miner, J. H., Lewis, R. M., and Sanes, J. R. (1995). Molecular cloning of a novel laminin chain, $\alpha5$, and widespread expression in adult mouse tissues. *J. Biol. Chem.* **270,** 28523–28526.

Miner, J. H., Patton, B. L., Lentz, S. I., Gilbert, D. J., Snider, W. D., Jenkins, N. A., Copeland, N. G., and Sanes, J. R. (1997). The laminin alpha chains: expression, developmental transitions, and chromosomal locations of $\alpha1$–5, identification of heterotrimeric laminins 8–11, and cloning of a novel $\alpha3$ isoform. *J. Cell Biol.* **137,** 685–701.

Miyamoto, S., Akiyama, S. K., and Yamada, K. M. (1995). Synergistic roles for receptor occupancy and aggregation in integrin transmembrane function. *Science* **267,** 883–885.

Murray, P., and Edgar, D. (2001). Regulation of laminin and COUP-TF expression in extraembryonic endodermal cells. *Mech. Dev.* **101,** 213–215.

Murshed, M., Smyth, N., Miosge, N., Karolat, J., Krieg, T., Paulsson, M., and Nischt, R. (2000). The absence of nidogen 1 does not affect murine basement membrane formation. *Mol. Cell. Biol.* **20,** 7007–7012.

Noakes, P. G., Gautam, M., Mudd, J., Sanes, J. R., and Merlie, J. P. (1995a). Aberrant differentiation of neuromuscular junctions in mice lacking s-laminin/laminin β2. *Nature* **374,** 258–262.

Noakes, P. G., Miner, J. H., Gautam, M., Cunningham, J. M., Sanes, J. R., and Merlie, J. P. (1995b). The renal glomerulus of mice lacking s-laminin/laminin β2: nephrosis despite molecular compensation by laminin β1. *Nat. Genet.* **10,** 400–6.

Ott, U., Odermatt, E., Engel, J., Furthmayr, H., and Timpl, R. (1982). Protease resistance and conformation of laminin. *Eur. J. Biochem.* **123,** 63–72.

Patton, B. L., Miner, J. H., Chiu, A. Y., and Sanes, J. R. (1997). Distribution and function of laminins in the neuromuscular system of developing, adult, and mutant mice. *J. Cell Biol.* **139,** 1507–1521.

Paulsson, M. (1988). The role of Ca^{2+} binding in the self-aggregation of laminin–nidogen complexes. *J. Biol. Chem.* **263,** 5425–5430.

Paulsson, M., Yurchenco, P. D., Ruben, G. C., Engel, J., and Timpl, R. (1987). Structure of low density heparan sulfate proteoglycan isolated from a mouse tumor basement membrane. *J. Mol. Biol.* **197,** 297–313.

Pesciotta Peters, D. M., and Mosher, D. F. (1994). Formation of fibronectin extracellular matrix. *In* "Extracellular Matrix Assembly and Structure," Biology of Extracellular Matrix (P. D. Yurchenco, D. E. Birk, and R. P. Mecham, eds.), pp. 315–350. Academic Press, San Diego.

Poschl, E., Fox, J. W., Block, D., Mayer, U., and Timpl, R. (1994). Two non-contiguous regions contribute to nidogen binding to a single EGF-like motif of the laminin γ1 chain. *EMBO J.* **13,** 3741–3747.

Poschl, E., Mayer, U., Stetefeld, J., Baumgartner, R., Holak, T. A., Huber, R., and Timpl, R. (1996). Site-directed mutagenesis and structural interpretation of the nidogen binding site of the laminin γ1 chain. *EMBO J* **15,** 5154–5159.

Prockop, D. J., and Hulmes, J. S. (1994). Assembly of collagen fibrils de novo from soluble precursors: polymerization and copolymerization of procollagen, pN-collagen, and mutated collagens. *In* "Extracellular Matrix Assembly and Structure," Biology of Extracellular Matrix (P. D. Yurchenco, D. E. Birk, and R. P. Mecham, eds.), pp. 47–90. Academic Press, San Diego.

Pulkkinen, L., Christiano, A. M., Airenne, T., Haakana, H., Tryggvason, K., and Uitto, J. (1994). Mutations in the γ2 chain gene (LAMC2) of kalinin/laminin 5 in the junctional forms of epidermolysis bullosa. *Nat. Genet.* **6,** 293–297.

Pulkkinen, L., Gerecke, D. R., Christiano, A. M., Wagman, D. W., Burgeson, R. E., and Uitto, J. (1995). Cloning of the β3 chain gene (LAMB3) of human laminin 5, a candidate gene in junctional epidermolysis bullosa. *Genomics* **25,** 192–198.

Rambukkana, A., Salzer, J. L., Yurchenco, P. D., and Tuomanen, E. I. (1997). Neural targeting of *Mycobacterium leprae* mediated by the G domain of the laminin-α2 chain. *Cell* **88,** 811–821.

Richards, A., Al-Imara, L., and Pope, F. M. (1996). The complete cDNA sequence of laminin α4 and its relationship to the other human laminin α chains. *Eur. J. Biochem.* **238,** 813–821.

Richards, A. J., al Imara, L., Carter, N. P., Lloyd, J. C., Leversha, M. A., and Pope, F. M. (1994). Localization of the gene (LAMA4) to chromosome 6q21 and isolation of a partial cDNA encoding a variant laminin A chain. *Genomics* **22,** 237–239.

Risteli, J., Bachinger, H. P., Engel, J., Furthmayr, H., and Timpl, R. (1980). 7-S collagen: characterization of an unusual basement membrane structure. *Eur. J. Biochem.* **108,** 239–250.

Rohde, H., Bachinger, H. P., and Timpl, R. (1980). Characterization of pepsin fragments of laminin in a tumor basement membrane. Evidence for the existence of related proteins. *Hoppe Seylers. Z. Physiol. Chem.* **361,** 1651–1660.

Rousselle, P., and Aumailley, M. (1994). Kalinin is more efficient than laminin in promoting adhesion of primary keratinocytes and some other epithelial cells and has a different requirement for integrin receptors. *J. Cell. Biol.* **125,** 205–214.

Ryan, M. C., Lee, K., Miyashita, Y., and Carter, W. G. (1999). Targeted disruption of the LAMA3 gene in mice reveals abnormalities in survival and late stage differentiation of epithelial cells. *J. Cell Biol.* **145,** 1309–1324.

Ryan, M. C., Tizard, R., VanDevanter, D. R., and Carter, W. G. (1994). Cloning of the LamA3 gene encoding the alpha 3 chain of the adhesive ligand epiligrin. Expression in wound repair. *J. Biol. Chem.* **269,** 22779–22787.

Sakashita, S., Engvall, E., and Ruoslahti, E. (1980). Basement membrane glycoprotein laminin binds to heparin. *FEBS Lett.* **116,** 243–246.

San Antonio, J. D., Karnovsky, M. J., Gay, S., Sanderson, R. D., and Lander, A. D. (1994). Interactions of syndecan-1 and heparin with human collagens. *Glycobiology* **4,** 327–332.

Sasaki, M., and Yamada, Y. (1987). The laminin B2 chain has a multidomain structure homologous to the B1 chain. *J. Biol. Chem.* **262,** 17111–17117.

Sasaki, M., Kato, S., Kohno, K., Martin, G. R., and Yamada, Y. (1987). Sequence of the cDNA encoding the laminin B1 chain reveals a multidomain protein containing cysteine-rich repeats. *Proc. Natl. Acad. Sci. USA* **84,** 935–939.

Sasaki, M., Kleinman, H. K., Huber, H., Deutzmann, R., and Yamada, Y. (1988). Laminin, a multidomain protein. The A chain has a unique globular domain and homology with the basement membrane proteoglycan and the laminin B chains. *J. Biol. Chem.* **263,** 16536–16544.

Schittny, J. C. (1994). Affinity retardation chromatography: characterization of the method and its application. The description of low affinity laminin self-interactions. *Anal. Biochem.* **222,** 140–148.

Schittny, J. C., and Schittny, C. M. (1993). Role of the B1 short arm in laminin self-assembly. *Eur. J. Biochem.* **216,** 437–441.

Schittny, J. C., and Yurchenco, P. D. (1990). Terminal short arm domains of basement membrane laminin are critical for its self-assembly. *J. Cell. Biol.* **110,** 825–832.

Schwarzbauer, J. E., and Sechler, J. L. (1999). Fibronectin fibrillogenesis: a paradigm for extracellular matrix assembly. *Curr. Opin. Cell Biol.* **11,** 622–627.

Siebold, B., Qian, R. A., Glanville, R. W., Hofmann, H., Deutzmann, R., and Kuhn, K. (1987). Construction of a model for the aggregation and cross-linking region (7S domain) of type IV collagen based upon an evaluation of the primary structure of the α1 and α2 chains in this region. *Eur. J. Biochem.* **168,** 569–575.

Smyth, N., Vatansever, H. S., Murray, P., Meyer, M., Frie, C., Paulsson, M., and Edgar, D. (1999). Absence of basement membranes after targeting the LAMC1 gene results in embryonic lethality due to failure of endoderm differentiation. *J. Cell Biol.* **144,** 151–160.

Sorokin, L., Sonnenberg, A., Aumailley, M., Timpl, R., and Ekblom, P. (1990). Recognition of the laminin E8 cell-binding site by an integrin possessing the α6 subunit is essential for epithelial polarization in developing kidney tubules. *J. Cell. Biol.* **111,** 1265–1273.

Stetefeld, J., Mayer, U., Timpl, R., and Huber, R. (1996). Crystal structure of three consecutive laminin-type epidermal growth factor-like (LE) modules of laminin gamma1 chain harboring the nidogen binding site. *J. Mol. Biol.* **257,** 644–657.

Sunada, Y., Bernier, S. M., Utani, A., Yamada, Y., and Campbell, K. P. (1995a). Identification of a novel mutant transcript of laminin α2 chain gene responsible for muscular dystrophy and dysmyelination in dy2J mice. *Hum. Mol. Genet.* **4,** 1055–1061.

Sunada, Y., Edgar, T. S., Lotz, B. P., Rust, R. S., and Campbell, K. P. (1995b). Merosin-negative congenital muscular dystrophy associated with extensive brain abnormalities. *Neurology* **45,** 2084–2089.

Sung, U., O'Rear, J. J., and Yurchenco, P. D. (1993). Cell and heparin binding in the distal long arm of laminin: identification of active and cryptic sites with recombinant and hybrid glycoprotein. *J. Cell. Biol.* **123,** 1255–1268.

Sung, U., O'Rear, J. J., and Yurchenco, P. D. (1997). Localization of heparin binding activity in recombinant laminin G domain. *Eur. J. Biochem.* **250,** 138–143.

Talts, J. F., and Timpl, R. (1999). Mutation of a basic sequence in the laminin α2LG3 module leads to a lack of proteolytic processing and has different effects on β1 integrin-mediated cell adhesion and α-dystroglycan binding. *FEBS Lett.* **458,** 319–323.

Talts, J. F., Andac, Z., Göhring, W., Brancaccio, A., and Timpl, R. (1999). Binding of the G domains of laminin α1 and α2 chains and perlecan to heparin, sulfatides, α-dystroglycan and several extracellular matrix proteins. *EMBO J.* **18,** 863–870.

Timpl, R., Johansson, S., van Delden, V., Oberbaumer, I., and Hook, M. (1983). Characterization of protease-resistant fragments of laminin mediating attachment and spreading of rat hepatocytes. *J. Biol. Chem.* **258,** 8922–8927.

Timpl, R., Martin, G. R., Bruckner, P., Wick, G., and Wiedemann, H. (1978). Nature of the collagenous protein in a tumor basement membrane. *Eur. J. Biochem.* **84,** 43–52.

Timpl, R., Rohde, H., Robey, P. G., Rennard, S. I., Foidart, J. M., and Martin, G. R. (1979). Laminin—a glycoprotein from basement membranes. *J. Biol. Chem.* **254,** 9933–9937.

Timpl, R., Wiedemann, H., van Delden, V., Furthmayr, H., and Kuhn, K. (1981). A network model for the organization of type IV collagen molecules in basement membranes. *Eur. J. Biochem.* **120,** 203–211.

Tisi, D., Talts, J. F., Timpl, R., and Hohenester, E. (2000). Structure of the C-terminal laminin G-like domain pair of the laminin $\alpha 2$ chain harbouring binding sites for α-dystroglycan and heparin. *EMBO J.* **19,** 1432–1440.

Veis, A., and George, A. (1994). Fundamentals of interstitial collagen self-assembly. *In* "Extracellular Matrix Assembly and Structure," Biology of Extracellular Matrix (P. D. Yurchenco, D. E. Birk, and R. P. Mecham, eds.), pp. 14–45. Academic Press, San Diego.

Vuolteenaho, R., Nissinen, M., Sainio, K., Byers, M., Eddy, R., Hirvonen, H., Shows, T. B., Sariola, H., Engvall, E., and Tryggvason, K. (1994). Human laminin M chain (merosin): complete primary structure, chromosomal assignment, and expression of the M and A chain in human fetal tissues. *J. Cell. Biol.* **124,** 381–394.

Weber, S., Dolz, R., Timpl, R., Fessler, J. H., and Engel, J. (1988). Reductive cleavage and reformation of the interchain and intrachain disulfide bonds in the globular hexameric domain NC1 involved in network assembly of basement membrane collagen (type IV). *Eur. J. Biochem.* **175,** 229–236.

Weber, S., Engel, J., Wiedemann, H., Glanville, R. W., and Timpl, R. (1984). Subunit structure and assembly of the globular domain of basement-membrane collagen type IV. *Eur. J. Biochem.* **139,** 401–410.

Xu, H., Wu, X. R., Wewer, U. M., and Engvall, E. (1994). Murine muscular dystrophy caused by a mutation in the laminin $\alpha 2$ (Lama2) gene. *Nat. Genet.* **8,** 297–302.

Yurchenco (1986), p. 8, Table I.

Yurchenco, P. D. (1994). Assembly of laminin and type IV collagen into basement membrane networks. *In* "Extracellular Matrix Assembly and Structure," Biology of Extracellular Matrix Series (P. D. Yurchenco, D. E. Birk, and R. P. Mecham, eds.), pp. 351–388. Academic Press, New York.

Yurchenco, P. D., and Cheng, Y. S. (1993). Self-assembly and calcium-binding sites in laminin. A three-arm interaction model. *J. Biol. Chem.* **268,** 17286–17299.

Yurchenco, P. D., and Furthmayr, H. (1984). Self-assembly of basement membrane collagen. *Biochemistry* **23,** 1839–1850.

Yurchenco, P. D., and Furthmayr, H. (1986). Type IV collagen "7S" tetramer formation: aspects of kinetics and thermodynamics. *Ann. N.Y. Acad. Sci.* **460,** 530–533.

Yurchenco, P. D., and O'Rear, J. J. (1994). Basement membrane assembly. *Methods Enzymol.* **245,** 489–518.

Yurchenco, P. D., and Ruben, G. C. (1987). Basement membrane structure in situ: evidence for lateral associations in the type IV collagen network. *J. Cell Biol.* **105,** 2559–2568.

Yurchenco, P. D., and Ruben, G. C. (1988). Type IV collagen lateral associations in the EHS tumor matrix. Comparison with amniotic and in vitro networks. *Am. J. Pathol.* **132,** 278–291.

Yurchenco, P. D., Cheng, Y. S., and Colognato, H. (1992). Laminin forms an independent network in basement membranes. *J. Cell. Biol.* **117,** 1119–1133.

Yurchenco, P. D., Cheng, Y. S., and Schittny, J. C. (1990). Heparin modulation of laminin polymerization. *J. Biol. Chem.* **265,** 3981–3991.

Yurchenco, P. D., Quan, Y., Colognato, H., Mathus, T., Harrison, D., Yamada, Y., and O'Rear, J. J. (1997). The α chain of laminin-1 is independently secreted and drives secretion of its β- and γ-chain partners. *Proc. Natl. Acad. Sci. USA* **94,** 10189–10194.

Yurchenco, P. D., Sung, U., Ward, M. D., Yamada, Y., and O'Rear, J. J. (1993). Recombinant laminin G domain mediates myoblast adhesion and heparin binding. *J. Biol. Chem.* **268,** 8356–8365.

Yurchenco, P. D., Tsilibary, E. C., Charonis, A. S., and Furthmayr, H. (1985). Laminin polymerization in vitro. Evidence for a two-step assembly with domain specificity. *J. Biol. Chem.* **260,** 7636–7644.

Zhong, C., Chrzanowska-Wodnicka, M., Brown, J., Shaub, A., Belkin, A. M., and Burridge, K. (1998). Rho-mediated contractility exposes a cryptic site in fibronectin and induces fibronectin matrix assembly. *J. Cell Biol.* **141,** 539–551.

CHAPTER 6

Preparation and Analysis of Synthetic Multicomponent Extracellular Matrix

Kim S. Midwood,* Iwona Wierzbicka-Patynowski,* and Jean E. Schwarzbauer

Department of Molecular Biology
Princeton University
Princeton, New Jersey 08544

I. Introduction to the ECM

All cells are surrounded by an extracellular matrix (ECM) which dynamically regulates cellular functions including adhesion, migration, growth, and differentiation. It also provides a stable structural support for cells and a framework for tissue architecture. Fibronectin (FN) is a multifunctional adhesive glycoprotein which is a major component of most matrices (Hynes, 1990; Mosher, 1989). Many cell types make FN, including fibroblasts, endothelial cells, myoblasts, and astrocytes, and incorporate this cellular FN into the ECM. The plasma form of FN (pFN) is synthesized by hepatocytes and released into the blood. The primary differences between pFN and cellular FN arise by alternative splicing (Schwarzbauer, 1991a). The composition of the ECM varies from tissue to tissue and undergoes continuous assembly and remodeling throughout the lifetime of an organism. Therefore, FN must be able to interact with different combinations of cells,

---

*These authors contributed equally to this manuscript.

145

collagens, proteoglycans, and other macromolecules. FN does this through functional domains for binding to fibrin, heparin, cells, collagen/gelatin, and FN itself. The matrix provides the cell with environmental information through interactions with a number of different types of cell-surface receptor, including integrins and proteoglycans. The integrins are a large family of related dimeric receptor complexes that are composed of α and β subunits (Hynes, 1992). Proteoglycans are a diverse set of macromolecules defined by the posttranslational addition of glycosaminoglycan side chains such as chondroitin or heparan sulfate (Lander, 1998). Integrins and transmembrane proteoglycans bind to a number of ligands within the ECM and on the surfaces of other cells. They also interact with cytoplasmic and cytoskeletal signaling proteins.

To address matrix assembly and function, a variety of experimental systems have been developed. One such system is based on the provisional matrix of the wound bed; another uses a cell-associated matrix.

The provisional matrix is synthesized in response to tissue injury and blood vessel damage. It is a covalently cross-linked network consisting predominantly of fibrin and pFN, which is remodeled over time to regain normal tissue structure and function (Clark, 1996). The provisional matrix is formed as the culmination of a series of enzymatic reactions. Upon vascular endothelial damage, blood comes into contact with subendothelial structures and other exposed injured tissues. This initiates a chain of reactions that ultimately results in the cleavage of the inactive precursor enzyme prothrombin, which is present in circulating blood, to active thrombin. Activated thrombin acts on soluble fibrinogen and, by cleaving off small fibrinopeptides, converts it to fibrin. This reveals previously cryptic self-assembly sites in the newly formed fibrin that allow spontaneous polymerization to occur. Following polymerization, factor XIII (transglutaminase) catalyzes the formation of intermolecular covalent cross-links in a calcium-dependent reaction. Factor XIII is present in plasma and is activated upon cleavage by thrombin. Factor XIII also catalyzes the formation of covalent cross-links between pFN and fibrin to form heterodimers and large molecular weight polymers. pFN is covalently cross-linked to lysine residues in the carboxy termini of fibrin α chains using glutamine residues in the amino terminus of FN (Mosher, 1989). This leaves other FN domains free to interact with cell surface receptors, such as integrins and proteoglycans, and other matrix proteins, including collagen and tenascin. This cascade of reactions results in the formation of a stable, semirigid, three-dimensional matrix. The immediate function of the provisional matrix is to fill the wound and maintain the integrity of the vascular system. It then initiates the wound healing process by providing a substratum that supports cell adhesion and migration into the injured tissues. The development of matrices that recapitulate the provisional matrix cell autonomously, *in vitro*, was first described by Mosher (1975).

Cell-mediated matrix formation is another model commonly used to study matrix function. Fibroblasts assemble the ECM into an extensive fibrillar network with FN as a major structural and functional component. FN matrices are vital to vertebrate development and wound healing, and modulate tumorigenesis. FN interacts with cells to control cell adhesion, cytoskeletal organization, and intracellular signaling (Hynes, 1990; Mosher, 1993). It is secreted as a disulfide-bonded dimer with subunits of 230–270 kDa and is assembled into matrix fibrils through a stepwise process (Schwarzbauer and Sechler, 1999). Initiation of assembly depends on the interactions between the cell-binding domain of FN and $\alpha5\beta1$ integrin receptor. These interactions induce receptor clustering and promote FN self-association. Fibril elongation is then propagated by continual addition of FN dimers to growing multimers. As fibrils mature, they are gradually converted into a detergent-insoluble high molecular weight form yielding stable matrix. Matrix assembly modulates cell growth and activates a number of intracellular signal transduction cascades. The $\alpha5\beta1$ integrin receptor is a major receptor for matrix assembly (Ruoslahti, 1991); however, three other integrins, $\alpha IIb\beta3$, $\alpha4\beta1$, and $\alpha v\beta3$ (Sechler et al., 2000; Wennerberg et al., 1996; Wu et al., 1995b) can substitute for $\alpha5\beta1$ to support fibril formation. Sechler (Sechler et al., 1996) developed a matrix assembly system using cell lines lacking an endogenous FN matrix but which are capable of assembling exogenously provided FN. This system has been used to study the involvement of different FN domains in the fibril assembly process.

II. Development of Synthetic Matrices

A. The Provisional Matrix

In this section we describe how to reconstitute the final stages of the blood-clotting cascade using purified components to form a three-dimensional fibrin-FN matrix.

1. Advantages

Three-dimensional matrix substrates reflect a physiologically relevant environment in which to analyze cell behavior. This overcomes the limitations of presenting cells with a single protein as a substratum in a planar organization. Since the provisional matrix is synthesized using purified components, it is easy to control the composition of the matrix and the proportion in which the matrix proteins are presented to cells. This system also allows for the use of recombinant FNs. The provisional matrix interacts with and regulates the behavior of many cells in vivo, including fibroblasts, blood platelets, endothelial cells, and smooth muscle cells (Clark, 1996). The formation of this matrix is not cell mediated; therefore it is possible to analyze the response of many different cells types, with known repertoires of receptors, to the matrix.

2. Disadvantages

By reducing the ingredients to include only the simplest necessary components, we can produce a matrix that enables the dissection of the function of a single protein. However, since not all components present at sites of wound healing are included, preparation of synthetic matrices of this kind in vitro may not recapitulate all of the functions of native matrix.

3. Protocols

a. Fibrinogen Preparation

Lyophilized human fibrinogen (98% clottable, American Diagnostica Inc., Greenwich, CT) is reconstituted in 0.15 M NaCl, 10 mM Tris-HCl, pH 7.4. These preps contain 15–20 μg endogenous human pFN per mg fibrinogen. Elimination of contaminating FN is achieved by batch incubation with gelatin-agarose beads for 1 h at room temperature, after which time the sample is centrifuged to pellet the beads. The supernatant should be collected carefully and recentrifuged to ensure that all agarose is removed. The fibrinogen is then aliquoted, stored at $-85°$C, and thawed immediately before use. FN specifically binds to the gelatin (denatured collagen) using its collagen-binding domain. Removal of FN can be monitored by SDS–PAGE and Western blotting with anti-human pFN monoclonal antibodies. This procedure results in 100-fold depletion of FN to 0.15 μg per mg fibrinogen (Corbett *et al.,* 1996; Wilson and Schwarzbauer, 1992) and is important to enable precise control of the concentration and type of FN in the synthetic matrices. This treatment eliminates the activity of endogenous factor XIII in the fibrinogen preps; therefore purified human factor XIII must be added to achieve levels of matrix cross-linking equivalent to untreated fibrinogen.

b. In Vitro Synthesis of Provisional Matrices

Bovine thrombin (96% clottable, Sigma Chemical Co., St. Louis, MO) is reconstituted in distilled water, aliquoted, stored at $-20°$C, and thawed immediately before use. Rat pFN is purified from freshly drawn rat plasma by gelatin agarose chromatography (Corbett *et al.,* 1996). Matrix components are mixed together in physiological proportions. A 10:1 mass ratio of fibrin : pFN (600 μg/ml fibrin and 60 μg/ml pFN) is incubated with 10 μg/ml human coagulation factor XIII (Calbiochem-Novabiochem Corp., La Jolla, CA), 50 mM CaCl$_2$, 0.15 M NaCl, 0.05 M Tris HCl, pH 7.5, in 0.1 to 1.5 ml volumes. The mixtures are vortexed and kept on ice for 10 min. Two U/ml thrombin are added to start the reaction, then the mixture is rapidly pipetted into 48-well non-tissue culture dishes or onto glass coverslips and allowed to incubate at 4°C overnight to allow maximum cross-linking (see Fig. 1). To inhibit fibrinolysis, 1 μg/ml aprotinin can be added to matrices intended for long time point experiments. Fibrin circulates in the blood at 3000 μg/ml and pFN at 300 μg/ml (Clark, 1996). These higher physiological concentrations can be used in the formation of the matrix. Alternatively, a mass ratio of 20:1 fibrin : pFN can be used with identical results to ratios of 10:1 to allow more conservative usage of matrix proteins (Wenk *et al.,* 2000). After polymerization, this matrix can be used for cell adhesion experiments.

c. Inclusion of Other Proteins

Full-length native pFN can be replaced with recombinant FNs to enable analysis of the specific effect of different domains of FN on matrix formation and cell phenotype (Corbett *et al.,* 1997; Corbett and Schwarzbauer, 1999; Wilson and Schwarzbauer, 1992). This synthetic matrix can also be built upon by the addition of relevant proteins to reconstitute increasingly physiological matrices. Other proteins that contact the provisional

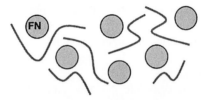

Fibrinogen and fibronectin in solution

thrombin

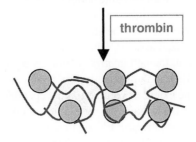

Soluble fibrin clot

Ca^{2+}
Factor XIIIa

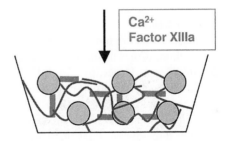

Covalently cross-linked provisional matrix

Fig. 1 Provisional matrix formation. The events common to provisional matrix formation both *in vivo* and *in vitro* are depicted. Fibrinogen (thin lines), the soluble precursor of fibrin, is mixed with FN. Polymerization is initiated by the addition of thrombin to the system, which cleaves fibrinogen to form fibrin. The resulting monomeric fibrin spontaneously polymerizes to form a clot. In the presence of Ca^{2+}, factor XIIIa covalently cross-links fibrin to itself and FN (thick gray lines), producing high molecular weight polymers. These matrices provide a physiologically relevant three-dimensional substrate for cell adhesion.

matrix *in vivo,* for example collagen and tenascin-C, can be incorporated into the matrix simply by adding them into the matrix mixture along with the basic components (Wenk *et al.,* 2000). Those that contain binding domains for fibrin or FN will be incorporated into the matrix via interactions with target proteins. Proteins that contain neither type of site will be incorporated into the matrix by being trapped into the network of fibrin-FN fibrils. After polymerization, the provisional matrix can be treated prior to the addition of cells to dissect matrix protein function. For example, small molecule inhibitors such as the RGD peptide or antibodies to domains of FN have been shown to interfere with cell–matrix interactions (Corbett and Schwarzbauer, 1999).

d. Matrix Contraction

To incorporate cells into the three-dimensional matrix, cells are added to the matrix proteins before the addition of thrombin (Corbett and Schwarzbauer, 1999). Once thrombin is added, the mixture should be incubated at 37°C for 30 min. The cells become evenly distributed within the matrix as it polymerizes.

Matrix contraction can be studied using this synthetic three-dimensional matrix in a well-characterized assay (Corbett *et al.*, 1997; Corbett and Schwarzbauer, 1999). After polymerization, the cell–matrix mixture is released from the sides of the dish and the ability of cells to contract the matrix is assessed by measuring the area of the matrix over time. The extent of contraction is dependent on matrix composition and cell type. For example, fibrin matrices will contract up to 60% over a 4-h time course, while fibrin matrices containing FN will contract 80% over the same time course (Corbett and Schwarzbauer, 1999). The concentration of cells added to achieve optimum cell density varies depending on the cell type; for example, for maximal contraction NIH3T3 fibroblasts are used at 1×10^6/ml and CHO cells at 2×10^6/ml.

e. Immunofluorescence

For cell adhesion on or culture within three-dimensional matrices, cells are released from tissue culture dishes using 0.2 mg/ml EDTA in PBS, washed with PBS, and resuspended in 0.025 M Hepes, pH 7.4, 0.13 M NaCl. TPCK trypsin (Sigma) may also be used to harvest cells from tissues culture dishes at a concentration of 0.1 mg/ml in Versene (Life technologies/Gibco-BRL). Cells are then washed in 0.5 mg/ml soybean trypsin inhibitor (Sigma) and resuspended in 0.025 M Hepes, pH 7.4, 0.13 M NaCl.

Cells within three-dimensional matrices can be stained with fluorescently labeled antibodies (Fig. 2). The cell–matrix mixture is fixed with 3.7% formaldehyde for 15 min at room temperature, permeabilized with ice-cold acetone for 5 min at −20°C, then incubated with primary or secondary antibody in 2% ovalbumin (Sigma) in PBS at 37°C for 1 h. The cell–matrix mixture is then removed from the 48-well dish and mounted onto slides with SlowFade Light Antifade Kit (Molecular Probes Inc., Eugene, OR).

To eliminate matrix thickness as a variable in the analysis of cell behavior, the provisional matrix can be carefully removed from the dish/coverslip by aspiration, leaving behind a visible fibrillar matrix. Cells respond identically to this fibrillar matrix as to a three-dimensional matrix (Corbett *et al.*, 1996). Matrices are prepared as before except cells are not added to the matrix components. Immediately after the addition of thrombin at 2 U/ml, the mixture is pipetted onto a glass coverslip (Fisher Scientific). After polymerization the matrix is aspirated, and the remaining matrix substratum blocked with 1% BSA in PBS. Cells are allowed to adhere to matrix-coated glass coverslips, then washed with PBS, fixed for 15 min at room temperature with 3.7% formaldehyde in PBS, and permeabilized for 15 min at room temperature with 0.5% NP-40 (Calbiochem) in PBS. Cells are

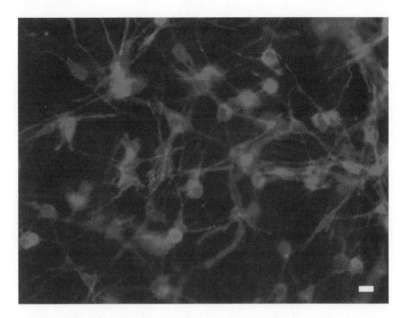

Fig. 2 Immunofluorescent staining of cells within a three-dimensional matrix. Cells were cultured for 18 h in a three-dimensional matrix consisting of fibrin and FN. After this time, cells were fixed and the actin cytoskeleton visualized using rhodamine-labeled phalloidin. Scale bar = 20 μm. (See Color Plate.)

incubated with primary or secondary antibody in 2% ovalbumin (Sigma) in PBS at 37°C for 1 h. Coverslips are mounted with SlowFade Light Antifade Kit (Molecular Probes).

Cells adherent to two-dimensional matrices can also be lysed using RIPA buffer (50 mM Tris-HCl, pH 7.5, 150 mM NaCl, 1% NP-40, 0.25% sodium deoxycholate, 1 mM PMSF, 1 mM NaVO$_4$, 1 mM EDTA, 50 mg/ml leupeptin, 0.5% aprotinin) on ice for 15 min (Kanner *et al.,* 1989). The cells are then scraped with a rubber policeman and the lysate collected and centrifuged for 10 min at 4°C. The cell lysates can be used in biochemical analyses such as immunoprecipitations to characterize levels of protein expression or activation.

4. Controls and Troubleshooting

One variable that affects cell behavior on this synthetic matrix is fibrin–FN cross-linking. The degree of cross-linking of matrix components can be monitored by adding an equal volume of solubilization buffer (8 M urea, 2% SDS, 2% 2-mercaptoethanol, 0.16 M Tris HCl, pH 6.8) to the matrix for 10 min at 100°C. Separation and identification of cross-linked products is performed by SDS–PAGE. Fibrin polymers can be detected on a 7% gel, visualized by silver staining, and pFN on a 5% gel, detected by Western blots probed with anti-FN monoclonal antibodies.

B. Cell–Assembled ECM

In this section we describe *de novo* matrix assembly by cells that do not produce an endogenous FN matrix. The mouse pituitary cell line AtT-20, which does not produce either the $\alpha5$ integrin subunit or endogenous FN, and Chinese hamster ovary (CHO)-K1 cells, which express low levels of hamster $\alpha5$ integrin, were transfected with human $\alpha5$ cDNA (Sechler *et al.,* 1996). The resulting transfected cells (AtT-20α5 and CHOα5) produced no or negligible endogenous FN, respectively, but both could assemble a fibrillar matrix using exogenous FN supplied in the culture medium.

1. Advantages

This matrix assembly system is not complicated by the presence of normal endogenous FN. It allows for the control of timing, rate, and amount of matrix assembly. One can plate the cells overnight to ensure appropriate adhesion and spreading, add known types and concentrations of FN, and follow the progression of fibril formation at specific time points after addition of FN. *In vivo* matrices are composed of more than one protein. The described method allows for the use of a mixture of different ECM proteins to create a more physiological composition of the matrix. Cells can be transfected with different types of receptors or combinations of receptors. One can also use different recombinant FNs to test for involvement of FN domains in the process of matrix assembly. A major advantage of this system is the ability to dissect the independent assembly of recombinant FNs from the earliest stages of matrix assembly.

2. Disadvantages

In a defined or reconstituted system cells will not necessarily have access to all naturally occurring ECM proteins. The composition of ECM is complex and not fully known; thus the absence of some proteins may affect matrix structure or assembly.

3. Protocols

a. Cells and Proteins

AtT-20α5 cells are grown in a 50:50 mixture of Ham's F12 and DMEM supplemented with 20 mM Hepes, pH 7.4, 4 mM L-glutamine, 10% fetal calf serum (FCS) (Hyclone Labs, Logan, UT), and Geneticin (Life Technologies/Gibco-BRL). CHOα5 cells are cultured in DMEM medium containing 10% FCS, 2 mM L-glutamine, 1% nonessential amino acids, and 100 μg/ml Geneticin. In a typical experiment, 1.5×10^5 cells per well are plated on a 24-well dish in 500 μl of medium. To avoid traces of FN present in serum, one can use medium containing FN depleted serum. FN depleted serum can be prepared by passing FCS over a gelatin–agarose column twice (Engvall and Ruoslahti, 1977). The cells are allowed to adhere and spread (for a few hours or overnight) and then incubated with 25–50 μg/ml of pFN added to the medium for different amounts of time. The extent of matrix assembly can be modulated by increasing the amount of

FN added. Adding 25–50 μg/ml of FN results in formation of an extensive FN matrix. At concentrations below 25 μg/ml, a sparser pattern of FN fibrils is observed, with no apparent matrix formation below 5 μg/ml.

Native FN can be purified from human, rat or bovine plasma by gelatin-agarose affinity chromatography (Engvall and Ruoslahti, 1977). Recombinant FNs are expressed using the baculovirus insect cell expression system (Aguirre *et al.*, 1994; Sechler *et al.*, 1996).

b. Immunofluorescence Staining

Cells are plated on glass coverslips in 24-well dish or Lab Tek Chamber Slides (Nunc Inc., Naperville, IL). Spread cells are then incubated with FN. After the desired incubation time, cells are washed with PBS + 0.5 mM MgCl$_2$ and then fixed with fresh formaldehyde solution (3.7% in PBS) for 15 min at room temperature. Coverslips are incubated in a moist chamber with primary anti-fibronectin antibody followed by incubation with fluorescently labeled secondary antibody, both diluted in 2% ovalbumin in PBS at 37°C for 30 min. Each step is proceeded by several gentle washes with PBS. Finally, coverslips are mounted with FluoroGuard Antifade Reagent (Bio-Rad, Hercules, CA). Fibrils are visualized with a Nikon Optiphot microscope with epifluorescence (Fig. 3A).

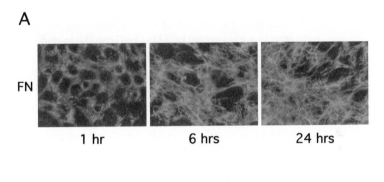

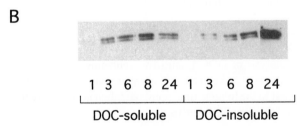

Fig. 3 Time course of FN fibril formation. (A) Analysis by immunofluorescence staining. CHOα5 cells were incubated with 50 μg/ml of rat pFN. At indicated times, the cells were fixed, stained with rat-specific monoclonal antibody IC3, and visualized with fluorescently labeled secondary antibody. (B) Analysis of DOC-soluble and DOC-insoluble material. Cells were lysed with DOC lysis buffer. Fractions (3 μg/lane) were separated by 5% SDS–PAGE and FN was detected with rat-specific monoclonal antibody IC3 and ECL reagents. (See Color Plate.)

c. Isolation and Detection of DOC-Soluble and -Insoluble Material

Cells are plated in a 24-well dish without coverslips and incubated with FN. After the desired incubation time, cells are gently washed with cold PBS buffer and then lysed with 200 μl of deoxycholate (DOC) lysis buffer (2% DOC, 0.02 M Tris-HCl, pH 8.8). Two mM PMSF, 2 mM EDTA, 2 mM iodoacetic acid, and 2 mM N-ethylmaleimide are added as protease inhibitors. Other inhibitors can be substituted. Cells are scraped with a rubber policeman and transferred with a 26G needle to tubes. To shear the DNA, the lysate should be passed through the needle 5 times. DOC-insoluble material is isolated by centrifugation at 14,000 rpm for 15 min at 4°C, and then solubilized in 25 μl of 1% SDS, 25 mM Tris-HCl, pH 8.0, plus protease inhibitors. Total protein concentration can be determined using BCA Protein Assay (Pierce, Rockford, IL). Equal aliquots or equal amounts of total protein of DOC-soluble and insoluble material is electrophoresed on a 5% polyacrylamide SDS gel nonreduced or reduced with 0.1 M DTT. Separated proteins are transferred to nitrocellulose (Sartorius Corp., Bohemia, NY) for immunodetection. Membranes are blocked overnight in buffer A (25 mM Tris-HCl, pH 7.5, 150 mM NaCl, 0.1% Tween-20) at room temperature followed by 1 h incubation with anti-fibronectin antibody in buffer A and 1 h incubation with secondary antibody in buffer A. Each incubation is followed by extensive washes with buffer A (3 times for 10 min). Finally, immunoblots are developed with chemiluminescence reagents (Pierce) (Fig. 3B).

To quantitate the amount of DOC-soluble and insoluble material, equal amounts of total protein are separated by SDS–PAGE and transferred to nitrocellulose. After blocking overnight with 5% BSA in TBS buffer (50 mM Tris-HCl pH 7.5, 200 mM NaCl), the membrane is incubated with anti-fibronectin antibody in blocking buffer for 1 h and then with 1 μg/ml of unconjugated secondary antibody in blocking buffer for 1 h. Each incubation is followed by extensive washes with TBS. Approximately 6 μCi ^{125}I-protein A (specific activity 10 mCi/mg, ICN Biomedicals Inc., Irvine, CA) in 10 ml of blocking buffer is then used in a 1-h incubation. After extensive washes with buffer A (until the background is minimal), the blot is exposed to a phosphor storage screen and analyzed using a Molecular Dynamics PhosphorImager (Sunnyville, CA). All of these steps are performed at room temperature.

d. Time Course

One can follow the matrix assembly process using both immunofluorescence staining and analysis of DOC lysates by incubating the cells with FN for different lengths of time. In order to keep the cell number similar for each time point, it is suggested to add FN first to the longest time point, and then accordingly to the next ones. All incubations should finish at the same time.

e. Serum Starvation

In some applications, such as analysis of protein phosphorylation or cell cycle progression, it is necessary to keep the cells in serum-free medium. CHOα5 cells can be serum starved for up to 24 h to enrich for a population of cells in G_0. Matrix assembly can then be followed by addition of FN in serum-free medium.

4. Controls and Troubleshooting

In experiments that test matrix assembly using recombinant FNs, one should always include a control of native FN purified from plasma. When analyzing a new cell type, it is advisable to compare FN matrix assembly between test cells and fibroblasts, a cell type that normally produces FN matrix. Fibroblasts use endogenously produced FN for the assembly; however, they are also capable of assembling exogenously provided FN. Thus the same amounts and types of FN can be added to parallel cultures and assembly monitored microscopically and biochemically.

Cell density is an important variable in determining the amount of fibril formation. Sparse cells form short fibrils, mainly between the cell surface and the substratum. Cell cultures with many cells in contact will assemble a denser matrix with fibrils extending between and over neighboring cells. When comparing different matrices, one should make sure that the cell density in the observed fields is comparable.

III. Discussion

A. Synthetic Provisional Matrix

The three-dimensional provisional matrix is an excellent model for deciphering the molecular events that regulate tissue injury and wound healing. Its applications are not limited to these events, however, and matrices of this kind can also be used to examine other processes which depend on a three-dimensional framework. Studies have focused on the organization of matrix architecture (Weisel, 1996) and how specific variations in the composition of the fibrin-FN matrix can be used as a mechanism to control cell behavior at sites of tissue repair (Wenk *et al.,* 2000). Using recombinant protein technology, it has also been determined precisely which domains of matrix proteins mediate specific cellular effects. For example, the alternatively spliced V region of FN is required for efficient incorporation into a fibrin matrix (Wilson and Schwarzbauer, 1992), and covalent cross-linking of fibrin and FN is needed for maximum cell adhesion to the matrix (Corbett *et al.,* 1997).

Three-dimensional synthetic matrices have also been used to study different stages of tissue injury including initial cell attachment to the matrix, migration through the matrix, and contraction of the matrix (Clark, 1996). For example, wound contraction has been implicated in the pathology of organ scarring in a variety of fibrotic diseases. Disruptions in the regulation of contraction can lead to undesirable cosmetic scarring; and body deformation and loss of joint motion have been observed in cases where contraction persists after wound closure. Contraction has been studied by culturing cells within a three-dimensional matrix and determining the ability of cells to exert force on and contract the matrix. Recent work has demonstrated that integrins are required to communicate signals from three-dimensional matrices to the cell, causing Rho activation and focal adhesion and stress fiber formation, which lead to contraction of the matrix (Corbett and Schwarzbauer, 1999; Grinnell, 2000; Hocking *et al.,* 2000; Yee *et al.,* 1998; Midwood, unpublished observations).

Cell migration into three-dimensional fibrin matrices has been used as a model for tumor metastasis and invasion. The ability of cells to cross barrier matrices gives an indication of the potency of tumor growth and spread. This type of system also allows analysis of the molecular basis of tumor progression, including identification of cell surface receptors used for attachment to and invasion into tissues (Svee *et al.*, 1996) and enzymes used to proteolytically forge paths through matrices for tumor cells and supporting vascular systems (Weidner *et al.*, 1993). Migration of cells is also important in the inflammatory response, and neutrophil migration through fibrin gels has been used as a model for events occurring in early inflammation (Schnyder *et al.*, 1999). The relationship between the migration of macrophages into three-dimensional fibrin matrices to the composition of the matrix, for example, fibrin concentration, glycosaminoglycan content, or the degree of cross-linking, has also been studied (Ciano *et al.*, 1986; Lanir *et al.*, 1988).

Matrix degradation is an essential part of wound repair. Multiple cell types must migrate through the matrix scaffolding, and migration is dependent on the activity of the fibrinolytic and proteolytic enzymes. Proteolysis has been analyzed *in vitro* using synthetic matrices. Migration of human keratinocytes into fibrin matrices occurs through tunnels of digested fibrin created by pericellular fibrinolysis. Formation of the tunnels requires that plasminogen activator be localized on the advancing surface of the keratinocyte (Ronfard and Barrandon, 2001).

Fibrin-based matrices have also been used in the study of angiogenesis: the formation of new blood vessels. The angiogenic response of human microvascular endothelial cells can be analyzed by seeding cells on top of a three-dimensional fibrin matrix, resulting in the in-growth of capillary-like tubular structures into the matrix (van Hinsbergh *et al.*, 2001). Similarly, endothelial cells resuspended in fibrin matrices form intracellular vacuoles that coalesce into lumenal structures, which process is regulated by integrins (Bayless *et al.*, 2000).

Tissue engineering combines cell biology, biomaterials science, and surgery, with a view to achieving tissue and organ replacement therapies using the patient's own cells. Fibrin matrices are commonly used as three-dimensional *in vitro* culture systems, consisting of single or multiple different cell populations, to develop vital tissue transplants before these preformed tissues are implanted into test subjects. Fibrin is an excellent biocompatible, biodegradable scaffold for cell anchorage, proliferation, and differentiation. It allows uniform cell distribution and quick tissue development, with none of the immunogenic effects of traditional scaffolds, which exhibit toxic degradation and inflammatory reactions (Ye *et al.*, 2000).

Tissues cultured in fibrin matrices have resulted in successful joint cartilage regeneration. Stable *in vivo* transplants which produce typical morphological tissue structure have been introduced into mice and rabbits (Sittinger *et al.*, 1999). Treatment of defects in dogs with exogenous fibrin clots promotes fibrocartilaginous repair and stimulates the regeneration of tissue with a normal histological appearance. Such therapy may be used in the arthroscopic treatment of injury in an effort to improve postoperative outcome (Arnoczky *et al.*, 1988). The attachment of endothelial cells after angioplasty can be greatly improved with fibrin glue matrix, resulting in a significant reduction of restenosis in atherosclerotic rabbits (Kipshidze *et al.*, 2000). Other applications include the long-term regeneration of human epidermis on third-degree burns transplanted with autologous cultured epithelium grown on a fibrin matrix (Ronfard *et al.*, 2000). Fibrin gels modified with covalently bound heparin-binding peptides have been studied as a therapeutic agent to enhance peripheral nerve regeneration by stimulating neurite extension (Sakiyama *et al.*, 1999).

The use of three-dimensional matrices in tissue engineering has progressed to the development of designer matrices. Similarly, in many fields manipulation of matrix composition and molecular interactions has been used to develop agents for *in vivo* use. In this way, analyses done with synthetic multicomponent provisional matrices encompass a wide range of biological and therapeutic applications.

B. Cell-Derived ECM

The deposition of FN into the ECM is an integrin-dependent, complex, and highly regulated process. Its involvement in many physiological and pathological processes encouraged scientists to study the assembly of FN matrices. Cell-assembled ECM systems have provided valuable insights into the role of matrix organization in the regulation of cell function. Different cell systems have been used to study how matrices are assembled. One of the first systems used fibroblasts, the cells that naturally assemble FN fibrils, to analyze the timing of incorporation of FN into existing matrix. Iodinated pFN or FN fragments such as the amino-terminal 70-kDa fragment were added to confluent fibroblasts that were surrounded by an endogenous FN matrix (McKeown-Longo and Mosher, 1983,1985). Incorporation of radiolabeled protein was followed over time, and this assay defined two pools of matrix FN. Pool I was soluble in buffered DOC and, over time, was converted into pool II, which was DOC-insoluble. The DOC-solubility assay became the standard for following FN assembly into matrix. A number of groups have combined this approach with inhibitory fragments, peptides, or antibodies to identify FN domains and receptors involved in the matrix assembly process (Chernousov *et al.*, 1991; Hocking *et al.*, 1996; McDonald *et al.*, 1987; Morla and Ruoslahti, 1992).

To avoid the complications inherent in using a cell system that produces an endogenous FN matrix, we have developed the CHO cell system described in this chapter (Sechler *et al.*, 1996). CHOα5 cells have been used to identify FN domains involved and to understand the mechanism and regulation of the assembly process. Using CHOα5 cells and recombinant FNs containing or lacking specific sequences, it has been demonstrated that RGD-dependent interactions with α5β1 are essential for the initiation of matrix assembly (Sechler *et al.*, 1996), that the cell-binding synergy site in FN is involved in α5β1-mediated accumulation of FN matrix (Sechler *et al.*, 1997), and that the first type III repeats play a regulatory role in conversion to DOC insolubility and in the rate of fibril formation (Sechler *et al.*, 1996).

Receptor requirements for FN assembly have also been dissected using CHO transfectants. CHO(B2) cells selected for deficiency in α5 integrin expression have been particularly useful for these purposes (Schreiner *et al.*, 1989). These cells require reintroduction of $\alpha 5$ by transfection in order to assemble FN matrix (Wu *et al.,* 1993). CHO(B2) cells engineered to express different integrin receptors have been used to demonstrate that $\alpha IIb\beta 3$ and $\alpha v\beta 3$ receptors provide alternative pathways for assembly of FN (Wu *et al.,* 1996, 1995b). Integrin subunits other than $\alpha 5$ can also be introduced into these cells. CHO B2 cells transfected with $\alpha 4$ integrin receptor adhere, spread, and migrate on FN but do not assemble FN into fibrils (Wu *et al.,* 1995a). Interestingly, however, CHO(B2)$\alpha 4$ cells are able to assemble FN matrix after integrin stimulation with Mn^{2+} or a $\beta 1$-stimulatory antibody (Sechler *et al.,* 2000).

Cell lines other than CHO and AtT-20 have been adapted for use in analyzing *de novo* FN matrix assembly and the effects of matrix on cell behavior. Sottile *et al.* (1998) isolated a FN-null cell line and showed that it is capable of assembling exogenous FN by a mechanism similar to fibroblasts. Oncogenically transformed cells, which often express reduced levels of FN, other matrix proteins, and their receptors, are also useful models for analyzing the molecules and intracellular pathways involved in these processes (Brenner *et al.,* 2000; Schwarzbauer, 1991b; Zhang *et al.,* 1997).

Cells that depend on exogenous sources of FN have been very useful in determining the contributions of cytoskeletal organization to FN assembly as well as the effects of FN matrix on intracellular signaling and cell cycle progression. CHO$\alpha 5$ cells show a rapid reorganization of actin into stress fibers, an accumulation of focal adhesion proteins, and activation of focal adhesion kinase (FAK) during assembly of native FN (Sechler and Schwarzbauer, 1997, 1998). A mutant recombinant FN lacking type III repeats 1-7 (FNΔIII_{1-7}) forms a structurally distinct fibrillar network. Cells assembling FNΔIII_{1-7} show mainly cortical actin organization and reduced activation of FAK. Whereas native FN matrix stimulates cell growth (Mercurius and Morla, 1998; Sechler and Schwarzbauer, 1998; Sottile *et al.,* 1998), this mutant FN has the opposite effect and inhibits cell growth (Sechler and Schwarzbauer, 1998). These observations show that modification of matrix architecture has profound effects on cells and may provide a novel approach to control cell proliferation.

The ECM has important effects on cell morphology, growth, and gene expression. Defects in matrix organization contribute to disease and developmental defects. Therefore, it is important to understand the mechanisms of ECM assembly and function as well as to decipher the compositions of matrices from different tissues. Using synthetic multicomponent matrices will allow us to draw a more complete picture of how cells are affected by their environment and will provide new insights into the control of cell phenotype by ECM interactions with cell surface receptors.

References

Aguirre, K. M., McCormick, R. J., and Schwarzbauer, J. E. (1994). Fibronectin self-association is mediated by complementary sites within the amino-terminal one-third of the molecule. *J. Biol. Chem.* **269,** 27863–27868.

Arnoczky, S. P., Warren, R. F., and Spivak, J. M. (1988). Meniscal repair using an exogenous fibrin clot. An experimental study in dogs. *J. Bone Joint Surg. Am.* **70,** 1209–1217.

Bayless, K. J., Salazar, R., and Davis, G. E. (2000). RGD-dependent vacuolation and lumen formation observed during endothelial cell morphogenesis in three-dimensional fibrin matrices involves the alpha(v)beta(3) and alpha(5)beta(1) integrins. *Am. J. Pathol.* **156,** 1673–1683.

Brenner, K. A., Corbett, S. A., and Schwarzbauer, J. E. (2000). Regulation of fibronectin matrix assembly by activated Ras in transformed cells. *Oncogene* **19,** 3156–3163.

Chernousov, M. A., Fogerty, F. J., Koteliansky, V. E., and Mosher, D. F. (1991). Role of the I-9 and III-1 modules of fibronectin in the formation of an extracellular fibronectin matrix. *J. Biol. Chem.* **266,** 10851–10858.

Ciano, P. S., Colvin, R. B., Dvorak, A. M., McDonagh, J., and Dvorak, H. F. (1986). Macrophage migration in fibrin gel matrices. *Lab. Invest.* **54,** 62–70.

Clark, R. A. F. (1996). "The Molecular and Cellular Biology of Wound Repair.". Plenum Press, New York.

Corbett, S. A., Lee, L., Wilson, C. L., and Schwarzbauer, J. E. (1997). Covalent cross-linking of fibronectin to fibrin is required for maximal cell adhesion to a fibronectin-fibrin matrix. *J. Biol. Chem.* **272,** 24999–25005.

Corbett, S. A., and Schwarzbauer, J. E. (1999). Requirements for alpha(5)beta(1) integrin-mediated retraction of fibronectin-fibrin matrices. *J. Biol. Chem.* **274,** 20943–20948.

Corbett, S. A., Wilson, C. L., and Schwarzbauer, J. E. (1996). Changes in cell spreading and cytoskeletal organization are induced by adhesion to a fibronectin–fibrin matrix. *Blood* **88,** 158–166.

Engvall, E., and Ruoslahti, E. (1977). Binding of soluble form of fibroblast surface protein, fibronectin, to collagen. *Int. J. Cancer* **20,** 1–5.

Grinnell, F. (2000). Fibroblast–collagen–matrix contraction: growth-factor signalling and mechanical loading. *Trends Cell Biol.* **10,** 362–365.

Hocking, D. C., Smith, R. K., and McKeown-Longo, P. J. (1996). A novel role for the integrin-binding III-10 module in fibronectin matrix assembly. *J. Cell Biol.* **133,** 431–444.

Hocking, D. C., Sottile, J., and Langenbach, K. J. (2000). Stimulation of integrin-mediated cell contractility by fibronectin polymerization. *J. Biol. Chem.* **275,** 10673–10682.

Hynes, R. O. (1990). "Fibronectins." Springer-Verlag, New York.

Hynes, R. O. (1992). Integrins: versatility, modulation, and signaling in cell adhesion. *Cell* **69,** 11–25.

Kanner, S. B., Reynolds, A. B., and Parsons, J. T. (1989). Immunoaffinity purification of tyrosine-phosphorylated cellular proteins. *J. Immunol. Methods* **120,** 115–24.

Kipshidze, N., Ferguson, J. J., 3rd, Keelan, M. H., Jr., Sahota, H., Komorowski, R., Shankar, L. R., Chawla, P. S., Haudenschild, C. C., Nikolaychik, V., and Moses, J. W. (2000). Endoluminal reconstruction of the arterial wall with endothelial cell/glue matrix reduces restenosis in an atherosclerotic rabbit. *J. Am. Coll. Cardiol.* **36,** 1396–1403.

Lander, A. D. (1998). "Extracellular Matrix, Anchor, and Adhesion Proteins." Oxford University Press, Oxford.

Lanir, N., Ciano, P. S., Van de Water, L., McDonagh, J., Dvorak, A. M., and Dvorak, H. F. (1988). Macrophage migration in fibrin gel matrices. II. Effects of clotting factor XIII, fibronectin, and glycosaminoglycan content on cell migration. *J. Immunol.* **140,** 2340–2349.

McDonald, J. A., Quade, B. J., Broekelman, T. J., LaChance, R., Forsman, K., Hasegawa, E., and Akiyama, S. (1987). Fibronectin's cell-adhesive domain and an amino-terminal matrix assembly domain participate in its assembly into fibroblast pericellular matrix. *J. Biol. Chem.* **262,** 2957–2967.

McKeown-Longo, P. J., and Mosher, D. F. (1983). Binding of plasma fibronectin to cell layers of human skin fibroblasts. *J. Cell Biol.* **97,** 466–472.

McKeown-Longo, P. J., and Mosher, D. F. (1985). Interaction of the 70,000-mol. wt. amino terminal fragment of fibronectin with matrix-assembly receptor of fibroblasts. *J. Cell Biol.* **100,** 364–374.

Mercurius, K. O., and Morla, A. O. (1998). Inhibition of vascular smooth muscle cell growth by inhibition of fibronectin matrix assembly. *Circ. Res.* **82,** 548–556.

Morla, A., and Ruoslahti, E. (1992). A fibronectin self-assembly site involved in fibronectin matrix assembly: reconstruction in a synthetic peptide. *J. Cell Biol.* **118,** 421–429.

Mosher, D. F. (1975). Cross-linking of cold-insoluble globulin by fibrin-stabilizing factor. *J. Biol. Chem.* **251,** 6614–6621.

Mosher, D. F. (ed.) (1989). "Fibronectin." Academic Press, New York.

Mosher, D. F. (1993). Assembly of fibronectin into extracellular matrix. *Curr. Opin. Struct. Biol.* **3,** 214–222.

Ronfard, V., and Barrandon, Y. (2001). Migration of keratinocytes through tunnels of digested fibrin. *Proc. Natl. Acad. Sci. USA* **98,** 4504–4509.

Ronfard, V., Rives, J. M., Neveux, Y., Carsin, H., and Barrandon, Y. (2000). Long-term regeneration of human epidermis on third degree burns transplanted with autologous cultured epithelium grown on a fibrin matrix. *Transplantation* **70,** 1588–1598.

Ruoslahti, E. (1991). Integrins. *J. Clin. Invest.* **87,** 1–5.

Sakiyama, S. E., Schense, J. C., and Hubbell, J. A. (1999). Incorporation of heparin-binding peptides into fibrin gels enhances neurite extension: an example of designer matrices in tissue engineering. *FASEB J.* **13,** 2214–2224.

Schnyder, B., Bogdan, J. A., Jr., and Schnyder-Candrian, S. (1999). Role of interleukin-8 phosphorylated kinases in stimulating neutrophil migration through fibrin gels. *Lab. Invest.* **79,** 1403–1413.

Schreiner, C. L., Bauer, J. S., Danilov, Y. N., Hussein, S., Sczekan, M. M., and Juliano, R. L. (1989). Isolation and characterization of Chinese hamster ovary cell variants deficient in the expression of fibronectin receptor. *J. Cell Biol.* **109,** 3157–3167.

Schwarzbauer, J. E. (1991a). Alternative splicing of fibronectin: three variants, three functions. *Bioessays* **13,** 527–533.

Schwarzbauer, J. E. (1991b). Identification of the fibronectin sequences required for assembly of a fibrillar matrix. *J. Cell Biol.* **113,** 1463–1473.

Schwarzbauer, J. E., and Sechler, J. L. (1999). Fibronectin fibrillogenesis: a paradigm for extracellular matrix assembly. *Curr. Opin. Cell Biol.* **11,** 622–627.

Sechler, J. L., Cumiskey, A. M., Gazzola, D. M., and Schwarzbauer, J. E. (2000). A novel RGD-independent fibronectin assembly pathway initiated by $\alpha4\beta1$ integrin binding to the alternatively spliced V region. *J. Cell Sci.* **113,** 1491–1498.

Sechler, J. L., Corbett, S. A., and Schwarzbauer, J. E. (1997). Modulatory roles for integrin activation and the synergy site of fibronectin during matrix assembly. *Mol. Biol. Cell* **8,** 2563–2573.

Sechler, J. L., and Schwarzbauer, J. E. (1997). Coordinated regulation of fibronectin fibril assembly and actin stress fiber formation. *Cell Adhes. Commun.* **4,** 413–424.

Sechler, J. L., and Schwarzbauer, J. E. (1998). Control of cell cycle progression by fibronectin matrix architecture. *J. Biol. Chem.* **273,** 25533–25536.

Sechler, J. L., Takada, Y., and Schwarzbauer, J. E. (1996). Altered rate of fibronectin matrix assembly by deletion of the first type III repeats. *J. Cell Biol.* **134,** 573–583.

Sittinger, M., Perka, C., Schultz, O., Haupl, T., and Burmester, G. R. (1999). Joint cartilage regeneration by tissue engineering. *Z. Rheumatol.* **58,** 130–135.

Sottile, J., Hocking, D. C., and Swiatek, P. J. (1998). Fibronectin matrix assembly enhances adhesion-dependent cell growth. *J. Cell Sci.* **111,** 2933–2943.

Svee, K., White, J., Vaillant, P., Jessurun, J., Roongta, U., Krumwiede, M., Johnson, D., and Henke, C. (1996). Acute lung injury fibroblast migration and invasion of a fibrin matrix is mediated by CD44. *J. Clin. Invest.* **98,** 1713–1727.

van Hinsbergh, V. W., Collen, A., and Koolwijk, P. (2001). Role of fibrin matrix in angiogenesis. *Ann. N. Y. Acad. Sci.* **936,** 426–437.

Weidner, N., Carroll, P. R., Flax, J., Blumenfeld, W., and Folkman, J. (1993). Tumor angiogenesis correlates with metastasis in invasive prostate carcinoma. *Am. J. Pathol.* **143,** 401–419.

Weisel, J. W. (1996). Fibrin clot architecture. *Ukr. Biokhim. Zh.* **68,** 29–30.

Wenk, M. B., Midwood, K. S., and Schwarzbauer, J. E. (2000). Tenascin-C suppresses Rho activation. *J. Cell Biol.* **150,** 913–920.

Wennerberg, K., Lohikangas, L., Gullberg, D., Pfaff, M., Johansson, S., and Fassler, R. (1996). Beta 1 integrin-dependent and -independent polymerization of fibronectin. *J. Cell Biol.* **132,** 227–238.

Wilson, C. L., and Schwarzbauer, J. E. (1992). The alternatively spliced V region contributes to the differential incorporation of plasma and cellular fibronectins into fibrin clots. *J. Cell Biol.* **119,** 923–933.

Wu, C., Bauer, J. S., Juliano, R. L., and McDonald, J. A. (1993). The $\alpha5\beta1$ integrin fibronectin receptor, but not the $\alpha5$ cytoplasmic domain, functions in an early and essential step in fibronectin matrix assembly. *J. Biol. Chem.* **268,** 21883–21888.

Wu, C., Fields, A. J., Kapteijin, B. A. E., and McDonald, J. A. (1995a). The role of $\alpha4\beta1$ integrin in cell motility and fibronectin matrix assembly. *J. Cell Sci.* **108,** 821–829.

Wu, C., Hughes, P. E., Ginsberg, M. H., and McDonald, J. A. (1996). Identification of a new biological function for the integrin $\alpha v\beta3$: Initiation of fibronectin matrix assembly. *Cell Adhes. Commun.* **4,** 149–158.

Wu, C., Keivens, V. M., O'Toole, T. E., McDonald, J. A., and Ginsberg, M. H. (1995b). Integrin activation and cytoskeletal interaction are essential for the assembly of a fibronectin matrix. *Cell* **83,** 715–724.

Ye, Q., Zund, G., Benedikt, P., Jockenhoevel, S., Hoerstrup, S. P., Sakyama, S., Hubbell, J. A., and Turina, M. (2000). Fibrin gel as a three dimensional matrix in cardiovascular tissue engineering. *Eur. J. Cardiothorac. Surg.* **17,** 587–591.

Yee, K. O., Rooney, M. M., Giachelli, C. M., Lord, S. T., and Schwartz, S. M. (1998). Role of $\beta 1$ and $\beta 3$ integrins in human smooth muscle cell adhesion to and contraction of fibrin clots in vitro. *Circ. Res.* **83,** 241–251.

Zhang, Q., Magnusson, M. K., and Mosher, D. F. (1997). Lysophosphatidic acid and microtubule-destabilizing agents stimulate fibronectin matrix assembly through Rho-dependent actin stress fiber formation and cell contraction. *Mol. Biol. Cell* **8,** 1415–1425.

CHAPTER 7

Analysis of Matrix Dynamics by Atomic Force Microscopy

Helen G. Hansma, * **Dennis O. Clegg,** [†] **Efrosini Kokkoli,** [‡,§]
Emin Oroudjev, * **and Matthew Tirrell** [‡]

*Department of Physics

[†] Department of Molecular, Cellular and Developmental Biology

[‡] College of Engineering
University of California
Santa Barbara, California 93106

[§] Current address: Department of Chemical Engineering, University of Massachusetts, Amherst, Amherst, Massachusetts 01003.

I. Introduction

Analysis of matrix dynamics is a new area of research for atomic force microscopy. It can be divided into three broad categories. First, atomic force microscope (AFM) imaging in fluid has revealed dynamic processes such as the movements of laminin arms and the real-time degradation of collagen. Second, molecular force spectroscopy, also in

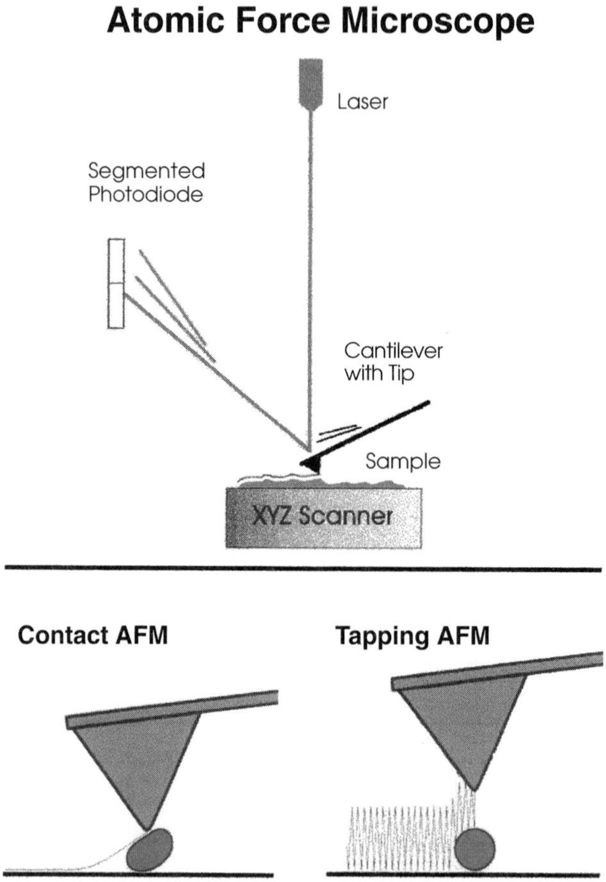

Atomic Force Microscope

Fig. 1 Probe microscopy. (*Top*) The AFM (also known as the scanning force microscope, SFM) images samples by raster-scanning a small tip back and forth over the sample surface. When the tip encounters features on the sample surface, the cantilever deflects. This deflection is sensed with an optical lever: a laser beam reflecting off the end of the cantilever onto a segmented photodiode magnifies small cantilever deflections into large changes in the relative intensity of the laser light on the two segments of the photodiode. (*Bottom*) Tapping AFM often induces less distortion and movement of soft biomolecules than Contact AFM. In tapping AFM, the cantilever oscillates as it scans the sample surface, which reduces lateral forces on the biomolecules, as illustrated in these cartoons. The molecular force probe (http://www.asylumresearch.com/) is similar to the AFM but is optimized for accurate measurements and precise control of movement in the Z-direction, enabling pulling on biomolecules to measure intermolecular and intramolecular interactions. These interactions are visualized as force–distance curves as in Figs. 6 and 10. (See Color Plate.)

fluid, has revealed intramolecular forces in the matrix protein tenascin and intermolecular forces between integrin and fibronectin. Third, AFM in air, the most common use of AFM for matrix analysis, has revealed static interactions between matrix macromolecules such as the binding sites of laminin and of human blood factor IX on collagen IV, and of heparin on fibronectin. Matrix research is about biological surfaces, so the AFM, which investigates surfaces, is a good instrument for matrix research. One might predict that biology in the coming century will evolve from research in test tubes to research on surfaces.

While the extracellular matrix (ECM) has traditionally been thought to play mostly a structural role, it has become clear that the ECM also has a crucial regulatory role, controlling cell adhesion, movement, and differentiation. The ECM consists of a complex array of fibrous proteins, growth factors, proteoglycans, and glycosaminoglycans that serves as nesting material for many cell types. Cells that reside in the matrix have the ability to synthesize, degrade, and reorganize matrix molecules as appropriate, during wound healing and development. However, intermolecular relationships in the ECM have been difficult to study because of the many covalent links between insoluble, fibrous components. Although rapid progress has resulted from the application of molecular biological approaches, molecular associations within the matrix and, particularly, the structural dynamics of ECM molecules are poorly understood.

The AFM (Binnig, 1992; Binnig *et al.,* 1987) (Fig. 1) is well suited to this exploration in at least two ways. The AFM images surfaces at resolutions between sub-nanometer to several nanometers and also senses the material properties of surfaces, through phase imaging in tapping mode AFM (Czajkowsky and Shao, 1998; Engel and Muller, 2000; Fisher *et al.,* 2000; Hansma and Pietrasanta, 1998). A literature search for AFM references on collagen and ECM ("matrix" specifically) gives more than 50 references already.

II. General Experimental Design

AFM of biomaterials is still such a new field that the experimental design varies from lab to lab, from one biomaterial to the next, and often from one experiment to the next experiment within the same lab. Therefore, we present several examples of experimental designs and the results obtained from them. As an example of the variability of experimental design, the careful reader will note that in some cases PBS is a useful buffer for matrix AFM, while in other cases it is quite inferior to other buffers.

AFM imaging is still something of an art, so it is useful to experiment with imaging parameters such as the scan speed, the imaging force or setpoint, and the integral gain, in order to optimize the image quality. In general, the imaging force needs to be higher when imaging a large scan area than when imaging a small scan area. Thus, when zooming in on an area of interest, it is often good to reduce the imaging force slightly beforehand, in order to avoid damaging or removing biomaterials on the sample surface.

General methods for biological AFM in fluid are presented in another article in this series (Kindt *et al.,* 2002). This article includes such details as imaging with or without an O-ring and changing fluid while imaging.

A. Mica

Mica is usually the preferred sample surface for AFM imaging because it is flatter than glass, silicon (Hansma *et al.,* 1996), and most other readily available sample surfaces. Good mica will be free from internal bubbles, which can be observed by holding the mica up to a light. Bubbles in mica seem to contain contaminants that can appear in the AFM as a lawn of particles on the mica surface.

Disks of mica (S&J Trading Co., Glen Oaks, NY) are glued to steel disks with 2-Ton epoxy (Devcon Corporation, Wood Dale, IL) and allowed to dry overnight or longer. The dried mica disks are cleaved with Scotch tape immediately before use. There are two schools of thought about cleaving mica. One school believes that it is best to remove only a very thin layer of firmly attached mica from the mica surface because the freshly cleaved mica surface will then be free of the contaminants that may be present between loosely bound mica layers. The other school believes that it is better to cleave the mica surface repeatedly until a thicker layer of mica is removed because this minimizes the likelihood that there will be incomplete layers or cracks on the resulting mica surface.

For non-imaging probe microscopy methods such as molecular pulling, the sample surface is often gold-coated glass, which seems to bind biomolecules better than uncoated glass.

B. Sample Preparation

Biomaterials in solution are deposited on mica under the various conditions described later. The solutions are incubated for times varying from 1 s to many hours, followed sometimes by rinsing and/or drying before imaging. Since the mica surface is negatively charged (Pashley and Israelachvili, 1984), divalent cations can be useful for binding acidic biomolecules to the surface (see, e.g., Hansma and Laney, 1996). Mica and some of its modifications are discussed more fully in Kindt *et al.* (2002).

One important aspect of sample preparation is the rinsing procedure. The method of rinsing can have extreme effects on the amount of biomaterial remaining on the surface. We (HH) typically rinse our samples with 1–2 ml Milli-Q water, dispensed with a Pasteur pipette across a tilted sample surface.

The rinse can be either gentle or vigorous. With a gentle rinse, fewer biomolecules are rinsed off the surface, so that a higher percentage of the molecules can be observed by AFM. With a vigorous rinse, all loosely bound biomolecules will be removed, so that the remaining biomolecules will all be firmly bound to the surface for stable imaging. A gentle way to rinse is to direct the water above the center of the sample so that it flows down over the sample. A less gentle way to rinse is to direct the water at the center of the sample. This rinse typically leaves many fewer biomolecules on the sample surface (R. Golan and H. Hansma, unpublished results with DNA on mica). A vigorous way to rinse the sample is to use a WaterPik (Teledyne Corp., Fort Collins, CO) (Bezanilla *et al.,* 1994).

Sometimes samples are dried with compressed air or gas. Samples are always dried if they are to be imaged in air. Samples to be imaged under fluid can also be dried and then rehydrated to facilitate the adhesion of the biomolecules to the surface, but this is not commonly done for fear of denaturing the biomolecules. Sample drying needs to be done gently, if one wants to avoid flow-induced rearrangements of molecules. In our hands, evaporative drying of samples produces drying artifacts, at least with DNA on mica; this is the area with which we have the most experience.

C. Probe Microscopy

1. Cantilevers and Tips

There are two general types of cantilevers in widespread use for probe microscopy. Both types of cantilevers have integrated tips at the end. Cantilever suppliers include Digital Instruments (Santa Barbara, CA), AsylumResearch (Santa Barbara, CA), and ThermoMicroscopes (Sunnyvale, CA).

One type of cantilever is a V-shaped silicon nitride cantilever with typical spring constants in the range of 0.06–0.4 N/m. These are used for contact mode AFM, tapping mode AFM in fluid, and for most non-imaging probe microscopy such as molecular pulling. For tapping mode AFM in fluid, the cantilevers are typically oscillated at frequencies in the range of ∼6–15 kHz, depending on where in this range there is a strong resonance peak. If the image quality is poor even after varying the scan speed, gains, and imaging force, then it is useful to try a slightly lower tapping frequency or a different resonance peak. With experience, one finds conditions that are reliably good for the specific type of cantilever and cantilever holder being used. For additional detail, see Kindt *et al.* (2002).

The second type of cantilever is a barlike ("diving board") silicon cantilever with a resonant frequency of 200–400 kHz. These are typically used for tapping-mode AFM in air and are operated at their resonant frequency or slightly below.

The tips on the cantilevers are sometimes modified by electron beam deposition (EBD) in a scanning electron microscope to produce EBD tips (Keller and Chou, 1992) with high aspect ratios and altered chemical properties; the radius of curvature of EBD tips is typically comparable to that of the cantilever's integrated tip. The typical tip of the future may be a nanotube, which has a smaller radius of curvature than most commercial AFM tips, as well as a high aspect ratio (Cheung *et al.,* 2000); for a discussion on nanotube tips, see Hansma (2001).

2. AFM Imaging

The samples discussed in this chapter were imaged by tapping mode, in air or fluid, with a MultiMode AFM with Nanoscope III (Digital Instruments, Santa Barbara, CA) unless stated otherwise. Samples can be imaged in fluid either with or without an O-ring, as described previously (Kindt *et al.,* 2002).

The AFM piezoelectric scanners had maximum scan ranges of 10–15 μm. The vertical-engage E scanner (Digital Instruments) is the most convenient because it does not need to be leveled manually. The specific imaging conditions and cantilevers are described for each experiment.

Images are typically processed by flattening to remove the background slope. Zeroth-order flattening is often best. Flattening modifies images such that each scan line has the same average height. Therefore, scan lines that cross high features will be flattened to generate artificial depressions in the scan lines to compensate for the high features. When high features cause the appearance of such low (dark) bands across the image, these features can be boxed, and the image can be flattened again to eliminate these dark artifacts.

3. Non-Imaging Probe Microscopy and Molecular Pulling

Two examples of non-imaging probe microscopy are described in Section IV. The best commercial instrument for such research is the Molecular Force Probe (MFP; Asylum-Research, Santa Barbara, CA), which is optimized for producing force–distance curves by having better control of tip movement, less drift, and a fivefold better sensitivity than the best commercial AFMs.

III. Probe Microscopy in Fluid: AFM Imaging

A. Dynamic Motion of Laminin-1

The cruciform structure of laminin-1, with binding sites for molecules and cells localized to various domains, has been well documented (Jones *et al.,* 2000). However, little is known about the dynamics and flexibility of the laminin-1 cross. We have employed AFM in fluid to examine laminin dynamics (Chen *et al.,* 1998). Movement of the laminin arms provided dramatic evidence of their flexibility, which may be important in laminins intermolecular interactions.

Several important experimental parameters were identified. The choice of good buffered solutions was the key to seeing this laminin motion—as well as a patient and talented AFM user. As always, it was challenging to find a buffered solution in which laminin bound to the mica well enough for stable imaging, but loosely enough to allow motion. This is a chronic challenge for AFM of biolomolecular processes (Hansma, 1999; Hansma *et al.,* 1999). The quality of imaging also varied somewhat from experiment to experiment, which is another typical experience.

1. Materials and Methods

Four different buffer solutions were used, all at pH 7.4: (1) high-salt Mops (20 mM Mops, 5 mM MgCl$_2$, 150 mM NaCl), a solution with physiologic ionic strength; (2) low-salt Mops (20 mM Mops, 5 mM MgCl$_2$, 25 mM NaCl); (3) PBS in 5 mM MgCl$_2$ (10 mM phosphate buffer, 2.7 mM KCl, 137 mM NaCl; PBS tablets, Sigma Chemical, St. Louis, MO); (4) Tris (50 mM Tris, 150 mM NaCl, 5 mM MgCl$_2$).

The first two buffers gave good results, as shown in Figs. 2 and 3. The second two buffers worked poorly, as shown in Fig. 4. These four solutions were used to prepare

Fig. 2 Laminin in motion. Successive images of a single molecule moving its arms in fluid in the AFM. The imaging solution was high-salt Mops (20 mM Mops, 5 mM MgCl$_2$, 150 mM NaCl). The elapsed time between these height images varied from 1.2 to 4.8 min. See Chen *et al.* (1998).

laminin samples (Laminin-1; Gibco-BRL, Grand Island, NY) at 0.002–0.01 μg/μl in two ways:

1. About 35–40 μl was pipetted onto freshly cleaved mica and left in a tightly sealed petri dish filled with water overnight. The micas were washed with Milli-Q water the next day and imaged in both water and buffers by adding about 100–150 μl of liquid directly onto the washed mica.

2. A Pap Pen (Electron Microscopy Sciences, Ft. Washington, PA) was used to create a hydrophobic boundary on the mica so that the sample (30–40 μl) stayed in a droplet form overnight. In this way, we could use a smaller sample volume. The micas were washed with Milli-Q water the next day and imaged in buffers.

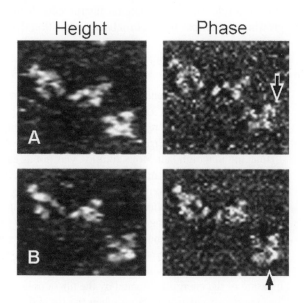

Fig. 3 Details of moving Ln-1 molecules can be seen in AFM phase images (right). A few molecules from the same field are shown in A-B in both height (left) and phase (right) images. Phase imaging provides additional information that is difficult or impossible to detect in the height mode (arrows). The images are 220 nm × 190 nm. See Chen *et al.* (1998).

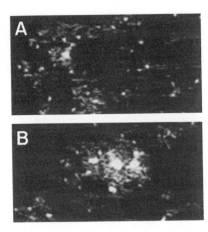

Fig. 4 Laminin molecules are aggregated and tend to move during imaging (as indicated by streaks in the images) under Tris buffer (A) and PBS (B). It is also possible that the laminin molecules adopt globular instead of cruciform conformations in these two solutions, since there are a number of "dots" in the image that are roughly the expected size of a laminin molecule but do not have the "arms" seen in Figs. 2 and 3. These height images are 1 μm $\times$ 0.5 μm. See Chen *et al.* (1998).

Samples were imaged at tapping frequencies of 10–20 kHz, and scan rates of 6–8 Hz. At these scan rates, image acquisition times were typically 45 s/image. Narrow 100-μm silicon-nitride cantilevers (Digital Instruments, Santa Barbara, CA) were modified by growing "supertips" by electron beam deposition (EBD).

2. Observations

Random arm movements of individual laminin molecules were often seen in the two Mops solutions (Fig. 2). This laminin molecule exhibits dynamic motion in fluid; other laminin molecules were relatively stable, showing only small changes in the conformations of the arms (e.g., Fig. 6 in Chen *et al.*, 1998).

Imaging in a high-salt Mops solution was not as easy as imaging in a low-salt Mops solution. Generally, the images appeared less well defined in high salt, perhaps because the laminin molecules were more weakly attached to the mica.

AFM phase images sometimes showed substructures in the laminin arms that were not visible in height images (Fig. 3, arrows). Phase images in tapping AFM show the phase difference between the oscillation driving the cantilever and the oscillation of the cantilever as it interacts with the sample surface (Argaman *et al.*, 1997; Babcock and Prater, 1995; Hansma *et al.*, 1997; Magonov *et al.*, 1997). In air, phase images are a measure of the energy dissipated by the tip-sample interaction (Cleveland *et al.*, 1998). The energy dissipation is large when the tip and sample stick to each other; this tip–sample adhesion can be due to such things as electrostatic interactions (for example, Hoh *et al.*, 1991; Radmacher *et al.*, 1994). The degree of adhesion between the tip and the sample correlates with the darkness in the phase image, if the tip is imaging at a force high enough to produce a repulsive interaction with the surface. The cantilever oscillation in air is sinusoidal, whereas in liquid the harmonics of cantilever motion are more complex. Therefore the interpretation of phase images in liquid is also more complex. It is quite possible, however, that for phase images in liquid the dark regions are also more adhesive than the light regions. Mica, which is generally more adhesive than biomolecules (Hansma, 1996; Radmacher et al., 1994), is darker in these phase images than the bulk of the laminin molecules. Therefore the substructures in phase images of laminin (Fig. 3, arrows) may be regions of the laminin arms that have greater and lesser adhesion to the AFM tip.

The other solutions, Tris and PBS, were not good for imaging with AFM in fluid. The laminin molecules tended to aggregate in these buffers, and the images showed streaks typical of poorly adsorbed molecules (Fig. 4). Since it was hard to image individual molecules, these two solutions are not recommended for this molecule.

B. Collagen Degradation

The ECM is a dynamic structure that is constantly being renovated and remodeled. While some ECM molecules are quite stable, others turn over rapidly. The proteolysis of ECM is thought to play important roles in cell migration, angiogenesis, inflammation, tumor cell metastasis, and other pathologies (McCawley and Matrisian, 2001; Vu and Werb, 2000).

Collagens are the major structural components of most ECMs, accounting for approximately 25% of total protein in mammals (Vanderrest and Garrone, 1991). The triple-helical structure of collagen has been studied with X-ray diffraction (Brodsky and Eikenberry, 1988) and single collagen molecules have been imaged with electron microscopy (Kielty et al., 1985; Mould et al., 1985). In general, steady-state levels of ECM components such as collagens are thought to be regulated by matrix metalloproteinases (MMPs) and their naturally occurring proteinaceous inhibitors (TIMPs). MMPs constitute a family of about 20 Zn^{2+}-dependent proteases that cleave extracellular matrix proteins at neutral pH (Parsons et al., 1997; Sethi et al., 2000). Small molecule MMP inhibitors are currently in clinical trials for treatment of cancer, macular degeneration, and other diseases. However, the molecular mechanisms of ECM turnover are just beginning to be uncovered.

Because AFM can be used for real-time and high-resolution imaging in aqueous fluids, it has tremendous potential for investigation of ECM dynamics. AFM has been used to image single collagen molecules, as well as the binding of human factor IX to collagen IV molecule at specific sites (Shattuck et al., 1994; Wolberg et al., 1997). We used AFM to image the real-time proteolysis of single triple helical collagen I molecules by Clostridium histolyticum collagenase on mica surfaces (Lin et al., 1999).

1. Imaging Individual Collagen I Molecules

As discussed earlier, the first requirement for imaging of any molecule is to determine the best concentration to use, which is done more or less empirically. Acid-solubilized collagen I extracted from rat tail tendon (Brodsky and Eikenberry, 1982) was diluted to 1–20 μg/ml in phosphate-buffered saline (PBS) (Life Technologies, Grand Island, NY) containing 1 mM Ca^{2+} and 1 mM Mg^{2+}, and various concentrations were absorbed to freshly cleaved mica surfaces at room temperature for 1–2 min. The mica was then thoroughly washed with PBS and imaged by AFM in PBS.

All samples were analyzed with silicon nitride cantilevers (200 μm long with nominal spring constant ~0.06 N/m). The tapping frequencies were selected at around 28 kHz, the scan rates were set between 2 and 3 Hz, and the proportional and integral gains were set between 1 and 3. Height and amplitude images, and sometimes phase images, of the samples were simultaneously recorded. Images shown herein were flattened off-line, and occasionally were magnified and filtered at low-pass using the Digital Instruments Nanoscope III software program.

When 1–2 μg/ml collagen I was adsorbed to the mica surface, single, isolated collagen I molecules could be visualized. Single collagen I molecules were approximately 300 nm long with a height (thickness) of 1–2 nm. These spatial features are consistent with measurements from previous studies (Brodsky and Eikenberry, 1988). The widths of imaged collagen I molecules are between 6 and 14 nm, which is wider than observed with EM. This widening is likely due to probe tip-induced broadening, commonly observed in AFM images of isolated macromolecules.

When 20 μg/ml collagen was absorbed, significantly more collagen molecules were attached to the mica in an overlapping, random meshwork one to two molecules deep. The collagen I molecules were firmly attached to the mica surface in PBS and could be imaged in tapping mode for up to 2 h without much alteration in the positioning of the collagen molecules. This concentration was used for imaging of proteolysis.

2. Real-Time Imaging of Proteolytic Digestion

To image the proteolysis of collagen by collagenase, PBS containing clostridium histolyticum collagenase (827 U/mg, purchased from Worthington Biochemical Co., Freehod, NJ) was perfused over collagen-coated mica with a peristaltic pump (Rainin Instruments, Emeryville, CA). Collagenase (2 μg/ml) in PBS was perfused for approximately 1 min at a rate of 2 ml/min before perfusion was stopped. This ensured that the fluid covering collagen on mica was replaced with the new perfusion buffer, since the chamber volume is less than 0.1 ml. These conditions were determined by numerous preliminary experiments where flow rates and concentrations were systematically altered. In most early experiments, all of the collagen rapidly disappeared.

The samples were continuously imaged during and after fluid change. Perfusion of fluid often introduced noise in the AFM image and sometimes caused tip to disengage; when this occurred the tip was immediately reengaged after the fluid change was completed. Of note, a computer-controlled perfusion system without these problems has now been developed (Kindt *et al.*, 2002). To distinguish the collagenase particles from the collagen, the enzyme was imaged by itself at the concentration used in the perfusion (Lin *et al.*, 1999).

Collagen I molecules were imaged prior to, during, and after collagenase was added. Four minutes after the addition of collagenase, some collagen I molecules (~40%) were already cleaved, and some smaller globular structures began to appear in the image. These structures are likely to be collagenase-digested collagen fragments and collagenase molecules bound to collagen molecules or deposited on mica. After incubation with collagenase for 20 min, the collagen I molecules were almost entirely digested. When higher concentrations of collagenase (10–100 μg/ml) were added, the collagen degradation was much faster: the collagen I molecules were almost completely degraded in 2 min or less, within a single image frame.

Figure 5 shows enlarged views of images during collagenase addition (Fig. 5A) and approximately 4 min after addition (Fig. 5B). The stop time of the perfusion is indicated by a horizontal line caused by tip disengagement (in Fig. 5A). Individual collagen I

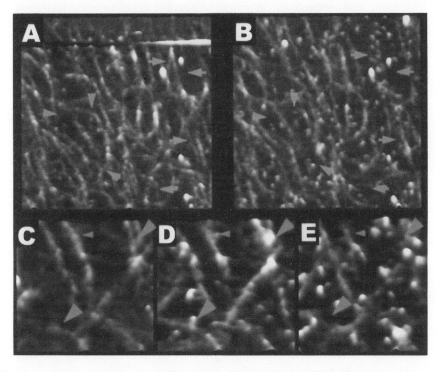

Fig. 5 Imaging collagenase digestion of collagen. Height images obtained using the tapping mode reveal individual collagen I fibers being digested by *C. histolyticum* collagenase. (A, B) Images were collected right after collagenase addition (A) or 4 min after enzyme addition (B). Arrows indicate collagen molecules that are cleaved during this time. The line across the image (A) indicates the end of the collagenase addition. Image area is 1 μm^2. (C–E) Images were collected before (C), immediately after (D), or 4 min after (E) collagenase addition. Arrows indicate sites on collagen that are bound by globular collagenase and then cleaved. Image area = 0.6 μm^2. (Reprinted with permission from *Biochemistry* **38,** 9956–9963. Copyright 1999 American Chemical Society.) (See Color Plate.)

molecules before and after cleavage are indicated by arrows. Images taken before (Fig. 5C), immediately after (Fig. 5D), and 4 min after (Fig. 5E) addition of collagenase show globular particles that bind to collagen fibers, which then appear to be cleaved at these sites (in most but not all cases). These globules were interpreted to be collagenase molecules because of the timing of their appearance and their similarity to the size of collagenase imaged alone. Furthermore, most of the collagen molecules subsequently broke at the globule binding sites.

There was some concern that the cleavage of collagen I plated on a mica surface may proceed differently from digestion in solution. However, when monomer cleavage was quantified in the AFM images, the data fit well with previously measured kinetic parameters for this collagenase (Lin *et al.,* 1999).

These images are the first of a single protein molecule being digested by a proteinase. By altering parameters, such as collagen plating density, collagenase concentration, scanning speed, etc., it should be possible to obtain better images of collagen/collagenase complexes. Rate constant measurements suggest that the reaction time for a single collagen cleavage ranges from 1.7 s to several minutes (Lin *et al.,* 1999). AFM should be capable of capturing multiple snapshots of some of the slower collagenases, which may reveal the molecular mechanisms of helix unwinding and provide further information about the reaction. It should also be possible to image proteolysis, synthesis, and reorganization of other ECM molecules as well as intact ECMs. For example, intact basal laminae might be imaged in regions invaded by metastasizing tumor cells.

C. Collagen III Assemblies

Collagen III, found in the basement membranes of endothelial cells, has also been investigated by AFM in fluid (Taatjes *et al.,* 1999). These investigators studied the fibrillar assemblies that formed after 2–4 h incubations. In this case, a pH 9.5 bicarbonate buffered solution was used, with a freshly cleaved mica substrate. Tapping mode AFM was performed using a Bioscope with Nanoscope IIIa (Digital Instruments, Santa Barbara, CA) mounted on an Olympus IX70 inverted light microscope (Olympus America, Inc., Melville, NY). This study demonstrates the feasibility of using AFM to study matrix assembly.

IV. Probe Microscopy in Fluid: Molecular Pulling/Force Spectroscopy

The last decade has seen novel non-imaging methods for studying mechanical forces within and between single macromolecules. Some of these methods for investigating protein and DNA folding/unfolding and ligand-receptor interactions have emerged from AFM techniques (Fisher *et al.,* 1999; Rief *et al.,* 1999). At present, the protocols for these non-imaging AFM methods are very well developed and could be easily applied to new model systems. With the appearance on the market of a commercial instrument, the Molecular Force Probe (MFP, Asylum Research, Santa Barbara, CA), that is specifically dedicated to such experiments, these studies soon will run as routine laboratory applications rather than unique and exotic experiments.

Molecular pulling has even provided a theory for the bulk materials properties of bone in a new paper: "Bone indentation recovery time correlates with bond reforming time" (Thompson *et al.*, 2001). The time dependence of the recovery of thoughness is similar for: (1) bone that has been indented, (2) molecules from bone that have been pulled, and (3) collagen molecules that have been pulled. This correlation in time dependence suggests that the refolding of collagen molecules in bone may be the molecular basis for the recovery of thoughness in bulk bone. This study highlights the potential importance of non-imaging AFM methods for future research on the extracellular matrix.

Two other recent molecular-pulling studies are presented here. These studies on tenascin (Oberhauser *et al.*, 1998) and molecular force measurements for interactions between integrin and the fibronectin peptide (Kokkoli, 2002) are chosen to represent the two basic types of molecular pulling. In the tenascin study, single molecules are stretched between the tip and the sample surface. In the integrin–fibronectin study, one member of the ligand–receptor pair is bound to the tip, and the other is bound to the surface. This permits the study of ligand–receptor interactions at the level of single molecules.

A. Unfolding Tenascin: An ECM Macromolecule

We present here the protocols that are based on molecular-elasticity studies of native and recombinant tenascin proteins (Oberhauser *et al.*, 1998). These protocols are applicable for studying the physicomechanical properties of structural/mechanical proteins in the ECM.

Native tenascin is a large ECM glycoprotein with both adhesive and anti-adhesive properties. It has a distinctive hexabrachion (six-armed) structure (Chiquetehrismann, 1995). Each arm in this structure is a single polypeptide folded into an irregular series of independent globular domains, as diagrammed in Fig. 6A. The globular domains in each arm can be separated into three different subclasses, depending on their homology to other proteins: EGF-type (epidermal growth factor), FN-III (fibronectin type III) and fbg (fibrinogen). The molecular elasticity experiments were performed on both native tenascin and recombinant proteins (Aukhil *et al.*, 1993) containing 7 or 15 FN-III type domains. Each of these FN-III domains folds into a β-barrel fold, made by β-strands forming a closed circular β-sheet (Bork *et al.*, 1996; Campbell and Spitzfaden, 1994).

1. Methods and Observations

The proteins were applied as a solution in phosphate-buffered saline (PBS), at concentrations of 10–100 μg/ml, on freshly gold-coated glass coverslips. After 10 min for protein absorption, the coverslip was washed with PBS and placed into a single-axis AFM, built as described (Oberhauser *et al.*, 1998). As mentioned earlier, there is now a commercial instrument, the MFP, that can perform this research (Best *et al.*, 2001; Carl *et al.*, 2001; Kindt *et al.*, 2002).

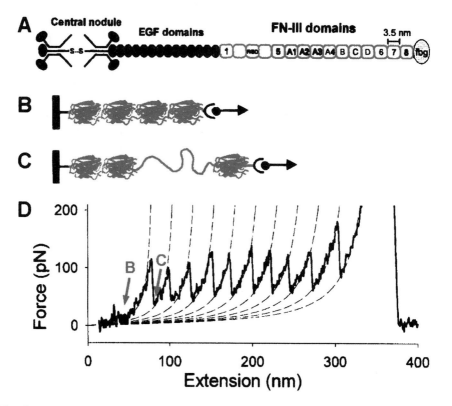

Fig. 6 Molecular pulling (force spectroscopy) of tenascin. (A) Diagram of a tenascin arm. (B, C) Diagram of four FN-III domains of tenascin, tethered to surface and tip (B) at start of pull and (C) after rupture of one FN-III domain. (D) Force–extension spectrum of the unfolding of a molecule containing at least 10 FN-III domains. B and C indicate stages of unfolding as diagramed in (B) and (C) above. Dashed lines show WLC fits. (Reprinted by permission from *Nature,* 393: 181–185. Copyright 1998 Macmillan Magazines Ltd. See also Erickson, 1997; Rief *et al.,* 1997.) (See Color Plate.)

The cantilever (silicon nitride, Digital Instruments, Santa Barbara, CA, USA) in the cantilever holder was covered with PBS and placed onto the sample. The spring constant for each individual cantilever was determined by equipartition method as described (Florin *et al.,* 1994); the cantilever spring constant is now measured automatically by the MFP.

The cantilever tip was brought to the surface and kept in contact under small force load for several seconds in order to allow a fraction of the large protein to adsorb onto the tip. After this, the tip was retracted with speed 200–600 nm/s and an extension curve was recorded, as in Fig. 6D. The majority of the force vs extension traces exhibited forces at tip–gold distances of more than few hundred nanometers. This indicates that long molecular structures bridged the tip and the gold surface, as diagrammed in Figs. 6B and 6C. In all these extension traces the start region was marked by high force load onto AFM tip. Authors attributed these high force peaks in the beginning to multiple molecular interactions between the tip and the gold surface, which rupture upon initial separation distances. On the larger extension distances, the traces both for native and recombinant tenascin proteins typically exhibited a sawtooth-like pattern with as many as 12 peaks with average force 137 ± 12 pN. These unfolding forces were related to pulling speed and increased when pulling speed was increasing.

The worm-like-chain (WLC) model (Bustamante *et al.,* 1994) was used to characterize these traces. The WLC was found to describe, with a good degree of accuracy, the relationship between the polymer extension and entropic force generated as a result of such extension. The persistence length was estimated in a 0.42 ± 0.22 nm range, which is in agreement with results obtained in a similar experiments on forced protein unfolding (Fisher *et al.,* 1999; Rief *et al.,* 1999). The contour length spacing between force peaks was measured to be 28.5 ± 4.0 nm. These measurements were in good agreement with predicted difference in length between the fully extended (near 31 nm) and folded (3.6 nm) single FN-III type domains from tenascin.

The unfolded FN-III domains can be refolded by allowing the AFM tip to travel partially back toward the surface, but not close enough to pick another protein molecule. Domain refolding is the process that occurs when moving from Fig. 6C to Fig. 6B; the direction of the force arrow is reversed during refolding. When the molecule was rapidly extended again to the same distance as in the previous pull, few unfolding events could be registered. This refolding was strictly time dependent and tens of seconds were needed for full refolding to occur.

B. Interactions Between an Integrin Receptor and a Fibronectin Peptide Ligand

A model biomimetic system has been designed that allows the investigation of collective unbinding processes between the integrin $\alpha_5\beta_1$ receptor and GRGDSP peptides found in fibronectin. An AFM was used to provide high-resolution images and direct adhesion measurements (Kokkoli, 2002).

Bioartificial membranes that mimic the tenth type III module of fibronectin (GRGDSP) are constructed from mixtures of peptide amphiphiles and polyethylene glycol (PEG) amphiphilic molecules. The peptide amphiphiles consist of a lipophilic tail, which is composed of dialkyl hydrocarbon chains, and a peptide headgroup that contains the bioactive sequence. The tail serves to align the peptide strands and provide a hydrophobic surface for self-association and interaction with other surfaces.

Peptide amphiphiles containing the GRGDSP sequence were deposited on a surface by the Langmuir–Blodgett technique. The purified $\alpha_5\beta_1$ integrin receptor was immobilized onto a polystyrene sphere using antibodies, and the sphere was then glued onto the AFM cantilever.

1. Preparation and AFM Characterization of Bioartificial Membranes

1,2-Distearoyl-*sn*-glycero-3-phosphatidylethanolamine (DSPE) and the PEG chains with molecular weight of 120 covalently linked to DSPE (DSPE-PEG-120) were obtained from Avanti Polar Lipids, Inc. (Alabaster, AL). $(C_{16})_2$-Glu-C_2-KAbuGRGDSPA buK (GRGDSP), shown in Fig. 7, was synthesized as described elsewhere (Berndt *et al.,* 1995; Yu *et al.,* 1997).

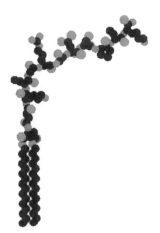

Fig. 7 The general structure of the peptide amphiphile (C_{16})$_2$-Glu-C_2-KAbuGRGDSPAbuK. (See Color Plate.)

The pure amphiphiles were dissolved at approximately 1 mg/ml in a 99:1 chloroform/methanol solution. The solution was stored at 4°C and heated to room temperature prior to use. The Langmuir–Blodgett (LB) technique was used to create supported bioactive bilayer membranes. LB film depositions were done on a KSV 5000 LB system (KSV Instruments, Helsinki, Finland). Prior to use the trough was thoroughly cleaned with a 9:1 chloroform/methanol solution (Fisher). The clean trough was filled with Milli-Q (Millipore) water and an amphiphile solution was spread on the air–water interface with a 100 μl microsyringe (Hamilton). The solvents were allowed to evaporate for 10 min after which the pressure vs molecular area isotherm was recorded. The barrier speed during compression of the layer varied from 50 to 5 mm/min. The surface pressure was recorded with a flame-cleaned platinum Wilhelmy plate (KSV Instruments, Monroe, CT). Thin 15 mm diameter mica disks were cleaved, rinsed with chloroform/methanol solution followed by Milli-Q water, and hung onto the dipper. All the depositions were done at a surface pressure of 41 mN/m, which is well below the collapse pressure of 60 mN/m. The deposition pressure was held for 10 min prior to deposition to equilibrate the film. The deposition speed for both the up and down strokes was 1 mm/min. The transfer ratio was 0.8–1 for all depositions. The first step in producing a supported bilayer membrane was to make the mica hydrophobic with a layer of DSPE in the up stroke. The second layer with the peptide amphiphiles was deposited in the down stroke. The resulting supported bilayer membranes were transferred into glass vials under water. Care was taken to avoid exposing the surfaces to air, as they rearrange and become disorganized (Fig. 8).

AFM images of pure GRGDSP and DSPE-PEG-120 amphiphile bilayer membranes and a 75 mol% mixture are shown in Fig. 9. Images of these LB films were obtained in tapping mode under water using standard 100 μm V-shaped silicon nitride AFM cantilevers with pyramidal tips (Digital Instruments). The O-ring was boiled in water

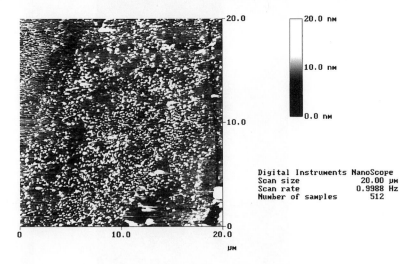

Fig. 8 An AFM topography image of a $(C_{16})_2$-Glu-C_2-KAbuGRGDSPAbuK supported bilayer membrane after being exposed to air for 3 h. The image was taken in air in tapping mode.

prior to use. The boiled O-ring was then placed on the sample surface, which was immersed in a water-filled vial. The sample was lifted from the water taking care not to disturb the O-ring so that a thin water meniscus, held by the O-ring, always covered the sample. Excess water outside the O-ring was removed, and the underside of the mica substrate was attached to a metal disk using double-stick tape and positioned on the piezo scanner. The AFM head was carefully lowered onto the O-ring.

2. $\alpha_5\beta_1$ Immobilization

Purified human $\alpha_5\beta_1$ protein was purchased from Chemicon International (Temecula, CA). The integrins were dialyzed overnight at $4°C$ against solution A, pH 7.2, containing 20 mM Tris-HCl (Sigma), 0.1% Triton X-100 (Sigma), 1 mM MgCl$_2$ (Aldrich), 1 mM MnCl$_2$ (Aldrich), and 0.5 mM CaCl$_2$ (Aldrich). The integrins were removed from the dialysis membrane, diluted to 5 μg/ml in solution A, aliquoted, and stored at $-80°C$.

A 16 μm polystyrene sphere (Bangs Laboratories, Fishers, IN) was attached on the apex of a standard 100 μm V-shaped silicon nitride AFM cantilever as described elsewhere (Kokkoli and Zukoski, 1998, 2001), washed with ethanol and phosphate buffered saline (PBS) (Life Technologies, Rockville, MD), and coated overnight at $37°C$ with goat anti-mouse IgG Fc (Cappel, Aurora, OH) at a final concentration of 8 μg/ml in PBS. The surface was washed with 1× PBS and nonspecific binding sites were blocked in 1× PBS with 1% BSA (Sigma) and 1:500 Tween-20 (Sigma) at $37°C$ for 1 h. The surface was washed with 1× PBS and incubated with purified monoclonal mouse anti-human TS2/16 (anti-β_1) antibody (Endogen, Woburn, MA) in 1× PBS (20 μg/ml) for 3 h at

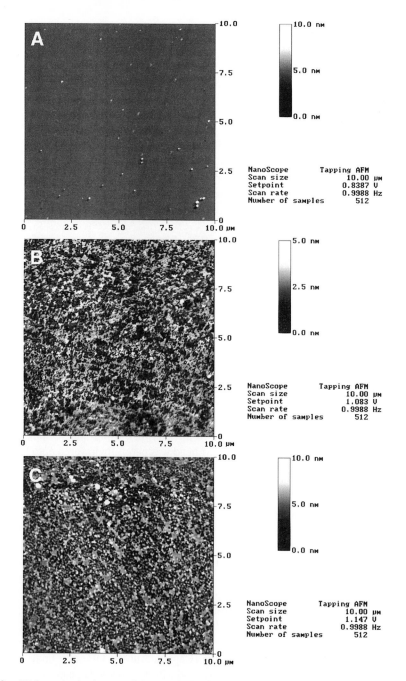

Fig. 9 AFM topography images of the pure DSPE-PEG-120 and $(C_{16})_2$-Glu-C_2-KAbuGRGDSPAbuK supported bilayer membranes and a mixture. The images were taken in water in tapping mode. (A) DSPE-PEG-120. (B) 75% $(C_{16})_2$-Glu-C_2-KAbuGRGDSPAbuK: 25% DSPE-PEG-120. (C) $(C_{16})_2$-Glu-C_2-KAbuGRGDSPAbuK. The PEG-lipid membrane in (A) is smooth, with a few raised particles (contaminants?); the membranes in (B) and (C), which contain the peptide amphiphile, show height variations of several nanometers.

37°C. After washing and blocking for another hour, purified human $\alpha_5\beta_1$ integrins were added and incubated at 37°C for 2 h. Surfaces were washed with PBS before use.

3. AFM Force Measurements

The force between the tip and the sample is calculated by multiplying the cantilever spring constant by the cantilever deflection. The spring constants of the cantilevers with functionalized 16 μm polystyrene probe tips were determined using the resonant frequency method (Cleveland *et al.*, 1993). An average value of 0.5855 N/m was used after calibrating 10 probe tips. Surface force measurements were performed using a Multi-Mode AFM , a Nanoscope III with MultiMode AFM and fluid cell (Digital Instruments, Santa Barbara, CA), in 1 mM MnCl$_2$, at a loading rate of 1.16–2.34 μN/s. All experiments were carried out at room temperature. In order to minimize the drift effects the AFM was warmed up for at least half an hour before an experiment. AFM force data were converted to force–distance curves using the method developed by Ducker *et al.* (1991).

A retraction AFM force–distance curve, after the two surfaces are in contact, is shown in Fig. 10. The separation of $\alpha_5\beta_1$-GRGDSP pairs produces a stepwise profile that is a combination of multiple unbinding and stretching events between the $\alpha_5\beta_1$-GRGDSP pairs. The ligand–receptor pairs do not break at once but in multiple steps, thus producing a stepwise return to zero force. As bonds break at the edge of the contact area, that induces

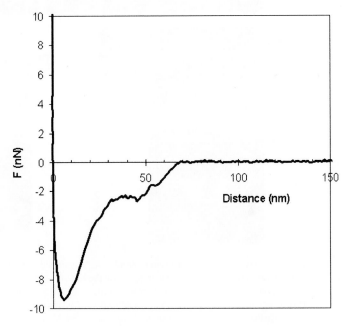

Fig. 10 Retraction force–distance curve between immobilized $\alpha_5\beta_1$ integrins on a polystyrene sphere and (C$_{16}$)$_2$-Glu-C$_2$-KAbuGRGDSPAbuK surface in 1 mM MnCl$_2$.

an increased stretching of bonds in the center, thus generating a series of unbinding and stretching events. Different models can be used to analyze the stretching events quantitatively as described elsewhere (Kokkoli *et al.,* 2002).

V. Probe Microscopy in Air: AFM Imaging

Probe microscopy in air includes imaging AFM but not molecular pulling. In air, there is a thin layer of fluid covering the sample. This thin layer of fluid creates a large meniscus force (Drake *et al.,* 1989). Therefore pulling experiments are dominated by the meniscus force and not by the molecular interactions. Imaging in dry gas reduces but does not eliminate the meniscus force.

Imaging in air is usually easier than imaging in fluid. In addition, imaging in air can give better resolution than imaging in fluid, as seen by the comparison of laminin in air (Fig. 11A) and in fluid (Figs. 2 and 3). However, hydrated molecules in fluid are in motion. Even if the individual molecules are well bound to the surface, the domains are able to flex and "breathe" when they are in fluid, and so matrix molecules are

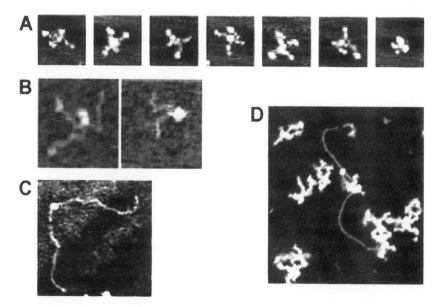

Fig. 11   Basement membrane macromolecules. AFM in air or dry He gas. (A) Laminin-1 molecules in varied conformations show the large globular domain at the end of the long arm of the cross and two small globular domains on the ends of many of the short arms. Images are 140 × 140 nm. (B) Single polysaccharide chains of HSPG are visible in these images, ~200 × 200 nm. (C) Collagen IV dimer. Globular C-terminal NC1 domains of collagen IV combine in dimerization. Images are 385 × 355 nm. (D) Interactions of laminin-1 (Ln-1) and collagen IV in AFM in air form the basis for proposed research on their interactions in near-physiological solutions. Image is 445 × 495 nm. (From Chen *et al.,* 1998; Chen and Hansma, 2000.)

assumed to be in a more physiological state when in aqueous fluids, since the matrix *in vivo* is not dry. Therefore the matrix AFM research in air will be reviewed more briefly.

A. Major Macromolecules of the Extracellular Matrix

Two highlights of this research are the visualization of single polysaccharide chains of hyaluron (Cowman *et al.*, 1998) and heparan sulfate from HSPG (Fig. 11B) (Chen and Hansma, 2000). Such chains are thinner even than single-stranded DNA.

In terms of "dynamics," AFM imaging in air can be done to investigate and map the interactions of matrix macromolecules. Similar research can be done also by EM; for example, laminin-binding sites on collagen IV have been mapped by both AFM (Chen and Hansma, 2000) and EM (Charonis *et al.*, 1986; Laurie *et al.*, 1986) and have been compared with the map (Yamada and Kuehn, 1993) of sequence imperfections in Col IV (Chen and Hansma, 2000). Site-specific binding on Col IV of human blood Factor IX shows specific binding sites 98 and 50 nm from the C-terminal end of Col IV (Wolberg *et al.*, 1997).

1. Materials

The following macromolecules were stored at −80°C: mouse laminin (laminin-1, 1 mg/ml) and mouse type IV collagen (0.5 mg/ml), both from Gibco-BRL (Grand Island, NY), and heparan sulfate proteoglycan (HSPG, 0.6 mg protein/ml) from Sigma (St. Louis, MO). The laminin may have contained some nidogen/entactin as well, according to the supplier.

The following four buffered solutions were used at pH 7.4: (1) high-salt Mops (20 mM Mops, 5 mM MgCl$_2$, 150 mM NaCl), (2) high-salt Mops with Zn (20 mM Mops, 5 mM MgCl$_2$, 150 mM NaCl, 15 μM ZnCl$_2$), (3) low-salt Mops with Zn (20 mM Mops, 5 mM MgCl$_2$, 25 mM NaCl, 15 μM ZnCl$_2$), and (4) Tris buffer (50 mM Tris, 150 mM NaCl, 5 mM MgCl$_2$) (Chen *et al.*, 1998; Chen and Hansma, 2000).

2. Sample Preparations

For sample containing only laminin, laminin-1 was diluted in PBS to 0.1-0.01 μg/μl. Approximately 10 μl diluted laminin was pipetted onto freshly cleaved mica. The sample was left on the mica for about 10 sec (Chen *et al.*, 1998).

For other samples, laminin (0.01 mg/ml), collagen IV (0.01–0.02 mg/ml), and HSPG (6–60 ng/μl) were prepared with various buffers. Laminin, collagen IV, and HSPG were mixed at different concentrations in order to find the best combination and conditions for AFM imaging. The reaction times allowed for the mixtures were 10 min to 2 h. Aliquots of approximately 1.5 μl were then pipetted on to freshly cleaved mica and were left on the mica for approximately 1–5 min (Chen and Hansma, 2000).

The micas were washed with Milli-Q water (Millipore Corp., Bedford, MA) and then dried immediately with compressed air. In some cases the samples were further dried in a vacuum desiccator over P_2O_5. AFM imaging was done at scan rates of 6–8 Hz with tapping frequencies of 270–330 kHz in air and in dry He gas. All imaging was done with 100-μm silicon cantilevers (Digital Instruments, Santa Barbara, CA).

3. Image Analysis

Relative molecular volumes of the HSPG protein core were measured with the Nanoscope Bearing software in two ways, which gave similar results. The two ways differed in the method for determining the base of the molecule. For some molecules, the Threshold feature of the software was used to visually determine the best area for the base of the molecule. For other molecules, the maximum height of the background was subtracted from the maximum height of the molecule to determine the base of the molecule. Volumes of the molecules were measured from the top of the molecule to the base, as determined by one of these two methods.

Molecular volumes measured from AFM images correlate roughly with the volumes calculated from their molecular weights, assuming a density of 1–1.3 g/ml (Golan *et al.,* 1999; Pietrasanta *et al.,* 1999; Schneider *et al.,* 1998).

4. Observations

It was difficult to prepare diluted samples of HSPG on mica, probably because the negative charges of HSPG's heparan sulfate tails did not bind to the negatively charged mica. One might be able to modify the mica with a divalent cation, as in Laney *et al.* (1997), but such an approach has not yet succeeded with HSPG. Preparing a concentrated sample brought another problem of aggregation. The individual molecules of HSPG were hardly seen among the aggregates (Chen and Hansma, 2000).

Individual HSPG molecules were observed (Fig. 11B) in mixed samples of HSPG with collagen IV and laminin. Molecular volumes were measured for the cores of molecules that appeared to be HSPG monomers, but in some cases these may have been small aggregates of HSPG. The histogram of molecular volumes showed peaks at $\sim$200 and $\sim$600 nm^3, corresponding to molecular weights of $\sim$150 and 400 kDa, respectively (Chen and Hansma, 2000). This is consistent with the heterogeneity observed in HSPGs (Hassell *et al.,* 1985; Fujuwara *et al.,* 1984; Paulsson *et al.,* 1987).

Collagen IV dimerized at the C-terminal globular domain (Fig. 11C) (Chen and Hansma, 2000) as expected (Yurchenco and O'Rear, 1993). Laminin-1 binding sites on collagen IV (Fig. 11D) (Chen and Hansma, 2000) mapped to positions correlating roughly with collagen IV sequence defects (Yamada and Kuehn, 1993) and with a map made from EM images (Laurie *et al.,* 1986).

5. Related Work

Two other labs have prepared matrix samples on freshly cleaved mica, rinsed with water and dried with compressed gas (dry N_2). For collagen IV and human blood Factor IX, 10 μl of sample in TBS-CM (20 mM Tris, 150 mM NaCl, 2 mM CaCl$_2$, and 1 mM MgCl$_2$, pH 8.0) was incubated for 5 s on mica (Wolberg *et al.*, 1997). Binding sites of human Blood Factor IX on collagen IV were mapped. For hyaluron, 4 μl of sample in 10 mM CaCl$_2$ was incubated for 1–2 min on mica (Cowman *et al.*, 1998). Various intramolecular associations of hyaluron were described.

B. Fibronectin–Heparin–Gold Complexes

A molecular understanding of the ECM will require dissection of the many complex intermolecular associations that occur as the matrix is assembled and remodeled. Most ECM proteins have a modular structure that allows multiple interactions with other components. Fibronectin (Fn) consists of repeating domains known as type I, type II, and type III repeats (Chothia and Jones, 1997; Hynes, 1990). These domains contain binding sites for collagens, fibrinogen, glycosaminoglycans, proteoglycans, bacteria, viruses, and cells. Fn binds to heparin, and this activity is required for formation of focal adhesions (Izzard *et al.*, 1986; Woods *et al.*, 1986). At least three heparin binding sites have been identified, and two of these, Hep I at the amino terminus, and Hep II at the C terminus, are thought to be the most physiologically relevant (Bultmann *et al.*, 1998; Hocking *et al.*, 1998). The Hep II site is thought to have the highest affinity for heparin, perhaps 100-fold higher than the affinity of the Hep I site (Benecky *et al.*, 1988). However, these studies have used fragments of fibronectin generated by proteolysis, and it is possible that binding sites may be altered during the preparation of fragments. We have analyzed the binding of heparin to whole intact Fn molecules using AFM. In contrast to previous studies, our results suggest that the Hep I site at the amino terminus has an affinity equal to or higher than that of Hep II (Lin *et al.*, 2000).

1. Preparation of Fn–Heparin–Gold Complexes

For imaging individual Fn molecules, bovine Fn (Life Technologies, Inc., Rockville, MD) was diluted to 1 μg/ml in phosphate-buffered saline (PBS, Life Technologies, Inc.) containing 1 mM Ca^{2+} and 1 mM Mg^{2+}. Diluted Fn was absorbed to a freshly cleaved mica surface and incubated at room temperature for 1–2 min. The mica was then washed with deionized water and dried using compressed argon gas.

To analyze Fn interactions with heparin, 2–5 μg Fn was incubated with 1 μl heparin-albumin coated gold particles (Sigma Chemical Co., St. Louis, MO) and albumin-coated gold particles (Electron Microscopy Sciences, Ft. Washington, PA) in 0.2 ml of PBS containing 1 mM Ca^{2+} and 1 mM Mg^{2+} for 2–12 h at 4°C. Heparin–gold or albumin–gold particles (diameter ~3.5–6.5 nm) were present at a concentration of 1×10^{12} ml^{-1}, and Fn was present at a concentration of 5×10^{13} ml^{-1}. Glutaraldehyde was then added to a final concentration of 0.1% and the solution was further incubated for 30 min. We empirically chose a higher Fn concentration compared to heparin–gold particles to ensure that most heparin–albumin–gold particles were bound by Fn.

To make analysis easier, we sought to examine only Fn–heparin–gold complexes and not unbound molecules. Fn–heparin–old complexes were separated from uncoupled Fn molecules by two methods. First, the mixture was diluted with PBS and centrifuged at 10,000*g* to pellet Fn–heparin–gold complexes, and the supernatant was discarded. This procedure was repeated at least three times for each sample preparation. Second, in some experiments, the Fn–heparin–gold complexes were separated from uncoupled Fn by size-exclusion chromatography on a Sephacryl S-500 column (6–10 ml volume, Pharmacia) in PBS. Fractions of 0.25 ml were collected and gold particles in each fraction were quantified by optical absorption at 520 nm. The fraction containing the highest gold particle concentration was analyzed by AFM.

2. Imaging Fn–Heparin–Gold Complexes

All samples were imaged in air using etched silicon cantilevers (125 μm in length, spring constant ∼2–10 N/m, resonance frequency ∼200–400 kHz, tip oscillation amplitude ∼10–20 nm). The scan rates were 3–15 Hz, and the proportional and integral gains were set between 0.2 and 0.6. Both height and phase images of the samples were recorded. As described previously, images were flattened, magnified, and filtered using the Digital Instruments Nanoscope III software program.

Images of individual Fn dimers showed the expected V-shape, where the two C termini of the monomers are connected by a disulfide bond at the bottom of the V. AFM images are similar to previous electron microscopy images where individual peptides are mostly extended with some globular domains visible (Engel *et al.,* 1981; Erickson *et al.,* 1981). Even though the samples were washed with water, the Fn molecules appeared to maintain their physiological conformation. Thus, it appears Fn absorption did not alter its conformation.

Fn molecules that were complexed with heparin–gold did not appear significantly different from individual, unfixed Fn dimers, indicating that fixation with glutaraldehyde and purification on sizing columns did not significantly alter the structure. The heparin–albumin–gold particles were 3–6 nm high, as expected, but their width was broadened to 8–20 nm because of the broadening effect of the tip.

Images of single heparin–albumin–gold particles bound to individual Fn molecules are shown in Fig. 12, with the dotted line indicating the inferred peptide backbone of the Fn dimmer. Binding between Fn and gold particles coated with only albumin was not detected, indicating the specificity of the heparin–Fn interaction.

One difficulty encountered when imaging complexes was that the heparin–gold particles have a much greater height than Fn, so they appeared much brighter in the height mode images. In some experiments, it was difficult to delineate the Fn molecules using the height mode. To circumvent this problem, we also used phase-mode imaging, which detects differences in stiffness of the sample (Young's modulus) and energy dissipation of sample–tip interactions (Cleveland *et al.,* 1998; Vesenka *et al.,* 1993). In most cases, it was possible to determine where the heparin bound along the length of the Fn dimer. Heparin–gold was detected either near the center of the Fn dimer (corresponding to the C termini of Fn monomers) or near one of the N termini.

Surprisingly, quantification of the data showed that N-terminal binding events, presumably to the Hep I site, were twofold greater than C-terminal binding events to the

Heparin–gold particles bound to Fn dimers

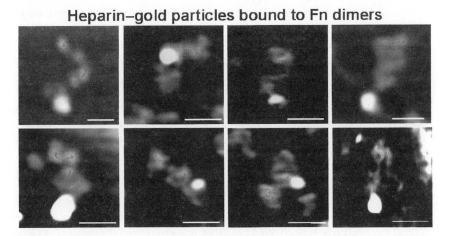

Fig. 12 Height mode images of heparin–gold particles bound to Fn dimers. The peptide backbone of Fn is indicated by a dotted line. Scale bars = 50 nm. (See Color Plate.)

Hep II site. This result is contrary to previous reports, but almost all of these studies were carried out with proteolytic fragments, which may have altered binding properties (Lin *et al.,* 2000). AFM allows analysis of binding events that occur with the intact molecule.

It should be possible to expand this analysis to image Fn complexes in fluid, to see how interactions with heparin might affect the movement and conformation of Fn. Evidence has been obtained to suggest that interactions between Fn and heparin are critical for self-assembly of Fn into fibrils (Bentley *et al.,* 1985). Self-assembly of Fn has been shown to require the Hep I and Hep II domains. Thus, heparin binding to the Hep I site may initiate conformational changes in Fn that are important in Fn–cell interactions and Fn fibrillogenesis.

C. Fibrillar Collagens

In addition to the work reported here, there has been considerable AFM research on the fibrillar collagens. For collagen I, this research includes imaging (Baselt *et al.,* 1993; Chernoff and Chernoff, 1992; Revenko *et al.,* 1994), stages in the assembly of fibers (Paige and Goh, 2001; Paige *et al.,* 1998, 2001), and the effects of diabetes (Odetti *et al.,* 2000). AFM of collagen II (Adachi *et al.,* 1999; Sun *et al.,* 2000) includes a method for stretching and aligning collagen II molecules (Sun *et al.,* 2000).

1. Aligning and Orienting Collagen Molecules

The last paper (Sun *et al.,* 2000) cites a number of methods used to align DNA molecules (Bensimon *et al.,* 1994; Pingoud and Jeltsch, 1997; Smith *et al.,* 1996, 1992; Washizu and Kurosawa, 1990; Zimmermann and Cox, 1994) and describes a relatively simple moving-meniscus method that was used to align collagen molecules. In this method, mica was rinsed successively with 50-μl volumes of water, 500-mM MgCl$_2$, and water. Ten microliters of collagen solution was dispensed between the treated mica surface and a double-edged razor blade placed at an angle of 45°. The razor blade was moved at 0.65 mm/s to align and orient the collagen molecules. The sample was then rinsed and air-dried (Sun *et al.*, 2000). In our hands, drying with compressed air generally gives better results than evaporative air-drying, but we have not experimented with aligning collagen molecules. In the foregoing method, however, it appears that evaporative drying may have been used, and the results were quite good.

2. Humidity Effects

One of the collagen papers mentions that the samples were imaged in air at relative humidities of 40–50% (Odetti *et al.*, 2000). At humidities above 60%, images were less clearly defined. AFM of DNA has given similar experiences (Bustamante *et al.*, 1992; Vesenka *et al.*, 1992), though humidity problems are less severe with tapping-mode AFM than with contact AFM.

VI. Conclusion

From the extent of this paper, it is clear that AFM and other probe microscopies are already beginning to have an impact on matrix research. One hope for the future is that probe microscopies will contribute to the understanding and treatment of matrix diseases, such as the kidney matrix diseases Alport syndrome, Goodpasture autoimmune disease, and Alport posttransplant nephritis (Borza *et al.*, 2000; David *et al.*, 2001; Rohrbach and Murrah, 1993; Tryggvason *et al.*, 1993). AFM imaging may have the resolution to detect abnormal binding patterns of matrix ligands to elongated matrix polymers such as collagen IV, while molecular pulling and force spectroscopy could be used to measure changes in affinities between matrix ligands and receptors, as has been done for avidin/streptavidin and several biotin-like ligands (Florin *et al.*, 1994).

Acknowledgments

This work was supported by an NSF MCB grant (9982743) to H.H., and grants from the NIH (EY09736) and the UC Tobacco Related Disease Program (9RT-0212) to D.O.C. It was also partially supported by the MRSEC Program of the National Science Foundation under Award No. DMR00-80034 at the University of California in Santa Barbara. Support from the University of Minnesota MRSEC Artificial Tissues Program (DMR99-0934) is also acknowledged with appreciation.

References

Adachi, E., Katsumata, O., Yamashina, S., Prockop, D. J., and Fertala, A. (1999). Collagen II containing a Cys substitution for Arg-alpha1-519. Analysis by atomic force microscopy demonstrates that mutated monomers alter the topography of the surface of collagen II fibrils. *Matrix Biol.* **18**, 189–96.

Argaman, M., Golan, R., Thomson, N. H., and Hansma, H. G. (1997). Phase imaging of moving DNA molecules and DNA molecules replicated in the atomic force microscope. *Nucleic Acids Res.* **25,** 4379–4384.

Aukhil, I., Joshi, P., Yan, Y. Z., and Erickson, H. P. (1993). Cell-binding and heparin-binding domains of the hexabrachion arm identified by tenascin expression proteins. *J. Biol. Chem.* **268,** 2542–2553.

Babcock, K. L., and Prater, C. B. (1995). Phase imaging: beyond topography. Digital Instruments, Santa Barbara, CA.

Baselt, D. R., Revel, J. P., and Baldeschwieler, J. D. (1993). Subfibrillar structure of Type-I collagen observed by atomic force microscopy. *Biophys. J.* **65,** 2644–2655.

Benecky, M. J., Kolvenbach, C. G., Amrani, D. L., and Mosesson, M. W. (1988). Evidence that binding to the carboxyl-terminal heparin-binding domain (Hep II) dominates the interaction between plasma fibronectin and heparin. *Biochemistry* **27,** 7565–7571.

Bensimon, A., Simon, A., Chiffaudel, A., Croquette, V., Heslot, F., and Bensimon, D. (1994). Alignment and sensitive detection of DNA by a moving interface. *Science* **265,** 2096–2098.

Bentley, K. L., Klebe, R. J., Hurst, R. E., and Horowitz, P. M. (1985). Heparin binding is necessary, but not sufficient, for fibronectin aggregation. A fluorescence polarization study. *J. Biol. Chem.* **260,** 7250–7256.

Berndt, P., Fields, G. B., and Tirrell, M. (1995). Synthetic lipidation of peptides and amino acids—monolayer structure and properties. *J. Am. Chem. Soc.* **117,** 9515–9522.

Best, R. B., Li, B., Steward, A., Daggett, V., and Clarke, J. (2001). Can non-mechanical proteins withstand force? Stretching barnase by atomic force microscopy and molecular dynamics simulation. *Biophys. J.* **81**(4), 2344–2356.

Bezanilla, M., Bustamante, C., and Hansma, H. G. (1994). Improved visualization of DNA in aqueous buffer with the atomic force microscope. *Scanning Microsc.* **7,** 1145–1148.

Binnig, G. (1992). Force microscopy. *Ultramicroscopy* **42–44,** 7–15.

Binnig, G., Gerber, C., Stoll, E., Albrecht, R. T., and Quate, C. F. (1987). Atomic resolution with atomic force microscope. *Europhys. Lett.* **3,** 1281–1286.

Bork, P., Downing, A. K., Kieffer, B., and Campbell, I. D. (1996). Structure and distribution of modules in extracellular proteins. *Quart. Rev. Biophys.* **29,** 119–167.

Borza, D. B., Netzer, K. O., Leinonen, A., Todd, P., Cervera, J., Saus, J., and Hudson, B. G. (2000). The goodpasture autoantigen. Identification of multiple cryptic epitopes on the NC1 domain of the alpha3(IV) collagen chain. *J. Biol. Chem.* **275,** 6030–6037.

Brodsky, B., and Eikenberry, E. F. (1982). *Methods Enzymol.* **82 Pt A,** 127–174.

Brodsky, B. T. S., and Eikenberry, E. F. (1988). "Collagen." CRC Press, Boca Raton, FL.

Bultmann, H., Santas, A. J., and Peters, D. M. P. (1998). Fibronectin fibrillogenesis involves the heparin II binding domain of fibronectin. *J. Biol. Chem.* **273,** 2601–2609.

Bustamante, C., Marko, J. F., Siggia, E. D., and Smith, S. (1994). Entropic elasticity of lambda-phage DNA. *Science* **265,** 1599–1600.

Bustamante, C., Vesenka, J., Tang, C. L., Rees, W., Guthold, M., and Keller, R. (1992). Circular DNA molecules imaged in air by scanning force microscopy. *Biochemistry* **31,** 22–26.

Campbell, I. D., and Spitzfaden, C. (1994). Building proteins with fibronectin type-III modules. *Structure* **2,** 333–337.

Carl, P., Kwok, C. H., Manderson, G., Speicher, D. W., and Discher, D. E. (2001). Forced unfolding modulated by disulfide bonds in the Ig domains of a cell adhesion molecule. *Proc. Natl. Acad. Sci. USA* **98,** 1565–1570.

Charonis, A. S., Tsilibary, E. C., Saku, T., and Furthmayr, H. (1986). Inhibition of laminin self-assembly and interaction with type IV collagen by antibodies to the terminal domain of the long arm. *J. Cell Biol.* **103,** 1689–1697.

Chen, C. H., Clegg, D. O., and Hansma, H. G. (1998). Structures and dynamic motion of laminin-1 as observed by atomic force microscopy. *Biochemistry* **37,** 8262–8267.

Chen, C. H., and Hansma, H. G. (2000). Basement membrane macromolecules: Insights from atomic force microscopy. *J. Struct. Biol.* **131,** 44–55.

Chernoff, E. A. G., and Chernoff, D. A. (1992). Atomic force microscope images of collagen fibers. *J. Vac. Sci. Technol. A* **10,** 596–599.

Cheung, C. L., Hafner, J. H., and Lieber, C. M. (2000). Carbon nanotube atomic force microscopy tips: Direct growth by chemical vapor deposition and application to high-resolution imaging. *Proc. Natl. Acad. Sci. USA* **97,** 3809–3813.

Chiquetehrismann, R. (1995). Tenascins, a growing family of extracellular matrix proteins. *Experientia* **51,** 853–862.

Chothia, C., and Jones, E. Y. (1997). The molecular structure of cell adhesion molecules. *Ann. Rev. Biochem.* **66,** 823–862.

Cleveland, J. P., Anczykowski, B., Schmid, A. E., and Elings, V. B. (1998). Energy dissipation with a tapping-mode atomic force microscope. *Appl. Phys. Lett.* **72,** 2613–2615.

Cleveland, J. P., Manne, S., Bocek, D., and Hansma, P. K. (1993). A nondestructive method for determining the spring constant of cantilevers for scanning force microscopy. *Rev. Sci. Instr.* **64,** 403–405.

Cowman, M. K., Min, L., and Balazs, E. A. (1998). Tapping mode atomic force microscopy of hyaluronan: extended and intramolecularly interacting chains. *Biophys. J.* **75,** 2030–2037.

Czajkowsky, D. M., and Shao, Z. (1998). Submolecular resolution of single macromolecules with atomic force microscopy. *FEBS Lett.* **430,** 51–54.

David, M., Borza, D. B., Leinonen, A., Belmont, J. M., and Hudson, B. G. (2001). Hydrophobic amino acid residues are critical for the immunodominant epitope of the Goodpasture autoantigen. A molecular basis for the cryptic nature of the epitope. *J. Biol. Chem.* **276,** 6370–6377.

Drake, B., Prater, C. B., Weisenhorn, A. L., Gould, S. A., Albrecht, T. R., Quate, C. F., Cannell, D. S., Hansma, H. G., and Hansma, P. K. (1989). Imaging crystals, polymers, and processes in water with the atomic force microscope. *Science* **243,** 1586–1589.

Ducker, W. A., Senden, T. J., and Pashley, R. M. (1991). Direct measurement of colloidal forces using an atomic force microscope. *Nature* **353**(6341), 239–241.

Engel, A., and Muller, D. J. (2000). Observing single biomolecules at work with the atomic force microscope. *Nat. Struct. Biol.* **7,** 715–718.

Engel, J., Odermatt, E., and Engel, A. (1981). Shapes, domain organizations and flexibility of laminin and fibronectin, two multifunctional proteins of the extracellular matrix. *J. Mol. Biol.* **150,** 97–120.

Erickson, H. P. (1997). Protein biophysics—Stretching single protein molecules: Titin is a weird spring. *Science* **276,** 1090–1092.

Erickson, H. P., Carrell, N., and McDonagh, J. (1981). Fibronectin molecule visualized in electron microscopy: a long, thin, flexible strand. *J. Cell Biol.* **91,** 673–678.

Fisher, T. E., Carrion-Vazquez, M., Oberhauser, A. F., Li, H., Marszalek, P. E., and Fernandez, J. M. (2000). Single molecular force spectroscopy of modular proteins in the nervous system. *Neuron* **27,** 435–446.

Fisher, T. E., Oberhauser, A. F., Carrion-Vazquez, M., Marszalek, P. E., and Fernandez, J. M. (1999). The study of protein mechanics with the atomic force microscope. *Trends Biochem. Sci.* **24,** 379–384.

Florin, E. L., Moy, V. T., and Gaub, H. E. (1994). Adhesion forces between individual ligand–receptor pairs. *Science* **264,** 415–417.

Fujiwara, S., Wiedemannn, H., Timpl, R., Lustig, A., and Engel, J. (1984). Structure and interactions of heparan sulfate proteoglycans from a mouse tumor basement membrane. *Eur. J. Biochem.* **143,** 145–157.

Golan, R., Pietrasanta, L. I., Hsieh, W., and Hansma, H. G. (1999). DNA toroids: stages in condensation. *Biochemistry* **38,** 14069–14076.

Hansma, H. G. (1996). Atomic force microscopy of biomolecules. *J. Vac. Sci. Tech. B* **14,** 1390–1394.

Hansma, H. G. (1999). Varieties of imaging with scanning probe microscopes. *Proc. Natl. Acad. Sci. USA* **96,** 14678–14680.

Hansma, H. G. (2001). Surface biology of DNA by atomic force microscopy. *Ann. Rev. Phys. Chem.* **52,** 71–92.

Hansma, H. G., and Laney, D. E. (1996). DNA binding to mica correlates with cationic radius: assay by atomic force microscopy. *Biophys. J.* **70,** 1933–1939.

Hansma, H. G., and Pietrasanta, L. (1998). Atomic force microscopy and other scanning probe microscopies. *Curr. Opin. Chem. Biol.* **2,** 579–584.

Hansma, H. G., Golan, R., Hsieh, W., Daubendiek, S. L., and Kool, E. T. (1999). Polymerase activities and RNA structures in the atomic force microscope. *J. Struct. Biol.* **127,** 240–247.

Hansma, H. G., Kim, K. J., Laney, D. E., Garcia, R. A., Argaman, M., Allen, M. J., and Parsons, S. M. (1997). Properties of biomolecules measured from atomic force microscope images: a review. *J. Struct. Biol.* **119,** 99–108.

Hansma, H. G., Revenko, I., Kim, K., and Laney, D. E. (1996). Atomic force microscopy of long and short double-stranded, single-stranded and triple-stranded nucleic acids. *Nucleic Acids Res.* **24,** 713–720.

Hassell, J. R., Leyshon, W. C., Ledbeter, S. R., Tyree, B., Suzuki, S., Kato, M., Kimata, K., and Kleinman, H. K. (1985). Isolation of two forms of basement membrane proteoglycans. *J. Biol. Chem.* **260,** 8098–8105.

Hocking, D. C., Sottile, J., and McKeownLongo, P. J. (1998). Activation of distinct alpha(5)beta(1)-mediated signaling pathways by fibronectin's cell adhesion and matrix assembly domains. *J. Cell Biol.* **141,** 241–253.

Hoh, J. H., Revel, J.-P., and Hansma, P. K. (1991). Tip–sample interactions in atomic force microscopy: I. Modulating adhesion between silicon nitride and glass. *Nanotechnology* **2,** 119–122.

Hynes, R. O. (1990). "Fibronectins." Springer-Verlag, New York.

Izzard, C. S., Radinsky, R., and Culp, L. A. (1986). Substratum contacts and cytoskeletal reorganization of BALB/c 3T3 cells on a cell-binding fragment and heparin-binding fragments of plasma fibronectin. *Exp. Cell Res.* **165,** 320–336.

Jones, J. C., Dehart, G. W., Gonzales, M., and Goldfinger, L. E. (2000). Laminins: an overview. *Microsc. Res. Tech.* **51,** 211–213.

Keller, D., and Chou, C. C. (1992). Imaging steep, high structures by scanning force microscopy with electron beam deposited tips. *Surface Sci.* **268,** 333–339.

Kielty, C. M., Kwan, A. P., Holmes, D. F., Schor, S. L., and Grant, M. E. (1985). Type X collagen, a product of hypertrophic chondrocytes. *Biochem. J.* **227,** 545–554.

Kindt, J. H., Sitko, J. C., Pietrasanta, L. I., Oroudjev, E., Becker, N., Viani, M. B., and Hansma, H. G. (2002). Methods for biological probe microscopy in aqueous fluids. *In* "Atomic Force Microscopy in Cell Biology" (B. P. Jena and H. Hoerber, eds.). Academic Press, New York.

Kokkoli, E., and Zukoski, C. F. (1998). Interactions between hydrophobic self-assembled monolayers. Effect of salt and the chemical potential of water on adhesion. *Langmuir* **14,** 1189–1195.

Kokkoli, E., and Zukoski, C. F. (2001). Surface pattern recognition by a colloidal particle. *Langmuir* **17,** 369–376.

Kokkoli, E., Ochsenhirt, S. E., and Tirrell, M. (2001). Direct adhesion measurements of integrin–ligand pairs by atomic force microscopy, submitted.

Laney, D. E., Garcia, R. A., Parsons, S. M., and Hansma, H. G. (1997). Changes in the elastic properties of cholinergic synaptic vesicles as measured by atomic force microscopy. *Biophys. J.* **72,** 806–813.

Laurie, G. W., Bing, J. T., Kleinman, H. K., Hassell, J. R., Aumailley, M., Martin, G. R., and Feldmann, R. J. (1986). Localization of binding sites for laminin, heparan sulfate proteoglycan and fibronectin on basement membrane (type IV) collagen. *J. Mol. Biol.* **189,** 205–216.

Lin, H., Clegg, D. O., and Lal, R. (1999). Imaging real-time proteolysis of single collagen I molecules with an atomic force microscope. *Biochemistry* **38,** 9956–9963.

Lin, H., Lal, R., and Clegg, D. O. (2000). Imaging and mapping heparin-binding sites on single fibronectin molecules with atomic force microscopy. *Biochemistry* **39,** 3192–3196.

Magonov, S. N., Elings, V., and Whangbo, M. H. (1997). Phase imaging and stiffness in tapping-mode atomic force microscopy. *Surf. Sci.* **375,** L385–L391.

McCawley, L. J., and Matrisian, L. M. (2001). Tumor progression: Defining the soil round the tumor seed. *Curr. Biol.* **11,** R25–R27.

Mould, A. P., Holmes, D. F., Kadler, K. E., and Chapman, J. A. (1985). Mica sandwich technique for preparing macromolecules for rotary shadowing. *J. Ultrastruct. Res.* **91,** 66–76.

Oberhauser, A. F., Marszalek, P. E., Erickson, H. P., and Fernandez, J. M. (1998). The molecular elasticity of the extracellular matrix protein tenascin. *Nature* **393,** 181–185.

Odetti, P., Aragno, I., Rolandi, R., Garibaldi, S., Valentini, S., Cosso, L., Traverso, N., Cottalasso, D., Pronzato, M. A., and Marinari, U. M. (2000). Scanning force microscopy reveals structural alterations in diabetic rat collagen fibrils: role of protein glycation. *Diabetes Metab. Res. Rev.* **16,** 74–81.

Paige, M. F., and Goh, M. C. (2001). Ultrastructure and assembly of segmental long spacing collagen studied by atomic force microscopy. *Micron* **32,** 355–361.

Paige, M. F., Rainey, J. K., and Goh, M. C. (1998). Fibrous long spacing collagen ultrastructure elucidated by atomic force microscopy. *Biophys. J.* **74,** 3211–3216.

Paige, M. F., Rainey, J. K., and Goh, M. C. (2001). A study of fibrous long spacing collagen ultrastructure and assembly by atomic force microscopy. *Micron* **32,** 341–353.

Parsons, S. L., Watson, S. A., Brown, P. D., Collins, H. M., and Steele, R. J. C. (1997). Matrix metalloproteinases. *Br. J. Surg.* **84,** 160–166.

Pashley, R. M., and Israelachvili, J. N. (1984). DLVO and hydration forces between mica surfaces in Mg^{2+}, Ca^{2+}, Sr^{2+}, and Ba^{2+} chloride solutions. *J. Colloid Interface Sci.* **97,** 446–455.

Paulsson, M., Yurchenco, P. D., Ruben, G. C., Engel, J., and Timpl, R. (1987). Structure of low density heparan sulfate proteoglycan isolated from a mouse tumor basement membrane. *J. Molec. Biol.* **197**(2), 297–313.

Pietrasanta, L. I., Thrower, D., Hsieh, W., Rao, S., Stemmann, O., Lechner, J., Carbon, J., and Hansma, H. G. (1999). Probing the *Sacchromyces cervisiae* CBF3-*CEN* DNA kinetochore complex using atomic force microscopy. *Proc. Natl. Acad. Sci. USA* **96,** 3757–3762.

Pingoud, A., and Jeltsch, A. (1997). Recognition and cleavage of DNA by type-II restriction endonucleases. *Eur. J. Biochem.* **246,** 1–22.

Radmacher, M., Cleveland, J. P., Fritz, M., Hansma, H. G., and Hansma, P. K. (1994). Mapping interaction forces with the atomic force microscope. *Biophys. J.* **66,** 2159–2165.

Revenko, I., Sommer, F., Minh, D. T., Garrone, R., and Franc, J. M. (1994). Atomic force microscopy study of the collagen fibre structure. *Biol. Cell* **80,** 67–69.

Rief, M., Gautel, M., Oesterhelt, F., Fernandez, J. M., and Gaub, H. E. (1997). Reversible unfolding of individual titin immunoglobulin domains by AFM. *Science* **276,** 1109–1112.

Rief, M., Pascual, J., Saraste, M., and Gaub, H. E. (1999). Single molecule force spectroscopy of spectrin repeats: low unfolding forces in helix bundles. *J. Mol. Biol.* **286,** 553–61.

Rohrbach, D. H., and Murrah, V. A. (1993). Molecular aspects of basement membrane pathology. *In* "Molecular and Cellular Aspects of Basement Membranes" (D. H. Rohrbach and R. Timpl, eds.), pp. 19–47. Academic Press, San Diego.

Schneider, S. W., Larmer, J., Henderson, R. M., and Oberleithner, H. (1998). Molecular weights of individual proteins correlate with molecular volumes measured by atomic force microscopy. *Pflugers Arch.* **435,** 362–367.

Sethi, C. S., Bailey, T. A., Luthert, P. J., and Chong, N. H. V. (2000). Matrix metalloproteinase biology applied to vitreoretinal disorders. *Br. J. Ophthalmol.* **84,** 654–666.

Shattuck, M. B., Gustafsson, M. G. L., Fisher, K. A., Yanagimoto, K. C., Veis, A., Bhatnagar, R. S., and Clarke, J. (1994). Monomeric collagen imaged by cryogenic force microscopy. *J. Microscopy—Oxford* **174,** RP1–RP2.

Smith, S. B., Cui, Y., and Bustamante, C. (1996). Overstretching B-DNA: the elastic response of individual double-stranded DNA molecules. *Science* **271,** 795–798.

Smith, S. B., Finzi, L., and Bustamante, C. (1992). Direct mechanical measurements of the elasticity of single DNA molecules by using magnetic beads. *Science* **258,** 1122–1126.

Sun, H. B., Smith, G. N., Jr., Hasty, K. A., and Yokota, H. (2000). Atomic force microscopy-based detection of binding and cleavage site of matrix metalloproteinase on individual type II collagen helices. *Anal. Biochem.* **283,** 153–158.

Taatjes, D. J., Quinn, A. S., and Bovill, E. G. (1999). Imaging of collagen type III in fluid by atomic force microscopy. *Microsc. Res. Tech.* **44,** 347–52.

Thompson, J. B., Kindt, J. H., Drake, B., Hansma, H. G., Morse, D. E., and Hansma, P. K. (2001). Bone indentation recovery time correlates with bond reforming time. *Nature,* **114,** 773–776.

Tryggvason, K., Zhou, J., and Hostikka, S. L. (1993). Alport syndrome and other inherited basement membrane disorders. *In* "Molecular and Cellular Aspects of Basement Membranes" (D. H. Rohrbach and R. Timpl, eds.), pp. 19–47. Academic Press, San Diego.

Vanderrest, M., and Garrone, R. (1991). Collagen family of proteins. *FASEB J.* **5,** 2814–2823.

Vesenka, J., Guthold, M., Tang, C. L., Keller, D., Delaine, E., and Bustamante, C. (1992). A substrate preparation for reliable imaging of DNA molecules with the scanning force microscope. *Ultramicroscopy* **42–44,** 1243–1249.

Vesenka, J., Manne, S., Giberson, R., Marsh, T., and Henderson, E. (1993). Collidal gold particles as an incompressible atomic force microscope imaging standard for assessing the compressibility of biomolecules. *Biophys. J.* **65,** 992–997.

Vu, T. H., and Werb, Z. (2000). Matrix metalloproteinases: effectors of development and normal physiology. *Genes Dev.* **14,** 2123–2133.

Washizu, M., and Kurosawa, O. (1990). Electrostatic manipulation of DNA in microfabricated structures. *IEEE Trans. Indust. Appl.* **26,** 1165–1172.

Wolberg, A. S., Stafford, D. W., and Erie, D. A. (1997). Human factor IX binds to specific sites on the collagenous domain of collagen IV. *J. Biol. Chem.* **272,** 16717–16720.

Woods, A., Couchman, J. R., Johansson, S., and Höök, M. (1986). Adhesion and cytoskeletal organisation of fibroblasts in response to fibronectin fragments. *EMBO J.* **5,** 665–670.

Yamada, Y., and Kuehn, K. (1993). Genes and regulation of basement membrane collagen and laminin synthesis. *In* "Molecular and Cellular Aspects of Basement Membranes" (D. H. Rohrbach and R. Timpl, eds.), pp. 189–210. Academic Press, San Diego.

Yu, Y.-C., Pakalns, T., Dori, Y., McCarthy, J. B., Tirrell, M., and Fields, G. B. (1997). Construction of biologically active protein molecular architecture using self-assembling peptide-amphiphiles. *Methods Enzymol.* **289,** 571–587.

Yurchenco, P. D., and O'Rear, J. (1993). Supramolecular organization of basement membranes. *In* "Molecular and Cellular Aspects of Basement Membranes" (D. H. Rohrbach and R. Timpl, eds.), pp. 19–47. Academic Press, San Diego.

Zimmermann, R. M., and Cox, E. C. (1994). DNA stretching on functionalized gold surfaces. *Nucleic Acids Res.* **22,** 492–497.

CHAPTER 8

Analysis of Matrix Degradation

H. G. Munshi* and M. Sharon Stack†

* Department of Medicine
Division of Hematology and Oncology

† Department of Cell and Molecular Biology
Northwestern University Medical School
Chicago, Illinois 60611

I. Introduction

The majority of cells in multicellular organisms are in contact with an extracellular matrix (ECM) composed of a complex variety of protein and polysaccharide molecules. By forming specialized structures such as cartilage, bone, tendon, and basal lamina, ECM macromolecules stabilize the physical structure of tissues. The physiocochemical properties of the major classes of ECM macromolecules, collagen, proteoglycan (discussed elsewhere in this volume), and glycoproteins, account for the unique mechanical properties of the connective tissue. Although the relative amounts and organization of matrix macromolecules vary among tissues, the collagens are the most abundant, comprising approximately one-third of the total body protein in humans. Collagens, in turn, interact with a number of matrix glycoproteins including fibronectin, laminins, and vitronectin, forming a three-dimensional multicomponent matrix (Yurchenco, 1990; Kleinman *et al.*, 1981). Degradation of the ECM and its component proteins by proteolytic enzymes is necessary for a number of physiologic and pathologic processes including embryonic development, tissue resorption and remodeling, angiogenesis, wound healing, and tumor invasion and metastasis (Lochter *et al.*, 1998; Werb and Chin, 1998; Ellerbroek and Stack, 1999).

A number of proteinases from distinct mechanistic classes contribute to ECM processing (Vu and Werb, 2000; Birkedal-Hansen, 1995). However, predominant among the matrix-degrading proteinases are the matrix metalloproteinases (MMPs) or matrixins (Nagase and Woessner, 1999; Ellerbroek and Stack, 1999). This family of secreted and membrane-bound zinc-dependent metalloendopeptidases is widely expressed and can cleave a variety of ECM components including, but not limited to, native fibrillar collagen, denatured collagen (gelatin), basement membrane collagen, and noncollagenous ECM glycoproteins such as fibronectin, laminin-1, and laminin-5. This chapter will focus predominantly on methods by which to assess proteolytic modification of ECM proteins using isolated proteinases or cell-associated enzymes.

II. Experimental Approaches and Protocols

A. Preparation and Cleavage of Multicomponent ECM

Cells in culture deposit ECM components onto the culture substratum. The composition of the subcellular matrix is dictated predominantly by cell type and may be influenced by cell culture conditions. A variety of cells have been utilized to deposit multicomponent matrices enhanced in specific proteins for further analysis of matrix composition and structure, relative proteinase susceptibility, and analysis of cell behavior. For example, the matrix deposited by endothelial cells resembles subendothelial basement membrane and contains type IV collagen, laminin-1, fibronectin, and various proteoglycans (Gospodarowicz *et al.,* 1980; Jaffe and Mosher, 1978; Kramer *et al.,* 1985). In contrast, smooth muscle cell ECM is composed primarily of types I and III collagen and elastin, with minor components including fibronectin, laminin-1, and type IV collagen (Werb *et al.,* 1980). Epithelial cells such as the rat bladder line 804G and the human mammary epithelial cell line MCF10A produce an ECM enriched in the glycoprotein laminin-5 (Langhofer *et al.,* 1993; Goldfinger *et al.,* 1998). The distinct protein composition of an ECM renders the protein components differentially susceptible to degradation by individual proteinases.

1. Matrix Preparation

The ECM deposited by postconfluent cells can be recovered for subsequent experimental analysis by removal of the cells, leaving behind the matrix components (Gospodarowicz *et al.,* 1980). Cells are grown in tissue culture dishes or microtiter plates for 5–10 days postconfluence to allow for matrix deposition followed by aspiration of the medium and removal of cells. For stabilization of collagen-rich matrices, cultures may be supplemented with ascorbic acid (50 μg/ml) (Jones *et al.,* 1979). To avoid processing of matrix proteins by trypsin, cells are exposed to either detergent or base to denude the ECM. Empirical results indicate that detergent treatment is preferred for cells that have been confluent for approximately 7 days, whereas base treatment is more efficient for cells that deposit a very thick ECM or have been confluent for longer time periods. The cells are solubilized with 0.5% Triton X-100 in PBS for 2–10 min followed by 3–5 PBS washes. Alternatively, cells can be removed by treatment for approximately 2–5 min with 20 mM NH$_4$OH followed by 3 rapid washes in sterile distilled water and 3 times in sterile PBS (Gospodarowicz, 1984). To determine the optimum duration of treatment, a control plate may be examined for cell lysis under phase-contrast microscopy. The plates can be stored for 7–10 days at 4°C by adding PBS supplemented with antimicrobial and antifungal agents. Radioactivity may be incorporated into the deposited matrices by culturing cells in medium supplemented with ^{3}H-proline (1 μCi/ml) to label collagen, or other amino acids to label matrix glycoproteins (Jones *et al.,* 1979).

2. Analysis of Matrix Degradation

Analysis of matrix proteolysis can be performed using purified enzymes (available from American Diagnostica, Greenwich, CT; Chemicon, Temecula, CA), proteinase-containing conditioned medium, or intact cells. In general, enzyme : substrate ratios should be maintained at approximately 1 : 20–1 : 50, although this is difficult to ascertain in nonpurified systems. Furthermore, it may be speculated that supraphysiologic enzyme : substrate ratios may prevail at proteinase-rich sites of cell–matrix contact. Serum contains numerous proteinase inhibitors, such that serum-free culture conditions should be utilized unless proteinase inhibition is desired.

a. Quantitative Determination of Matrix Degradation

For quantitative determination of matrix degradation, use of a radioactively labeled matrix is preferred. The purified proteinase, conditioned medium, or proteinase-expressing cell is added to labeled matrix for an empirically determined time period, followed by collection of the conditioned medium and scintillation counting. As a positive control, matrix samples can be incubated with Dispase (48 h at 37°C, Gibco BRL, Baltimore, MD) to fully solubilize the matrix and provide a numerical value of 100% degradation (Young *et al.,* 1996). As an additional control, an appropriate proteinase inhibitor should be included in the experiment to demonstrate that inhibition of proteinase activity blocks matrix protein solubilization. If the degrading proteinase is known, specific proteinacious or small molecule inhibitors may be incorporated into the assay (Table I). If the identity of the matrix-degrading proteinase is unknown, an initial approach is to include mechanistic class-specific proteinase inhibitors in the assay to first identify the type of proteinase(s) involved in matrix processing (Table I). An extensive listing of

Table I
Commonly Used Low Molecular Weight Proteinase Inhibitors[a]

Serine proteinases	3,4-Dichloroisocoumarin at 5–100 μM
Cysteine proteinases	E-64 (L-*trans*-epoxysuccinylleucylamide-(4-guanidino)butane, *N*-[*N*-(L-3-*trans*-carboxyirane-2-carbonyl)-L-leucyl]agmatine) at 1–10 μM
Aspartic proteinases	Pepstatin (isovaleryl-Val-Val-AHMHA-Ala-AHMHA where AHMHA = (3S,4S)-4-amino-3-hydroxy-6-methylheptanoic acid) at 1 μM
Metalloproteinases	1,10-Phenanthroline at 1–10 mM

[a]Adapted from *Proteolytic Enzymes, a Practical Approach* by R. J. Beynon and J. S. Bond, IRL Press, Oxford, 1989. Compounds available from Sigma Chemical Co., St. Louis, MO.

class-specific proteinase inhibitors and their appropriate working concentrations is contained in Beynon and Bond (1989). It should be noted that more than one proteinase may participate in matrix solubilization. For example, native triple helical type I collagen is resistant to cleavage by gelatinolytic proteinases. However, following initial cleavage by an interstitial collagenase such as MMP-1, localized denaturation of the helix renders type I collagen susceptible to gelatinases (Cawston *et al.,* 2001; Ellerbroek *et al.,* 2001, Fig. 2A). Proteinases with gelatinolytic activity are not limited to the MMP family and may include broad spectrum serine proteinases such as plasmin, as well as other enzymes. Thus, release of radioactivity from a type I collagen-containing matrix may reflect the sequential activity of several distinct proteinases.

b. Qualitative Determination of Matrix Breakdown

In addition to the quantitative determination of radioactive matrix cleavage, nonlabeled matrices may be analyzed qualitatively for evidence of matrix breakdown by solubilization of the matrix followed by electrophoretic analysis and visualization of degradation products. Matrices may be solubilized by incubating with Laemmli sample buffer (62.5 mM Tris-Hcl, pH 6.5, 10% glycerol, 5% β-mercaptoethanol) containing 2.5% SDS, followed by scraping and boiling (Laemmli, 1970). Alternatively, treated matrices may be solubilized in sample buffer consisting of 8 M urea, 1% SDS, 15% β-mercaptoethanol in 10 mM Tris-HCl, pH 6.8, followed by scraping and boiling (Langhofer *et al.,* 1993). The sample is then passed through a 26-gauge syringe to make it less viscous and easier to load onto the gel. Electrophoretic separation of matrix breakdown products on SDS–polyacrylamide gels (Laemmli, 1970) may be followed by protein staining of the gel to visualize overall degradation patterns. To evaluate cleavage of specific protein components, Western blotting with protein-specific antibodies is advantageous. Care should be taken in data interpretation, however, as proteolytic cleavage may solubilize the epitope recognized by a monoclonal antibody. Thus, polyclonal antibodies or a mixture of monoclonal antibodies directed against multiple distinct epitopes should be employed. However, Western blotting using a monoclonal antibody generated against a defined epitope can also be used to characterize loss of the epitope by proteolytic processing. For example, the laminin-5-rich subepithelial matrix produced by 804G cells contains a 190-kDa α3 subunit. Laminin-5 is susceptible to limited proteolytic cleavage by plasmin, releasing the COOH-terminal G5 subdomain of the α3 subunit. Using a monoclonal antibody directed against the G5 subdomain, loss of antibody reactivity in the solubilized matrix can be observed following plasmin incubation (Goldfinger *et al.,* 1998).

3. Matrix Invasion Assays

Cellular penetration of a three-dimensional ECM barrier is thought to require both proteolytic activity and cell motility. The ability of cells to invade an artificial matrix protein barrier, such as the matrix protein extract Matrigel, has been shown to correlate with cellular invasive activity in many model systems (Kleinman *et al.,* 1986). To evaluate *in vitro* invasion of an ECM extract, porous polycarbonate filters (6–12 μm pore size, Becton-Dickinson, Bedford, MA) are coated on the upper surface with Matrigel (10–15 μg/filter,

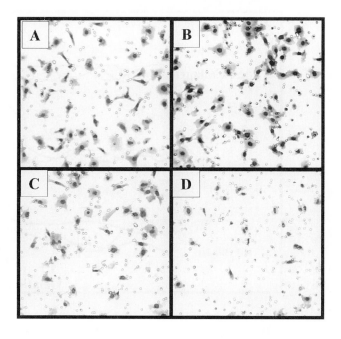

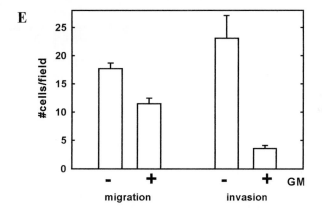

Fig. 1 Proteinase dependence of migration and Matrigel invasion. Oral squamous cell carcinoma cells (SCC25, 10^5) were added to porous polycarbonate filters (8 μm pore) that were either coated on the underside with type I collagen (A, B, migration) or coated on the upper surface with Matrigel (10 μg) (C, D, invasion) in the presence (B, D) or absence (A, C) of the MMP inhibitor GM6001 (25 μM) as indicated. After incubation for 6 h (migration) or 36 h (invasion), non-migrating cells were removed from the upper chamber, filters were fixed and stained, and invading cells enumerated using an ocular micrometer and counting a minimum of 10 high powered fields (E).

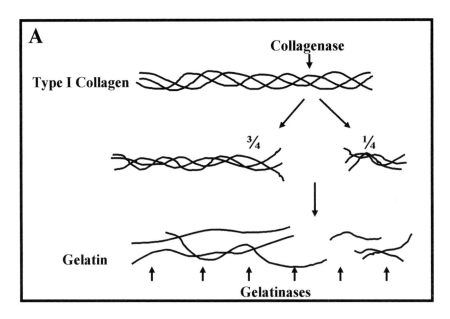

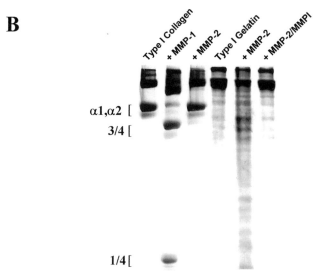

Fig. 2 Cleavage of triple helical type I collagen and gelatin. (A) Diagram of collagen cleavage. Following processing of collagen at the Gly^{775}-Leu/Ile^{776} bond by an interstitial collagenase such as MMP-1, localized denaturation of the helix renders collagen susceptible to gelatinolytic proteinases. (B) Type I collagen (10 μg, first three lanes) was incubated with MMP-1 or MMP-2 (50 n*M*, as indicated) at 25°C for 18 h followed by electrophoretic separation of the degradation products on 8–15% gradient SDS–polyacrylamide gels and staining with Coomassie blue. The migration position of the uncleaved $\alpha1(I)$ and $\alpha2(I)$ chains as well as the 3/4 and 1/4 degradation products are indicated. Note that interstitial collagenase (MMP-1) catalyzed collagen cleavage (lane designated "+MMP-1") whereas the gelatinase MMP-2 is ineffective against native collagen

Becton-Dickinson) and placed in a transwell invasion chamber (Becton-Dickinson). Cells (approximately 10^5) are placed in the upper chamber in a small volume (200–500 μl) while culture medium (750 μl) is added to the lower well. A chemoattractant may also be added to the lower chamber. Following incubation at 37°C for 8–72 h, the filter is removed from the transwell chamber. Noninvading cells adherent to the upper surface of the filter are removed with a cotton swab, and the filters are then fixed and stained with a Wright–Giemsa-based stain such as Diff-Quik (Fisher). Invading cells adherent to the lower surface of the filter can then be enumerated using an inverted phase microscope with a 20× objective equipped with an ocular micrometer. The proteinase dependence of invasion may be assessed by inclusion of proteinase inhibitors in the upper well. For example, invasion of Matrigel by the oral squamous cell carcinoma cell line SCC25 requires MMP activity and is significantly inhibited by the broad spectrum hydroxamic acid-based peptide inhibitor GM6001 (Fig. 1). In contrast, migration of SCC25 cells through an uncoated Transwell filter is MMP-independent.

B. Analysis of Individual Matrix Protein Components

In addition to analysis of multicomponent ECMs, proteolytic cleavage of individual matrix protein components may also be characterized. Protein cleavage may be evaluated in solution or on solid phase in the presence or absence of radioactive labeling.

1. Protein Preparation

Matrix protein components may be passively adsorbed to tissue culture plastic by solubilizing at a concentration of 5–10 μg/ml in 0.1 M sodium carbonate buffer, pH 9.6, and incubating overnight at 4°C or for 2 h at 37°C. The amount of protein deposited varies depending on the protein and the type of plate and can be determined empirically using radioactively labeled protein of a defined specific activity. The method of radiolabeling is dependent on available functional groups in the protein, but may include ^{125}I labeling with Bolton–Hunter reagent (Kanemoto et al., 1990) or [14]C- or ^{3}H-acetylation (Harris and Krane, 1972). Alternatives to radiolabeling include the use of heavily fluoresceinated proteins, such as commercially available labeled collagen and gelatin (DQ-gelatin, Molecular Probes, Eugene, OR) that release fluorescent peptides when cleaved. Both labeled and nonlabeled proteins may also be analyzed in solution phase using the buffer of choice. An exception is analysis of cleavage of native type I collagen (Fig. 2). In this case, lyophilized collagen is dissolved to a concentration of 1–2 mg/ml in 20 mM

(lane designated "+MMP-2"). In the last three lanes, type I gelatin (10 ug) (prepared by thermal denaturation of type I collagen at 60°C for 20 min) was incubated with MMP-2 (50 nM, as indicated) at 25°C for 18 h followed by electrophoretic separation of the multiple lower molecular weight degradation products as described above. Note that incubation with the broad-spectrum MMP inhibitor GM6001 (10 μM) effectively blocks the gelatinase activity of MMP-2 (compare the absence of multiple low molecular weight degradation products in lane designated "+MMP-2/MMPI" to the products present in lane designated "+MMP-2").

acetic acid overnight at 4°C and diluted 1 : 3 into Tris–glucose buffer (50 mM Tris, 0.2 M glucose, 0.2 M NaCl, pH 7.4) to prevent collagen fibril formation (Terato *et al.,* 1976).

2. Analysis of Protein Degradation

Similar to the procedures described earlier for intact matrices, radioactive and non-radioactive detection methods can be utilized to detect matrix protein cleavage by purified enzymes, proteinase-containing conditioned medium, or intact cells. For immobilized, radiolabeled protein, the procedure is identical to that in Section II.A.2, wherein solubilized radioactivity is used as a quantitative measure of protein cleavage. For solution-phase radiolabeled proteins, uncleaved substrate is removed by precipitation with 15% trichloroacetic acid, and soluble radioactivity determined as described previously. When unlabeled protein is utilized, electrophoretic separation of protein breakdown products on SDS–polyacrylamide gels (Laemmli, 1970) may be followed by protein staining (Fig. 2) or Western blotting, as described in Section II.A.2. In a single component protein system, a semiquantitative analysis of the extent of protein cleavage can be estimated using densitometry (Darlak *et al.,* 1990).

3. Substrate Gel Electrophoresis

Substrate gel electrophoresis, or zymography, is commonly utilized as a rapid screen for matrix-degrading proteinases and can provide information on substrate cleavage as well as approximate size of the proteinase. In this technique a substrate, commonly 0.1% gelatin or casein, is copolymerized into an SDS–polyacrylamide gel (Heussen and Dowdle, 1980). Prior to electrophoresis, proteinase-containing samples (pure enzyme, conditioned medium, etc.) are incubated with nonreducing Laemmli sample dilution

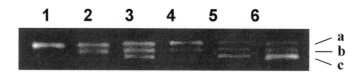

Fig. 3 Calcium-induced changes in proMMP-2 activation as evaluated by substrate gel electrophoresis. SCC25 cells (10^5) were cultured on a thin layer of type I collagen (lanes 1–3, prepared as described in Section II.B.1) or on a three-dimensional collagen gel (lanes 4–6, prepared by diluting acid solubilized type I collagen to a concentration of 1 mg/ml in cold DMEM, neutralizing with NaOH, adding to 12-well plates in a 700 μl volume, and allowing to gel at 37°C for 30 min prior to plating of cells) in serum-free medium with varying [Ca^{2+}] for 24 h. Conditioned media were removed, diluted 1 : 2 with nonreducing Laemmli sample dilution buffer for 30 min at room temperature, and electrophoresed on an 8% SDS–polyacrylamide gel containing copolymerized gelatin. Following electrophoresis, the gel was washed twice in Triton X-100 as described in Section II.B.3, incubated overnight in the buffer indicated, and stained with Coomassie blue. Regions of gelatinolytic activity are shown as clear zones against a dark background. Lanes 1, 4: 0.09 mM [Ca^{2+}]; lanes 2, 5: 0.4 mM [Ca^{2+}]; lanes 3,6: 1.2 mM [Ca^{2+}]. Note that as [Ca^{2+}] is increased and as cells are cultured on a three-dimensional matrix protein substratum, the zymogen form of MMP-2 (a,66 kDa) is converted through an intermediate species (b,64 kDa) to the mature form of the proteinase (c,62 kDa). Although all three forms are visible by zymography, only the mature enzyme is proteolytically competent in solution.

buffer containing 2.5% SDS for 30 min at room temperature without boiling. Following electrophoresis, proteinase activity is renatured by washing the gel twice for 10 min with 2.5% Triton X-100 to remove SDS. The gel can then be incubated for 6–48 h at 37°C in the buffer of choice, for example 0.1 M glycine, 10 mM CaCl$_2$, 1 μM ZnCl$_2$, pH 8.3, for analysis of MMP activity. Following the incubation period, gels are stained with Coomassie blue and regions of proteolytic activity are visualized as clear zones against a blue background (Fig. 3). Class-specific proteinase inhibitors can be added to the sample prior to electrophoresis and incorporated into the incubation buffer to elucidate the mechanistic class of the proteinase. Although not quantitative, this method provides a rapid and sensitive initial evaluation of proteinase profiles. Data obtained using zymography to evaluate MMP activity are frequently misinterpreted, however, as (a) MMP zymogens attain proteolytic activity in the presence of sodium dodecyl sulfate *without* propeptide cleavage and (b) MMP complexes with tissue inhibitors of metalloproteinases (TIMPs) are noncovalent and are thus dissociated during electrophoresis, leading to the potential for overestimation of enzymatic activity. Thus, preliminary results obtained using zymography should be confirmed by other methods.

4. Collagen Invasion Assays

Similar to the Matrigel invasion assay described in Section II.A.3, cellular penetration of a three-dimensional collagen gel can be utilized to evaluate cellular collagenolytic potential (Ellerbroek *et al.,* 2001). To prepare three-dimensional collagen gels for invasion assays, type I collagen is dissolved in 0.5 M acetic acid at a concentration of 2 mg/ml and neutralized with 100 mM Na$_2$CO$_3$ (pH 9.6) to a final concentration of

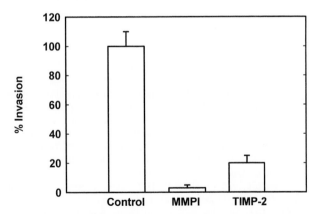

Fig. 4 Invasion of three-dimensional collagen gels. DOV13 ovarian cancer cells (10^5) were added to porous polycarbonate filters (Transwell chambers) coated with type I collagen (20 μg) and allowed to invade for 48 h at 37°C. Following invasion, nonmigrating cells were removed from the upper chamber with a cotton swab, filters were fixed and stained, and invading cells were enumerated using an ocular micrometer and counting a minimum of 10 high-powered fields. Results are expressed as percent of control invasion (designated 100%). Both the broad-spectrum hydroxamic acid peptide inhibitor MMPI (10 μM, INH-3850-PI, Peptides International, Louisville, KY) and TIMP-2 (10 nM) effectively block invasion of collagen gels.

0.4 mg/ml. Collagen gels are then prepared in the upper well of Transwell inserts by adding 50 μl of collagen (20 μg) at room temperature and allowing gels to air dry overnight. Collagen-coated inserts are then washed with culture medium three times to remove buffer salts and used immediately. Preparation of cells, quantitation of collagen invasion, and determination of proteinase dependence of invasion are determined as described in Section II.A.3. For example, invasion of three-dimensional collagen gels by DOV13 ovarian carcinoma cells is dependent on MMP activity and is blocked by both a broad-spectrum MMP inhibitor and by the proteinacious inhibitor TIMP-2 (Fig. 4).

Acknowledgment

This work is supported by grants RO1 CA86984 and RO1 CA 85870 from the National Cancer Institute and PO1 DE12328 from the National Institute of Dental and Craniofacial Research. The authors thank Yueying Liu and Yi Wu for assistance with figure preparation.

References

Birkedal-Hansen, H. (1995). Proteolytic remodeling of extracellular matrix. *Curr. Opin. Cell Biol.* **7,** 728–735.

Beynon, R. J., and Bond, J. S. (1989). "Proteolytic Enzymes, a Practical Approach." IRL Press, Oxford, pp. 241–249.

Cawston, T. E., Koshy, P., and Rowan, A. D. (2001). Assay of matrix metalloproteinases against matrix substrates. *In* "Matrix Metalloproteinase Procotols" (Clark, I. M., ed.). Humana Press, Totowa, NJ, pp. 389–398.

Darlak, K., Miller, R. B., Stack, M. S., Spatola, A. F., and Gray, R. D. (1990). Thiol-based inhibitors of mammalian collagenase. *J. Biol. Chem.* **265,** 5199–5205.

Ellerbroek, S. M., and Stack, M. S. (1999). Membrane associated matrix metalloproteinases in metastasis. *Bioessays* **21,** 940–949.

Ellerbroek, S. M., Wu, Y. I., Overall, C. M., and Stack, M. S. (2001). Functional interplay between type I collagen and cell surface matrix metalloproteinase activity. *J. Biol. Chem.* **276,** 24833–24842.

Goldfinger, L. E., Stack, M. S., and Jones, J. C. R. (1998). Processing of laminin-5 and its functional consequences: role of plasmin and tissue-type plasminogen activator. *J. Cell Biol.* **41,** 255–265.

Gospodarowicz, D. (1984). Preparation of extracellular matrices produced by cultured bovine corneal endothelial cells and PF-HR-9 endodermal cells: Their use in cell culture. *In* "Methods for Preparation of Media, Supplements and Substrata for Serum-Free Animal Cell Culture." AR Liss Inc., New York, pp. 275–293.

Gospodarowicz, D., Delgado, D., and Vlodavsky, I. (1980). Permissive effect of the extracellular matrix on cell proliferation in vitro. *Proc. Natl. Acad. Sci. USA* **77,** 4094–4098.

Harris, E. D., and Krane, S. M. (1972). An endopeptidase from rheumatoid synovial tissue culture. Biochimica et Biophysica Acta 258: 566–576.

Heussen, C., and Dowdle, E. B. (1980). Electrophoretic analysis of plasminogen activators in polyacrylamide gels containing sodium dodecyl sulfate and copolymerized substrates. *Anal. Biochem.* **102,** 196–202.

Jaffe, E. A., and Mosher, D. (1978). Synthesis of fibronectin by cultured endothelial cells. *J. Exp. Med.* **147,** 1779–1781.

Jones, P. A., Scott-Burden, T., and Gevers, W. (1979). Glycoprotein, elastin and collagen secretion by rat smooth muscle cells. *Proc. Natl. Acad. Sci. USA* **76,** 353–357.

Kanemoto, T., Reich, R., Royce, L., Greatorex, D., Adler, S. H., Shiraishi, N., Martin, G. R., Yamada, Y., and Kleinman, H. K. (1990). Identification of an amino acid sequence from the laminin A chain that stimulates metastasis and collagenase IV production. *Proc. Natl. Acad. Sci. USA* **87,** 2279–2283.

Kleinman, H. K., Klebe, R. J., and Martin, G. R. (1981). Role of collagenous matrices in the adhesion and growth of cells. *J. Cell Biol.* **88,** 473–485.

Kleinman, H. K., McGarvey, M. L., Hassell, J. R., Star, V. L., Cannon, F. B., Laurie, G. W., and Martin, G. R. (1986). Basement membrane complexes with biological activity. *Biochemistry* **25,** 312–318.

Kramer, R. H., Fuh, G. M., and Karasek, M. A. (1985). Type IV collagen synthesis by cultured human microvascular endothelial cells and its deposition into the subendothelial basement membrane. *Biochemistry* **24,** 7423–7430.

Laemmli, U. K. (1970). Cleavage of structural proteins during the assembly of the head of bacteriophage T4. *Nature* **227,** 680–685.

Langhofer, M., Hopkinson, S. B., and Jones, J. C. R. (1993). The matrix secreted by 804G cells contains laminin-related components that participate in hemidesmosome assembly in vitro. *J. Cell Sci.* **105,** 753–764.

Lochter, A., Sternlicht, M. D., Werb, Z., and Bissell, M. J. (1998). The significance of matrix metalloproteinases during early stages of tumor progression. *Ann. NY Acad. Sci.* **857,** 180–193.

Nagase, H., and Woessner, J. F. (1999). Matrix metalloproteinases. *J. Biol. Chem.* **274,** 21491–21494.

Terato, K., Nagai, Y., Kawanishi, K., and Yamamoto, S. (1976). A rapid assay method of collagenase activity using ^{14}C-labeled soluble collagen as substrate. *Biochim. Biophys. Acta* **445,** 753–762.

Vu, T. H., and Werb, Z. (2000). Matrix metalloproteinases: effectors of development and normal physiology. *Genes Dev.* **14,** 2123–2133.

Werb, Z., and Chin, J. R. (1998). Extracellular matrix remodeling during morphogenesis. *Ann. NY Acad. Sci.* **857,** 110–118.

Werb, A., Banda, M. J., and Jones, P. A. (1980). Degradation of connective tissue matrices by macrophages. I. Proteolysis of elastin, glycoproteins and collagen by proteinases isolated from macrophages. *J. Exp. Med.* **152,** 1340–1357.

Young, T. N., Rodriguez, G. C., Rinehart, A. R., Bast, R. C., Pizzo, S. V., and Stack, M. S. (1996). Characterization of gelatinases linked to extracellular matrix invasion in ovarian adenocarcinoma: Purification of matrix metalloproteinase 2. *Gyn. Onc.* **62,** 89–88.

Yurchenco, P. D. (1990). Assembly of basement membranes. *Ann. NY Acad. Sci.* **580,** 195–213.

PART III

Adhesion Receptor Methodologies

CHAPTER 9

Expression Cloning Strategies for the Identification of Adhesion Molecules

Joe W. Ramos* and Mark Ginsberg†

* Rutgers, The State University of New Jersey
Piscataway, New Jersey 08854

† The Scripps Research Institute
La Jolla, California 92037

I. Introduction

Over the past 30 years considerable progress has been made in identifying molecules responsible for mediating and regulating cell adhesion. This includes the identification of families of adhesion proteins such as the integrins, cadherins, and syndecans and of proteins that bind directly to these receptors and regulate their functions. Aside from their role in development, adhesion molecules play important roles in diverse pathological conditions including cancer, arthritis, and other inflammatory diseases and cardiovascular disease. Thus, identification of molecules involved in the regulation of cell adhesion continues to be an area of intense research activity.

For the purposes of this review, we will confine our discussion to the use of expression cloning to identify proteins that modulate integrin-mediated adhesion. With an appropriate selection strategy, these methods are generally applicable to the study of other adhesion systems as well. Integrins mediate cell–cell and cell–extracellular matrix adhesion and are transmembrane heterodimers consisting of one α and one β subunit (Hynes, 1992). The affinity of some integrins for ligand is regulated by "inside-out" cell signaling cascades (Ginsberg, 1995; Hughes and Pfaff, 1998). This dynamic regulation of integrin affinity for ligand (activation) is important in cell migration (Huttenlocher *et al.,* 1996), fibronectin matrix assembly (Wu *et al.,* 1995), hemostasis and thrombosis (Shattil *et al.,* 1998), and morphogenesis (Martin-Bermudo *et al.,* 1998; Ramos *et al.,* 1996). Conversely, integrin binding to ligands initiates an "outside-in" signaling cascade that modifies cell shape, gene expression, growth, and survival (Schwartz *et al.,* 1995). Inside-out signaling is cell type specific, energy dependent, and requires the cytoplasmic domains of both the α and β subunits (Williams *et al.,* 1994). Changes in integrin affinity that result from inside-out signaling can be measured by adhesion of ligands such as fibrinogen, or by activation specific antibodies such as PAC1. Hence one can use inside-out signaling as a selective marker in expression cloning strategies.

The advent of highly efficient transient transfection and expression methodologies for mammalian cells in combination with improved cDNA library construction set the stage for the design of expression cloning screens. Sensitive assays such as immunoselection by panning were used to isolate cDNA clones for the cell surface proteins CD2 and CD28 (Seed and Aruffo, 1987; Aruffo and Seed, 1987). These methods have since been adapted to identify changes in receptors for matrix ligands such as collagen (Pullman and Bodmer, 1992), P-selectin (Sako *et al.,* 1993), and thrombospondin (Adams *et al.,* 1998). They have also been used in studies not directly related to adhesion to identify sulfotransferases (Ill *et al.,* 1985) and proteins that complement overexpressed signaling mutants (Han and Colicelli, 1995; Akiyama and Ito, 1995). Moreover, we have been successful with two distinct screens to identify proteins that regulate inside-out signaling to integrins (Fenczik *et al.,* 1997; Ramos *et al.,* 1998).

In both screens we took advantage of the use of the PAC1 antibody to measure integrin activation by flow cytometry (Shattil *et al.,* 1985). In the first such strategy we suppressed integrin activation by overexpressing the cytoplasmic domain of the integrin $\beta1$ subunit as a chimera fused with the extracellular and transmembrane domains of the IL2 receptor (Tac-$\beta1$) (Chen *et al.,* 1994). The inhibition of integrin activation is sequence specific because the isolated cytoplasmic tail of the α subunit or certain mutants of the $\beta1$ tail have no effect on integrin activation. The Tac-$\beta1$ effect is also cell autonomous (Chen *et al.,* 1994). The presumption was that the $\beta1$ integrin tail may titer out the proteins that signal to the integrin. We reasoned that overexpression of these proteins would compensate the effect and overcome the Tac-$\beta1$ suppression. Hence we coexpressed the Tac-$\beta1$ with a cDNA library in Chinese hamster ovary (CHO) cells. The cells were sorted by flow cytometry to isolate cells in which integrins were active in the presence of high levels of Tac-$\beta1$. These cDNAs were further enriched by retransfection of cells in batches and reanalysis by flow cytometry. This process was repeated until a single cDNA was isolated that would block the Tac-β1 inactivation of integrins. In this way we identified CD98, an early T-cell activation antigen, as an integrin regulatory protein (Fenczik *et al.*, 1997). Subsequent studies have documented that CD98 binding to integrin β1A cytoplasmic domains accounts for its activities in this screen (Zent *et al.*, 2000; Fenczik *et al.*, 2001). A summary of this expression cloning method is shown in Fig. 1.

In a separate strategy, we devised an expression cloning screen to identify proteins that negatively regulate the suppression of integrin activation (Ramos *et al.*, 1998). This screen was based on the observation that activation of the small GTP-binding protein H-Ras,

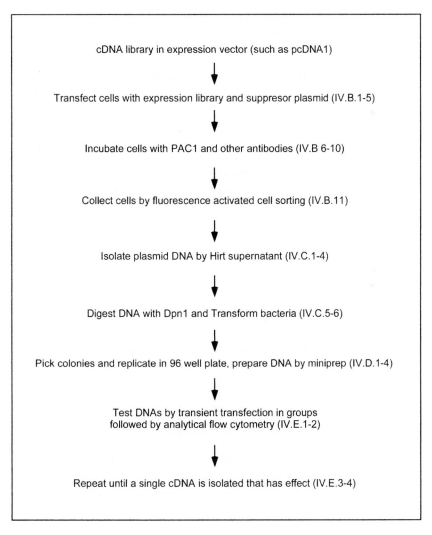

Fig. 1 Flow chart summarizing expression cloning strategy.

or its effector kinase, c-Raf-1, initiates a MAP kinase-dependent signaling pathway that suppresses integrin activation (Hughes *et al.,* 1997) and again PAC1 was used to monitor integrin activation state by flow cytometry (O'Toole *et al.,* 1994). The strategy consisted of transfecting a CHO cell line with activated H-Ras and a CHO cDNA expression library. The cells were sorted by flow cytometry and only cells that had active integrins in the presence of activated H-Ras were collected. The cDNAs were isolated from these cells and amplified in bacteria. These cDNAs were further enriched as above until a single cDNA was isolated that would block the Ras/ERK signal to the integrins. By this method, we identified PEA-15 as a candidate novel regulator of the ERK to integrin signal (Ramos *et al.,* 1998).

What follows is a step-by-step description of the Tac-β1 cloning strategy in detail along with a commentary on the advantages and disadvantages of the method. We also provide alternative approaches to various steps where appropriate.

II. Importance of Methodology

Expression cloning has been used to investigate two types of questions in cell adhesion. One is to identify receptors for new ligands or regions of a known ligand. The second is to identify proteins that affect the adhesive activity of a receptor such as an integrin or cadherin. The latter approach can be used to identify secreted proteins and intracellular proteins and is not biased by any preconceived targets. Consequently the approach offers the possibility of uncovering new directions in cell adhesion research.

Expression cloning has many advantages for the identification of adhesion molecules and their regulators. It provides a non-biased approach to identify proteins that affect adhesion. Every protein expressed in the library is tested in the screen. Another advantage is that this cloning method can identify proteins that have a functional relationship to adhesion regardless of whether they interact directly with adhesion molecules. This sets this screen apart from binding based methods such as the yeast two-hybrid screen.

The primary disadvantage of this approach is that functional alterations that require more than one protein may be impossible to uncover. Of course this is a limitation of most cloning strategies and is not peculiar to expression cloning. These limitations aside, expression cloning has proven to be a valuable tool in analysis of cell adhesion and new screens will likely continue to provide new insight into adhesive protein function and mechanism.

III. General Considerations

There are four factors that must be considered before embarking on an expression cloning project: (1) the quality of the library to be screened, (2) the transfection efficiency of the cells used, (3) autonomous replication of the cDNA library in the cells, and (4) appropriate controls for critical steps. The particulars of each step may differ, but they will have certain features in common.

Of critical importance is the kind and quality of the cDNA library used. An appropriate library should be in a plasmid with a promoter such as CMV that drives high levels of protein expression in mammalian cells. The plasmid must also contain an origin for episomal replication such as SV40 or polyoma that matches the elements used to drive replication (e.g., SV40 large T or polyoma large T, respectively). The plasmid pCDM8 was originally designed for expression cloning (Seed and Aruffo, 1987) and pcDNA1 was derived from this. Neither of these plasmids is currently available commercially. We chose to use pcDNA1 because it contains a CMV promoter, both the SV40 and polyoma origins of replication, the supF suppressor tRNA for selection in *Escherichia coli* strains that carry the P3 episome, and the ColE1 origin for maintenance in *E. coli*. The supF provides Tet resistance to the bacteria and thus helps us to select specifically for the transfected library plasmids, as only bacteria carrying library cDNAs as opposed to other transfected plasmids are Tet resistant. Episomal replication also provides a means to help confine isolated plasmids to those derived from the cDNA library. This is because plasmids replicated episomally in mammalian cells by SV40 large T are not methylated. After isolation of the plasmid DNA from the cells by Hirt supernatant (Hirt, 1967), we treat the plasmid with *Dpn*I, which is a four-base specific restriction enzyme that digests only methylated DNA, thus removing any contaminant bacterial plasmid in the prep. More importantly, episomal replication is necessary to amplify the transfected plasmid DNA in the mammalian cells so that the odds of isolating all the plasmids by Hirt supernatant are improved.

The quality of the library is also an important issue. Ideally the library should not have been amplified or at worst amplified only once. This is because with each amplification step rare cDNAs and cDNAs that affect bacterial growth may be lost. This diminishes the complexity of the library and may result in the loss of cDNAs that affect the function of interest. As a guide, the library we used was a pcDNA1 library purchased from Invitrogen that had been amplified once and had 1.8×10^7 primary recombinants. To insure the presence of full-length clones in the library it was oligo dT primed and size selected for clones larger than 0.5 kb. Full-length clones are crucial in approaches where cell-surface expression is needed, whereas intracellular molecules need not be full-length to be isolated in these screens. Random-primed libraries contain antisense and clone fragments that may be beneficial in particular strategies.

Expression cloning is, to a large extent, a numbers game. Thus, transfection efficiency and episomal replication of the library are of key importance. Lipofection such as with Lipofectamine (LifeTechnologies) can achieve high transfection efficiencies in CHO cells. On average we obtain transfection efficiencies above 60%. Other methods of transfection with similar efficiency can be used and should be chosen to best suit the cell line used in the screen. Episomal replication is important because it results in multiple copies of each cDNA in an individual cell, which improves the chances of recovering the relevant cDNA in the Hirt supernatant. Once the cDNAs have been extracted it is important that they all be recovered during the transformation procedure. For this reason the transformation competence of the bacteria should be tested prior to transforming with the Hirt supernatant cDNAs. Only cells with a competency level of $\geq 10^9$ cfu/μg DNA should be used.

The sensitivity and specificity of the screen must also be considered. Flow cytometry or cell panning screens allow millions of cells to be screened at a time. From a typical screen by flow cytometry we can sort at least 5×10^6 cells and collect a target number of 50,000 cells in the gate. Thus a major advantage of the flow cytometry method is that it is very sensitive, allowing us to examine function quantitatively at the level of a single cell millions of times.

IV. Protocols

A. Materials and Equipment

Flow cytometer (FACScalibur, Becton Dickinson)

Fluorescence activated cell sorter (FACSTAR, Becton Dickinson)

Cell line: $\alpha\beta$py cells, a CHO-K1 cell line (ATCC) stably transfected with the polyoma large T antigen and αIIβ6A (Hughes *et al.,* 1997). These cells may be requested from M. H. Ginsberg.

cDNA library: CHO cell library directionally cloned into the mammalian expression vector, pcDNA1.

Other cDNAs: Tac-β1 (suppressor); Tac-α5 (transfection marker); pcDNA1 (control vector)

Negative control: Competitive integrin αIIbβ3 antagonist (e.g. lamifiban (Roche), eptifibatide (Cor Therapeutics); inhibits PAC1 binding to αIIbβ3).

Primary and secondary antibodies: PAC1 IgM ascites (Shattil *et al.,* 1985) available from Becton Dickinson; 7G7B6 (ATCC HE 8784) biotinylated with biotin *N*-hydroxysuccinimide (Pierce) according to manufacturer's instructions; FITC-anti-mouse IgM Mu chain (BioSource).

Streptavidin-phycoerythrin (Molecular Probes) (used to visualize biotinylated 7G7B6 binding)

Transfection reagent: Lipofectamine (LifeTechnologies)

Cell Dissociation Buffer (LifeTechnologies)

Propidium iodide (Sigma)

Phosphate-buffered saline (PBS)

0.6% SDS/10 m*M* EDTA/5 *M* NaCl (Hirt supernatant cell lysis buffer; make fresh)

Phenol:chloroform

Glycogen

3 *M* sodium acetate

*Dpn*I restriction enzyme (Promega)

MC1061/P3 cells $\geq 10^9$ cfu/μg DNA (Invitrogen)

60% glycerol

96-well microtiter plates

B. Transfection and Flow Cytometry

1. We use a CHO cell line called $\alpha\beta$py maintained in DMEM containing 10% serum, 1% nonessential amino acids, 1% glutamine, and 1% penicillin/streptomycin. To prepare cells for transfection, plate at 2×10^6 cells per 100-mm tissue culture plate 24 h in advance. This should be done in accordance with protocols for the transfection reagent. We transfect between 12 and 18 plates per screen.

2. Transfect cells using 20 μl of Lipofectamine in serum-free media as follows: control plate 1:Tac-α5 (2 μg) and CHO library (5 μg); control plate 2: Tac-α5 (2 μg) and pcDNA1 (5 μg); control plate 3: Tac-β1 (2 μg) and pcDNA1 (5 μg); and sort plates (12–18 plates): Tac-β1 (2 μg) and CHO library (5 μg). Follow transfection reagent protocols for specifics of transfection (e.g., serum-free media, and amounts of reagent).

3. Twenty-four hours after the start of the transfection, wash plates with complete DMEM medium containing serum in accordance with transfection protocol.

4. Collect cells: Forty-eight hours after the start of transfection, remove the medium and wash cells once with $1\times$ PBS, then add 3 ml of Cell Dissociation Buffer to the plates. Let this sit 5 min. Use a transfer pipette to collect the cells and put them into the appropriately labeled test tubes. At this point, combine the cells to be sorted (Tac-β1+lib cells) in one tube. Set aside Control Plate 1 (Tac-α5+cDNA library), to be used later as a Hirt supernatant control.

5. Spin cells down at 1000 rpm in tabletop centrifuge at room temperature. Aspirate off supernatant.

6. Resuspend the cells to be sorted (Tac-β1+cDNA library) in 1.4 ml complete DMEM total. Divide the sort cells into six tubes (Falcon 2054) of 225 μl each. Add 25 μl of diluted PAC1 ascites (PAC1 should be titered in each lab and the appropriate dilution determined). Resuspend cells transfected with Tac-β1+pcDNA1 in 45 μl and add 5 μl of the diluted PAC1 ascites. Resuspend the Tac-α5 control (Tac-α 5+pcDNA1) in 200 μl complete DMEM; aliquot into four separate tubes labeled 1, A, B, and C. Add 5 μl of diluted PACl to tubes 1 and B, and nothing to the other two. The cells in tube 1 will be stained with both PACl and 7G7B6 to measure transfection levels and to determine the sort box for FACS. The cells in the next three tubes will be used as compensation controls for the flow cytometry: Tube A is the no-antibody control; B, the FITC-only control; C, the streptavidin–phycoerythrin only control. Incubate at room temperature for 30 min.

7. Add 2 ml complete DMEM to each tube and centrifuge for 5 min at 1000 rpm. Decant liquid and resuspend cells to be sorted in 250 μl DMEM/4% (v/v) biotinylated 7G7B6. For controls, resuspend in 50 μl DMEM/4% (v/v) biotinylated 7G7B6. Incubate on ice for 30 min.

8. Wash cells with 2 ml complete DMEM, centrifuge for 5 min at 1000 rpm, then decant liquid. Resuspend sort cells in 250 μl DMEM/4% streptavidin–phycoerythrin/4% FITC-labeled anti-mouse IgM Mu chain antibodies. For controls, final resuspend in 50 μl instead of 250 μl. Incubate on ice in the dark for 30 min.

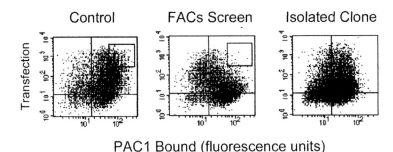

Fig. 2 Setting the gate and the final isolated clone. $\alpha\beta$py cells were cotransfected with cDNA encoding Tac-α5 alone (Control) or with Tac-β1 in combination with the CHO cDNA library (FACS Screen). After 48 h the cells were stained for Tac expression (ordinate) and PAC1 binding (abscissa). The box represents the gate set for cell collection. Note that the gate on the sort screen contains very few cells while this window contains a large number of cells with active integrins in the control. A gate set in this way will collect cells in which a library cDNA has recovered integrin binding activity as reflected by PAC1 binding. After subsequent analytical flow cytometry of groups of cDNAs from the screen, you should obtain a single cDNA that will rescue integrin binding (Isolated Clone).

9. Add propidium iodide (0.1 μg to 0.5 μg/1 × 10^5 cells) during the last 5 min of PE/FITC antibody incubation. This stains the dead cells, which are permeable, so they can be removed from the sort gate.

10. Wash cells with 2 ml of cold 1× PBS. Spin cells down at 1000 rpm, resuspend and combine cells to be sorted in 4 ml of 1× PBS, and resuspend controls in 500 μl 1× PBS each.

11. This step should be done with the guidance of someone trained in FACS. Use FACS to collect PE/FITC positive cells. Collect cells into 3 ml complete DMEM. We set our collection window as shown in Fig. 2. Generally, we sorted above 5 × 10^6 cells and collected at least 50,000 cells. Make a note of the number of cells sorted and collected for comparison between multiple screen runs.

C. Recovery of cDNAs by Hirt Supernatant

1. To isolate cDNA containing plasmids: Pipette 1.5 ml of cells collected from the sort into an Eppendorf tube and spin down for 5 min at 1000 rpm in a centrifuge. Remove supernatant and pipette remaining 1.5 ml into the same tube and spin the cells onto the previous cell pellet. Remove supernatant. All cells should be collected at the bottom of the Eppendorf tube.

2. Add 400 μl of 0.6% SDS/10 mM EDTA and let sit at room temperature for 20 min. At the same time remove the DMEM from the Tac-α 5+lib plate, wash with PBS, and add 800 μl of 0.6% SDS/10 mM EDTA to the saved plate. Also let this sit at room temperature for 20 min. Collect lysate from plate into Eppendorf tube with cell scraper. Add 100 μl (or 200 μl to plate lysate) of 5 M NaCl, mix by inversion, and leave at 4°C overnight.

3. Spin in microfuge on high for 5 min. Carefully remove supernatant to a new tube. Phenol : chloroform extract the supernatant twice. Transfer the supernatant to new tube and add 1 μl of glycogen (MolBiol grade). Fill the tube to the top with 100% ethanol to precipitate DNA. Spin down for 5 min at full speed in microfuge. Remove supernatant. Resuspend pellet in 100 μl of double-distilled H_2O.

4. Add 10 μl of 3 M sodium acetate and 300 μl of 100% ethanol and repeat precipitation spinning down in a microcentrifuge for 20 min in cold room. Wash pellet with 70% ethanol.

5. Resuspend the pellet in 8 μl H_2O. Resuspend the control Tac-α5+lib pellet in 35 μl H_2O. Digest the DNAs overnight at 37°C with *Dpn*I.

6. Transformation: Incubate each entire 10 μl digestion with 100 μl of highly competent MC10B/P3 cells on ice for 20 min in Falcon 2059 tubes. Transfer tubes to 42°C bath for 1 min; return to ice. Add 1 ml SOC medium and incubate on the shaker at 37°C for 1 h. Plate entire transformation of each on Amp (12.5 mg/ml)/Tet (7.5 mg/ml) plates. Leave overnight at 37°C.

D. Identification and Isolation of Candidate Clones

1. Count and record the number of colonies that grow. A typical number of colonies in our screens ranged from 200 to 500. One successful screen had only 72 colonies.

2. Pick the colonies from the sort plate and inoculate a 2 ml miniprep grow for each. Let grow overnight at 37°C.

3. Prepare 96-well microtiter plates with 50 μl of 60% glycerol in each well. Prepare a frozen glycerol stock of each miniprep by adding 150 μl of the miniprep to a numbered microtiter plate. The coordinates of the microtiter plate will serve as the name of each colony/clone (i.e., clone lAl for the first clone picked and placed in plate 1, etc.). Seal the micro titer plates with Parafilm and store at −80°C for later use.

4. After you prepare the glycerol stock, pool the minipreps in groups of no more than 20. Carefully note the plate and coordinates corresponding to each set of pooled mini preps. Prepare DNA from each of the pools by Qiagen prep or CsCl preparation according to standard protocols. The advantage in pooling the cDNAs at this point, rather than at a later stage, is to prevent the need for subsequent mass cell sorts, which can get costly.

E. Rescreening and Identification of Positive Clones

1. To identify pools containing cDNAs that prevent Tac-β1 suppression, cotransfect 8 μg of each pool of DNA into cells with the appropriate test plasmids. Look for rescue by analytical FACS staining with PACl and 7G7B6. The rescue can be very subtle at this point. Pay particular attention to any increase in cell number in the upper right-hand quadrant. There may be an increase of as little as 10% in this region over control experiments. Such an increase should be pursued.

2. Once a candidate batch has been identified, go back to the saved microtiter plate corresponding to that pool and start a miniprep grow from each colony in the pool. Prepare DNA by standard methods.

3. To test DNAs from the pool, combine them into groups of 4 to 5 and cotransfect them with the appropriate test plasmids as above. Look for rescue by analytical FACS. The rescue should be more obvious at this step. Next, test each DNA from the candidate pool individually by analytical flow cytometry.

4. Sequence the positive cDNAs and begin characterization of the protein structure and function.

V. Points of Variation

There are several possible points of variation from the protocol as presented here. A less expensive and specialized approach is to replace the flow cytometry steps (IV.B.4) with panning. With panning you coat the antibody (PAC1 here) or ligand on a petri dish and apply the transfected cell population to the plates and allow them to adhere (Seed and Aruffo, 1987). You then wash away the nonadherent population and harvest the adherent cells for Hirt supernatant and subsequent rescreening. This cell-panning screen has been very successful and constitutes a large portion of the published expression cloning methods. See *Current Protocols in Molecular Biology* for a detailed protocol (Hollenbaugh and Aruffo, 1998).

Episomal replication of the library cDNA is critical for success. If a cell line stably expressing EBV, SV40 large T antigen, or polyoma large T is not available, one can cotransfect one of these along with the library (step IV.B.2) (Ong *et al.,* 1998). An example of a plasmid containing polyoma large T antigen is pPSVE1-PyE (Bierhuizen and Fukuda, 1992). Alternatively, some cell lines can replicate episomal DNA, the most commonly used being COS7 cells that contain SV40 large T. The important point is that the cells be able to exogenously replicate the transfected cDNA plasmid library.

In our protocol we have performed only one round of selection. We follow this by painstakingly isolating all of the subsequent cDNAs for rescreening (steps IV.D.1–4). A common alternative is to follow the first round of selection with two or three subsequent rounds of enrichment. In these instances the cDNAs that were isolated from the first round of selection are retransfected and enriched. The key to this is that the cDNAs must be transfected for enrichment by means that limit the number of plasmids introduced into each individual cell. Two methods to do this are by spheroplast fusion or by electroporation (Hollenbaugh and Aruffo, 1998). It is important to limit the number of cDNAs per cell so that each cell identified in the second and third rounds of screens will contain primarily multiple copies of the cDNA of interest. In this way almost all of the cDNAs in the final enrichment screen will be those of interest and multiple copies of each cDNA will be present. The disadvantage here is that if an enrichment step goes poorly, one must start again from the beginning.

If one follows our protocol without enrichment, a final point of difference in our screens is how the cDNAs are managed (step IV.D.2). Here we isolate each colony from the primary screen and preserve it as a glycerol stock in a microtiter plate. This has the advantage that each colony is safely stored away as subsequent screens are pursued. Its main disadvantage is that it is the most labor-intensive part of the strategy and many colonies of no interest are preserved. With the advent of colony-picking robots this step can be adjusted for high throughput. However, many labs do not have access to this technology. An alternative method is to replica plate the original cDNA colonies for sib-selection (Ong *et al.,* 1998). In this approach lifts are done on these plates and plasmid DNA is isolated and retested in the screen. A plate that has a plasmid cDNA with the desired activity is then lifted again with each filter being subdivided and the plasmid from the colonies from a single division being isolated and retested in the screen. A positive section is then further subdivided and retested until a single colony with the desired function is identified. The disadvantage here is that the rescreening must be done quickly to avoid contamination or death of the colonies on the plates. Also, it is nearly as labor-intensive as our method using microtiter plates.

VI. Critical Controls

We have included several controls throughout our protocols. To test the efficiency of transfection and the activity of the integrins we include cells that have been transfected with the transfection marker Tac-α5 alone. We also include cells that have been transfected with the suppressor Tac-β1 alone as a control to verify that in a particular assay the Tac-β1 continues to suppress integrin activation. Moreover, by comparing PAC1 binding in cells highly transfected with Tac-α5 to cells transfected with Tac-β1 we can draw a gate to collect cells in the presence of Tac-β1 in the screen that have regained integrin activation (see Fig. 2 and step IV.B.11).

In order to be certain that the cells replicated the plasmid library episomally, we also include a control plate transfected with Tac-α5 and plasmid library (see steps IV.B.4 and IV.C.1–6). We take advantage of the fact that while DNA replicated in bacterial cells is methylated, DNA replicated in mammalian cells is not. We isolate DNA from this control plate by Hirt supernatant (Hirt, 1967) and digest half of the DNA with the methylation-specific restriction enzyme DpnI. Equal amounts of digested and undigested DNAs are then transformed into bacteria and plated. The next day we count colonies, usually getting between 300 and 1000 colonies. However, in the screen that identified PEA-15 we had only 72 colonies. There should be substantially more colonies on the nondigested plate than the *Dpn*I-digested one, indicating that plasmid was replicated. If the plasmid was not replicated, the *Dpn*I digest plate will not have any colonies or only very few and the screen should not be continued further as success is unlikely.

VII. Potential Use in Disease/Drug Screening Applications

The power of expression cloning is to isolate novel cDNAs that have functional consequences in cell adhesion. These new molecules may then be evaluated as potential drug targets in diseases that stem from defects in the adhesion systems being evaluated. Moreover, the contribution of the new molecules themselves to specific disease states can be determined by further analysis. For example, we isolated PEA-15 as a protein that alters ERK MAP kinase signaling to integrins. Other screens uncovered roles for PEA-15 in type II diabetes (Condorelli *et al.*, 1998), breast cancer (Bera *et al.*, 1994), and glioma (Hao *et al.*, 2001). Our functional screen has now provided a possible molecular mechanism to explain how PEA-15 may affect these diverse diseases through its activity on ERK and integrin signaling.

Acknowledgments

Correspondence should be addressed to M.H.G. (Ginsberg@scripps.edu). We thank Dr. M. L. Matter for critical evaluation of the manuscript. This work was supported by grants from the NIH. This is manuscript #14162-VB from T.S.R.I.

References

Adams, J. C., Seed, B., and Lawler, J. (1998). Muskelin, a novel intracellular mediator of cell adhesive and cytoskeletal responses to thrombospondin-1. *EMBO J.* **17,** 4964–4974.

Akiyama, Y., and Ito, K. (1995). A new *Escherichia coli* gene, fdrA, identified by suppression analysis of dominant negative FtsH mutations. *Mol. Gen. Genet.* **249,** 202–208.

Aruffo, A., and Seed, B. (1987). Molecular cloning of a CD28 cDNA by a high efficiency COS cell expression system. *Proc. Natl. Acad. Sci. USA* **84,** 8573–8577.

Bera, T. K., Guzman, R. C., Miyamoto, S., Panda, D. K., Sasaki, M., Hanyu, K., Enami, J., and Nandi, S. (1994). Identification of a mammary transforming gene (MAT1) associated with mouse mammary carcinogenesis. *Proc. Natl. Acad. Sci. USA* **91,** 9789–9793.

Bierhuizen, M. F., and Fukuda, M. (1992). Expression cloning of a cDNA encoding UDP-GlcNAc:Gal beta 1-3-GalNAc-R (GlcNAc to GalNAc) beta 1-6GlcNAc transferase by gene transfer into CHO cells expressing polyoma large tumor antigen. *Proc. Natl. Acad. Sci. USA* **89,** 9326–9330.

Chen, Y.- P., O'Toole, T. E., Shipley, T., Forsyth, J., LaFlamme, S. E., Yamada, K. M., Shattil, S. J., and Ginsberg, M. H. (1994). "Inside-out" signal transduction inhibited by isolated integrin cytoplasmic domains. *J. Biol. Chem.* **269,** 18307–18310.

Condorelli, G., Vigliotta, G., Iavarone, C., Caruso, M., Tocchetti, C. G., Andreozzi, F., Cafieri, A., Tecce, M. F., Formisano, P., Beguinot, L., and Beguinot, F. (1998). PED/PEA-15 gene controls glucose transport and is overexpressed in type 2 diabetes mellitus. *EMBO J.* **17,** 3858–3866.

Fenczik, C. A., Sethi, T., Ramos, J. W., Hughes, P. E., and Ginsberg, M. H. (1997). Complementation of dominant suppression implicates CD98 in integrin activation. *Nature* **390,** 81–85.

Fenczik, C. A., Zent, R., Dellos, M., Calderwood, D. A., Satriano, J., Kelly, C., and Ginsberg, M. H. (2001). Distinct domains of CD98hc regulate integrins and amino acid transport. *J. Biol. Chem.* **276,** 8746–8752.

Ginsberg, M. H. (1995). Cell adhesion: the cytoplasmic face. *Biochem. Soc. Trans.* **23,** 439–446.

Han, L., and Colicelli, J. (1995). A human protein selected for interference with Ras function interacts directly with Ras and competes with Raf1. *Mol. Cell Biol.* **15,** 1318–1323.

Hao, C., Beguinot, F., Condorelli, G., Trencia, A., Van Meir, E. G., Yong, V. W., Parney, I. F., Roa, W. H., and Petruk, K. C. (2001). Induction and intracellular regulation of tumor necrosis factor-related apoptosis-inducing ligand (TRAIL) mediated apotosis in human malignant glioma cells. *Cancer Res.* **61,** 1162–1170.

Hirt, B. (1967). Selective extraction of polyoma DNA from infected mouse cell cultures. *J. Mol. Biol.* **26,** 365–369.

Hollenbaugh, D., and Aruffo, A. (1998). Specialized strategies for screening libraries. *In* "Current Protocols in Molecular Biology" (F. M. Ausebel, R. Brent, R. Kingston, D. Moore, J. Seidman, J. Smith, and K. Struhl eds.). John Wiley and Sons, New York, pp. 6.11.1–6.11.16.

Hughes, P. E., and Pfaff, M. (1998). Integrin affinity modulation. *Trends Cell Biol.* **8,** 359–364.

Hughes, P. E., Renshaw, M. W., Pfaff, M., Forsyth, J., Keivens, V. M., Schwartz, M. A., and Ginsberg, M. H. (1997). Suppression of integrin activation: A novel function of a Ras/Raf-initiated MAP kinase pathway. *Cell* **88,** 521–530.

Huttenlocher, A., Ginsberg, M. H., and Horwitz, A. F. (1996). Modulation of cell migration by integrin mediated cytoskeletal linkages and ligand binding affinity. *J. Cell Biol.* **134,** 1551–1562.

Hynes, R. O. (1992). Integrins: Versatility, modulation, and signalling in cell adhesion. *Cell* **69,** 11–25.

Ill, C. R., Engvall, E., and Ruoslahti, E. (1985). Adhesion of platelets to laminin in the absence of activation. *J. Cell Biol.* **99,** 2140–2145.

Martin-Bermudo, M. D., Dunin-Borkowski, O. M., and Brown, N. H. (1998). Modulation of integrin activity is vital for morphogenesis. *J. Cell Biol.* **141,** 1073–1081.

Ong, E., Yeh, J. C., Ding, Y., Hindsgaul, O., and Fukuda, M. (1998). Expression cloning of a human sulfotrans-ferase that directs the synthesis of the HNK-1 glycan on the neural cell adhesion molecule and glycolipids. *J. Biol. Chem.* **273,** 5190–5195.

O'Toole, T. E., Katagiri, Y., Faull, R. J., Peter, K., Tamura, R. N., Quaranta, V., Loftus, J. C., Shattil, S. J., and Ginsberg, M. H. (1994). Integrin cytoplasmic domains mediate inside-out signal transduction. *J. Cell Biol.* **124,** 1047–1059.

Pullman, W. E., and Bodmer, W. F. (1992). Cloning and characterization of a gene that regulates cell adhesion. *Nature* **356,** 529–532.

Ramos, J. W., Kojima, T. K., Hughes, P. E., Fenczik, C. A., and Ginsberg, M. H. (1998). The death effector domain of PEA-15 is involved in its regulation of integrin activation. *J. Biol. Chem.* **273,** 33897–33900.

Ramos, J. W., Whittaker, C. A., and DeSimone, D. W. (1996). Integrin-dependent adhesive activity is spatially controlled by inductive signals at gastrulation. *Development* **122,** 2873–2883.

Sako, D., Chang, X. J., Barone, K. M., Vachino, G., White, H. M., Shaw, G., Veldman, G. M., Bean, K. M., Ahern, T. J., and Furie, B. (1993). Expression cloning of a functional glycoprotein ligand for P-selectin. *Cell* **75,** 1179–1186.

Schwartz, M. A., Schaller, M. D., and Ginsberg, M. H. (1995). Integrins: emerging paradigms of signal transduction. *Ann. Rev. Cell Dev. Biol.* **11,** 549–599.

Seed, B., and Aruffo, A. (1987). Molecular cloning of the CD2 antigen, the T cell erythrocyte receptor by a rapid immunoselection procedure. *Proc. Natl. Acad. Sci. USA* **84,** 3365–3369.

Shattil, S. J., Hoxie, J. A., Cunningham, M., and Brass, L. F. (1985). Changes in the platelet membrane glycoprotein IIb-IIIa Complex during platelet activation. *J. Biol. Chem.* **260,** 11107–11114.

Shattil, S. J., Kashiwagi, H., and Pampori, N. (1998). Integrin signaling: the platelet paradigm. *Blood* **91,** 2645–2657.

Williams, M. J., Hughes, P. E., O'Toole, T. E., and Ginsberg, M. H. (1994). The inner world of cell adhesion: integrin cytoplasmic domains. *Trends Cell Biol.* **4,** 109–112.

Wu, C., Keivens, V. M., O'Toole, T. E., McDonald, J. A., and Ginsberg, M. H. (1995). Integrin activation and cytoskeletal interaction are essential for the assembly of a fibronectin matrix. *Cell* **83,** 715–724.

Zent, R., Fenczik, C. A., Calderwood, D. A., Liu, S., Dellos, M., and Ginsberg, M. H. (2000). Class- and splice variant-specific association of CD98 with integrin beta cytoplasmic domains. *J. Biol. Chem.* **275,** 5059–5064.

CHAPTER 10

Purification of Integrins and Characterization of Integrin-Associated Proteins

Johannes A. Eble* and Fedor Berditchevski[†]

*Institute of Physiological Chemistry and Pathobiochemistry
Universität Münster
48149 Münster, Germany

[†]CRC Institute for Cancer Studies
The University of Birmingham
Edgbaston, Birmingham, B15 2TA United Kingdom

I. Introduction

By binding to extracellular matrix (ECM) proteins with their ectodomains and to cytoskeletal proteins with their cytoplasmic domains integrins form transmembrane linkages that connect the ECM with the cytoskeleton. The structure of the integrins has been recently reviewed (Humphries, 2000). About 25 different integrins with different ligand binding capabilities are known so far (Plow *et al.,* 2000). Either for structural studies, such as X-ray crystallography, circular dichroism, or proteolytic susceptibility assays, or to study interactions of integrins with their respective ligands or association partners, integrins need to be isolated from cells. Purification protocols of integrins will be discussed in Section II of this chapter.

Integrin-mediated cell adhesion triggers a network of signaling pathways that control cellular proliferation, differentiation, motility, and anchorage-dependent cell survival. While our knowledge about the down-stream signalling events that are associated with integrin activation has significantly widened in recent years (Schwartz *et al.,* 1995; Giancotti and Ruoslahti, 1999), less is known about how the signals are initiated by the ligand-bound receptors. Integrins do not possess intrinsic enzymatic activities themselves but, instead, function as "scaffolding platforms" that initiate assembly of signaling aggregates that include a wide array of cytoskeletal and signaling proteins. Hence, to understand how integrin-dependent signaling pathways are triggered, one has to establish spatial organization of the signaling clusters, and, more importantly, to identify integrin-associated protein partners that affect the assembly of these aggregates (and, therefore, signaling capacities of integrins). Over the past few years the list of integrin-associated proteins has grown significantly and it now includes cytoplasmic, transmembrane, and peripheral membrane proteins (reviewed in Hemler, 1998, and Liu *et al.,* 2000). Some of the associated proteins were copurified with integrins as "contaminants" using standard protocols for integrin purification (see Section II) (Altruda *et al.,* 1989). In addition, a range of biochemical and molecular biology approaches, which are commonly used in the studies of protein–protein interactions, have been successfully applied for characterization of integrin-associated proteins. These include affinity chromatography purification using immobilized cytoplasmic tails of integrin subunits (Otey *et al.,* 1990, 1993; Schaller *et al.,* 1995), monoclonal antibody (mAb) production against the affinity purified integrin complexes (Berditchevski *et al.,* 1997) and the yeast two-hybrid system (Shattil *et al.,* 1995; Kolanus *et al.,* 1996; Hannigan *et al.,* 1996; Chang *et al.,* 1997). Finally, biological assays could be used for identifying integrin-associated proteins (Fenczik *et al.,* 1997). The methodology of yeast two-hybrid system and wide variations of affinity chromatography protocols were extensively reviewed elsewhere (Frederickson, 1998; Vidal and Legrain, 1999; Scopes, 1994). In Section III of this chapter, we will focus on the details of the procedure that allows identification of the integrin-associated proteins through the production of the mAbs.

II. Isolation of Integrins: General Considerations and Points of Variation

General biochemical protocols are applied to purify integrins. However, as integrins possess characteristic properties, e.g., their requirement for the presence or absence of specific cofactors, the common protein purification procedures will be discussed in the light of integrin isolation. Being transmembrane proteins, integrins need to be extracted from the plasma membrane with detergents, a procedure that will be discussed first (Section II.B.1). Recombinant expression of soluble integrins obviates detergent extraction (Section II.B.2). Detergent-solubilized or soluble integrins can be purified by different means, such as affinity chromatography on their natural ligands (Section II.B.1), on lectin columns (Section II.B.2), or on antibody immune matrices (Section II.B.3). Recombinant manipulation of integrins also allows the addition of tag sequences, which facilitates their purification (Section II.B.4). Being acidic proteins of comparatively high molecular mass, integrins have also been purified by "classical" protein purification procedures, such as ion exchange chromatography (Section II.B.5) and gel filtration chromatography.

A. Extraction and Solubilization of Integrins

1. Extraction of Cell Membrane–Anchored Wild-Type Integrins from Tissues and Cells

Because integrins vary in their tissue distribution, the choice of certain tissues or cells is of paramount importance as a starting point to isolate a specific integrin in high yield and without major contamination of other integrins or other cell proteins. The integrin $\alpha_{IIb}\beta_3$ accounts for about 17% of total membrane protein on thrombocytes, and can be isolated in high amounts from platelets (Calvete, 1994). Yet, depending on the isolation procedure, platelets are also a good source to purify the laminin-binding $\alpha_6\beta_1$ and the collagen-binding $\alpha_2\beta_1$ integrin (Sonnenberg et al., 1991; Vandenberg et al., 1991). Integrin $\alpha_2\beta_1$ is the only integrin collagen receptor on the platelet surface. In addition, platelets possess other collagen receptors, such as the glycoprotein (gp) VI, which interacts with collagen directly, or gp Ib-V-IX, which binds collagen indirectly via von Willebrand factor. These non-integrin receptors show distinct binding characteristics and requirements for divalent cations. Hence, they can be easily separated from $\alpha_2\beta_1$ integrin. Leukocytes are the preferred starting material to isolate $\alpha_4\beta_1$ integrin (Makarem et al., 1994). Placenta is a good source to purify $\alpha_5\beta_1$ (Mould et al., 1996) and $\alpha_1\beta_1$ integrin (Vandenberg et al., 1991; Eble et al., 1993; Kern et al., 1993). However, tissue or cells rarely express only one integrin species. Therefore, when using tissue or cells as source for integrin isolation, one has to be aware that the integrin preparation may contain various integrin species.

As transmembrane glycoproteins, integrins are released from the cell surface by extraction with detergents. Various detergents have been used to solubilize integrins: 1%

Nonidet P-40 (NP40) was used for the extraction of several β_1 integrins (Takada *et al.,* 1987), whereas Triton X-100 (1–4%(w/v)) was utilized for the extraction of platelet integrin $\alpha_{IIb}\beta_3$ (Calvete *et al.,* 1992; Gulino *et al.,* 1995), $\alpha_4\beta_1$, and $\alpha_5\beta_1$ integrin (Makarem *et al.,* 1994). Octyl-β-D-glucopyranoside or octyl-β-D-thioglucopyranoside are the detergents most commonly used to extract integrins from the plasma membrane. The latter one is preferred because it is more stable than octyl-β-D-glucopyranoside. Both detergents can be removed by dialysis because of their convenient biophysical properties, such as low critical micelle concentration (CMC) and aggregation number. However, their comparatively high prices increase costs of large-scale preparations immensely. Concentrations of octyl-β-D-glucopyranoside used for extraction of integrins from tissue and cells vary from 100 to 200 mM (Smith and Cheresh, 1988; Sonnenberg *et al.,* 1991; Song *et al.,* 1992), but a concentration as low as 25 mM seems to suffice for integrin solubilization (Gehlsen *et al.,* 1988; von der Mark *et al.,* 1991). Octyl-β-D-glucopyranoside and its sulfur homolog have a higher probability than Triton X-100 or NP-40, thus increasing the probability that the native conformation of the integrin will remain intact.

Pytela *et al.* (1987) studied how various detergents affect the yield of biologically active fibronectin receptor, $\alpha_5\beta_1$ integrin, and vitronectin receptor, $\alpha_V\beta_3$ integrin, when used for integrin extraction from placental tissue (Table I). Octyl-β-D-glucoside has been proven superior in its properties and is nowadays most commonly used for integrin extraction from tissue and cells. Interestingly, even Tween-20, which is also considered a mild detergent, did not allow extraction of either integrin in a native and ligand-binding conformation. Whereas $\alpha_{IIb}\beta_3$ integrin remains biologically active when extracted with Triton X-100 even at higher concentration (Calvete *et al.,* 1992; Gulino *et al.,* 1995), both the fibronectin receptor $\alpha_5\beta_1$ and the vitronectin receptor $\alpha_V\beta_3$ partially lose ligand binding activity in the presence of Triton X-100 (Pytela *et al.,* 1987). Instead of

Table I

Yields of Fibronectin and Vitronectin Receptor, $\alpha_5\beta_1$ and $\alpha_V\beta_3$ Integrin, Respectively, from Placental Tissue Extracted with Various Detergents after Purification by Affinity Chromatography (according to Pytela *et al.,* 1987)

Detergent used	Concentration	Yield of $\alpha_5\beta_1$ (%)	Yield of $\alpha_V\beta_3$ (%)
Octylglucoside	100 mM	100	100
Triton X-100	0.5%	40	23
NP-40	0.5%	43	29
Tween-20	0.5%	0	0
Zwittergent 3–12	0.5%	32	13
Brij 35	0.5%	0	0
Sodium deoxycholate	50 mM	0	0

The yields were determined by densitometric scanning of stained receptor polypeptide bands from SDS–PAGE using bovine serum albumin as standard. The yield obtained with octylglucoside was set as 100%.

This table was taken from Pytela *et al.* (1987).

Triton X-100, reduced Triton X-100 is usually utilized since the cyclohexyl ring of reduced Triton X-100 does not absorb light at 280 nm, unlike the aromatic ring of Triton X-100. Reduction of background absorption caused by detergent helps to detect protein photometrically, facilitating the isolation of detergent extracted integrins in the later purification step of affinity chromatography. In contrast, $\alpha_4\beta_1$ integrin is very sensitive toward dissociation and inactivation. Its subunits are easily dissociated, unless very mild detergents, such as CHAPS at no more than 0.3%, is used (Hemler et al., 1990).

Solubilization of integrins from the plasma membrane with detergents does not only release the integrins but also sets free various proteases. To inhibit these proteases, protease inhibitors, e.g., aprotinin, leupeptin, phenylmethylsulfonyl fluoride (PMSF), or N-ethylmaleinimide (NEM), are added to the extraction buffer. Integrins are not very sensitive to proteases, unless their native conformation is partially or entirely destroyed or unless they lack N-glycosidic sugar side chains. However, reducing agents may loosen the rigid conformation of the cysteine-rich region within the integrin β subunit, thereby making the protein highly accessible and sensitive toward proteolytic degradation. The native conformation of integrins is stablized by divalent cations. For some integrins, such as $\alpha_{IIb}\beta_3$, Ca^{2+} ions are essential in keeping the two subunits together (McKay et al., 1996). Furthermore, divalent cations are essential cofactors for integrins to bind to their ligands. Binding to ligands can be affected by both species and concentration of the respective cation (Gailit and Ruoslahti, 1988; Sonnenberg et al., 1988; Mould et al., 1998). Usually, Mg^{2+} ions are added in a concentration of 1–2 mM. Interestingly, Mn^{2+} ions as nonphysiological substitute for Mg^{2+} increase stability and binding affinity to extracellular matrix ligands significantly. Especially if integrins are isolated by ligand binding affinity chromatography, Mn^{2+} ions are supplied at a concentration of 1 mM. On the other hand, Ca^{2+} at the physiological concentration of 1–2 mM in the extracellular medium, decreases binding affinity of β_1-integrins. However, at micromolar concentrations, the destabilizing effect of Ca^{2+} ions is reversed and increases binding affinities (Onley et al., 2000).

2. Production of Recombinant, Soluble Integrins

Recombinant expression of soluble integrin ectodomain heterodimers lacking the transmembrane domains obviates the extraction of integrin from the plasma membrane by using detergents. Furthermore, as the soluble integrins are secreted by the transfected cells only one integrin species is found in the cell culture supernatant, thus avoiding contamination of soluble integrin with other cell membrane-bound integrins (Smith et al., 1990; Busk et al., 1992). Hence, a secreted soluble integrin can be specifically isolated from the supernatant of transfected cells by affinity chromatography on resin-immobilized fibronectin or vitronectin (Nishimura et al., 1994; Weinacker et al., 1994). Such integrin ectodomain heterodimers have proven to be ideal tools to study interaction of integrins with the extracellular matrix ligands.

Two major strategies have been utilized to produce recombinant soluble integrin heterodimers. In the first strategy both integrin subunits were truncated at the junction between the extracellular and the transmembrane domains. When expressed together the truncated subunits were able to form heterodimers and were secreted into the cell culture medium. Soluble truncated forms have been described for the integrins $\alpha_{IIb}\beta_3$ (Wippler *et al.*, 1994; Gulino *et al.*, 1995; McKay *et al.*, 1996), $\alpha_V\beta_3$ (Mehta *et al.*, 1998), $\alpha_V\beta_6$ (Kraft *et al.*, 1999), $\alpha_L\beta_2$ (Dana *et al.*, 1991), $\alpha_4\beta_1$ (Clark *et al.*, 2000), $\alpha_4\beta_7$ (Higgins *et al.*, 1998), $\alpha_8\beta_1$ (Denda *et al.*, 1998), and $\alpha_1\beta_1$ (Briesewitz *et al.*, 1993). However, yields vary substantially among different integrins. The purification yields (around 1 mg/liter) were comparable for double-truncated soluble $\alpha_8\beta_1$, $\alpha_V\beta_3$, and $\alpha_V\beta_6$ and the intact heterodimers when the integrins were expressed in insect cells (Denda *et al.*, 1998; Mehta *et al.*, 1998; Kraft *et al.*, 1999). In contrast the corresponding truncations of the $\alpha_1\beta_1$, $\alpha_{IIb}\beta_3$, and $\alpha_M\beta_2$ yielded relatively low amounts (around 10 μg/liter) of soluble integrins (Dana *et al.*, 1991; Briesewitz *et al.*, 1993; Bennett *et al.*, 1993; Wippler *et al.*, 1994; Gulino *et al.*, 1995).

The second strategy to produce soluble integrins involves replacing the transmembrane and cytoplasmic domains of both integrin subunits by dimerization motifs, such as the ones of the transcription factors Fos and Jun. This not only helps the heterodimerization of integrin subunits but also prevents their homodimerization, thereby significantly increasing the yields of the soluble products (Eble *et al.*, 1998). Interestingly, as shown for the soluble $\alpha_3\beta_1$ integrin, the Fos and Jun dimerizing motifs can be cleaved off proteolytically, without any loss of heterodimer stability, once the heterodimer is formed. Yet, heterodimer formation may differ from one integrin α subunit to another. Therefore, the requirement for dimerizing motifs for the association of soluble integrin ectodomains should be tested for each individual α–β pair. An alternative way to dimerize the integrin α–β ectodomains is to express the recombinant subunits in which the integrin sequences are fused with F_c-fragments of immunoglobulins (Coe *et al.*, 2001).

For some integrin α subunits shorter fragments of their ectodomains could be expressed as recombinant soluble proteins. Importantly, these "integrin fragments" can retain their ligand binding capability. The high yield-production of A-domains of the integrin α_L, α_M, and α_2 subunits by bacterial expression systems are examples (Michishita *et al.*, 1993; Zhou *et al.*, 1994; Qu and Leahy, 1995; Tuckwell *et al.*, 1995; Calderwood *et al.*, 1997). All three A-domains were expressed as fusion proteins with glutathione-thionyltransferase (GST), which facilitated their purification by affinity chromatography on a glutathione column. However, although able to bind to their ligands, the isolated A-domains may differ in their binding characteristics from the intact ectodomain (Zhou *et al.*, 1994). Among the integrin α subunits lacking A-domains, the α_5 and α_{IIb} subunit have been truncated to minimal ligand-binding integrin fragments (Gulino *et al.*, 1995; McKay *et al.*, 1996; Banéres *et al.*, 1998, 2000).

B. Purification of Solubilized Integrin by Chromatographic Means

1. Purification of Integrin using Affinity Chromatography on Natural Ligands

Whether as detergent-solubilized wild-type forms or as soluble ectodomain heterodimers, the integrins have to be isolated from the crude cellular (or tissue) extract or from the cell culture media. High purification factors are achieved by binding the integrin to an immobilized ligand on an affinity chromatography column. The immobilization of the extracellular matrix proteins to the resin may encounter problems in that extracellular matrix proteins are rather insoluble or only soluble under conditions that are inappropriate for the coupling reaction. For example, collagen is highly insoluble at the slightly alkaline pH that is used for coupling to cyanogen bromide-activated Sepharose. To circumvent the solubility problem, the pH of the collagen solution must be adjusted at the time of adding the cyanogen-bromide activated sepharose. Furthermore, the coupling reaction may block residues essential for integrin binding or lead to steric hindrance and thus inaccessibility to the integrin.

After nonbound contaminants are washed off the column, the bound integrin can be selectively eluted either with a ligand mimetic, such as linear RGD peptide, or by depriving essential divalent cations from the integrin with chelator substances, such as EDTA. Binding of integrins to ligands depends on both the species and the concentration of divalent cations used as cofactors (Gailit and Ruoslahti, 1988; Sonnenberg et al., 1988). Accordingly, to elute integrins bound to an immobilized ligand, one has to apply EDTA, which chelates all divalent cations with high affinity. Immediately after elution, the integrin needs to be stabilized by addition of divalent cations to the eluate (e.g., Mg^{2+} ions).

A potential problem for the purification of integrins on the ligand-immobilized resin is that many extracellular matrix integrin ligands (such as collagen, laminin, and fibronectin) are multifunctional proteins that interact with one another and various other cell-surface receptors (Timpl, 1996). This may result in contamination of the purified integrin with other cell-surface receptors or extracellular matrix molecules (Dufour et al., 1988). One route around this problem is to use chemical or proteolytic fragments of the large extracellular matrix proteins to identify smaller integrin-binding fragments that can be used for the affinity matrix. Such fragments include the cyanogen bromide-derived fragment CB3[IV] of type IV collagen and the chymotrypsin-derived 120-kDa fragment of fibronectin (Pierschbacher et al., 1981; Vandenberg et al., 1991; Kern et al., 1993). However, fragmentation of extracellular matrix proteins can affect their ternary structure and destroy the integrin-binding sites. In this regard, the CB3[IV] fragment retains its triple helical conformation and supports $\alpha_1\beta_1$ and $\alpha_2\beta_1$-mediated cell adhesion. Furthermore, this fragment has been successfully used to purify both of these collagen binding integrins (Vandenberg et al., 1991; Kern et al., 1993).

The integrins $\alpha_3\beta_1$, $\alpha_6\beta_1$, $\alpha_7\beta_1$ and $\alpha_6\beta_4$ can be isolated on laminins (Gehlsen et al., 1988; Kramer et al., 1989; Elices et al., 1991; von der Mark et al., 1991; Lee et al., 1992; Song et al., 1993). Proteolytic fragmentation of laminin-1 produces a C-terminal fragment, E8, which contains the binding site for integrins $\alpha_3\beta_1$, $\alpha_6\beta_1$, and $\alpha_7\beta_1$ (Aumailley et al., 1987; Goodman et al., 1987), but is not recognized by $\alpha_6\beta_4$ integrin (Sonnenberg et al., 1990), nor by $\alpha_1\beta_1$ and $\alpha_2\beta_1$ integrins (Hall et al., 1990; Pfaff et al., 1994; Ettner et al., 1998). This fragment has also been used to purify laminin-binding integrins (Sonnenberg et al., 1991; von der Mark et al., 1991). Furthermore, a peptide derived from the C-terminal G-domain of laminin-1 has been used as an affinity matrix to isolate $\alpha_3\beta_1$ integrin (Gehlsen et al., 1992). Importantly, neither the collagen-binding nor the laminin-binding integrins can be eluted from the ligand affinity column by RGD peptides or other short peptide mimetics, but are eluted from their affinity columns by chelating agents, such as EDTA.

The integrin recognition sites of several other extracellular matrix proteins (e.g., fibronectin, fibrinogen, vitronectin) are less restricted by the conformation of the molecules. They include a linear RGD peptide that protrudes from the rigid molecular framework as a rather flexible looplike structure, which is accessible to the integrin receptor (Leahy et al., 1996). The RGD-containing sequences are recognized by the fibronectin receptor $\alpha_5\beta_1$, the platelet fibrinogen receptor $\alpha_{IIb}\beta_3$, the vitronectin receptor $\alpha_V\beta_3$, and other RGD-dependent integrins of the integrin α_V subfamily, including $\alpha_V\beta_1$, $\alpha_V\beta_5$, $\alpha_V\beta_6$, and $\alpha_V\beta_8$. For the RGD-dependent integrins, the ligand-binding site can be reduced to the linear RGD-peptide sequence GRGDSP. The platelet integrin $\alpha_{IIb}\beta_3$ and the vitronectin receptor $\alpha_V\beta_3$ were isolated with high purity from platelet, placental, and cellular extracts, respectively, using immobilized GRGDSP peptide (Pytela et al., 1985b, 1986). Yields were increased when a spacer sequence between the RGD peptide and the resin was included to increase accessibility of the peptide to integrins. Although it belongs to increase the group of RGD-dependent integrins, the fibronectin receptor $\alpha_5\beta_1$ cannot be isolated by affinity chromatography with immobilized RGD peptide. This is probably because, in addition to the RGD loop within the 10th type III repeat, the so-called synergy site in the ninth type III repeat of fibronectin is also required for avid $\alpha_5\beta_1$ integrin binding (Aota et al., 1994). Although the RGD peptide does not suffice to mediate high-affinity binding of $\alpha_5\beta_1$ integrin, it is still a potent inhibitor of the $\alpha_5\beta_1$ integrin interaction with fibronectin. It can therefore be utilized for the purification step. For example, $\alpha_5\beta_1$ integrin from the placental extract is eluted specifically and with high purity from the immobilized 120-kDa integrin binding fragment of fibronectin with RGD peptide (Pytela et al., 1985b, 1987).

The LDV/IDS amino acid sequence of fibronectin, VCAM (vascular cell adhesion molecule), MadCAM (mucosal addressin cell adhesion molecule), and ICAMs (intercellular cell adhesion molecules) is known to be part of the recognition sites for the α_4-integrins, $\alpha_4\beta_1$ and $\alpha_4\beta_7$, and the leukocyte integrins, $\alpha_M\beta_2$ and $\alpha_L\beta_2$, respectively (Mould and Humphries, 1991; Osborn et al., 1994; Renz et al., 1994; Vonderheide et al., 1994). The binding sites for $\alpha_4\beta_1$ integrin within fibronectin had been delineated to two distinct peptide sequences CS1 and CS5 within the alternatively spliced, so-called connecting segment (Humphries et al., 1987). Only the high-affinity binding site of CS-1, which contains the LDV sequence, can be used as immobilized ligand on an affinity chromatography column (Guan and Hynes, 1990; Mould et al., 1990). In contrast, the integrin–ligand interface is more complex for the immunoglobulin family members (VCAM, MadCAM, and the ICAMs), as further residues apart from the loop-based LDV/IDS-sequence contribute to their interaction with the integrins. Consequently, LDV- or IDS-containing peptides do not support the avid integrin binding necessary for purification by affinity chromatography and, therefore, have not been used for integrin isolation.

Integrins are also recognized by viruses and pathogenic microorganisms, which subvert integrins as docking sites for invasion into eukaryotic cells. Examples are the foot-and-mouth disease virus, which bears an RGD sequence within its capsid protein and binds to RGD-dependent integrins, and bacterial strains of *Shigella, Yersinia,* and enteropathogenic *E. coli,* which bind with their surface proteins, invasin and intimin, respectively, to the eukaryotic host cell (Isberg and Leong, 1990; Frankel *et al.*, 1996). Invasin binds very avidly to most β_1 integrins, except for those ones that contain an A-domain within their α subunit (Isberg and Leong, 1990). Because of its divalent cation-dependent high binding affinity to β_1 integrins, invasin has proved to be a valuable tool for purifying integrin by affinity chromatography. Integrins bound to the resin-immobilized invasin or its cell-binding domain (Hamburger *et al.*, 1999) can be eluted with EDTA under mild conditions (Eble *et al.*, 1998).

2. Purification of Integrins Using Affinity Chromatography of Lectin Columns

Integrins are glycoproteins bearing several N-linked carbohydrate side chains on both subunits. N-linked glycoconjugates of integrins isolated from mammalian tissue and cells generally belong to the complex-type N-coupled sugar chains. Therefore, resin-immobilized wheat-germ agglutinin or ricin can be used to isolate integrins from mammalian cells and tissues (Hemler *et al.*, 1987a; Takada *et al.*, 1987; Hemler *et al.*, 1989; Smith *et al.*, 1990; Eble *et al.*, 1993). However, integrins recombinantly expressed in insect cells, such as *Drosophila* Schneider S2 cells or lepidopteran Sf9 and High Five cells (BTI-Tn5B1-4 cells), usually bear N-linked carbohydrate side chains of the high-mannose or paucimannose type, since insect cells fail to process N-linked high-mannose type carbohydrate conjugates into complex-type sugar chains (Altmann *et al.*, 1995; Hollister and Jarvis, 2001; Marchal *et al.*, 2001). Integrins recombinantly expressed by insect cells can therefore be purified on concanavalin A-column (Wippler *et al.*, 1994). Elution of bound integrins is achieved by applying N-acetylglucosamine or α-D-methylmannoside, respectively, at high concentrations of 200–300 mM under native conditions. These conditions are sufficiently mild to retain the ligand-binding activities of the integrins. After the elution, the low molecular weight sugars can be removed by dialysis. However, as most extracellular matrix proteins or other cell surface proteins are N-glycosylated, lectin affinity chromatography of tissue or cell extracts by itself does not provide a very pure integrin preparation.

3. Purification of Integrins Using Affinity Chromatography on Antibody Columns

Unlike matrix affinity chromatography columns, resin-coupled monoclonal antibodies have the advantage of high specificity and affinity toward a distinct integrin. Thus, a particular integrin can be isolated with high purity even from tissues that contain many different integrins (Kraft *et al.*, 1999). By use of immune-affinity chromatography on monoclonal antibodies directed against the integrin β_1 subunit, highly enriched integrin preparations were obtained from different tissues, such as placenta, leukocytes, and platelets (Takada *et al.*, 1987; Makarem *et al.*, 1994; Mould *et al.*, 1996). The integrin $\alpha_6\beta_4$ was captured from placental extract using a monoclonal antibody against the integrin β_4 subunit (Hemler *et al.*, 1989). The separation of pure $\alpha_V\beta_3$ integrin from other members of the α_V integrin subfamily, which show similar ligand binding specificity and RGD dependence, was only possible by using the monoclonal antibody LM609, which is specifically directed against the whole $\alpha_V\beta_3$ integrin (Smith and Cheresh, 1988). Purification of other α_V integrins has been achieved by combining recombinant expression as secreted heterodimers with antibody affinity chromatography using a monoclonal antibody specific to the integrin α_V subunit (Smith *et al.*, 1990; Nishimura *et al.*, 1994; Weinacker *et al.*, 1994).

A wide range of monoclonal antibodies directed against integrins and their individual subunits have been developed (e.g., described in Takada *et al.*, 1987; Hall *et al.*, 1990) and are commercially available. To generate the immunoaffinity matrix, antibodies are coupled to cyanogen bromide-activated or *N*-hydroxysuccinimidyl-activated agarose resins according to the manufacturers' instructions. General protocols are also provided by Harlow and Lane (1988). After binding to the immobilized antibody, the integrin attached on the column can be washed extensively to remove all contaminants. Smith and Cheresh (1988) described a washing step even at pH 4.5 without losing the integrin from the column. The bound integrin is generally eluted from the antibody column with a pH shift to pH 3.2–3.5, but elution under strong alkaline conditions of pH 11.5 has been described as well (Hemler *et al.*, 1987a; Takada *et al.*, 1987; Takagi *et al.*, 2001). After the elution, the integrin solution must be neutralized immediately to avoid inactivation of the integrin and its biological activity should be tested. For integrins that are less stable and dissociate under the pH extremes used for elution from the immunoaffinity resin (such as $\alpha_4\beta_1$ integrin, Hemler *et al.*, 1987b; 1990), elution can also be achieved using a peptide that mimics the antibody epitope (Clark *et al.*, 2000).

4. Purification of Integrins Using Tag Sequences

Recombinant expression of integrins offers the opportunity to add or insert tag sequences at or into the integrin protein, which facilitate detection and isolation of the tag-labeled protein either with antibodies or nonimmunologically. Various tag sequences (e.g., the epitope of myc, fusion protein moieties of thioredoxin, and glutathionyl-S-transferase (GST)) had been utilized to ease detection and purification of recombinantly expressed integrins by affinity chromatographic approaches (Denda *et al.*, 1998; Banéres *et al.*, 1998; Michishita *et al.*, 1993; Zhou *et al.*, 1994; Tuckwell *et al.*, 1995). Oligohistidine tags, which allow easy purification of the labeled integrin on a Ni-NTA-column, are also in common use (Banéres *et al.*, 1998; Denda *et al.*, 1998; Banéres *et al.*, 2000). Another interesting tag that allows easy integrin detection is a fusion protein of soluble $\alpha_8\beta_1$ integrin with alkaline phosphatase (Denda *et al.*, 1998).

5. Purification by Ion-Exchange Chromatography

Most integrins are acidic proteins with isoelectric points around pH 5 (Hemler *et al.*, 1987a), suggesting that they can be purified by anion exchange chromatography under physiological pH conditions. Indeed, detergent-extracted $\alpha_{IIb}\beta_3$ was isolated on DEAE-anion exchange resin (Calvete *et al.*, 1992). Soluble $\alpha_2\beta_1$ integrin was also separated from contaminating proteins using a Mono-Q chromatography column (Eble *et al.*, 2001). Integrin A-domains, such as the A-domain of the integrin α_L-subunit, also have acidic character and bind to anion exchange resins under physiological conditions (Qu and Leahy, 1995). The integrins or domains that have been bound to the anion-exchange resin are eluted with a gradient of increasing NaCl concentration. Both binding and elution of integrin occur under native conditions.

C. Protocols

1. Extraction of Integrins from Tissue or Cells

Cells detached from the culture dish by EDTA-treatment, leukocytes, or minced tissue are washed several times with Tris-buffered saline (TBS: 50 mM Tris/HCl, pH 7.4, 150 mM NaCl) containing 2 mM MgCl$_2$, 1 mM PMSF, and 1 mM NEM by centrifugation and resuspension. After a final washing step, the suspension is centrifuged (5 min, 4°C, 1000g). The pellet is resuspended in about 10 times the volume of extraction buffer (100 mM octyl-β-D-glucoside, 2 mM MgCl$_2$, 1 mM of each phenylmethylsulfonyl fluoride (PMSF) and N-ethylmaleinimide (NEM), 1 μg/ml of each aprotinin, leupeptin, antipain, and pepstatin in TBS, pH 7.4). Integrins are extracted by the detergent by slow rotation of the tube at 4°C overnight. The suspension is then centrifuged at 25,000g at 4°C for 30 min. The supernatant contains the detergent-solubilized integrin.

2. Isolation of Solubilized or Soluble Integrin by Ligand Affinity Chromatography

To immobilize the integrin ligand on a column resin, dissolve the ECM protein or its fragment in a basic, nonnucleophilic buffer, such as 0.1 M NaHCO$_3$, pH 8.3, containing 0.5 M NaCl. Add the cyanogen-bromide activated or N-hydroxysuccinimidyl (NHS)-activated agarose resin (Biorad Affi-Gel 10) to the solution of the ECM-protein. Incubate the tube on a slowly turning rotation disk overnight at 4°C to avoid sedimentation. Excess coupling sites on the resin are blocked by adding a nucleophilic compound, such as Tris or ethanolamine at 0.1 M and pH 8.0. ECM proteins that are insoluble under these conditions, such as collagen, are dissolved in a nonnucleophilic buffer, such as 0.1 M acetic acid. Upon addition of the cyanogen bromide- or NHS-activated resin, the pH is adjusted with a concentrated solution of sodium hydroxide. The ligand-loaded resin is thoroughly washed to remove nonimmobilized ligand.

To isolate the integrin, the resin is loaded into a column and equilibrated with TBS, pH 7.4, containing 2 mM MgCl$_2$ and 1 mM MnCl$_2$ (wash buffer). Mn^{2+} ions are added as a general activator of integrins to increase their binding affinity toward the ligand. If detergent-solubilized integrin is purified, octyl-β-D-glucoside (OG) is added to the buffer (wash buffer + OG). The tissue extract with the detergent-solubilized integrin or cell supernatant with soluble integrin is adjusted to pH 7.5 and to 1 mM Mn^{2+} and loaded onto the resin-coupled ligand. Then, the column is washed with wash buffer with or without detergent, respectively, until the photometrically detected baseline is reached. Bound integrin is generally eluted with a 10 mM EDTA solution in wash buffer. Immediately after elution from the column, integrin-containing eluate fractions are adjusted to 20 mM MgCl$_2$ and pH 7.4.

3. Isolation of Integrin by Affinity Chromatography on a Lectin Column

> To isolate detergent-solubilized integrin or recombinantly expressed integrin, the lectin-column (Sigma, L1882 or C9017) is equilibrated with wash-buffer with or without detergent, respectively. Note that some lectins require Ca^{2+} ions to bind their sugar ligands avidly. The adjusted extraction solution or cell supernatant is loaded onto the lectin column. After washing the column, until baseline absorbance is achieved, bound integrin is eluted from the wheat germ agglutinin or concanavalin A column with wash-buffer containing the competing saccharide, N-acetylglucosamine or α-D-methyl-mannoside, respectively, at a concentration of 300 mM. The integrin-containing eluate fractions are dialyzed to remove the low molecular sugars.

III. Characterization of Integrin–Associated Proteins through mAb Production: General Considerations and Points of Variation

A. Detection of Integrin–Associated Proteins by Immunoprecipitation

> The immunoprecipitation protocol is as a principal method used for the detection of integrin-associated proteins. We did not find any differences when function-blocking or non-function-perturbing integrin antibodies were used for the primary precipitation. A wide range of the immunoprecipitation conditions should be tested at the preliminary stage to optimize the detection of the complexes while minimizing the coimmunoprecipitation of nonspecific proteins. The choice of detergent is critical, as not only does this influence the efficiency of integrin solubilisation itself (see Pytela et al., 1987, and Table 1), but it may also affect the stability of the association. Traditionally, certain nonionic (various polyoxyethylene ethers (Brij compounds), Digitonin) and zwitterionic (CHAPS) detergents are considered as "mild" and widely used in studying interactions involving membrane proteins. However, the stability of any given integrin-containing protein complex in the presence of a particular detergent may vary significantly and has to be established empirically in the initial "screening" experiments.
>
> Presence of divalent cations (Ca^{+2} or Mg^{+2}) may be important to stabilize the integrin heterodimer and to maintain it in the ligand-binding competent conformation. However, if divalent cations are included in the immunoprecipitation buffer, special precaution should be taken to inactivate divalent cation-dependent proteases (e.g., calpains, matrix metalloproteases). Hence, in addition to the obligatory inhibitors of aspartic, serine, and cysteine proteases (PMSF, leupeptin, aprotinin, pepstatin), the buffer should contain inhibitors of metalloproteases (e.g., calpeptin, bestatin). Finally, the presence of wide-range phosphatase inhibitors, including sodium orthovanadate, sodium pyrophosphate, and sodium fluoride, could also improve the recovery of the integrin-containing complexes (Mainiero et al., 1995).
>
> Using established cell lines allows for easy labeling and detection of the protein complexes by immunoprecipitation. Cellular proteins can be labeled either metabolically with (^{35}S-methionine, ^{35}S-cysteine) or by using a surface-labeling procedure (biotin, Na^{125}I).

Both protocols have their pros and cons: (A) metabolic labeling allows detection of the association of integrins with both intracellular and transmembrane proteins, but the labeling efficiency depends on protein turnover and the number of cysteine and methionine residues in the proteins. Furthermore, immunoprecipitation experiments using metabolically labeled cellular extracts often result in a higher general radiolabeled background than the immunoprecipitation of surface-labeled proteins. Shortening the time of labeling to 1–2 h can reduce the background. We found that this was a sufficient time interval to achieve good integrin labelling in many different adherent cells. However, short labelling may require additional "chase" time to account for the proteins with slow turnover. B) immunoprecipitation of surface-labelled protein complexes generates lower background and, in the case of biotinylation, allows for fast signal detection (in hours versus days when radioactivity is used) but it will only permit the detection of membrane proteins.

B. Characterization of Integrin–Associated Proteins by Monoclonal Antibody Production

If the association is robust, it may be possible to scale up the immunopurification protocol. This will typically involve prolonged preclearance and more extensive washing steps. The integrin-associated proteins can be further purified using preparative polyacrylamide gel electrophoresis and, if produced in sufficient quantities (picomoles), characterized by sequencing. Alternatively, the complexes purified by immunoprecipitation can be used as immunogens to generate mAbs against the component molecules (e.g., integrin subunits and integrin-associated proteins). It is advisable not to elute the complexes from the anti-integrin antibody prior to the injection into animals because dissociation of the complexes may change the protein conformation and preclude generation of mAbs with functional activities. Furthermore, no harmful side effects were noticed when mice (or rats) were injected with up to 100 μl of agarose matrices per animal. Because many commonly used detergents are quite toxic, the resin-immobilized complexes should be extensively washed with PBS prior to the injection. The injection of the matrix-immobilized complexes does not require addition of adjuvant. The immune responses were comparable when mice were injected subcutaneously and intraperitoneally. Typically, three injections with 3-week intervals suffice to generate strong immune responses to the integrin-associated proteins. Test bleeds should be assayed by immunoprecipitation.

C. Screening Hybridoma Clones for the Production of Monoclonal Antibodies Directed against Integrin–Associated Proteins

Choosing an appropriate screening protocol is the most important step in the methodology. Two scenarios are feasible: (i) an integrin-associated protein is expressed on the cell surface; (ii) an integrin-associated protein is not exposed to the extracellular space. If the target integrin-associated proteins are expressed on the cell surface (Fig. 1, step 1), analysis by flow cytometry can be used as a first step for the selection of positive clones. Positive clones (or rather supernatants) are taken for further analysis. Integrins are strong immunogens when delivered as purified complexes. Consequently, a significant

1. <u>Selection for mAbs recognizing cell surface proteins.</u>

Flow cytometry: human cells that were used as a source for purification of the integrin complexes.

2. <u>Selection against mAb recognizing integrins.</u>

Flow cytometry: non-human cells expressing human $\alpha\beta$ subunits and the parental non-human cell line (e.g CHO cells and CHO cells expressing human α and β subunits).

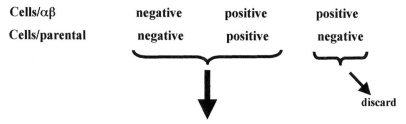

3. <u>Selection for mAbs recognizing integrin-associated proteins.</u>

Differential immunoprecipitation : immunoprecipitation is performed under both "mild" and "stringent" conditions followed by Western blotting with anti-integrin Ab.

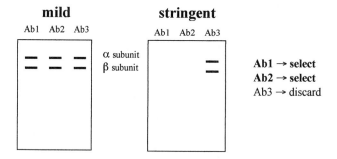

Fig. 1 Screening protocol for integrin-associated proteins exposed on the cell surface.

proportion of hybridoma clones will secrete mAbs to integrin subunits. From our experience, mAbs against α subunits are more frequent than against $\beta1$ subunit. To eliminate these clones, the mAbs are analyzed by a second round of flow cytometry experiments using a pair of nonhuman cell lines (preferably of rodent origin), one of which expresses the human integrin to which the mAbs were raised. For example, for the mAbs directed to the $\alpha3\beta1$-associated proteins we used CHO cells that expressed human $\alpha3$ and human $\beta1$ integrin subunits and the parental CHO cells. Any mAb that stains the cells expressing the human integrin and is negative on the parental cell line is discarded from further analysis (Fig. 1, step 2).

However, mAbs that give either positive or negative results on both cell lines are selected for use in subsequent immunoprecipitation screening steps. Initially, the interaction is analyzed under "mild" immunoprecipitation conditions (see earlier discussion), and the presence of an integrin in the immunoprecipitates is detected by a standard Western blotting protocol using specific anti-integrin Ab (rabbit polyclonal Abs for most integrin subunits are available from several commercial sources including Chemicon, Santa Cruz, or Pharmingen). Alternatively, cells can be surface biotinylated and precipitated integrins, which are well labeled with biotin, are then easily visualized as two high molecular weight bands in a range of 100–150 kDa using streptavidin-conjugated horseradish peroxidase (Fig. 1, step 3).

Positive hybridoma supernatants (e.g., those that coimmunoprecipitate an integrin) should be characterized in further immunoprecipitation experiments. At this stage selected clones combine two categories of mAbs: (i) directed to the integrin-associated proteins; (ii) directed to the integrins or the associated proteins and cross-reacting with their nonhuman orthologs. In most cases, the anti-integrin mAbs can be distinguished by performing the immunoprecipitation experiments under "more stringent" conditions (e.g., using higher concentrations of NaCl, or more hydrophobic detergent). The mAbs that precipitate integrins only under "mild" conditions are considered to be against the associated proteins (Fig. 1, step 3). The corresponding hybridoma clones are expanded and after two rounds of subcloning, each mAb can be purified and used for further characterization of its target antigen.

A variation of the foregoing protocol is used for screening the mAbs for which the target integrin-associated proteins are not exposed on the cell surface (Fig. 2). This screening scheme relies almost entirely on the immunoprecipitation experiments. Significant workload in the initial screening steps (it may be necessary to carry out up to 1000 immunoprecipitation reactions during 10–20 days) can be eased slightly by initial flow cytometry analysis. The supernatants that stain the cells used as a source of purification of the integrin complexes are excluded from further analysis (Fig. 2, step 1). The initial large-scale screening by immunoprecipitation (using only the "mild" conditions) and the subsequent comparative immunoprecipitation analysis using "more stringent" conditions are carried out as described above (Fig. 2, steps 2 and 3). The selected hybridoma clones are then used individually to characterize the corresponding target antigens. This includes not only analyzing the association with an integrin over a wide range of experimental conditions (by approaches similar to the initial characterization of the integrin-associated proteins described above) but also identifying cells (or tissues) that express the highest level of the protein in question. These experiments are necessary to establish optimal conditions for subsequent immunoaffinity purification of the integrin-associated proteins.

D. Immunoaffinity Purification of Integrin-Associated Proteins

As described in Section II.C.3 immunoaffinity purification offers a highly specific and relatively quick way to, obtain sufficient amounts of pure material to permit sequence analysis. For most mAbs, an optimized immunoprecipitation protocol could be adapted for the large-scale immunoaffinity purification of integrin-associated proteins.

1. <u>Selection for mAbs recognizing cell surface proteins.</u>

<u>Flow cytometry:</u> **human cells that were used as a source for purification of the integrin complexes.**

negative   positive

discard

2. <u>Selection for mAbs recognizing integrin-associated proteins</u>.

<u>Immunoprecipitation:</u> **immunoprecipitation under "mild" conditions followed by Western blotting with anti-integrin Ab.**

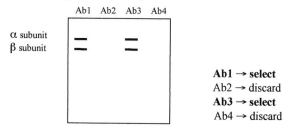

α subunit
β subunit

Ab1 → **select**
Ab2 → discard
Ab3 → **select**
Ab4 → discard

3. <u>Selection against mAbs recognizing cytoplasmic /transmembrane domains of integrin subunits.</u>

<u>Differential immunoprecipitation:</u> **parallel immunoprecipitation under "mild" and "stringent" conditions followed by Western blotting with anti-integrin Ab.**

mild   **stringent**

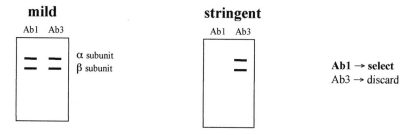

α subunit
β subunit

Ab1 → **select**
Ab3 → discard

Fig. 2   Screening protocol for integrin-associated intracellular proteins.

Occasionally, covalent coupling of the purified mAbs to solid matrix resin reduces the efficiency of antibody binding to the target proteins. Hence, preliminary small-scale purification experiments with the coupled Ab should be performed to make appropriate adjustments to the protocol. We routinely use a two-step preclearance procedure before applying the lysate to the matrix with the immobilized target mAb: we incubate the lysate with the BSA-conjugated matrix (4–6 h) and, subsequently, with the matrix-conjugated irrelevant Ab (4–16 h). Both preclearance incubations and subsequent incubation with the conjugated target mAb (8–16 h) are performed in batch. For the washing step the collected beads are loaded on the column. The elution of the antigen from the column is achieved by pH-shifting, as described in Section II.C.3.

E. Protocols

Detailed protocols for animal immunization and hybridoma production are described by Harlow and Lane (1988).

1. Cell Labeling

Metabolic labeling with ^{35}S-methionine and ^{35}S-cysteine. Before labeling, cells (2×10^6–10^7) are washed twice in PBS followed by incubation in the serum-/methionine-/cysteine-free medium for 1.5–2 h. Fresh serum-/methionine-/cysteine-free medium containing 0.5–1 mCi of EXPRE^{35}S^{35}S mix (NEN, >1000 Ci/mmol) is added to the cells for 1–2 h. After the removal of the radioisotope-containing medium, cells are incubated for an additional 2–4 h in the complete growth medium before the lysis.

Cell surface biotinylation. Cells (2×10^6–10^7) are washed twice with PBS containing 2 mM MgCl$_2$ and 0.5 mM CaCl$_2$ (PBS/Mg/Ca). Sulfo-NHS-LC-Biotin (Pierce) is dissolved in PBS/Mg/Ca at 0.2 mg/ml and the solution is added to the cells for 1–2 h at room temperature. Incubation is carried out on the rotating platform/orbital shaker set up at slow motion. Cells are washed three times with PBS/Mg/Ca before the lysis. When labeling is performed on the adherent cells, post-labeling washing steps have to be carried out with care as some cells become susceptible to shear force after prolonged incubation in PBS/Mg/Ca.

Cell surface iodination. Detached cells (2–5×10^7) are washed twice with PBS/Mg/Ca and subsequently resuspended in 1 ml PBS/Mg/Ca supplemented with 50 mM glucose. For labeling of adherent cells, 3–4 ml of PBS/Mg/Ca/glucose per 90 mm-diameter tissue culture dish is used. Bovine lactoperoxidase (Sigma, L8257), *Aspergillus* glucose oxidase (Sigma, G9010) (at 1 U/ml each), and, subsequently, 0.5–1 mCi of Na^{125}I (NEN, $\sim$17 Ci/mg) are added to the cells for 20 min. Labeling is performed at room temperature with occasional agitation. The labeling reaction is terminated by addition of NaN$_3$ (0.1%). The cells are washed three times with PBS/Mg/Ca/NaN$_3$ solution before lysis in the immunoprecipitation buffer.

2. Immunoprecipitation

Labeled cells are mechanically scraped from the dish into the lysis buffer (in the case of nonadherent cells the lysis buffer is added to the cellular pellet after centrifugation). We have used lysis buffer of the following composition for initial immunoprecipitations: 1% Brij 96 in PBS, containing 2 mM MgCl$_2$, 2 mM phenylmethylsulfonyl fluoride, 10 μg/ml aprotinin, 10 μg/ml leupeptin, 1 μg/ml pepstatin, 1 mM Na$_3$VO$_4$, 5 mM NaF, 10 mM Na$_3$P$_3$O$_7$. However, various other detergents can be tested as well (for example,

Brij 35, Brij 58, Brij 98, digitonin, CHAPS, Triton X-100—these detergents can be obtained from Sigma, Pierce, or Roche). All Brij compounds are solid at room temperature and need to be melted at 60°C for preparation of the lysis buffer. To prepare 1% digitonin lysis buffer, 1 g of digitonin is added to 50 ml H_2O and stirred at 37°C for 1 h. The solution is then boiled for 30 min and, subsequently, cooled to room temperature. The solution is supplemented with Tris-HCl (buffered at pH 7.5) and NaCl to the required concentrations (20–100 mM and 150–500 mM, respectively), and then filtered through a 0.45 μm filter.

If cells are lysed in the immunoprecipitation buffer at the ratio of 2–5 × 10^6 cells per ml, almost complete solubilization of integrins can be achieved during 2–4 h at 4°C. If the cell number is higher in proportion to the buffer volume, a longer solubilization time (6–8 h) may be required. We have also noticed that for certain cells (K562, MDA-MB-435) the extraction of integrin by 1% Brij 98 at the suggested cell number/volume ratio is less efficient. Hence, one may need to increase the volume of the immunoprecipitation buffer and prolong the time of the extraction to improve the yield.

After extraction, the insoluble material is pelleted at 12,000 rpm for 10 min at 4°C and the cell lysate is transferred into the preclearance tubes containing the agarose beads conjugated with goat anti-mouse antibodies (Sigma, A-6531). The volume of the packed beads for the preclearance should exceed the combined volume of the beads used in all immunoprecipitation reactions by 3- to 5-fold (e.g., if the lysate is used for four separate immunoprecipitation reactions with 10 μl of the packed beads for each of them, then the volume of the agarose beads taken into the preclearance reaction should be 120–200 μl). The preclearance is carried out with constant agitation for 0.5–4 h at 4°C.

In parallel with carrying out the cellular lysis, the mAbs to be tested in the immunoprecipitation reactions are mixed with the slurry of goat anti-mouse agarose beads. We typically use 10 μl of the packed beads (resuspended in 300 μl of the immunoprecipitation buffer without protease inhibitors) to precipitate integrin complexes from 5 × 10^5–2 × 10^6 cells. The volume of mAb to be added to the beads for immobilization depends on the concentrations in the original source: we typically use either 200–800 μl of undiluted hybridoma culture supernatant, 2–5 μl of ascites, or 2–10 μg of purified antibodies. Possible alternatives to goat anti-mouse agarose beads are protein G- and protein A-conjugated matrices (both are available from several commercial sources, e.g., Sigma, Pierce). We have consistently found that using protein G matrix generates higher background especially when the immunoprecipitation was performed in the presence of Brij detergents. Thus, a modified preclearance procedure involving protein G beads may be required. In contrast, using protein A matrices is preferable as it results in a relatively low background. However, as mouse mAbs of certain isotypes have poor binding affinity to protein A, precoating the protein A beads with anti-mouse Ab is required before the mAbs under test are applied to the beads.

The mAb immobilization step is continued throughout the time of cell lysis and preclearance. The immobilized mAbs are then washed three times (12,000 rpm, 5 s, 4°C) with the immunoprecipitation buffer before mixing with the precleared lysate. The immune complexes are collected onto the beads for 6–18 h at constant agitation at 4°C followed by four washes with the immunoprecipitation buffer. Immune complexes are eluted from beads with Laemmli sample buffer at 95°C for 5 min and resolved by SDS–PAGE. Alternatively, the immunoprecipitated material can be eluted from the beads by 0.1 M glycine, pH 2.7 : 50 μl of glycine is added to the beads for 5 min at 4°C, and the eluted material is neutralized by transferring the supernatant into 10 μl Tris-HCl, pH 8.0. This elution procedure is repeated twice to achieve complete recovery of the material from the beads. Immunoprecipitated proteins are then analyzed on SDS–PAGE gels or by Western blot according to standard procedures.

3. Preparation of Cells for Flow Cytometry Analysis

Cells are washed with ice-cold serum-free growth medium or PBS and resuspended at 10^6/ml in ice-cold 3% BSA/PBS/0.1% NaN$_3$. Aliquots of cells (100 μl) are pipetted into a round-bottom 96-well tissue culture plate (NUNCLON, Life Technologies) that has been precoated with 3% BSA/PBS for 30 min at 4°C. Undiluted hybridoma supernatant (100 μl) is added to the wells and the plate is incubated on ice for 1 h. The primary Ab supernatants are removed after centifugation at 1200 rpm for 3 min at 4°C, and the cells are washed three times with 150 μl ice-cold PBS. Cells are resuspended in 100 μl 1% BSA/PBS/0.1% NaN$_3$ supplemented with fluorescein-conjugated goat anti-mouse Ab (Sigma) at 1 : 75 dilution. The plate is incubated at 4°C for 1 h in the dark (e.g., wrapped in aluminium foil) and subsequently washed with PBS as above. After the final wash cells are resuspended in ice-cold PBS and transferred for fixation into 0.5 ml 2% paraformaldehyde aliquoted into tubes suitable for use in flow cytometry analysis. As a negative control, samples should be stained with an irrelevant mAb (e.g., a mAb directed to an antigen that is not expressed on the target cells). As a positive control, cells could be stained with an mAb directed against the β1 integrin subunit.

References

Altmann, F., Schwihla, H., Staudacher, E., Glössö, J., and März, L. (1995). Insect cells contain an unusual, membrane-bound β-N-actelyglucosaminidase propably involved in the processing of protein N-glycans. *J. Biol. Chem.* **270**, 17344–17349.

Altruda, F., Cervella, P., Gaeta, M. L., Daniele, A., Giancotti, F. G., Tarone, G., Stefanuto, G., and Silengo, L. (1989). Cloning of cDNA for a novel mouse membrane glycoprotein (gp42): shared identity to histocompatibility antigens, immunoglobulins and neural-cell adhesion molecules. *Gene* **85**, 445–451.

Aota, S.-I., Nomizu, M., and Yamada, K. M. (1994). The short amino acid sequence pro-his-ser-arg-asp in human fibronectin enhances cell-adhesive function. *J. Biol. Chem.* **269**, 24756–24761.

Aumailley, M., Nurcombe, V., Edgar, D., Paulsson, M., and Timpl, R. (1987). The cellular interactions of laminin fragments; Cell adhesion correlates with two fragment-specific high-affinity binding sites. *J. Biol. Chem.* **262**, 11532–11538.

Banéres, J.-L., Roquet, F., Green, M., LeCalvez, H., and Parello, J. (1998). The cation-binding domain from the α subunit of integrin α5β1 is a minimal domain for fibronectin recognition. *J. Biol. Chem.* **273**, 24744–24753.

Banéres, J.-L., Roquet, F., Martin, A., and Prello, J. (2000). A minimized human integrin $\alpha_5\beta_1$ that retains ligand recognition. *J. Biol. Chem.* **275**, 5888–5903.

Bennett, J. S., Kolodziej, M. A., Vilaire, G., and Poncz, M. (1993). Determinants of the intracellular fate of truncated forms of the platelet glycoprotein IIb and IIIa. *J. Biol. Chem.* **268**, 3580–3585.

Berditchevski, F., Chang, S., Bodorova, J., and Hemler, M. E. (1997). Generation of monoclonal antibodies to integrin-associated proteins. Evidence that $\alpha3\beta1$ complexes with EMMPRIN/basigin/OX47/M6. *J.Biol.Chem.* **272,** 29174–29180.

Briesewitz, R., Epstein, M. R., and Marcantonio, E. E. (1993). Expression of native and truncated forms of the human integrin $\alpha1$ subunit. *J. Biol. Chem.* **268,** 2989–2996.

Busk, M., Pytela, R., and Sheppard, D. (1992). Characterization of the integrin $\alpha v\beta6$ as a fibronectin-binding protein. *J. Biol. Chem.* **267,** 5790–5796.

Calderwood, D. A., Tuckwell, D. S., Eble, J., Kühn, K., and Humphries, M. J. (1997). The integrin $\alpha1$ A-domain is a ligand binding site for collagens and laminin. *J. Biol. Chem.* **272,** 12311–12317.

Calvete, J. J. (1994). Clues for understanding the structure and function of a prototypic human integrin: the platelet glycoprotein IIb/IIIa complex. *Thrombosis Haemostasis* **72,** 1–15.

Calvete, J. J., Schäfer, W., Mann, K., Henschen, A., and González-Rodríguez, J. (1992). Localization of the cross-linking sites of RGD and KQAGDV peptides to the isolated fibrinogen receptor, the human platelet integrin glicoprotein IIb/IIIb: Influence of peptide length. *Eur. J. Biochem.* **206,** 759–765.

Chang, D. D., Wong, C., Smith, H., and Liu, J. (1997). ICAP-1, a novel beta1 integrin cytoplasmic domain-associated protein, binds to a conserved and functionally important NPXY sequence motif of beta1 integrin. *J. Cell Biol.* **138,** 1149–1157.

Clark, K., Newham, P., Burrows, L., Askari, J. A., and Humphries, M. J. (2000). Production of recombinant soluble integrin $\alpha4\beta1$. *FEBS Lett.* **471,** 182–186.

Coe, A. P. F., Askari, J. A., Kline, A. D., Robinson, M. K., Kirby, H., Stephens, P. E., and Humphries, M. J. (2001). Generation of a minimal $\alpha5\beta1$ integrin-F_c fragment. *J. Biol. Chem.* **276,** 35854–35866.

Dana, N., Fathallah, D. M., and Arnaout, M. A. (1991). Expression of a soluble and functional form of the human $\beta2$ integrin CD11b/CD18. *Proc. Natl. Acad. Sci. USA* **88,** 3106–3110.

Denda, S., Müller, U., Crossin, K. L., Erickson, H. P., and Reichardt, L. F. (1998). Utilization of a soluble integrin–alkaline phosphatase chimera to characterize integrin $\alpha8\beta1$ receptor interactions with tenascin: murine $\alpha8\beta1$ binds to the RGD site in tenascin-C fragments, but not to native tenascin-C. *Biochemistry* **37,** 5464–5474.

Dufour, S., Duband, J.-L., Humphries, M. J., Obara, M., Yamada, K. M., and Thiery, J. P. (1988). Attachment, spreading and locomotion of avian neural crest cells are mediated by multiple adhesion sites on fibronectin molecules. *EMBO J.* **7,** 2661–2671.

Eble, J. A., Beermann, B., Hinz, H.-J., and Schmidt-Hederich, A. (2001). $\alpha_2\beta_1$ integrin is not recognized by rhodocytin but is the specific high affinity target of rhodocetin, an RGD-independent disintegrin and potent inhibitor of cell adhesion to collagen. *J. Biol. Chem.* **276,** 12274–12284.

Eble, J. A., Golbik, R., Mann, K., and Kühn, K. (1993). The $\alpha1\beta1$ integrin recognition site of the basement membrane collagen molecule [$\alpha1$(IV)]$_2\alpha2$(IV). *EMBO J.* **12,** 4795–4802.

Eble, J. A., Wucherpfennig, K. W. L. G., Dersch, P., Krukonis, E., Isberg, R. R., and Hemler, M. E. (1998). Recombinant soluble human $\alpha_3\beta_1$ integrin: purification, processing, regulation, and specific binding to laminin-5 and invasin in a mutually exclusive manner. *Biochemistry* **37,** 10945–10955.

Elices, M. J., Urry, L. A., and Hemler, M. E. (1991). Receptor functions for the integrin VLA-3: fibronectin, collagen, and laminin binding are differentially influenced by arg-gly-asp peptide and by divalent cations. *J. Cell. Biol.* **112,** 169–181.

Ettner, N., Göhring, W., Sasaki, T., Mann, K., and Timpl, R. (1998). The N-terminal globular domain of the laminin $\alpha1$ chain binds to the $\alpha1\beta1$ and $\alpha2\beta1$ integrin and to the heparan sulfate-containing domains of perlecan. *FEBS Lett.* **430,** 217–221.

Fenczik, C. A., Sethi, T., Ramos, J. W., Hughes, P. E., and Ginsberg, M. H. (1997). Complementation of dominant suppression implicates CD98 in integrin activation. *Nature* **390,** 81–85.

Frankel, G., Lider, O., Hershkoviz, R., Mould, A. P., Kachalsky, S. G., Candy, D. C. A., Cahalon, L., Humphries, M. J., and Dougan, G. (1996). The cell-binding domain of intimin from enteropathogenic *Escherichia coli* binds to β_1 integrin. *J. Biol. Chem.* **271,** 20359–20364.

Frederickson, R. M. (1998). Macromolecular matchmaking: advances in two-hybrid and related technologies. *Curr. Opin. Biotechnol.* **9,** 90–96.

Gailit, J., and Ruoslahti, E. (1988). Regulation of the fibronectin receptor affinity by divalent cations. *J. Biol. Chem.* **263,** 12927–12932.

Gehlsen, R. K., Dillner, L., Engvall, E., and Ruoslahti, E. (1988). The human laminin receptor is a member of the integrin family of cell adhesion receptors. *Science* **241,** 1228–1229.

Gehlsen, K. R., Sriramarao, P., Furcht, L. T., and Skubitz, A. P. N. (1992). A synthetic peptide derived from the carboxy terminus of the laminin A chain represents a binding site for the $\alpha_3\beta_1$ integrin. *J. Cell Biol.* **117,** 449–459.

Giancotti, F. G., and Ruoslahti, E. (1999). Integrin signaling. *Science* **285,** 1028–1032.

Goodman, S. L., Deutzmann, R., and von der Mark, K. (1987). Two distinct cell-binding domains in laminin can independently promote nonneuronal cell adhesion and spreading. *J. Cell Biol.* **105,** 589–598.

Guan, J.-L., and Hynes, R. O. (1990). Lymphoid cells recognize an alternatively spliced segment of fibronectin via the integrin receptor $\alpha_4\beta_1$. *Cell* **60,** 53–61.

Gulino, D., Martinez, P., Delachanal, E., Concord, E., Duperray, A., Alemany, M., and Marguerie, G. (1995). Expression and purification of a soluble functional form of the platelet $\alpha_{IIb}\beta_3$ integrin. *Eur. J. Biochem.* **227,** 108–115.

Hall, D. E., Reichardt, L. F., Crowley, E., Holley, B., Moezzi, H., Sonnenberg, A., and Damsky, C. H. (1990). The α_1/β_1 and α_6/β_1 integrin heterodimers mediate cell attachment to distinct sites on laminin. *J. Cell Biol.* **110,** 2175–2184.

Hamburger, Z. A., Brown, M. S., Isberg, R. R., and Bjorkman, P. J. (1999). Crystal structure of invasin: a bacterial integrin-binding protein. *Science* **286,** 291–295.

Hannigan, G. E., Leung-Hagesteijn, C., Fitz-Gibbon, L., Coppolino, M. G., Radeva, G., Filmus, J., Bell, J. C., and Dedhar, S. (1996). Regulation of cell adhesion and anchorage-dependent growth by a new beta1-integrin-linked protein kinase. *Nature* **379,** 91–96.

Harlow, E., and Lane, D. (1988). Immunoaffinity purification. *In* "Antibodies, a Laboratory Manual" (E. Harlow and D. Lane, eds.), pp. 511–552. Cold Spring Harbor Laboratory Press, Cold Spring Harbor, NY.

Hemler (1988). p. 3.

Hemler, M. E. (1998). Integrin associated proteins. *Curr. Opin. Cell Biol.* **10,** 578–585.

Hemler, M. E., Crouse, C., and Sonnenberg, A. (1989). Association of the VLA α^6 subunit with a novel protein; a possible alternative to the common VLA β_1 subunit on certain cell lines. *J. Biol. Chem.* **264,** 6529–6535.

Hemler, M. E., Elices, M. J., Parker, C., and Takada, Y. (1990). Structure of the integrin VLA-4 and its cell–cell and cell–matrix adhesion functions. *Immunol. Rev.* **114,** 45–65.

Hemler, M. E., Huang, C., and Schwarz, L. (1987a). The VLA protein family; characterization of five distinct cell surface heterodimers each with a common 130,000 molecular weight β subunit. *J. Biol. Chem.* **262,** 3300–3309.

Hemler, M. E., Huang, C., Takada, Y., Schwarz, L., Strominger, J. L., and Clabby, M. L. (1987b). Characterization of the cell surface heterodimer VLA-4 and related peptides. *J. Biol. Chem.* **262,** 11478–11485.

Higgins, J. M. G., Mandlebrot, D. A., Shaw, S. K., Russell, G. J., Murphy, E. A., Chen, Y.-T., Nelson, W. J., Parker, C. M., and Brenner, M. B. (1998). Directed and regulated interaction of integrin $\alpha_E\beta_7$ with E-Cadherin. *J. Cell Biol.* **140,** 197–210.

Hollister, J. R., and Jarvis, D. L. (2001). Engineering lepidopteran insect cells for sialoglycoprotein production by genetic transformation with mammalian β-1,4-galactosyltransferase and α2,6-sialylstransferase genes. *Glycobiology* **11,** 1–9.

Humphries, M. J. (2000). Integrin structure. *Biochem. Soc. Trans.* **28,** 311–339.

Humphries, M. J., Komoriya, A., Akiyama, S. K., Olden, K., and Yamada, K. M. (1987). Identification of two distinct regions of the type III connecting segment of human plasma fibronectin that promote cell type-specific adhesion. *J. Biol. Chem.* **262,** 6886–6892.

Isberg, R. R., and Leong, J. M. (1990). Multiple β_1 chain integrins are receptors for invasin, a protein that promotes bacterial penetration into mammalian cells. *Cell* **60,** 861–871.

Kern, A., Eble, J., Golbik, R., and Kühn, K. (1993). Interaction of type IV collagen with the isolated integrins $\alpha 1\beta 1$ and $\alpha 2\beta 1$. *Eur. J. Biochem.* **215,** 151–159.

Kolanus, W., Nagel, W., Schiller, B., Zeitlmann, L., Godar, S., Stockinger, H., and Seed, B. (1996). $\alpha L\beta 2$ integrin/LFA-1 binding to ICAM-1 induced by cytohesin-1, a cytoplasmic regulatory molecule. *Cell* **86,** 233–242.

Kraft, S., Diefenbach, B., Mehta, R., Jonczyk, A., Luckenbach, G. A., and Goodman, S. L. (1999). Definition of an unexpected ligand recognition motif for $\alpha v\beta 6$ integrin. *J. Biol. Chem.* **274,** 1979–1985.

Kramer, H. R., McDonald, K. A., and Vu, M. P. (1989). Human melanoma cells express a novel integrin receptor for laminin. *J. Biol. Chem.* **264,** 15642–15649.

Leahy, D. J., Aukhil, I., and Erickson, H. P. (1996). A crystal structure of a fourdomain segment of human fibronectin encompassing the RGD loop and synergy region. *Cell* **84,** 155–164.

Lee, E. C., Lotz, M. M., Steele, G. D., and Mercurio, A. M. (1992). The integrin $\alpha6\beta4$ is a laminin receptor. *J. Cell Biol.* **117,** 671–678.

Liu, S., Calderwood, D. A., and Ginsberg, M. H. (2000). Integrin cytoplasmic domain-binding proteins. *J. Cell Sci.* **113,** 3563–3571.

Mainiero, F., Pepe, A., Wary, K. K., Spinardi, L., Mohammadi, M., Schlessinger, J., and Giancotti, F. G. (1995). Signal transduction by the alpha6 beta4 integrin: distinct beta4 subunit sites mediate recruitment of Shc/Grb2 and association with the cytoskeleton of hemidesmosomes. *EMBO J.* **14,** 4470–4481.

Makarem, R., Newham, P., Askari, J. A., Green, L. J., Clements, J., Edwards, M., Humphries, M. J., and Mould, A. P. (1994). Competitive binding of vascular cell adhesion molecule-1 and the HepII/IIICS domain of fibronectin to the integrin $\alpha4\beta1$. *J. Biol. Chem.* **269,** 4005–4011.

Marchal, I., Jarvis, D. L., Cacan, R., and Verbert, A. (2001). Glycoproteins from insect cells: sialylated or not?. *Biol. Chem.* **382,** 151–159.

McKay, B. S., Annis, D. S., Honda, S., Christie, D., and Kunicki, T. J. (1996). Molecular requirements for assembly and function of a minimized integrin $\alpha_{IIb}\beta_3$. *J. Biol. Chem.* **271,** 30544–30547.

Mehta, R. J., Diefenbach, B., Brown, A., Cullen, E., Jonczyk, A., Güssow, D., Luckenbach, G. A., and Goodman, S. L. (1998). Transmembrane-truncated $\alpha v \beta 3$ integrin retains high affinity for ligand binding: evidence for an "inside-out" supressor? *Biochem. J.* **330,** 861–869.

Michishita, M., Videm, V., and Arnaout, M. A. (1993). A novel divalent cation-binding site in the A-domain of the $\beta2$ integrin CR3 (CD11b/CD18) is essential for ligand binding. *Cell* **72,** 857–867.

Mould, A. P., and Humphries, M. J. (1991). Identification of a novel recognition sequence for the integrin $\alpha4\beta1$ in the CooH-terminal heparin binding domain of fibronectin. *EMBO J.* **10,** 4089–4095.

Mould, A. P., Akiyama, S. K., and Humphries, M. J. (1996). The inhibitory anti-$\beta1$ integrin monoclonal antibody 13 recognizes an epitope that is attenuated by ligand occupancy. Evidence for allosteric inhibition of integrin function. *J. Biol. Chem.* **271,** 20365–20374.

Mould, A. P., Garratt, A. N., Puzon-McLaughlin, W., Takada, Y., and Humphries, M. J. (1998). Regulation of integrin function: evidence that bivalent-cation-induced conformational changes lead to unmasking of ligand-binding sites within integrin $\alpha5\beta1$. *Biochem. J.* **331,** 821–828.

Mould, A. P., Komoriya, A., Yamada, K. M., and Humphries, M. J. (1991). The CS5 peptide is a second site in the IIICS region of fibronectin recognized by the integrin $\alpha_4\beta_1$. Inhibition of the $\alpha_4\beta_1$ function by RGD peptide homologues. *J. Biol. Chem.* **266,** 3579–3585.

Mould, A. P., Wheldon, L. A., Komoriya, A., Wayner, E. A., Yamada, K. M., and Humphries, M. J. (1990). Affinity chromatography isolation of the melanoma adhesion receptor for the IIICS region of fibronectin and its identification as the integrin $\alpha_4\beta_1$. *J. Biol. Chem.* **265,** 4020–4024.

Nishimura, S. L., Sheppard, D., and Pytela, R. (1994). Integrin $\alpha v \beta 8$: Interaction with vitronectin and functional divergence of the $\beta8$ cytoplasmic domain. *J. Biol. Chem.* **269,** 28708–28715.

Onley, D. J., Knight, C. G., Tuckwell, D. S., Barnes, M. J., and Farndale, R. W. (2000). Micromolar Ca^{2+} concentrations are essential for Mg^{2+}-dependent binding of collagen by the integrin $\alpha_2\beta_1$ in human platelets. *J. Biol. Chem.* **275,** 24560–24564.

Osborn, L., Vassallo, C., Browning, B. G., Tizard, R., Haskard, D. O., Benjamin, C. D., Dougas, I., and Kirchhausen, T. (1994). Arrangement of domains, and amino acid residues required for binding of vascular cell adhesion molecule-1 to its counter-receptor VLA-4 ($\alpha_4\beta_1$). *J. Cell Biol.* **124,** 601–608.

Otey, C. A., Pavalko, F. M., and Burridge, K. (1990). An interaction between α-actinin and the $\beta1$ integrin subunit in vitro. *J. Cell Biol.* **111,** 721–729.

Otey, C. A., Vasquez, G. B., Burridge, K., and Erickson, B. W. (1993). Mapping of the α-actinin binding site within the β_1 integrin cytoplasmic domain. *J. Biol.Chem.* **268,** 21193–21197.

Pfaff, M., Göhring, W., Brown, J. C., and Timpl, R. (1994). Binding of purified collagen receptors ($\alpha1\beta1$, $\alpha2\beta1$) and RGD-dependent integrins to laminins and laminin fragments. *Eur. J. Biochem.* **225,** 975–984.

Pierschbacher, M. D., Hayman, E. G., and Ruoslahti, E. (1981). Location of the cell-attachment site in fibronectin with monoclonal antibodies and protelytic fragments of the molecule. *Cell* **26**, 259–267.

Plow, E. F., Haas, T. A., Zhang, L., Loftus, J., and Smith, J. W. (2000). Ligand binding to integrins. *J. Biol. Chem.* **275**, 21785–21788.

Pytela, R., Pierschbacher, M. D., Argraves, S., Suzuki, S., and Ruoslahti, E. (1987). Arginine-glycine-aspartatic acid adhesion receptors. *Methods Enzymol.* **144**, 475–489.

Pytela, R., Pierschbacher, M. D., Ginsberg, M. H., Plow, E. F., and Ruoslahti, E. (1986). Platelet membrane glycoprotein IIB/IIIA: member of a family of arg-gly-asp-specific adhesion receptors. *Science* **231**, 1559–1562.

Pytela, R., Pierschbacher, M. D., and Ruoslahti, E. (1985a). A 125/115-kDa cell surface receptor specific for vitronectin interacts with the arginine-glycine-aspartic acid adhesion sequence derived from fibronectin. *Proc. Natl. Acad. Sci. USA* **82**, 5766–5770.

Pytela, R., Pierschbacher, M. D., and Ruoslahti, E. (1985b). Identification and isolation of a 140 kd cell surface glycoprotein with properties expected of a fibronectin receptor. *Cell* **40**, 191–198.

Qu, A., and Leahy, D. J. (1995). Crystal structure of the I-domain from the CD11a/CD18 (LFA-1, $\alpha_L\beta2$) integrin. *Proc. Natl. Acad. Sci. USA* **92**, 10277–10281.

Renz, M. E., Chiu, H. H., Jones, S., Fox, J., Kim, K. J., Presta, L. G., and Fong, S. (1994). Structural requirements for adhesion of soluble recombinant murine vascular cell adhesion molecule-1 to $\alpha4\beta1$. *J. Cell. Biol.* **125**, 1395–1406.

Schaller, M. D., Otey, C. A., Hilderbrand, J. D., and Parsons, J. T. (1995). Focal adhesion kinase and paxillin bind to peptides mimicking β integrin cytoplasmic domains. *J. Cell Biol.* **130**, 1181–1187.

Schwartz, M. A., Schaller, M. D., and Ginsberg, M. H. (1995). Integrins: emerging paradigms of signal transduction. *Ann. Rev. Cell Dev. Biol.* **11**, 549–599.

Scopes, R. P. (1994). "Protein Purification. Principles and Practice," 3rd ed. Chapters 5–7. Springer-Verlag, New York,

Shattil, S. J., O'Toole, T., Eigenthaler, M., Thon, V., Williams, M., Babior, B. M., and Ginsberg, M. H. (1995). β_3-Endonexin, a novel polypeptide that interacts specifically with the cytoplasmic tail of the integrin β_3 subunit. *J. Cell Biol.* **131**, 807–816.

Smith, J. W., and Cheresh, D. A. (1988). The arg-gly-asp binding domain of the vitronectin receptor: Photoaffinity cross-linking implicates amino acid residue 61-203 of the β subunit. *J. Biol. Chem.* **263**, 18726–18731.

Smith, J. W., Deborah, J., Irwin, S. V., Burke, T. A., and Cheresh, D. A. (1990). Purification and functional characterization of integrin $\alpha_v\beta_5$: An adhesion receptor for vitronectin. *J. Biol. Chem.* **265**, 11008–11013.

Song, W. K., Wang, W., Sato, H., Bielser, D. A., and Kaufman, S. J. (1993). Expression of α_7 integrin cytoplasmic domains during skeletal muscle development: alternative forms, conformational change, and homologies with serine/threonine kinases and tyrosine phosphatases. *J. Cell Sci.* **106**, 1139–1152.

Song, W. K., Weigwang, W., Foster, R. F., Bielser, D. A., and Kaufman, S. J. (1992). H36-α7 is a novel integrin alpha chain that is developmentally regulated during skeletal myogenesis. *J. Cell Biol.* **117**, 643–657.

Sonnenberg, A., Gehlsen, K. G., Aumailley, M., and Timpl, R. (1991). Isolation of $\alpha6\beta1$ integrins from platelets and adherent cells by affinity chromatography on mouse laminin fragment E8 and human laminin pepsin fragment. *Exp. Cell Res.* **197**, 234–244.

Sonnenberg, A., Linders, C. J. T., Modderman, P. W., Damsky, C. H., Aumailley, M., and Timpl, R. (1990). Integrin recognition of different cell-binding fragments of laminin (P1, E3, E8) and evidence that $\alpha6\beta1$ but not $\alpha6\beta4$ functions as a major receptor for fragment E8. *J. Cell Biol.* **110**, 2145–2155.

Sonnenberg, A., Modderman, P. W., and Hogervorst, F. (1988). Laminin receptor on platelets is the integrin VLA-6. *Nature* **336**, 487–489.

Takada, Y., Strominger, J. L., and Hemler, M. E. (1987). The very late antigen family of heterodimers is part of a superfamily of molecules involved in adhesion and embryogenesis. *Proc. Natl. Acad. Sci. USA* **84**, 3239–3243.

Takagi, J., Erickson, H. P., and Springer, T. A. (2001). C-terminal opening mimics "inside-out" activation of integrin $\alpha5\beta1$. *Nature Struct. Biol.* **8**, 412–416.

Timpl, R. (1996). Binding of laminins to extracellular matrix components. *In* "The Laminins" (P. Ekblom and R. Timpl, eds.), Vol. 2, pp. 97–125. Harwood Academic Publishers.

Tuckwell, D., Calderwood, D. A., Green, L. J., and Humphries, M. J. (1995). Integrin alpha2 I-domain is a binding site for collagens. *J. Cell Sci.* **108,** 1629–1637.

Vandenberg, P., Kern, A., Ries, A., Luckenbill-Edds, L., Mann, K., and Klaus, K. (1991). Characterization of a type IV collagen major cell binding site with affinity to the $\alpha 1\beta 1$ and $\alpha 2\beta 1$ integrin. *J. Cell Biol.* **113,** 1475–1483.

Vidal, M., and Legrain, P. (1999). Yeast forward and reverse "n"-hybrid systems. *Nucleic Acids Res.* **27,** 4919–4929.

von der Mark, H., Dürr, J., Sonnenberg, A., von der Mark, K., Deutzmann, R., and Goodman, S. L. (1991). Skeletal myoblasts utilize a novel $\beta 1$-series integrin and not $\alpha 6\beta 1$ for binding to the E8 and T8 fragments of laminin. *J. Biol. Chem.* **266,** 23593–23601.

Vonderheide, R. H., Tedder, T. F., Springer, T. A., and Staunton, D. E. (1994). Residues within a conserved amino acid motif of domains 1 and 4 of VCAM-1 are required for binding to VLA-4. *J. Cell Biol.* **125,** 215–222.

Weinacker, A., Chen, A., Agrez, M., Cone, R. I., Nishimura, S., Wayner, E., Pytela, R., and Sheppard, D. (1994). Role of the integrin $\alpha v\beta 6$ in cell attachment: Heterologous expression of intact and secreted forms of the receptor. *J. Biol. Chem.* **269,** 6940–6948.

Wippler, J., Kouns, W. C., Schlaeger, E.-J., Kuhn, H., Hadvary, P., and Steiner, B. (1994). The integrin α_{IIb}-β_3, platelet glycoprotein IIb-IIIa, can form a functionally active heterodimer complex without the cysteine-rich repeats of the β_3 subunit. *J. Biol. Chem.* **269,** 8754–8761.

Zhou, L., Lee, D. H. S., Plescia, J., Lau, C. Y., and Altieri, D. C. (1994). Differential ligand binding specificities of recombinant CD11b/CD18 integrin I-domain. *J. Biol. Chem.* **269,** 17075–17079.

CHAPTER 11

Methods for Analysis of the Integrin Ligand Binding Event

Jeffrey W. Smith

The Burnham Institute
La Jolla, California 92037

I. Introduction

The primary objective in making measurements of the integrin ligand binding event is to link structure with function, or to apply biochemistry to gain an understanding of regulation at the cellular level. In the case of integrins, biochemical information can be particularly important because their regulation appears to be complex (Hynes, 1992; Plow *et al.,* 2000). Many integrins contain multiple ligand binding sites that are influenced by allosteric effectors and inhibitors (Hu *et al.,* 1996; Kallen *et al.,* 1999). Unraveling the interplay between these regulatory sites and attributing their effects to specific protein domains requires an in-depth analysis of the ligand binding event.

METHODS IN CELL BIOLOGY, VOL. 69

The study of the integrin ligand binding event is relatively mature, with studies having begun in the 1980s (for reviews see Plow *et al.,* 2000; Leitinger *et al.,* 2000; van Kooyk and Figdor, 2000). Nevertheless, a number of questions on the biological regulation of this binding event and its structural basis remain. For example, many integrins bind to more than one ligand, so questions arise regarding the relationship and interplay between the ligand binding pockets. How are they structurally connected, and how does ligand binding at one site influence binding at the second site? Moreover, papers are published every year reporting the identification of additional integrin ligands. Consequently, approaches to gauge the significance of the affinity for these new ligands are important. Another pressing issue is to understand how the integrin ligand binding site is exposed, or turned on and off. Although it is very clear that part of this on/off switch resides in the cytoplasm (Hynes, 1992; Hughes *et al.,* 1996), the way in which conformational changes in the cytoplasm ultimately influence the exposure of the ligand binding site(s) remains unclear.

This chapter describes four separate methods for measuring parameters relevant to the integrin ligand binding event. The first section provides a discussion of cross-linking technology, which has been applied to map the topology of the integrin ligand binding domain. The second section provides detailed methods and a practical discussion of microtiter-based assays for measuring ligand binding. The third section discusses the application of surface plasmon resonance (SPR) to dissect the overall ligand binding reaction into its individual steps. The final section briefly introduces methods for a relatively new area, the study of integrin redox states. In each case I have attempted to provide sufficient theoretical background and experimental detail, and to disclose any caveats that we are aware of.

II. Application of Affinity Cross-Linking to Study of the Integrin Ligand Binding Pocket

A discussion of affinity cross-linking is provided because this technique laid the foundation for the study of the integrin ligand binding event. Therefore, it has great historical importance. In addition, the study of integrins is a fantastic illustration of advances in cross-linking methodology that can now be applied to gain incredibly precise information on protein topology. Although a specific method is not presented, key aspects of the considerations associated with cross-linking and their importance to the study of integrins are discussed. General descriptions of affinity cross-linking procedures can be found in protein methods guides such as *Protein Function, a Practical Approach* (Creighton, 1997, OUP) or the Molecular Probes methods handbook (www.probes.com/handbook).

The interpretation of results from affinity cross-linking studies depends on an understanding of proximity, particularly as it is defined by the structure of the cross-linking reagents that are employed. Affinity cross-linking reagents contain two key elements: a ligand that brings the cross-linking group in proximity to the protein, and the chemically reactive group, which covalently tags the receptor. The tagging of the receptor can be achieved either with chemical cross-linking reagents or with photolabile reagents. By changing the cross-linking strategy, one can alter the parameters that define the spacing between the ligand and the affinity tag. Consequently, different cross-linking strategies alter the constraints that define proximity.

Prior to the study of integrins, and even in the very initial searches for the fibronectin receptor, cross-linking studies were aimed primarily at proving the existence of a cellular receptor or assisting with its identification. In these analyses, a large protein ligand was used to tag a cellular receptor at any position. The position of such a tag was often unimportant, as the method was being applied to gain basic information on the size and molecular identity of the putative receptor. Indeed, Gardner and Hynes used this approach to cross-link the β subunit of the fibronectin receptor (Gardner and Hynes, 1985).

As the field progressed, however, affinity cross-linking was applied for an entirely different purpose. The ability of integrins to bind to small peptide ligands, such as RGD (Ruoslahti, 1996), allowed cross-linking to be employed to obtain information of much higher resolution. Because affinity cross-linking reagents were linked to small peptide ligands, these reagents facilitated the tagging of the integrin at sites proximal to the ligand binding pocket.

Santoro and Lawing were the first to apply photoaffinity cross-linking to investigate the structural basis of ligand binding to an integrin. They studied the platelet integrin αIIbβ3 (Santoro and Lawing, 1987), which binds to ligands containing the RGD motif and also to a peptide derived from the C-terminal region of the gamma chain of fibrinogen. The relationship between these two binding sites has been quite controversial because RGD peptides will compete for the binding of fibrinogen to αIIbβ3, but the nature of this competition is not classical competitive inhibition (Lam *et al.,* 1987; Hu *et al.,* 1999). Santoro and Lawing coupled synthetic peptides, representing the two types of ligands, to commercially available photoaffinity reagents. These photoaffinity derivatives were found to cross-link to distinct sites on αIIbβ3. These observations were really the first indication of the complex kinetic interplay between the two ligand binding sites (see below).

The first information on the position of the integrin ligand binding pocket came from closely related studies performed by D'Souza *et al.* (1988a,b, 1990) and by myself while I was in Cheresh's group (Smith and Cheresh, 1988, 1990). We each applied different strategies to identify the regions of the integrin proximal to the ligand binding sites. The methodological differences between the two approaches are worthy of discussion. The primary difference in approach relates to the type of cross-linking reagent that was used. Our group used photoaffinity derivatives of RGD peptides. The reasoning behind this choice centered on the fact that cross-linking efficiency is known to be very low, and the final objective was to use peptide mapping to identify proteolytic fragments of the integrin that contained the affinity label. Therefore, we reasoned that a photoaffinity derivative of an RGD peptide, which would tag the integrin with a small chemical moiety linked to ^{125}I, and not to the entire RGD peptide, would have less of an impact on the separation of proteolytic fragments containing the tag. Thus, we were using the photoaffinity-tagged region of the receptor as a tracer, to track and identify the nature of the RGD binding pocket on the αvβ3 integrin (Smith and Cheresh, 1988, 1990).

D'Souza *et al.* used chemical cross-linking of RGD and fibrinogen gamma chain peptides to identify the ligand binding site(s) on the platelet integrin αIIbβ3. They found the efficiency of chemical cross-linking to be relatively high, and also wisely chose to study the most abundant integrin, so that starting material was not a limitation (D'Souza *et al.,* 1988a,b, 1990). Results from both studies were similar and laid the groundwork for a far more in-depth analysis of this integrin's ligand-binding sites.

Although more than a decade has passed since these early cross-linking studies, the method appears to still have something to offer the integrinologist. Bitan *et al.* have synthesized custom RGD cross-linking probes in an attempt to further constrain the sites of cross-linking. This extension of the method refines the meaning of "proximal," making it a term that implies very precise positioning. Bitan synthesized a series of affinity tags of cyclic RGD peptides which incorporated a photoreactive benzophenone moiety (Bitan *et al.,* 2000). The position of the photoreactive group was varied in relation to the RGD sequence, yielding probes in which the reactive group is positioned at either the N or the C terminus of the peptide. Interestingly, the two affinity probes labeled the integrin at distinct sites. By precisely mapping the position of the cross-linking sites within the integrin, the authors were able to arrive at spatial constraints that define the topology of the RGD binding site. The results from this analysis challenge the prevailing models of the ligand binding domain, arguing against an I-domain structure within the N terminus of the integrin $\beta 3$ subunit. Given recent advances in mass spectrometry, one might anticipate that ligand affinity cross-linking methods could be coupled with nonspecific chemical cross-linking of integrin to gain an even higher resolution view of integrin topology.

III. Microtiter–Based Integrin Binding Assays

A microtiter-based assay for measuring integrin ligand interactions has also found widespread use. The assay was originally performed by Nachman and Leung in 1982, using purified platelet integrin $\alpha IIb\beta 3$ (Nachman and Leung, 1982). Since then, we and several other groups have employed this assay to measure binding parameters between integrins and a number of ligands (Smith *et al.,* 1990, 1994a,b; Scarborough *et al.,* 1991; Mould *et al.,* 1995). Others have used a similar assay to measure the binding of ligands to individual integrin domains (Dickeson *et al.,* 1997, 1999). The assay can provide relative measures of ligand binding affinity and assign a rank-order potency to compounds that block ligand binding. The approach has also helped define much of the allosteric regulation of the ligand binding event, including the effects of divalent cations (Mould *et al.,* 1995; Dickeson *et al.,* 1997).

A. General Considerations

The type and concentration of detergent used to purify the integrin is a primary consideration in the microtiter-based ligand binding assay. Although the integrin is coated onto microtiter plates by diluting it into buffer lacking detergent, significant amounts of detergent can be carried through this dilution. This can interfere with the coating of the integrin onto the microtiter wells. This problem is illustrated in Fig. 1, where $\alpha IIb\beta 3$ purified in either 1% or 0.1% Triton X-100 was immobilized into microtiter wells. Each integrin preparation was coated onto the microtiter wells across a range of

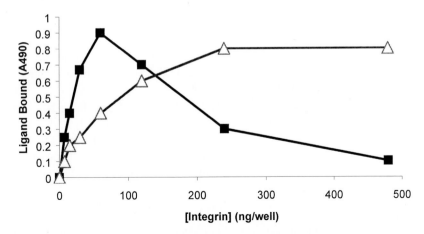

Fig. 1 Effects of integrin coating concentration on ligand binding in microtiter-based binding assays. The effect of coating integrin at different protein:detergent ratios on ligand binding in microtiter based assays was examined. The αIIbβ3 integrin was purified from human platelets on RGD-affinity resins in either 0.1% (triangles) or 1% Triton X-100 (squares). Purified protein was then concentrated to 1.0 mg/ml by incubating a dialysis bag of integrin in a bed of Aquacide. The concentration of integrin was measured with the BCA assay (Pierce Chemical Co.), and then different concentrations of integrin were used to coat microtiter wells. The binding of biotinylated human fibrinogen to wells was quantified as described in the text. The binding of ligand to the integrin purified in high concentrations of detergent exhibits a biphasic response across the range of coated protein. We attribute this to the fact that detergent interferes with coating of the integrin to microtiter wells.

concentrations, and the binding of fibrinogen was measured. The binding of fibrinogen to the integrin purified in 0.1% Triton X-100 exhibits an increase in ligand binding, which reaches a plateau just above 100 ng/well of integrin. In contrast, binding to the integrin purified in 1% Triton X-100 reaches a peak and then sharply drops. We have traced this effect to reduced coating of the integrin onto microtiter wells, and to the effect of detergent concentration present during coating. In general, we have found it desirable to achieve integrin concentrations of near 1 mg/ml in either 0.1% Triton X-100 or 50 mM octylglucoside. These stocks of integrin can be frozen at $-80°$C until use, when they can be diluted for coating of microtiter plates. In cases where this concentration of integrin cannot be achieved, the optimal coating concentration of integrin should be determined in a preliminary titration of the coating concentration of integrin, as illustrated in Fig. 1.

B. General Method

The following protocol is suggested for purified integrins containing their transmembrane domains. It has also been the experience of our group that all buffers should be made fresh on the day of use. Although concentrated stocks of Tris buffer are routinely used, they are not suitable for this assay and will invariably lead to failure of the assay.

1. Buffers

Coating Buffer: 20 mM Tris · HCl, pH 7.4, containing 150 mM NaCl and 1 mM each of CaCl$_2$ and MgCl$_2$. Divalent cations should be made as 100 mM stocks in distilled water and diluted into Coating Buffer. Stocks of divalent cations can be stored at room temperature for up to 1 month.

Binding Buffer: 50 mM Tris · HCl pH 7.4 containing 100 mM NaCl and 1 mg/ml of bovine serum albumin (Sigma, RIA grade). Typically MgCl$_2$ and CaCl$_2$ are added to the Binding Buffer to achieve a final concentration of 1 mM (diluted from 100 mM stocks). However, we and others have often varied the concentration and type of divalent ion in the binding reaction to gain insight into the mechanisms behind the effects of divalent ions.

2. Method

1. Purified integrin is diluted to a concentration of 0.5 μg/ml in Coating Buffer. Of this solution 100 μl is added to each well of a microtiter plate (LinBro Titertek 96-well flat-bottom plates are optimal). Integrin is allowed to coat the wells for 14–18 h at 4°C. Buffer is removed from the well by inverting.

2. The remaining nonspecific binding sites on the plate are blocked with Coating Buffer containing 10 mg/ml of bovine serum albumin (Sigma, RIA grade).

3. The blocking solution is removed by inversion, and the plate is washed three times with Binding Buffer (100 μl/well).

4. The binding assay is initiated by adding ligand to the wells of the plate. Ligand (100 μl/well) is diluted to a suitable concentration in Binding Buffer and is allowed to incubate with integrin for 1–3 h at 37°C. Most natural adhesive proteins bind to integrin with very slow association rates, so the binding time should be extended to 3 h. In contrast, many antibodies and optimized ligands bind rapidly so equilibrium is achieved within 1 h.

5. Following the binding period, the unbound ligand is removed by inversion or aspiration, and the plate is washed three times with Binding Buffer (100 μl per well).

6. The amount of bound ligand is then quantified. In cases where radioiodinated ligand is used, the ligand is released from the well by addition of 100 μl of boiling 0.2 N NaOH. For safety reasons, the sodium hydroxide is brought to boiling by heating in a beaker of water. The NaOH is incubated in the wells for 20 min and then removed individually with a Gilson P200 pipette. We have found that ejection of both the NaOH sample and the pipette tip into the tubes for gamma counting is efficient.

If precise quantification is not necessary, then biotinylated ligand can be used and detected with secondary antibody. In instances where ligand binding to integrin is irreversible, we have also been able to measure the bound ligand with antibody against ligand, followed by a suitable secondary antibody, much like a sandwich ELISA.

In most cases nonspecific binding in this assay is less than 10%. Nevertheless, accurate measures are made only when specific binding is determined by subtracting nonspecific binding from the total binding. Nonspecific binding should be measured for every concentration of added ligand and can be assessed by competition with a 300-fold excess of unlabeled ligand.

For the $\alpha IIb\beta3$ integrin and the αv-integrins, nonspecific binding can be measured by competition with a vast excess of the peptide GRGDSP. If the amount of ligand is limiting, then nonspecific binding can often be determined with the use of EDTA at a final concentration 10-fold above the concentration of $CaCl_2$ and $MgCl_2$. Typically, we use a final concentration of 20 mM EDTA, added directly to the binding reaction prior to addition of ligand. A 200 mM stock of EDTA is used (10 μl is added per 100 μl of reaction volume).

C. Interpretation of Binding Data

There are some aspects of the integrin ligand binding event that can complicate the interpretation of binding data. In particular, many integrins bind their ligands in an irreversible manner (Orlando and Cheresh, 1991; Muller *et al.*, 1993). The molecular

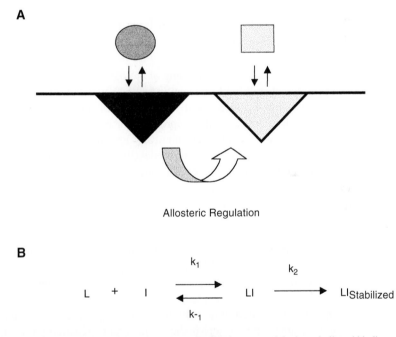

Fig. 2 Allosteric regulation of integrins. (A) A simplified cartoon of the integrin ligand binding pocket is shown with one allosteric regulatory site. The ligand binding site is depicted as a lightly shaded triangle with the ligand as a square, and the allosteric effector site as a dark triangle with the ligand as an oval. The influence of binding interactions at the allosteric regulatory site is shown by an arrow below the binding pockets. (B) An equation that describes the general features of the integrin ligand binding event is shown. In cases where the binding of macromolecular adhesive ligands (fibronectin, vitronectin, etc.) to the entire integrin is measured, the ligand binding reaction is described by a two-step reaction. The second step in this reaction, described by the constant k_2, describes the transition of the ligand : integrin complex to an irreversibly bound state. It should be emphasized that the binding of most model ligands, including RGD peptides, peptide mimetics, and engineered antibodies, is entirely reversible. Therefore, the second step in the reaction needs to be considered for these interactions.

basis for the transition from reversible binding to irreversible binding is still not entirely clear, but this parameter must be considered when analyzing binding data. In the case of most integrins, we are aware that most natural ligands bind irreversibly in the microtiter-based assay. This property is likely to have physiologic significance, and it has been studied in great detail (Orlando and Cheresh, 1991; Muller *et al.*, 1993). An equation describing the steps leading to this irreversible binding interaction is shown in Fig. 2. Binding data for ligands that bind irreversibly cannot be analyzed with typical methods, such as Scatchard plots (Scatchard, 1949). Rather, we described a constant, which we termed the observed association rate constant k_{1obs} (Smith *et al.*, 1994a), to describe binding kinetics. Others have done a more elaborate analysis, yielding even richer detail (Muller *et al.*, 1993). Some ligands, particularly small peptide ligands or antibodies engineered to bind the ligand binding site, bind to integrins in an irreversible manner. Therefore, if a new ligand is being examined, the reversible nature of its binding to integrin should be tested in preliminary experiments.

IV. Measuring Ligand Binding to Integrins with Surface Plasmon Resonance

To gain a deeper understanding of biological regulation of the integrin ligand binding event, the overall binding reaction can be separated into its individual steps, and each can be analyzed independently. A number of important aspects of integrin regulation have been revealed by this approach. For example, consider a model of the integrin containing two separate but kinetically linked ligand binding sites, as we reported for $\alpha IIb\beta3$ (Hu *et al.*, 1999). In this model (Fig. 2), the two ligand binding sites are allosteric to one another (physically separate), yet they do interact. Based on our findings, the RGD binding site remains accessible to its ligands, even when fibrinogen is bound to its binding pocket on the integrin. More importantly, RGD ligands that contact an existing complex between integrin and fibrinogen cause the ejection of fibrinogen from its binding site. Another example of the significance of understanding kinetic detail comes from our work aimed at understanding how divalent ions regulate ligand binding. We identified a class of calcium binding sites that inhibit ligand binding to the $\beta3$ integrins (Hu *et al.*, 1996). Interestingly, binding of Ca^{2+} to this site(s) influences ligand dissociation, without effect on ligand association. These insights can only be obtained by measuring ligand binding prior to equilibrium or in real time. By making these types of measurements, one seeks to measure the initial steps in the ligand binding reaction shown in Fig. 2. More recently, an examination of the kinetics of binding between the I domain of CD11b and ligand helped to establish residues that are involved in maintaining this domain in an open vs closed conformation (Li *et al.*, 1998).

SPR is a preferred method for measuring the kinetic aspects of the integrin ligand binding event. SPR arises within thin metal films under conditions of total internal reflection. The magnitude of the signal depends on the refractive index of solutions in contact with the surface. Typically, a protein of interest is coated onto such a surface, establishing a baseline. Then, when a ligand for this protein binds to the coated surface, the refractive index is altered, giving rise to an SPR signal. For reviews of the method and discussion of practical considerations see Rich and Myszka (2000); O'Shannessy (1994); O'Shannessy and Winzor (1996); and Morton *et al.* (1995). The most widely used SPR instrument is the BIAcore. The BIAcore makes use of "chips" to which proteins are immobilized. These chips are placed into the BIAcore instrument, where they are subject to the flow of an analyte. Here, I adopt BIAcore terminology and refer to the ligand that is present in solution during the SPR assay as the "analyte." Hence, changes to SPR of the chip occur when the analyte binds to immobilized protein. These changes can be monitored as a function of time.

A. General Considerations

Binding can be measured with either the integrin or the ligand immobilized to the SPR chips. CM5 chips from BIAcore are an appropriate choice for linking integrins. However, the best success has resulted when the ligand is coupled to the chip and integrin is used as the analyte. We have observed that the choice of an immobilization format often depends upon the individual properties of the integrin and/or ligand. For example, we have been unable to reproducibly immobilize the $\alpha v \beta 5$ integrin onto chips surfaces, yet both $\beta 3$ integrins are easy to immobilize onto SPR chips. Other issues can also arise. We found that osteopontin, which has a very low isoelectric point, could only be linked to the chips when it is expressed and used as a fusion protein with glutathione *S*-transferase (Hu *et al.,* 1995). Consequently, some aspects of this approach depend on empirical observations.

In general, the detergents that are used for purification of integrin are not compatible with SPR. When integrin is used in the analyte, we have found that a high concentration of detergent in the stock integrin solution (>1% along with an integrin concentration of less than 1 mg/ml) interferes with the analysis. Detergent is also a factor when one wishes to immobilize integrin onto the CM5 chips. We have found that elevated levels of detergent prevent proper linking of the integrin to the chips. As discussed earlier for the microtiter-based assay, best results are obtained when integrin concentrations are 1 mg/ml and the detergent concentration is limited to 0.1%. The P20 surfactant, which is used in all binding assays with the BIAcore, is sufficient to maintain whole integrins in solution during SPR assays.

B. General Method

1. Integrin or ligand is immobilized, according to the manufacturer's specifications, onto CM5 chips. Reactive esters on the chip surface are quenched by addition of ethanolamine. A reference channel lacking any conjugated protein is prepared simultaneously.

2. Following coupling, the surface of the chip is "regenerated" prior to use. This regeneration/washing step eliminates any nonspecifically bound material and ensures that protein linked to the chip is stable. We have found that different adhesive proteins respond to regeneration differently. Consequently, we found that 0.5 *M* L-arginine will

regenerate chips linked to GST-osteopontin without deleterious effects on the proteins. In contrast, though, we have found 0.1 N NaOH to be best for regenerating chips coupled to fibrinogen, and 0.5 M CaCl$_2$ or small RGD peptide are best when regenerating chips linked to integrin.

3. A series of binding isotherms are generated across a range of analyte (ligand) to determine the appropriate concentration for use in assessing kinetic parameters. When integrin is the analyte, we typically dilute purified integrin into phosphate-buffered saline containing 0.005% P20 surfactant (BIAcore) to achieve final concentration of integrin ranging from 0.1 to 10 μg/ml. We have found that integrin should be diluted just prior to injection onto the surface of the chip. We do not recommend storing the diluted integrin beyond 30 min on ice.

4. Once an appropriate concentration of analyte is established, a series of binding studies can be performed. The reader is referred to BIAcore manuals for specific detail, but the software provided with the instrument is capable of independently deriving both the association rate constant (k_1) and the dissociation rate constant (k_{-1}). In all cases, the instrument subtracts the binding of analyte to the reference channel (nonspecific binding) from the binding observed in the test channel. Most of the analyses are auto-mated.

Potential caveats: One should be aware of unstable baselines, and any aberrant peaks that occur in either the association or dissociation phases of the binding reactions. These are indicative of an unstable chip surface. We have found CM 5 chips linked to integrin to be stable for about 4 days at 4°C. Chips linked to adhesive ligands, such as vitronectin and fibrinogen, appear to be stable for about 10 days.

V. Analyzing Integrin Redox Status

Another important aspect of integrin function involves what is commonly referred to as activation state. Activation state describes the ligand binding affinity of integrins. Most integrins can adopt at least two conformations that have different affinities for ligand. For example, the platelet integrin αIIbβ3 exists in a dormant state on resting platelets. In this conformation, it has an almost undetectable binding affinity for its ligand fibrinogen. However, when platelets are activated by physiologic agonists such as ADP or thrombin, the integrin undergoes a series of conformational changes that expose the ligand binding site. Exposure of the ligand binding site leads to a dramatic increase in ligand binding affinity (Marguerie *et al.,* 1980). Considerable effort has been expended trying to understand the factors that govern the interconversion between resting and active integrin. Clearly, signals generated from within the cytoplasm enact changes to an integrins ligand binding affinity (inside-out signaling; Hynes, 1992), but these signals must be transmitted through the length of the integrin's extracellular face to the ligand binding pocket that resides in the N terminus of the protein.

We have discovered a redox site on the extracellular face of the αIIbβ3 integrin that appears to be involved in regulating the changes that lead to activation state (Yan and

Smith, 2000). This redox site comprises several unpaired cysteine residues that appear to reshuffle during activation. At this juncture, we do not know if other integrins also contain such a redox site, but there are hints that they do. Consequently, a general method for site specific modification and detection of unpaired cysteine residues within integrin is presented.

A. General Considerations

Unpaired cysteines within integrin can be detected with the use of sulfhydryl modification reagents that are linked to biotin. In our experience, only reagents that span at least 29 Å from the biotin to the reactive moiety are able to label the redox site on the two $\beta 3$-integrins. This limitation may be due to the inherent depth of the site. Two biotinylated modification reagents have yielded good results. These are 1-biotinamido-4-[4'(maleimidomethyl)-cyclohexane-carboximido]butane (biotin-BMCC) and N-[6-biotinamido)hexyl]-3'-(2'-pyridyldithio)propionamide (biotin-HPDP), both purchased from Pierce Chemical Co. Biotin-BMCC generates a maleimide link with free sulfhydryl that is insensitive to reduction, whereas biotin-HPDP creates a disulfide link with free sulfhydryls that is readily released by reduction.

This labeling strategy can also be adapted to examine the redox site within integrin on the surface of cells. For this application, the sulfhydryl modification reagents are added directly to 10^6 cells held in suspension, making sure that the final concentration of dimethyl sulfoxide does not exceed 5%. Best results are obtained when the modified integrin is subsequently immunoprecipitated from cell lysates and then analyzed by SDS–PAGE, blotting to PVDF, and detecting with avidin-HRP. Integrins are known to have a number of regulatory divalent cation binding sites, but we have yet to perform rigorous experiments to determine if the occupation of these sites influences the accessibility of the redox site to site-specific modification reagents. Therefore, to reduce variability in outcome, particular attention should be to maintaining consistent levels and types of divalent cations.

B. General Method

We have found the following method to reproducibly modify the redox site within $\alpha IIb\beta 3$, and $\alpha v\beta 3$, and consequently expect that the same approach could be applied to examine the redox site in other integrins.

1. Biotin-BMCC is made fresh as a 4 mM stock in dimethyl sulfoxide. Biotin-HPDP should be made in dimethylformamide. Either reagent can be added directly to purified integrin (1–10 μg) in 20 mM Tris · HCl, pH 7.2, containing 150 mM NaCl and containing 0.1% Triton X-100.

2. Modification is allowed to proceed for 1 h at ambient temperature. We have found that the labeling of the redox site of $\alpha IIb\beta 3$ saturates at approximately 100 μM of the sulfhydryl modification reagent. The reaction can be quenched by addition of a 100-fold molar excess of reduced glutathione or L-cysteine.

3. Following modification, the integrin can be separated by SDS–PAGE and transferred to PVDF membranes, and the modification can be assessed by probing the membrane with HRP-avidin (Sigma). It is conceivable that an ELISA could also be devised to quantify the extent of modification with biotin-BMCC or biotin-HPDP.

Acknowledgments

Work leading to development of these methods was funded by grants HL 58925, CA 69306, and AR 42750 from the NIH.

References

Bitan, G., Scheibler, L., Teng, H., Rosenblatt, M., and Chorev, M. (2000). Design and evaluation of benzophenone-containing conformationally constrained ligands as tools for photoaffinity scanning of the integrin $\alpha v \beta 3$-ligand bimolecular interaction. *J. Peptide Res.* **55,** 181–194.

Dickeson, S. K., Walsh, J. J., and Santoro, S. A. (1997). Contributions of the I and EF hand domains to the divalent cation-dependent collagen binding activity of the $\alpha 2 \beta 1$ integrin. *J. Biol. Chem.* **272,** 7661–7668.

Dickeson, S. K., Mathis, N. L., Rahman, M., Bergelson, J. M., and Santoro, S. A. (1999). Determinants of ligand binding specificity of the $\alpha 1 \beta 1$ and $\alpha 2 \beta 1$ integrins. *J. Biol. Chem.* **274,** 32182–32191.

D'Souza, S. E., Ginsberg, M. H., Lam, S. C., and Plow, E. F. (1988a). Chemical cross-linking of arginyl-glycyl-aspartic acid peptides to an adhesion receptor on platelets. *J. Biol. Chem.* **263,** 3943–3951.

D'Souza, S. E., Ginsberg, M. H., Burke, T. A., Lam, S. C., and Plow, E. F. (1988b). Localization of an Arg-Gly-Asp recognition site within an integrin adhesion receptor. *Science* **242,** 91–93.

D'Souza, S. E., Ginsberg, M. H., Burke, T. A., and Plow, E. F. (1990). The ligand binding site of the platelet integrin receptor GPIIb-IIIa is proximal to the second calcium binding domain of its α subunit. *J. Biol. Chem.* **265,** 3440–3446.

Gardner, J. M., and Hynes, R. O. (1985). Interaction of fibronectin with its receptor on platelets. *Cell* **42,** 439–448.

Hu, D. D., Hoyer, J. R., and Smith, J. W. (1995). Ca^{2+} suppresses cell adhesion to osteopontin by attenuating binding affinity for integrin $\alpha v \beta 3$. *J. Biol. Chem.* **270,** 9917–9925.

Hu, D. D., Barbas, C. F. I., and Smith, J. W. (1996). An allosteric Ca^{2+} binding site on the β_3-integrins that regulates the dissociation rate for RGD ligands. *J. Biol. Chem.* **271,** 21745–21751.

Hu, D. D., White, C. A., Panzer-Knodle, S., Page, J. D., Nicholson, N., and Smith, J. W. (1999). A new model of dual interacting ligand binding sites on integrin $\alpha IIb \beta 3$. *J. Biol. Chem.* **274,** 4633–4639.

Hughes, P. E., Diaz-Gonzalez, F., Leong, L., Wu, C., McDonald, J. A., Shattil, S. J., and Ginsberg, M. H. (1996). Breaking the integrin hinge. A defined structural constraint regulates integrin signaling. *J. Biol. Chem.* **271,** 6571–6574.

Humphries, M. J. (2000). Integrin structure. *Biochem. Soc. Trans.* **28,** 311–339.

Hynes, R. O. (1992). Integrins: versatility, modulation and signaling in cell adhesion. *Cell* **69,** 11–25.

Kallen, J., Welzenbach, K., Ramage, P., Geyl, D., Kriwacki, R., Legge, G., Cottens, S., Weitz-Schmidt, G., and Hommel, U. (1999). Structural basis for LFA-1 inhibition upon lovastatin binding to the CD11a I-domain. *J. Mol. Biol.* **292,** 1–9.

Lam, S. C.-T., Plow, E. F., Smith, M. A., Andrieux, A., Ryckwaert, J.-J., Marguerie, G., and Ginsberg, M. H. (1987). Evidence that Arginyl-Glycyl-Aspartate peptides and fibrinogen gamma chain peptides share a common binding site on platelets. *J. Biol. Chem.* **262,** 947–950.

Leitinger, B., McDowall, A., Stanley, P., and Hogg, N. (2000). The regulation of integrin function by Ca^{2+}. *Biochim. Biophys. Acta* **1498,** 91–98.

Li, R., Rieu, P., Griffith, D. L., Scott, D., and Arnaout, M. A. (1998). Two functional states of the CD11b A-domain: correlations with key features of two Mn^{2+}-complexed crystal structures. *J. Cell. Biol.* **143,** 1523–1534.

Marguerie, G. A., Edgington, T. S., and Plow, E. F. (1980). Interaction of fibrinogen with its platelet receptor as part of a multistep reaction in ADP-induced platelet aggregation. *J. Biol. Chem.* **255,** 154–161.

Morton, T. A., Myszka, D. G., and Chaiken, I. M. (1995). Interpreting complex binding kinetics from optical biosensors: a comparison of analysis by linearization, the integrated rate equation, and numerical integration. *Anal. Biochem.* **227,** 176–185.

Mould, A. P., Akiyama, S. K., and Humphries, M. J. (1995). Regulation of integrin $\alpha 5\beta 1$-fibronectin interactions by divalent cations. Evidence for distinct classes of binding sites for Mn^{2+}, Mg^{2+}, and Ca^{2+}. *J. Biol. Chem.* **270,** 26270–26277.

Muller, B., Zerwes, H.-G., Tangemann, K., Peter, J., and Engel, J. (1993). Two-step binding mechanism of fibrinogen to $\alpha IIb\beta 3$ integrin reconstituted into planar lipid bilayers. *J. Biol. Chem.* **268,** 6800–6808.

Nachman, R. L., and Leung, L. L. (1982). Complex formation of platelet membrane glycoproteins IIb and IIIa with fibrinogen. *J. Clin. Invest.* **69,** 263–269.

Orlando, R. A., and Cheresh, D. A. (1991). Arginine-glycine-aspartic acid binding leading to molecular stabilization between integrin $\alpha v\beta 3$ and its ligand. *J. Biol. Chem.* **266,** 19543–19550.

O'Shannessy, D. J. (1994). Determination of kinetic rate and equilibrium binding constants for macromolecular interactions: a critique of the surface plasmon resonance literature. *Curr. Opin. Biotechnol.* **5,** 65–71.

O'Shannessy, D. J., and Winzor, D. J. (1996). Interpretation of deviations from pseudo-first-order kinetic behavior in the characterization of ligand binding by biosensor technology. *Anal. Biochem.* **236,** 275–283.

Plow, E. F., Haas, T. A., Zhang, L., Loftus, J., and Smith, J. W. (2000). Ligand binding to integrins. *J. Biol. Chem.* **275,** 21785–21788.

Rich, R. L., and Myszka, D. G. (2000). Advances in surface plasmon resonance biosensor analysis. *Curr. Opin. Biotechnol.* **11,** 54–61.

Ruoslahti, E. (1996). RGD and other recognition sequences for integrins. *Ann. Rev. Cell Dev. Biol.* **12,** 697–715.

Santoro, S. A., and Lawing, W. J., Jr. (1987). Competition for related but nonidentical binding sites on the glycoprotein IIb–IIIa complex by peptides from platelet adhesive proteins. *Cell* **48,** 867–873.

Scarborough, R. M., Rose, J. W., Hsu, M. A., Phillips, D. R., Fried, V. A., Campbell, A. M., Nannizzi, L., and Charo, I. F. (1991). Barbourin. A GPIIb-IIIa-specific integrin antagonist from the venom of *Sistrurus M. Barbouri*. *J. Biol. Chem.* **266,** 9359–9362.

Scatchard, G. (1949). The attraction of proteins for small molecules and ions. *Ann. N.Y. Acad. Sci.* **51,** 660–672.

Smith, J. W., and Cheresh, D. A. (1988). The Arg-Gly-Asp binding domain of the vitronectin receptor. *J. Biol. Chem.* **263,** 18726–18731.

Smith, J. W., and Cheresh, D. A. (1990). Integrin($\alpha v\beta 3$)–ligand interaction. *J. Biol. Chem.* **265,** 2168–2172.

Smith, J. W., Vestal, D. J., Irwin, S. V., Burke, T. A., and Cheresh, D. A. (1990). Purification and functional characterization of integrin $\alpha v\beta 5$. *J. Biol. Chem.* **265,** 11008–11013.

Smith, J. W., Piotrowicz, R. S., and Mathis, D. M. (1994a). A mechanism for divalent cation regulation of $\beta 3$-integrins. *J. Biol. Chem.* **269,** 960–967.

Smith, J. W., Hu, D., Satterthwait, A., Pinz-Sweeney, S., and Barbas, C. F. (1994b). Building synthetic antibodies as adhesive ligands for integrins. *J. Biol. Chem.* **269,** 32788–32795.

van Kooyk, Y., and Figdor, C. G. (2000). Avidity regulation of integrins: the driving force in leukocyte adhesion. *Curr. Opin. Cell Biol.* **12,** 542–547.

Yan, B., and Smith, J. W. (2000). A redox site involved in integrin activation. *J. Biol. Chem.* **275,** 39964–39972.

CHAPTER 12

Intracellular Coupling of Adhesion Receptors: Molecular Proximity Measurements

Maddy Parsons and Tony Ng

Richard Dimbleby Department of Cancer Research/Cancer Research UK Labs
The Rayne Institute
St. Thomas Hospital
London SE1 7EH, United Kingdom

I. Introduction and Applications

The control of cellular function and intracellular signaling is vastly complex and highly regulated. Many of the key players involved in the generation of functional signal transduction pathways have been characterized, and the relationship between these proteins has been partially delineated using biochemical techniques such as immunoblotting, immunoprecipitation, and *in vitro* phosphorylation assays. These techniques, although invaluable in the dissection of the hierarchy of protein–protein interactions, do not deliver information on the true spatial and temporal nature of these complexes at the single-cell level, which is essential for understanding the dynamic regulation of cell signaling.

In the context of cell adhesion and migration, stimulatory signals are generated by the binding of cells to extracellular matrix (ECM) proteins, cytokines/chemoattractants, or ligands that are expressed on the surface of target tissues or vessels. The "sensing" of an extracellular signal results in cell polarization, which may be manifested by a redistribution and trafficking of membrane receptors (Bailly *et al.*, 2000) and/or intracellular signaling proteins, as well as the localized formation of protein complexes that can be stable or transient. Monitoring these protein–protein associations during cell adhesion or motility would be difficult with conventional biochemical analyses because of the transient nature of some of the protein complexes formed, and depending on the stability of the association, it may be difficult to preserve the complex following cell lysis prior to detection *ex vivo*.

Immunocytochemical techniques coupled with confocal microscopy allow us to preserve and image the relative localization of signaling molecules in cells that have been stimulated to adhere or migrate. The basis of this technique is to utilize specific antibodies to the target proteins, which are subsequently directly or secondarily labeled with fluorescent tags, thereby allowing visualization by wide-field epifluorescence or confocal microscopy. Although this provides a certain amount of information about spatial localization of proteins, this is only on a micrometer scale and does not provide the resolution required to "specifically" detect direct binding between two proteins on a nanometer scale.

Fluorescence resonance energy transfer (FRET) by fluorescence lifetime imaging microscopy (FLIM) is a recently described approach that provides the means to visualize individual proteins and quantitate various biochemical reactions which include phosphorylation, proteolysis, oligomerization, and conformational changes at a single-cell level, in cells challenged with different physiological stimuli, in either cell culture or tissue environments. This technique has been successfully employed to look at signal transduction processes in live and fixed cells as well as in archived pathological material (Bastiaens and Jovin, 1996; Gadella and Jovin, 1995; Harpur *et al.*, 2001; Murata *et al.*, 2000; Ng *et al.*, 2001, 1999a,b; Pepperkok *et al.*, 1999; Verveer *et al.*, 2000b; Wouters and Bastiaens, 1999).

Labeling target proteins within a cell with specific chromophores lends these proteins a fluorescence lifetime. Fluorescence lifetime, defined as the average amount of time a molecule spends in an excited state upon absorption of a photon of light, can range from picoseconds to nanoseconds. It is sensitive to excited state reactions such as FRET (which results in fluorescence lifetime shortening) and is independent of chromophore concentration and light path length. FRET occurs when two different spectrally overlapping chromophores are in close proximity to one another (generally <10 nm apart). Dipole–dipole coupling causes energy to be transferred from donor to acceptor fluorophore, thereby producing emission from the acceptor. The occurrence of FRET results in a shortening of fluorescence lifetime.

There are two main methods for measuring fluorescence lifetime, namely time-domain and frequency-domain-based fluorescence lifetime sensing (Draaijer *et al.*, 1995). In 1993, Clegg, Jovin, and others published the development of the first frequency domain fluorescence lifetime imaging microscope (Clegg *et al.*, 1992; Gadella *et al.*, 1993). The principles of frequency-domain-based fluorescence lifetime sensing, which is based on monitoring the extent of demodulation and phase shift between a sinusoidally

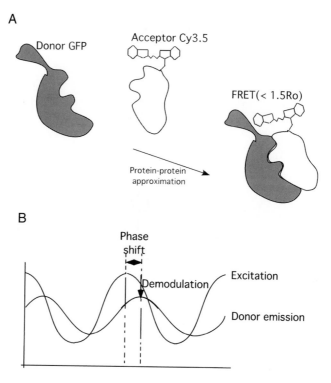

Fig. 1 (A) Representation of principles of FRET. Two proteins are fluorescently labeled with a pair of donor and acceptor fluorophores, chosen on the basis of a "proper" spectral overlap determined mathematically by the overlap interval $J(\lambda)$ (see text). The presence or absence of FRET is determined by the separating distance and the relative orientation of the chromophores' transition dipoles (Lakowicz, 1999). (B) Fluorescent lifetime measurement by frequency domain FLIM. Sinusoidally modulated light is used to excite the sample. Relative to this sinusoidally modulated excitation light (labeled "Excitation"), the donor emits at the same frequency but with a phase shift and reduced modulation depth/amplitude (labeled "Donor emission"). The phase (τ_p) and modulation (τ_m) fluorescence lifetimes can then be calculated from this phase shift and demodulation.

modulated excitation light and the donor emission consequent upon its excitation, are explained in Fig. 1B. In time-domain FLIM, the sample is excited with a short pulse, and the subsequent fluorescence is integrated into separate time windows. The fluorescence lifetime is then calculated by fitting an exponential model to the measured intensities. In this review we will focus on the frequency-domain-based method using a conventional inverted microscope configuration that employs phase-sensitive detection of fluorescence emission (Fig. 2, and Squire and Bastiaens, 1999).

A different methodology to monitor FRET is by the detection of acceptor-sensitized emission (Day, 1998; Kraynov *et al.*, 2000; Li *et al.*, 2001; Mahajan *et al.*, 1998, 1999; Mas *et al.*, 2000) or by monitoring donor : acceptor intensity ratios (Adams *et al.*, 1991; Honda *et al.*, 2001; Miyawaki *et al.*, 1997, 1999; Tyas *et al.*, 2000). An example of this approach is the intracellular monitoring of Rac1 activation (an assay known as

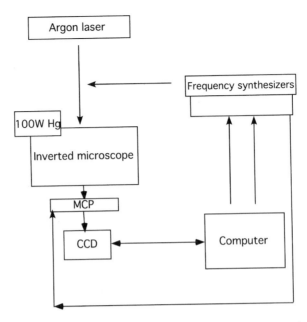

Fig. 2 Schematic representation of FLIM setup. System components of a frequency-domain-based FLIM apparatus. Refer to text for details.

FLAIR, fluorescence activation indicator for Rho proteins, Kraynov *et al.,* 2000). One major difference between the two methodologies is that using FLIM, provided that the acceptor fluorophore is present in excess, the extent of FRET measured by monitoring changes in donor fluorescence lifetime is independent of the subcellular distribution of protein concentrations. On the other hand, acceptor-sensitized emission is based upon the measurement of acceptor emission intensity, which is protein concentration-dependent. This will be discussed further in Section II.D.

New advances in genetic technology have allowed tagging of DNA with fluorophores, which in turn allows a direct visualization of nucleic acid hybridisation (reviewed by Glazer and Mathies, 1997). Similarly, in combination with the use of RNAs that are dually labeled with a pair of donor and acceptor fluorophores, FRET measurements can be used to detect the thermodynamic stability of their tertiary structures (Klostermeier and Millar, 2000, 2001). The FRET/FLIM approach is thus in many respects complementary to the more conventional structural biological approaches in this capacity. At an industrial level, this technology can be adapted to provide a valuable high-throughput screen for small-molecule inhibitors of specific protein interactions and/or phosphorylation processes.

In recent years, we have exploited FRET/FLIM techniques to monitor protein phosphorylation and protein–protein interactions in the cell context, using antibodies that label endogenous or nonfluorescently tagged, overexpressed proteins and more often, in combination with the use of a fluorescently tagged (such as GFP, YFP, and CFP) plasmid encoding the protein of interest. Figure 1 shows the basic principles of how this works. A pair of donor (e.g., GFP) and acceptor (e.g., Cy3.5) fluorophores is chosen on the basis that there is a "proper" degree of overlap between the emission spectrum of the donor and the excitation spectrum of the acceptor, determined mathematically by the overlap interval $J(\lambda)$, which is dependent upon the acceptor extinction coefficient $\varepsilon(\lambda)$ (refer to Harpur and Bastiaens, 2001, for a more mathematical definition). The spatial separation (R_0) between GFP and Cy3.5, for instance, at which the resonance energy transfer efficiency is 50% is 5.7 nm. Detectable energy transfer can, however, occur up to $1.5 \times R_0$ (i.e., 8.6 nm) (Fig. 1A; Ng *et al.,* 1999b). FRET is detected by virtue of a shortening of the fluorescence lifetime τ of the donor (defined as the average amount of time that the donor fluorophore stays in the excited state upon absorption of a photon of light). For protein–protein approximation, one of the experimental model systems we have frequently employed is to label the putative binding partner of a GFP-tagged protein of interest with a Cy3 or Cy3.5-conjugated antibody (acceptor antibody) (Ng *et al.,* 1999a, 2001). In this case, FRET has to reach across three intervening proteins between the donor and acceptor fluorophores. Only the acceptor antibodies in close proximity to the donor will be detectable through FRET (dictated by factors such as R_0 and the relative conformation of the interacting partners) and henceforth specificity of the acceptor antibody is not a limiting factor in the experiment. Protein phosphorylation can similarly be studied by detecting FRET across two intervening proteins (GFP-tagged donor protein and a Cy3- or Cy3.5-conjugated antibody that recognizes the donor protein only when it is phosphorylated) (Ng *et al.,* 1999b; Verveer *et al.,* 2000b; Wouters and Bastiaens, 1999).

II. Fluorescent Lifetime Imaging Microscopy (FLIM)

A. Basic Biophysical Principles of the FLIM System Setup

The method described here is frequency-domain FLIM, which utilizes a sinusoidally modulated laser excitation source to excite the donor fluorophore (Fig. 1B). A frequency-domain fluorescence lifetime imaging microscope has a number of components. Figure 2 shows a simplified schematic of our FLIM system setup. An inverted microscope with an argon laser light source (Coherent, 200 mW per line output typically) forms the basis of the instrument. The microscope is equipped with standing wave acoustooptic modulators (AOMs, Polytec) for the periodic modulation of a laser excitation source. As an example, for GFP donor lifetime measurements, the excitation source (488 nm line) is modulated typically at 80 MHz as a single frequency. An iris diaphragm, placed about 1.5 m from the AOMs, selects the central beam (zero order) and excludes the higher-order diffracted beams. Sample illumination is achieved by incorporating a lens at the epiillumination port of the microscope, which serves to collect and collimate the laser beam. A 100-W mercury arc lamp is used as a further illumination source for imaging (and photobleaching) the acceptor, whereas the laser is used for lifetime imaging of the donor fluorophore.

All images are taken using a Zeiss Plan-APOCHROMAT 100X/1.4NA phase 3 oil objective. The two commonly used donors, GFP and Cy3, are excited using the 488-nm and 514-nm lines, respectively, of a argon/krypton laser and the resultant fluorescence separated using a combination of dichroic beamsplitter (Q 495 LP and HQ 565 LP for GFP and Cy3, respectively) and emitter filter (BP514/10 and HQ 610/75 for GFP and Cy3, respectively). Acceptor images are recorded using a high Q filter set (for Cy3, exciter:HQ 535/50, dichroic:Q 565 LP, emitter:HQ 610/75 LP; for Cy5, exciter:HQ 620/60, dichroic:HQ 660 LP, emitter:HQ 700/75). Filters and dichroics are from Chroma or Delta Light & Optics.

Light from the sample on the slide is directed onto the photocathode of the image intensifier head, which in turn directs photoelectrons at the microchannel plate (MCP). The high-frequency fluorescence signal is mixed with the electronic gain of the MCP image intensifier modulated on the photocathode at the same frequency (80 MHz). The electron image generated by the MCP strikes a phosphor screen and generates an intensified light image, which is then projected onto the chip of a scientific grade CCD camera (LaVision or Hamamatsu Photonics). Two-by-two binning is used, and generally 16 phase-dependent images are acquired during a FLIM sequence. High-speed shutters are used during acquisition in order to minimize bleaching of the sample. Specific algorithms are used to retrieve lifetime information from the images taken. Phase shift and modulation of the donor emission at each pixel are extracted in order to calculate fluorescence lifetime. Detailed data analysis will be discussed at a later stage in this chapter.

B. Preparation of Samples for FLIM

1. Generation of Fab Fragments and Labeling of Antibodies

As an example of intramolecular FRET, the detection of the activated form of PKCα in human breast cancers *in situ* (Ng *et al.,* 1999b) can be achieved by costaining the tumor section with two IgGs labeled with different fluorophores (i.e., a pan-PKCα mAb labeled with a donor fluorophore and a phospho-T(P)250 rabbit polyclonal IgG labeled with an acceptor fluorophore). For intermolecular FRET, additional optimization is required by using a donor fluorophore-labeled IgG in combination with an acceptor fluorophore-conjugated IgG Fab fragment. The use of Fab fragments in these analyses serves to optimize the specificity of antibody binding as well as to reduce levels of steric hindrance often generated by whole antibody binding. Antibodies are concentrated (for instance, using centrifugation in a Centricon) to 20 mg/ml in 20 mM sodium phosphate/10 mM EDTA (pH 7.0). A 500-μl aliquot of this preparation is added to 500 ml of the same buffer containing 20 mM cysteine/HCl buffer (pH 7) and papain immobilized agarose beads (Pierce, in a 50:50 solution with buffer). Antibodies are then digested for around 8 h at 37°C with shaking. Longer digestion periods may be used, but there is a risk of overdigestion and hence generation of smaller peptide fragments. Fragments are purified by removing the Fc portion of the antibody using protein A–Sepharose beads and collecting 1 ml fractions of Fab from the column. The OD280 of each fraction is determined to identify those containing Fab fragments. The fractions containing Fab are pooled, reconcentrated, and dialyzed against a non-amino buffer. Samples are finally purified on a low molecular weight exclusion column and then reconcentrated to determine the final concentration.

Resultant Fab fragments may be directly labeled using fluorescent dyes, such as Cy3, Cy5, and Oregon Green (Molecular Probes). The fluorescent dye in each vial is resuspended in N,N-dimethylformamide (DMF) to a final concentration of 10 mM. Labeling of the antibody is carried out in a buffer containing 20 mM bicine and 50 mM NaCl, at pH 8.5 (labeling of Fab fragments is carried out at pH 9.0). The dye is added to the antibody sample (final volume of dye should not exceed 10% of total volume) at around a 10-fold molar excess and left for 30 min at room temperature. The labeling reaction is stopped by adding Tris buffer to a final concentration of 10 mM. To remove free dye, samples are then separated on a gel filtration column, and the protein-bound dye is eluted into PBS. Dilute a 10-μl aliquot of the labeled antibody 10–20 times and measure absorption at 280 nm and 496 nm for Oregon Green, 280 nm and 554 nm for Cy3, or 280 nm and 650 nm for Cy5.

Estimate the labeling ratio by the following formulas:

$$\text{Oregon Green: } A_{496} \times M/[(A_{280} - 0.12 \times A_{496}) \times 50]$$
$$\text{Cy3: } A_{554} \times M/[(A_{280} - 0.05 \times A_{554}) \times 150]$$
$$\text{Cy5: } A_{650} \times M/[(A_{280} - 0.05 \times A_{650}) \times 250]$$

where A_x is the absorbance at wavelength x and M is the molecular weight of the protein in kilodaltons.

The final ratio of protein : dye should be aimed at around 1:3 for the acceptor, but will vary depending upon the protein labeled and reaction times. The amount of dye used for the labeling the donor antibody should be reduced accordingly to achieve a ratio of protein : dye of 1:1.

2. Cell Transfection, Fixation, and Preparation

a. Transfection

Adherent cells are seeded onto glass coverslips and transfected with the relevant plasmids, and proteins are allowed to express for 36 h. Where possible, the transfection is optimized so that acceptor protein levels are in excess and therefore are not limiting in the detection of FRET. Typically, cells are transfected using a donor : acceptor plasmid ratio of 1:2.

b. Microinjection

For observing live cell FLIM, cells may need to be microinjected with plasmid DNA or antibodies/proteins prior to visualization. It is important to ensure the purity of the DNA and protein preparations prior to injection to avoid contamination and blocking of needles. Coverslips are marked prior to injection to assist in the subsequent location of injected cells. Standard microinjection procedure is used while the coverslips are immersed in a CO_2-independent medium. Postinjection, cells are returned to the incubator and allowed to quiesce or express the protein of interest for 2–3 h (1 h in the case of

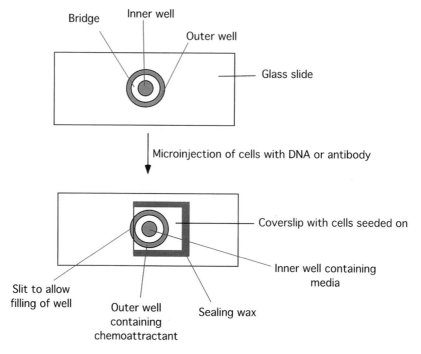

Fig. 3 Dunn chemotaxis chamber. Diagram depicts a Dunn chamber setup as described for chemotaxis assays using cells seeded onto a 22-mm coverslip. Cells are transfected or microinjected prior to setting up the chamber. The inner well is filled with the control medium, and the outer well with medium containing the chemoattractant. According to the diffusion theory (Zicha *et al.,* 1991), a 10- to 20-kDa protein will form a close approximation to a linear gradient within about 30 min of assembling the chamber and the $t_{1/2}$ of the gradient will be about 30 h. The times of gradient formation and decay are approximately proportional to the cube root of the molecular weight of the chemoattractant.

cytoplasmic injection of fluorescently labeled antibodies) before fluorescence lifetime measurements are taken.

c. Dunn Chemotaxis Chamber Setup

Figure 3 shows a schematic diagram of a Dunn chemotaxis chamber setup. The principle of this system is to directly track and measure motility of cells toward a chemotactic gradient (Zicha *et al.,* 1991, 1997). The chamber (Weber Scientific International, Teddington, Middlesex, UK) consists of a central well that is filled with control medium, an outer well into which the chemoattractant is placed, and a "bridge" that separates the two. Microinjection can be performed on cells seeded at subconfluence onto a coverslip, prior to chamber assembly. Coverslips are placed cell-side-down onto the chamber, allowing a small gap at one side for filling of the outer well. The other three sides of the coverslip are sealed to the glass chamber slide using a solution of hot wax. The outer well is drained of control medium, refilled via the "slit" with medium containing the chemoattractant, and then sealed with hot wax. A "gradient" then forms across the "bridge" section, and cell movement in response to this linear gradient is monitored using time-lapse microscopy. Live cell FLIM measurements can also be made using this setup.

d. Antibody Staining

For fixed preparations, medium is removed and cells washed in PBS before fixing in a solution of 4% paraformaldehyde for 15 min at room temperature. Following this, the cells are washed in PBS twice and permeabilized in 0.2% Triton X-100/PBS for 5 min at room temperature. Samples are washed in PBS again and then any background fluorescence is quenched by incubating coverslips in 1 mg/ml fresh sodium borohydride/PBS for 10 min. Cells are rewashed in PBS and blocked in a solution of 1% BSA. Remove BSA and incubate the cells in PBS–1% BSA containing 10 μg/ml fluorophore-conjugated antibody for 1 h to overnight (in a humidified chamber). To minimize the amount of antibody or Fab used, coverslips are placed onto Parafilm before the antibody is applied. Finally, coverslips are washed thoroughly in PBS , then fixed again in 4% paraformaldehyde to preserve the bound antibody. After washing with PBS and then distilled water, the coverslips are mounted onto a standard microscope slide using Mowiol (Calbiochem) mounting medium containing 2.5% (w/v) 1,4-diazabicyclo[2.2.2]octane as an antifade, and allowed to solidify at 4°C overnight before imaging.

3. Image Acquisition

The image acquisition and analysis algorithms (Squire and Bastiaens, 1999) we currently employ were written by Dr. Anthony Squire (EMBL, Heidelberg) as scripts in a copy of IPLab Spectrum v. 3.1f. To obtain a zero lifetime reference image, a piece of aluminium foil adhered to a glass microscope slide is used as a strong light scatterer. A cycle of 16 phase-dependent images is acquired (separated by 22.5°) with the incident excitation source dimmed, ensuring that the image is not overexposed or saturated. Foil images are taken approximately every 30 min to provide a reference for subsequent donor lifetime calculations, by monitoring fluctuations in the intensity modulation of the laser beam (which is sensitive to small temperature changes) throughout the experiment.

A cycle of phase-dependent donor emission images (in the absence of an acceptor; separated by 45°) is acquired first with the excitation source set at maximum and using the appropriate filter sets, to enable calculation of the control lifetime (see next section). It is important that the signal : background ratio be high in the donor image and that there be ≤15% bleaching of the sample between images 1 and 16. Samples with both donor and acceptor present can then be imaged. The acceptor is visualized using the 100-W mercury lamp (with adjustable percent output) as a source of illumination with the appropriate excitation and emission filters. Exposure times and percent lamp output are carefully recorded for comparison.

C. Data Analysis

Once the data have been analyzed, a representative field of cell(s) from each experiment or condition is selected for data presentation. An example is presented in Fig. 4A, which shows a Cdc42-PAK (p21-activated kinase) association in porcine aortic endothelial cells detected by FLIM. For each representative field, the following information is presented:

1. The fluorescence image from the donor (GFP-Cdc42)
2. The fluorescence image from the acceptor (anti-Myc-PAK mAb-Cy3)
3. The donor fluorescence lifetime $\langle \tau \rangle$ (the average of τ_p and τ_m) pseudocolor cell map
4. Eff, the pixel-by-pixel FRET efficiency represented on a pseudocolor scale (Eff $= 1 - \tau_{da}/\tau_d$, where τ_{da} is the lifetime map of the donor in the presence of acceptor and τ_d is the average lifetime $\langle \tau \rangle$ of the donor in the absence of acceptor, numerically taken as the average of mean $\langle \tau \rangle$ of ≥ 6 donor-alone control cells). Eff only applies to the donor + acceptor (anti-Myc-PAK mAb-Cy3) cells

For statistical analyses, all the FRET efficiency data (at each pixel of the cell from ≥ 6 donor + acceptor cells) are presented in a cumulative pixel counts ($\times 10^3$) vs Eff profiles. Most Eff histograms follow a Gaussian distribution pattern; sometimes a bimodal distribution showing two distinct (complexed and uncomplexed) populations can be observed, as seen in the case of wild-type (WT) GFP-Cdc42 in Fig. 4A. In this pilot study, an overall analysis of cumulative results from six fields of cells confirmed that a subpopulation of WT GFP-Cdc42 (but not the N17 inactive variant) associated with Myc-PAK1 following platelet-derived growth factor (PDGF) treatment.

The upper panels in Fig. 4B provide an alternative statistical representation of the cumulative lifetime data, i.e., the cumulative τ_p vs τ_m lifetimes (at each pixel of the cell from ≥ 6 cells) of donor alone (green) and that measured in the presence of the acceptor fluorophore (red) plotted on 2D histograms. The left histogram shows a slight separation of the donor-alone lifetime population (green; GFP-PKCα) and that of donor lifetimes measured in the presence of the acceptor (red; the receptor in this example is a Cy-3 labeled IgG that recognizes a PKCα-interacting protein), with an overlapping yellow area, suggesting that a small proportion of the donor molecules are complexed to the acceptor. Upon phorbol ester stimulation (right histogram), there is an increased proportion of the donor molecules complexed to the acceptor, giving rise to a greater separation of the green and red dots. A greater heterogeneity of the donor lifetimes in the presence of the acceptor is observed, indicating the increase in PKC/binding protein association occurs asynchronously between individual cells, or indeed between different subcellular compartments.

D. Optimization and Experimental Design

Several key criteria ought to be fulfilled before conclusive information can be derived from the use of FRET/FLIM analysis:

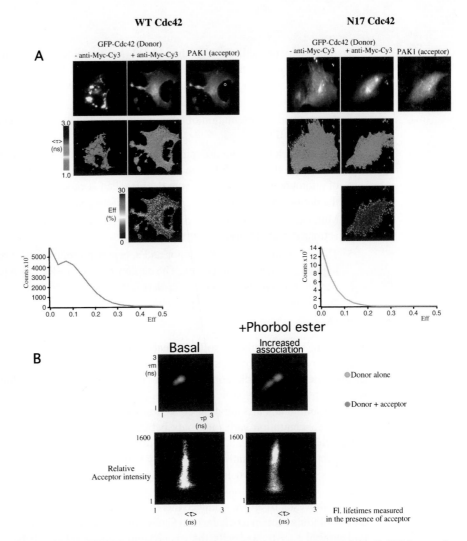

Fig. 4 (A) To determine the spatiotemporal distribution of the activated form of Cdc42, FLIM was undertaken to determine the extent of FRET between GFP-Cdc42 (donor) and Myc-tagged PAK1 stained with a Cy3 (acceptor) conjugated antibody 9E10 to the Myc epitope. Porcine aortic endothelial (PAE) cells stably expressing the platelet-derived growth factor (PDGF) PDGF β receptor were dually transfected with both a GFP-Cdc42 (wild-type WT or inactive N17 mutant form) and a Myc-tagged PAK1 construct for 36 h, then stimulated with PDGF (50 ng/ml) for 45 min before fixation. In a third of the WT Cdc42/PAK-cotransfected cells stimulated with PDGF and stained with an anti-Myc-Cy3 post-fixation, it was evident that fluorescence lifetime, $\langle \tau \rangle$, for GFP was decreased at the cell periphery and filopodial extensions. Because only the activated form of the small GTPases binds PAK, the localization of a reduced $\langle \tau \rangle$ (compared to the GFP lifetime in control cells that were not stained with the Cy3-labeled anti-Myc antibody) indicates the subcellular distribution of the activated form of Cdc42. (B) The detection of a phorbol ester-enhanced association between GFP-PKCα and its binding partner by FLIM. Please refer to text for a detailed description. (See Color Plate.)

1. A diversity of methods other than FLIM (such as biochemical pulldown/immuno-precipitation, electron microscopy, sucrose gradient fractionation, etc) should be employed wherever possible to provide corroborative evidence in support of the FLIM data.

2. A statistical representation of cumulative results should be presented (please refer to Section II.C).

3. Photobleaching of the acceptor fluorophore should partially reverse the lifetime shortening due to FRET. One of the caveats for this analysis is the accumulation of photobreakdown product from the bleached acceptor that may bleed through to the donor channel, giving rise to aberrant lifetimes.

4. Mutational analysis should provide the appropriate negative controls for the interactions detected through lifetime imaging.

5. The proximity between the donor and acceptor proteins may be regulated physiologically or pharmacologically. The demonstration of FRET as a regulated process partly circumvents the common criticism that the detected lifetime shortening might be a result of protein overexpression.

6. Donor/acceptor levels.

The dogma we have upheld so far is that, using FLIM, provided that the acceptor fluorophore is present in excess, the extent of FRET measured by FLIM is independent of the subcellular distribution of protein concentrations. This is achieved by labeling the "acceptor antibody" at a dye : protein molar ratio of 3:1 and using the conjugated acceptor antibodies at saturating concentrations. The reality is that the acceptor fluorophore concentration varies not only between cells, but also between different pixels within each individual cell. The pixel-to-pixel variation of acceptor availability is virtually impossible to quantify or control by experimental design. While performing the FLIM experiments, we therefore routinely check for any dependency of fluorescence lifetimes on acceptor fluorescence intensity. Figure 4B shows the basal (unstimulated) and phorbol ester treatment-induced increase in the association between GFP-PKCα and a PKC-binding protein labeled with a Cy3-conjugated antibody. In both the unstimulated and phorbol ester-treated cells, there is no significant correlation between donor lifetime $\langle\tau\rangle$ (the average of τ_p and τ_m) and the relative intensity of acceptor at each pixel of the cell, although the $\langle\tau\rangle$ values are systematically shifted to lower values after phorbol treatment. When the lifetime data from all the cells that do not show a lifetime dependency on local acceptor fluorophore concentrations are analyzed, there is a clear-cut increase in FRET after PDBu treatment (see comparison of 2D τ_p vs τ_m histograms before and after phorbol treatment in Fig. 4B, upper panel). The conclusion regarding the increase in association by phorbol ester treatment therefore cannot be due to protein overexpression.

E. Experimental Variations

1. Live Cell and Confocal FLIM

The "sensing" of an extracellular signal that stimulates in a polarized fashion both dissolution of focal adhesions and the subsequent cell movement is a poorly understood process. Some clues have emerged from the study of surface receptor redistribution and trafficking in cells exposed to spatial gradients of growth factors applied using a pipette (Bailly *et al.,* 2000). As an example, we present the first evidence that epidermal growth factor (EGF) stimulates a redistribution of its receptor ErbB1 to the leading edge, in MCF7 cells polarizing in response to the EGF gradient in a Dunn motility chamber. ErbB1 is known to interact both physically and functionally with β1 integrins in breast epithelial cells (Wang *et al.,* 1998). Given the changes in the distribution of ErbB1 and ligand-occupied (activated) β1 integrins during cell polarization and migration (Fig. 5), it would be intriguing to establish whether the two receptors come into close proximity during cell migration and the time course of such association, in MCF-7 cells responding to a growth factor gradient established in the Dunn chamber. GFP-ErbB1 (GFP at the C terminus), coupled with a Cy3-conjugated polyclonal IgG that recognizes the cytoplasmic domain of β1 integrin, provides the reagents necessary for the real-time analysis of ErbB1/integrin association by live cell FLIM analysis (Ng and Wouters, unpublished data). A large number of cells on the same coverslip would, however, have to be analyzed (as discussed under "Data Analysis") to provide a statistically relevant result, because of the lack of synchronicity of the protein–protein association event within the cell population.

Intermolecular FRET can be monitored in live cells using a cell line stably expressing a non-fluorescently tagged acceptor protein in combination with a transient expression of the GFP-tagged donor. A saturating amount of a Cy3-labeled IgG that recognizes the nonfluorescent tag is microinjected into the cytoplasm at least 1 h before FLIM measurements are taken.

A potential alternative to the use of stable transfectants is to tag the acceptor protein of interest with a tetracysteine motif (CCXXCC) (within an α-helical conformation). Such motif has been to shown to bind to FlAsH (fluorescein containing two arsenoxides), which is membrane permeant and nonfluorescent until it binds with high affinity and specificity to the tetracysteine motif (Griffin *et al.,* 1998, 2000). A red variant of FlAsH has been developed (Griffin *et al.,* 2000). This provides a potential FRET acceptor for GFP, which can be used in live cell FLIM analysis.

Certain protein contacts may be restricted spatially to specific sites in cells. A cumulative lifetime analysis of all the donor molecules through the depth of the cell may not detect such association, if the proportion of donor molecules that are complexed to acceptor is small. The detection of localized FRET can be achieved by confocal FLIM. A confocal, two-photon time-domain-based FLIM system has been used to measure the fluorescence lifetimes of GFP and DsRed (a red fluorescent GFP homolog) in *Escherichia coli* (Jakobs *et al.,* 2000). Adaptation of such an approach to study changes in fluorescence lifetimes in live mammalian cells will provide a better spatial resolution of the biochemical events that ensue as a direct consequence of cell adhesion/movement.

2. Multiple-Frequency FLIM

All of our published work to date has been carried out using a single modulation frequency (80 MHz). In principle the measurements are valid for most applications where one can assume a monoexponential decay of donor fluorescence. Some fluorescence lifetime measurements taken on a fluorescently labeled toxin in PC12 cells suggest

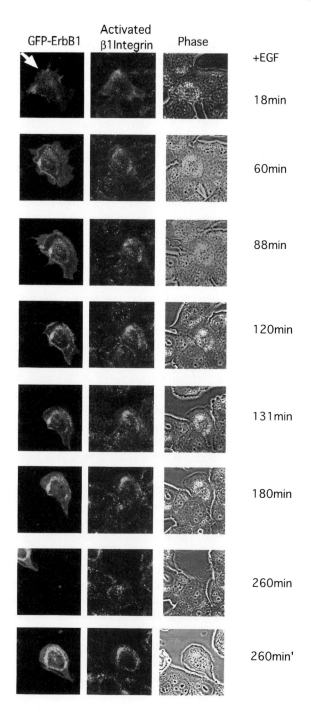

that the donor fluorescence of proteins that are bound to receptors in membrane raft structures (rich in cholesterol and polar sphingolipids) may undergo multiexponential decay, giving rise to distinct donor subpopulations with fast and slow decay rates, even in the absence of FRET (Herreros, Ng, and Schiavo, unpublished observations). For these biological situations, it is important to measure fluorescence lifetimes at multiple harmonic modulation frequencies (Squire *et al.,* 1999) in order to distinguish the fast and slow components in the control lifetimes (without acceptor) before the extent of FRET can be calculated accurately.

3. Global Analysis

Using global FLIM analysis (Verveer *et al.,* 2000a,b), not the conventional semi-quantitative FRET efficiency approach discussed here, one is able to calculate the exact proportion of the donor molecules that is complexed to the acceptor. In global analysis, populations are the fractions of donor molecules that transfer energy. Global analysis techniques exploit the prior knowledge that only a limited number of fluorescent molecule species, with spatially invariant fluorescence lifetimes, exist in the cell sample. This approach simultaneously fits the spatially invariant lifetimes and fractional contributions (populations) from each lifetime species in all pixels. The population derived in this way is different mathematically from the conventional efficiency (Eff), which is approximately proportional to the fraction of the total steady-state fluorescence that originates from donor molecules in their bound state. Since the donor-tagged molecules transfer energy to the acceptor when they are in a complex, their lifetimes are reduced and hence the number of photons emitted would be fractionally reduced. Mathematically, there is no linear relationship between the two quantities (i.e., population vs Eff); hence, the contrast in the Eff pseudocolor cell plots may appear different from the population cell images.

Fig. 5 MCF-7 cells on a coverslip were transiently transfected with a GFP-tagged ErbB1 construct (Verveer *et al.,* 2000b; Wouters and Bastiaens, 1999). The endogenous activated integrins were prelabeled with Cy3-conjugated mAb 12G10 by pulse-chase as described (Ng *et al.,* 1999a). 12G10 recognizes the extracellular domain of the ligand-occupied, activated form of $\beta 1$ integrins and is kindly donated by Prof. Martin Humphries (University of Manchester) (Mould *et al.,* 1995). Cells were then stimulated with an epidermal growth factor (EGF) (500 n*M* in the outer well) gradient in a Dunn motility chamber at 37°C. Each image represents a two-dimensional projection of 2–3 slices in the Z-series, taken across the mid-depth of the cell at 0.2-μm intervals. GFP-ErbB1 was initially distributed in numerous fine vesicular/endosomal structures throughout the cytoplasm (18 min). Redistribution of GFP-ErbB1 to the cell cortex (60 min) facing the EGF gradient (indicated by an arrow, i.e., from high to low EGF concentration) as well as a perinuclear ring structure was closely followed by the appearance of membrane protrusions/filopodia toward the gradient (88 min). After the dissociation of the apposing neighbour cell (131 min phase image), the GFP-ErbB1-expressing cell retracted its filopodia and eventually moved away from its location (260 min). A subsequent image was captured using a different field of view (260 min'). The 12G10-reactive $\beta 1$ integrins were distributed in a polarized manner in two populations situated at the front and tail ends of the cell throughout the experiment, until 260 min after the addition of EGF when both ErbB1 and activated integrins adopted a perinuclear/recycling endosomal localization within which there was a significant colocalization between these receptors.

In practice, the biological conclusions drawn from the cumulative/statistical analysis of these two parameters are generally identical (Ng *et al.,* 2001). Conventional Eff analysis therefore provides a good semiquantitative representation of protein–protein association.

Acknowledgments

We thank Dr. Jonathan Chernoff (Fox Chase Cancer Center, Philadelphia) and James Monypenny (Imperial Cancer Research Fund, London) for providing the PAK-1 and GFP-tagged Cdc42 constructs, respectively. We also thank Dr. Arne Ostman (Ludwig Institute, Uppsala) and Dr. Fred Wouters, (EMBL, Heidelberg) for kindly donating porcine aortic endothelial (PAE) cells stably expressing the platelet-derived growth factor (PDGF) β receptor and a C-terminally tagged ErbB1 construct, respectively. We are indebted particularly to Dr. Peter Verveer (EMBL, Heidelberg) for critical reading of the manuscript and helpful discussions. We are grateful for the generous support from the UK Medical Research Council (in the form of a Clinician Scientist Grant awarded to T.N.).

References

Adams, S. R., Harootunian, A. T., Buechler, Y. J., Taylor, S. S., and Tsien, R. Y. (1991). Fluorescence ratio imaging of cyclic AMP in single cells. *Nature* **349,** 694–697.

Bailly, M., Wyckoff, J., Bouzahzah, B., Hammerman, R., Sylvestre, V., Cammer, M., Pestell, R., and Segall, J. E. (2000). Epidermal growth factor receptor distribution during chemotactic responses. *Mol. Biol. Cell* **11,** 3873–3883.

Bastiaens, P. I., and Joviyn, T. M. (1996). Microspectroscopic imaging tracks the intracellular processing of a signal transduction protein: fluorescent-labeled protein kinase C beta I. *Proc. Natl. Acad. Sci. USA* **93,** 8407–8412.

Clegg, R. M., Feddersen, B. A., Gratton, E., and Jovin, T. M. (1992). Time-resolved imaging fluorescence microscopy. *Proc. SPIE* **1640,** 448–460.

Day, R. N. (1998). Visualization of Pit-1 transcription factor interactions in the living cell nucleus by fluorescence resonance energy transfer microscopy. *Mol. Endocrinol.* **12,** 1410–1419.

Draaijer, A., Sanders, R., and Gerritsen, H. C. (1995). Fluorescence lifetime imaging, a new tool in confocal microscopy. *In* "Handbook of Biological Confocal Microscopy" (Pawley, J. B., ed.), pp. 491–505. Plenum Press, New York.

Gadella, T. W., and Jovin, T. M. (1995). Oligomerization of epidermal growth factor receptors on A431 cells studied by time-resolved fluorescence imaging microscopy. A stereochemical model for tyrosine kinase receptor activation. *J. Cell Biol.* **129,** 1543–1558.

Gadella, T. W. J., Jovin, T. M., and Clegg, R. M. (1993). Fluorescence lifetime imaging microscopy (FLIM)— spatial resolution of microstructures on the nanosecond time-scale. *Biophys. Chem.* **48,** 221–239.

Glazer, A. N., and Mathies, R. A. (1997). Energy-transfer fluorescent reagents for DNA analyses. *Curr. Opin. Biotechnol.* **8,** 94–102.

Griffin, B. A., Adams, S. R., Jones, J., and Tsien, R. Y. (2000). Fluorescent labeling of recombinant proteins in living cells with FlAsH. *Methods Enzymol.* **327,** 565–578.

Griffin, B. A., Adams, S. R., and Tsien, R. Y. (1998). Specific covalent labeling of recombinant protein molecules inside live cells. *Science* **281,** 269–272.

Harpur, A. G., and Bastiaens, P. I. H. (2001). "Probing Protein interactions using GFP and Fluorescence Resonance Energy Transfer." Cold Spring Harbor Laboratory Press, New York.

Harpur, A. G., Wouters, F. S., and Bastiaens, P. I. (2001). Imaging FRET between spectrally similar GFP molecules in single cells. *Nat. Biotechnol.* **19,** 167–169.

Honda, A., Adams, S. R., Sawyer, C. L., Lev-Ram, V., Tsien, R. Y., and Dostmann, W. R. (2001). Spatiotemporal dynamics of guanosine $3',5'$-cyclic monophosphate revealed by a genetically encoded, fluorescent indicator. *Proc. Natl. Acad. Sci. USA* **98,** 2437–2442.

Jakobs, S., Subramaniam, V., Schonle, A., Jovin, T. M., and Hell, S. W. (2000). EFGP and DsRed expressing cultures of *Escherichia coli* imaged by confocal, two-photon and fluorescence lifetime microscopy. *FEBS Lett.* **479,** 131–135.

Klostermeier, D., and Millar, D. P. (2000). Helical junctions as determinants for RNA folding: origin of tertiary structure stability of the hairpin ribozyme. *Biochemistry* **39,** 12970–12978.

Klostermeier, D., and Millar, D. P. (2001). RNA conformation and folding studied with fluorescence resonance energy transfer. *Methods* **23,** 240–254.

Kraynov, V. S., Chamberlain, C., Bokoch, G. M., Schwartz, M. A., Slabaugh, S., and Hahn, K. M. (2000). Localized rac activation dynamics visualized in living cells. *Science* **290,** 333–337.

Lakowicz, J. R. (1999). "Principles of Fluorescence Spectroscopy." Kluwer Academic/Plenum Publishers, New York.

Li, H. Y., Ng, E. K., Lee, S. M., Kotaka, M., Tsui, S. K., Lee, C. Y., Fung, K. P., and Waye, M. M. (2001). Protein–protein interaction of FHL3 with FHL2 and visualization of their interaction by green fluorescent proteins (GFP) two-fusion fluorescence resonance energy transfer (FRET). *J. Cell. Biochem.* **80,** 293–303.

Mahajan, N. P., Harrison-Shostak, D. C., Michaux, J., and Herman, B. (1999). Novel mutant green fluorescent protein protease substrates reveal the activation of specific caspases during apoptosis. *Chem. Biol.* **6,** 401–409.

Mahajan, N. P., Linder, K., Berry, G., Gordon, G. W., Heim, R., and Herman, B. (1998). Bcl-2 and Bax interactions in mitochondria probed with green fluorescent protein and fluorescence resonance energy transfer. *Nat. Biotechnol.* **16,** 547–552.

Mas, P., Devlin, P. F., Panda, S., and Kay, S. A. (2000). Functional interaction of phytochrome B and cryptochrome 2. *Nature* **408,** 207–211.

Miyawaki, A., Griesbeck, O., Heim, R., and Tsien, R. Y. (1999). Dynamic and quantitative Ca^{2+} measurements using improved cameleons. *Proc. Natl. Acad. Sci. USA* **96,** 2135–2140.

Miyawaki, A., Llopis, J., Heim, R., McCaffery, J. M., Adams, J. A., Ikura, M., and Tsien, R. Y. (1997). Fluorescent indicators for Ca^{2+} based on green fluorescent proteins and calmodulin. *Nature* **388,** 882–887.

Mould, A. P., Garratt, A. N., Askari, J. A., Akiyama, S. K., and Humphries, M. J. (1995). Identification of a novel anti-integrin monoclonal antibody that recognizes a ligand-induced binding site epitope on the beta1 subunit. *FEBS Lett.* **363,** 118–122.

Murata, S., Herman, P., Lin, H. J., and Lakowicz, J. R. (2000). Fluorescence lifetime imaging of nuclear DNA: effect of fluorescence resonance energy transfer. *Cytometry* **41,** 178–185.

Ng, T., Parsons, M., Hughes, W. E., Monypenny, J., Zicha, D., Gautreau, A., Arpin, M., Gschmeissner, S. Verveer, P. J., Bastiaens, P. I. H., and Parker, P. J. (2001). Ezrin is a downstream effector of trafficking PKC/integrin complexes involved in the control of cell motility. *EMBO J.* (in press).

Ng, T., Shima, D., Squire, A., Bastiaens, P. I. H., Gschmeissner, S., Humphries, M. J., and Parker, P. J. (1999a). PKCa regulates b1 integrin-dependent motility, through association and control of integrin traffic. *EMBO J.* **18,** 3909–3923.

Ng, T., Squire, A., Hansra, G., Bornancin, F., Prevostel, C., Hanby, A., Harris, W., Barnes, D., Schmidt, S., Mellor, H., Bastiaens, P. I. H., and Parker, P. J. (1999b). Imaging PKC alpha activation in cells. *Science* **283,** 2085–2089.

Pepperkok, R., Squire, A., Geley, S., and Bastiaens, P. I. H. (1999). Simultaneous detection of multiple green fluorescent proteins in live cells by fluorescence lifetime imaging microscopy. *Curr. Biol.* **9,** 269–272.

Squire, A., and Bastiaens, P. I. (1999). Three dimensional image restoration in fluorescence lifetime imaging microscopy. *J. Microsc.* **193,** 36–49.

Squire, A., Verveer, P. J., and Bastiaens, P. I. H. (1999). Multiple frequency fluorescence lifetime imaging microscopy. *J. Microsc.* **197,** 136–149.

Tyas, L., Brophy, V. A., Pope, A., Rivett, A. J., and Tavare, J. M. (2000). Rapid caspase-3 activation during apoptosis revealed using fluorescence-resonance energy transfer. *EMBO Rep.* **1,** 266–270.

Verveer, P. J., Squire, A., and Bastiaens, P. I. (2000a). Global analysis of fluorescence lifetime imaging microscopy data. *Biophys. J.* **78,** 2127–2137.

Verveer, P. J., Wouters, F. S., Reynolds, A. R., and Bastiaens, P. I. (2000b). Quantitative imaging of lateral ErbB1 receptor signal propagation in the plasma membrane. *Science* **290,** 1567–1570.

Wang, F., Weaver, V. M., Petersen, O. W., Larabell, C. A., Dedhar, S., Briand, P., Lupu, R., and Bissell, M. J. (1998). Reciprocal interactions between b1-integrin and epidermal growth factor receptor in three-dimensional basement membrane breast cultures: a different perspective in epithelial biology. *Proc. Natl. Acad. Sci. USA* **95,** 14821–14826.

Wouters, F. S., and Bastiaens, P. I. (1999). Fluorescence lifetime imaging of receptor tyrosine kinase activity in cells. *Curr. Biol.* **9,** 1127–1130.

Zicha, D., Dunn, G. A., and Brown, A. F. (1991). A new direct-viewing chemotaxis chamber. *J. Cell Sci.* **99,** 769–775.

Zicha, D., Dunn, G., and Jones, G. (1997). Analyzing chemotaxis using the Dunn direct-viewing chamber. *In* "Basic Cell Culture Protocols" (Pollard, J. W. and Walker, J. M., eds.), Vol. 75, pp. 449–457. Humana Press Inc., Totowa, NJ.

PART IV

Functional Applications
of Cell–Matrix Adhesion
in Molecular Cell Biology

CHAPTER 13

Functional Analysis of Cell Adhesion: Quantitation of Cell–Matrix Attachment

Steven K. Akiyama

Laboratory of Molecular Carcinogenesis
National Institute of Environmental Health Sciences
National Institutes of Health
Research Triangle Park, North Carolina 27709

I. Introduction

Cell–extracellular matrix interactions contribute to normal processes such as differentiation, embryonic development, and wound healing as well as to the progression of diseases and pathological conditions. It has become clear that cell adhesion occurs through the binding of specific adhesion receptors to matrix proteins. Ligand binding, in turn, initiates signaling pathways that not only result in regulation of the cell–matrix interactions themselves but also stimulate a wide range of cellular responses including proliferation, migration, and modulation of gene expression. Thus cell adhesion assays can provide important insights into many aspects of cell physiology and regulation in addition to the mechanism of the adhesive response itself.

The best characterized of the cell adhesion receptors are the integrins, a family of at least 24 widely expressed, noncovalent, transmembrane, heterodimeric complexes each consisting of an α subunit and a β subunit (Hynes, 1992; Akiyama *et al.*, 1995b; Huttenlocher *et al.*, 1995; Schwartz *et al.*, 1995; Howe *et al.*, 1998; Ruoslahti, 1999). At least 16 different α subunits and eight different β integrin subunits have been identified. Most α and β subunits are 140–180 kDa and 120–140 kDa, respectively. The combination of α and β subunits in the integrin dimers determines ligand specificity, although most integrins bind to several ligands. Posttranslational modifications, growth factors, divalent cations, and cytoplasmic signals all can modulate integrin function. The specific effects of integrin modulation can be readily analyzed with quantitative cell adhesion assays. Integrins containing a β_1, β_3, β_4, or β_5 subunit generally include adhesion receptors for the extracellular matrix proteins.

There is a large number and variety of extracellular matrix molecules that can play important roles in cell adhesive processes including the collagens, laminin, fibronectin, vitronectin, thrombospondin, tenascin, fibrin, osteopontin, von Willebrand factor, and proteoglycans. Because of space limitations, only a few of these will be discussed in detail here.

Fibronectin is found primarily as dimers in blood and as higher multimers in extracellular matrices (Mosher, 1989; Hynes, 1990). This multifunctional glycoprotein is composed of three different types of homologous repeating units. Human fibronectin contains 12 type I repeats, two type II repeats, and up to 18 type III repeats. In addition, one particular region, designated IIICS (also known as the V region), can be inserted intact or in part by alternative splicing.

Fibronectin contains at least two independent cell adhesive regions. The site that binds the $\alpha_5\beta_1$ integrin is defined by an Arg-Gly-Asp (RGD) amino acid sequence located in the 10th type III repeat. The 10th type III module alone and even small (five- or six-residue) synthetic peptides containing the RGD sequence are effective inhibitors of fibronectin-mediated cell adhesion (Pierschbacher *et al.*, 1984; Yamada and Kennedy, 1984). However, full cell adhesive activity requires an additional site defined by a Pro-His-Ser-Arg-Asn (PHSRN) peptide sequence located in the ninth type III repeat (Dufour *et al.*, 1988; Obara *et al.*, 1988; Aota *et al.*, 1991, 1994; Nagai *et al.*, 1991). This sequence acts synergistically with the RGD site. Cell adhesion assays with chemically modified and engineered fragments of fibronectin suggest that the relative positioning or orientation of the PHSRN and RGD sites may be important for cell adhesive function.

The alternatively spliced IIICS connecting segment contains a cell adhesive region that acts independently from the RGD-synergistic regions (Humphries *et al.*, 1987a,c) and is recognized by the $\alpha_4\beta_1$ integrin (Humphreys *et al.*, 1987b; Mould and Humphreys, 1991; Guan and Hynes, 1990; Mould *et al.*, 1991). Two nonadjacent peptide sequences within the IIICS region contain cell adhesive activity: a Leu-Asp-Val (LDV) near the amino terminus and an Arg-Glu-Asp-Val (REDV) near the carboxy-terminus (Humphries *et al.*, 1987c).

Laminin is a major component of basement membrane consisting of three non-identical polypeptide chains, α, β, and γ (Martin and Timpl, 1987; Mercurio, 1990; Engvall and Wewer, 1996). Each subunit exists in multiple isoforms that can combine to produce more than 10 variant forms of laminin. These variants include "classical" murine laminin derived from the Englebreth–Holm–Swarm (EHS) tumor, merosin, s-laminin, and kalinin/epiligrin. Laminin has multiple cell adhesive regions. There is an RGD sequence on the amino-terminal half of the α chain. The β chain contains a Tyr-Ile-Gly-Ser-Arg (YIGSR) sequence that promotes adhesion and migration, modulates morphological differentiation, and inhibits angiogenesis and tumor growth (Graf *et al.,* 1987). A third cell adhesive site near the carboxy terminus of the α chain with the sequence Ser-Ile-Lys-Val-Ala-Val (SIKVAV) induces neurite outgrowth and formation of tubelike networks by endothelial cells *in vitro* and enhances tumor cell growth, metastasis, and angiogenesis *in vivo* (Grant *et al.,* 1989; Tashiro *et al.,* 1989).

Vitronectin is a relatively small, multifunctional adhesive glycoprotein (Preissner, 1991; Tomasini and Mosher, 1991). The polypeptide chains of vitronectin are often present as two molecular species of approximately 60 and 75 kDa, the smaller of which is truncated at the carboxy terminus. This protein is found mostly in blood but is also found in tissues and other body fluids. It is the major cell adhesion protein in serum-containing tissue culture medium. Cell adhesion to vitronectin occurs by binding to an RGD sequence (Suzuki *et al.,* 1985).

Collagen is a highly abundant, extremely diverse family of multichain glycoproteins, all of which contain repeating-Gly-X-Y-sequences (Prockop and Kivirikko, 1995). To date, there are at least 19 types of collagens. The collagens are often divided into classes on the basis of structural features. The fibril forming collagens (types I, II, III, V, and XI) are the most abundant and, along with the network-forming collagens (types IV, VIII, and X), are the key cell-adhesive collagens.

II. General Concepts in Cell Adhesion Assays

There is a wide array of quantitative and semiquantitative assays for the interaction of cells with extracellular matrix proteins, and many variations on each method. This chapter will cover in depth assays for cell attachment, cell spreading, direct binding, and cell migration. It must be noted that the assays described may not be optimal for every situation. Because of space limitations, only a few specific methods are discussed in detail, but suggestions for alternative approaches are provided. These methods were selected, in part, because they represent common practices and because they minimize the use of highly specialized equipment or techniques. There are some general concepts and techniques that are important for most of the assays.

A. Preparation of Adhesive Substrata

Quantitative cell adhesion and migration assays require the preparation of an adhesive substratum consisting of an extracellular matrix molecule immobilized onto a solid support. The support is typically a 96-well tissue culture plate, although virtually every format of tissue culture plates and dishes, ELISA plates, bacteriological dishes, or glass coverslips can be used. ELISA plates and 96-well tissue culture clusters have the advantages of minimizing the amount of protein, cells, and reagents that will be needed for each assay and of allowing attachment assays to be quantitated using an ELISA plate reader. Other chapters in this volume should cover three-dimensional matrices and multicomponent mixtures. This section will focus on the preparation of two-dimensional adhesive surfaces consisting of a single active component.

Intact extracellular matrix proteins and large fragments adsorb well to plastics and glass. To do this, dilute the extracellular matrix protein to the appropriate concentration (usually 0.5–50 μg/ml) in a suitable buffer, usually Dulbecco's phosphate-buffered saline (D-PBS), and incubate in the wells overnight at 4°C or for 1–2 h at 37°C. Creating an adhesive surface with small protein fragments and peptides may require special treatment. Most small protein fragments and synthetic peptides either do not adsorb well to plastic to begin with, or may not retain cell adhesive activity when adsorbed (e.g., see Nagai *et al.,* 1991; Akiyama *et al.,* 1995a) and, therefore, require conjugation to larger carrier proteins or covalent linkage to the cluster. The coupling of peptides containing a terminal cysteine to bovine serum albumin (BSA) via heterobifunctional cross-linking agents has been very well described by Humphries *et al.* (1994). This class of reagents contains two different reactive groups and, therefore, allows initial coupling of the cross-linker to the carrier protein, removal of unreacted material, and subsequent coupling of the peptide. This multistep process maximizes cross-linking of peptide to carrier by minimizing the chance of forming cross-linked carrier protein. Cross-linkers, such as *N*-succinimidyl 3-(2-pyridyldithio)propionate and *m*-maleimidobenzoyl-*N*-hydroxysuccinimide ester, will covalently link lysines on the carrier protein to terminal cysteines on peptides. An alternative cross-linking approach is to couple peptides or fragments directly to chemically modified 96-well clusters as described by Nagai *et al.* (1991). Amines grafted onto such products as Nunc CovaLink plates take the place of the carrier protein.

Regardless of the method chosen to create the adhesive substratum, the concentration dependence of adhesion should be determined experimentally for each cell type and each extracellular matrix protein. In almost all cases, the concentration dependence of adhesion will have a linear region at relatively low matrix protein concentrations and then reach a plateau of maximal cell adhesion at high concentrations. Selection of an appropriate concentration of matrix protein may depend on the goal of the experiment. For maximal sensitivity to inhibitors, it is usually best to select the lowest possible concentration of matrix protein that still yields maximal cell adhesion. If higher concentrations are used, it might not be possible to strongly inhibit cell adhesion. If both inhibition and activation of adhesion is expected, the concentration of matrix protein that yields half-maximal adhesion might be a good choice.

B. Blocking Nonspecific Adhesion

Most cells attach to tissue culture plastic independently of extracellular matrix proteins. This nonspecific interaction is due to the highly hydrophilic nature of tissue culture plastic (Barker and LaRocca, 1994). Even uniformly adsorbing a matrix protein onto the plastic may not provide enough coverage to prevent a nonspecific component of adhesion.

Thus, most quantitative cell adhesion assays require blocking the plastic with a nonadhesive protein. A most effective blocking agent for most cells is 10 mg/ml heat-denatured BSA, which usually should allow attachment of <2–3% of most cells. BSA in calcium- and magnesium-free Dulbecco's phosphate-buffered saline (CMF-PBS) is first filtered and then incubated 10 min at 85°C. The incubation time is critical, so the heat denaturing is best performed with small volumes (usually ≤25 ml) to ensure uniform and rapid heating of the sample. Properly prepared denatured BSA should yield a very slightly hazy, almost pearlescent solution that is obviously homogeneous. A clear solution after heating indicates that denaturation did not occur. A very cloudy, white suspension indicates that the BSA has been overheated and aggregated. PBS containing divalent cations should not be used to prepare heat-denatured BSA. In the presence of divalent cations, the result is most often achieved is the very cloudy suspension of large, aggregated particles of BSA. Heat-denatured BSA can be stored for up to a week in the refrigerator.

C. Preparation of Cells

Either adherent or nonadherent cells can be used in adhesion assays. Cells growing in suspension culture can simply be pelleted by mild centrifugation, washed once to remove serum, and resuspended at an appropriate concentration and used in an assay. It may initially sound illogical to try to quantitate the adhesion of nonadherent cells to matrix proteins, but several points should be kept in mind. First, the major adhesion factors in serum-containing tissue culture medium are usually vitronectin and fibronectin. Cells that do not have active receptors for these proteins may readily attach when presented with other matrix proteins. Second, normally nonadherent cells may be expressing adhesion receptors in inactive form. Adhesion assays can be used to characterize mechanisms of activation for these receptors. Third, the concentration of matrix proteins in culture medium may be too low to promote adhesion.

It is critical that the procedure used to harvest adherent cells yield individual, viable cells that have suffered minimal damage. The cells should be detached as gently as possible. Either EDTA (Versene) alone, trypsin alone, or trypsin–EDTA can be used to detach cells. These agents should have little effect of the ability of the cells to adhere in the assay, provided the cells were treated for the briefest possible time and then given time to recover. Prolonged treatment of cells with either trypsin or EDTA can cause the cells to aggregate, a condition that is one of the most common causes of irreproducible results in adhesion assays. Treatment time during harvesting can be minimized by augmenting the detachment reagent with tapping the sides of the culture flask to dislodge cells. The trypsin (or EDTA) is then quenched using medium supplemented with either 0.5 mg/ml soybean trypsin inhibitor or 10% fetal bovine serum. Cells should be pelleted by gentle centrifugation (generally, $700g$–$1000g$) for 3–5 min, resuspended in serum-free medium, and incubated at 37°C for 10–20 min to allow them to recover. While the cells are incubating an aliquot should be removed to measure cell number. After the incubation, the cells are again gently pelleted and resuspended to the appropriate concentration in serum-free medium. Mild conditions are used to pellet and resuspend cells. Again, overly harsh treatment can cause many cell types to aggregate.

Only medium that is prewarmed to 37°C and preequilibrated to the correct pH should be used. A simple way to achieve these conditions is to preequilibrate medium in a sterile dish in a tissue culture incubator for several hours before use. Finally, some published procedures specify the use of simple salt solutions in adhesion assays. Although this may be necessary for some experiments, the use of serum-free medium usually gives more consistent and reproducible results.

III. Cell Attachment Assays

The most commonly used assays for quantitating cell adhesion fall into two categories: cell attachment assays and cell spreading assays. Cell attachment assays quantitate the fraction of cells that attach to matrix-coated surfaces and are resistant to gentle washing. An outline of the general procedure is provided in Table I. Briefly, cells are added to prepared matrix-coated surfaces and allowed to attach (usually for 15–90 min) in serum-free medium. The unattached cells are gently washed away and the attached cells are then quantitated.

Attached cells can be quantitated by directly counting the attached and unattached cells, by labeling cells with radioactivity, by staining attached cells, by using metabolic assays (e.g., the MTT assay), or by prelabeling cells with fluroescent dyes. If any of the last three methods are used, assays performed in 96-well clusters can be quantitated using an appropriate ELISA plate reader. The simplest method of quantitating cells is to fix them and then stain, usually with 0.1% crystal violet. When the stain is added to the wells, great care should be taken not to allow droplets of stain to stick to the sides of the wells. These droplets can dry, leaving extra dye in the wells and giving artifactually high readings. Since the amount of stain per cell may vary from experiment

Table I
Cell Attachment Assay

1. Preload matrix-coated and blocked wells of a 96-well cluster with 50 μl of serum-free medium alone or with inhibitors/activators.
2. Harvest cells.
3. Quench trypsin and allow cells to recover for 20 min at 37°C.
4. Count the cells, centrifuge, and resuspend in serum-free medium at a concentration of 2×10^5/ml.
5. Add 50 μl of cells to wells prefilled with 50 μl of serum-free medium and incubate in a humidified tissue culture incubator.
6. Remove nonadherent cells by gently washing with serum-free medium and aspirating the unbound cells.
7. Fix attached cells in 5% (w/v) glutaraldehyde in CMF-PBS for 30 min.
8. Wash cells three times with CMF-PBS and stain cells, 0.1% crystal violet for 30–60 min.
9. Destain by washing extensively with deionized water. Use at least 200 μl water per well.
10. Solubilize the bound dye with 10% (w/v) acetic acid or 2% SDS in DPBS.
11. Read absorbance at 560–590 nm.

to experiment, known numbers of cells (usually aliquots of 25, 50, 75, and 100 μl) are added to triplicate control wells of the same cluster and fixed without washing away unattached cells. This control will allow the conversion of absorbance readings to actual numbers of cells attached and will also ensure that the absorbance increases linearly with the number of cells attached.

An alternative approach for quantitating attached cells is to prelabel cells with fluorescent dyes. Cells can be prelabeled with 5-chloromethylfluorescein diacetate (CMFDA; Molecular Probes, Inc., Eugene, OR) for 20 min during the recovery step after detachment. If this approach is used, medium without phenol red should be used for the assay to minimize background fluorescence. As with the staining approach, the total number of cells adhered per well is determined by normalizing fluorescent readings to those of a standard curve containing predetermined numbers of cells per well.

If radioactivity is used, the cells must be prelabeled, usually overnight (but at least several hours) with 1–10 μCi/ml [^{3}H]thymidine or radioactive amino acids (e.g., [^{35}S] methionine or [^{3}H]leucine), followed by a chase period of 1–2 h to deplete unincorporated label. After the unattached cells are washed away, the remaining, attached cells can be solubilized in 50 mM NaOH and 2% SDS and counted by liquid scintillation spectrometry. Either unattached or attached cells can also be quantitated by direct counting using a cell sorter or even a hemocytometer. This approach may require the use of 6-well clusters in order to get enough cells to count accurately.

The incubation time during which the cells are attaching to the matrix protein should be experimentally determined for each cell type and matrix protein. In order for the assay to be an accurate measure of simple cell attachment, this time should be long enough to yield maximal cell adhesion but not so long that the cells begin to spread. This can be checked microscopically using phase contrast optics. For fibroblastic cells, the incubation time for cell attachment is usually 15–30 min.

A critical step in this assay is the washing step, which may be required to reduce the background level of unbound cells in contact with the matrix-coated surface. When washing, medium at 37°C and the correct pH must be added gently to the side of wells in order to minimize turbulent flow which, if high enough, could dislodge even well-attached cells. Similarly the wash medium must be removed with gentle aspiration. Preliminary experiments should be performed to determine the number of washes that yields the highest possible ratio of cell attachment to the extracellular matrix protein divided by background cell attachment to wells coated with only BSA. It should be possible to obtain less than 2–3% (and often <1%) cell adhesion to BSA-coated plastic and 40–50% maximal cell adhesion (and often >60%) on the highest concentration of matrix protein. On occasion, even the gentlest turbulent flow during washing removes all the cells, even those that are specifically bound. A possible solution is presented by Channavajjala *et al.* (1997). These authors describe the use of a Percoll-induced density inversion to lift unbound cells from the matrix-coated surface very gently.

Cell attachment assays of the type described in Table I relate cell–matrix interactions to the relative numbers of cells that attach to the matrix protein. Assays have also been developed to quantitate directly the strength of cell–matrix interactions by measuring the force needed to detach cells in a controlled manner. Most of these assays control the detachment force through centrifugation (Lotz *et al.,* 1989; Channavajjala *et al.,* 1997) or hydrodynamic flow (Garcia *et al.,* 1998). These force quantitation assays usually require highly specialized equipment and/or difficult techniques. Therefore, they are mentioned here only as possible alternative approaches and will not be discussed further.

IV. Cell Spreading Assays

Another approach to measuring cell adhesion is to determine how well cells spread on extracellular matrix proteins. Although cell spreading to normal morphology is a more complex process than simple attachment, there are a number of advantages to this approach: (1) Cell spreading assays tend to be more specific with lower background levels of spreading. This is because although many proteins and nonphysiological molecules (e.g., polylysine or concanvalin A) can promote cell attachment nonspecifically, few of these can promote morphological spreading. (2) Because of the lower background, spreading assays can be more sensitive and are usually the methods of choice with which to test soluble inhibitors of cell adhesion. (3) Cell spreading is much more useful for initial screening experiments or for obtaining detailed dose-response curves. Cell spreading assays can provide reliable results without the use of multiple wells for each concentration or condition because replicate determinations can be obtained from sampling different locations in the same well. Therefore, it is possible to get detailed dose-response curves, accurate determinations of IC_{50} values for inhibitors, or screening of a large number of inhibitors in a single experiment. (4) Spread cells can be observed with a quick look under a microscope. Therefore, it is possible to "eyeball" qualitative results quickly. (5) Finally, since all the cells that are added to wells are fixed to the extracellular matrix protein, washing artifacts are eliminated.

A general protocol for carrying out cell spreading assays is provided in Table II. This assay is usually easier to perform than is the cell attachment assay, and there are usually only a few parameters that need to be adjusted, none of which require truly fine-tuning. One such parameter is density of cells. It is important to seed enough cells to count,

Table II
Cell Spreading Assay

1. Preload matrix-coated and blocked wells of a 96-well cluster with 50 μl of serum-free medium alone or with inhibitors.
2. Harvest cells.
3. Quench and allow cells to recover for 20 min at 37°C.
4. Count the cells, centrifuge, and resuspend in serum-free medium at a concentration of 2×10^5/ml.
5. Add 50 μl cells to the previously prepared matrix-coated surface and incubate in a tissue culture incubator.
6. Terminate the assay by fixing the cells with 100 μl 6% glutaraldehyde/6% formaldehyde in CMF-DPBS for 60 min.
7. Count sets of 100 cells from 3–8 randomly chosen fields. Alternatively, the surface areas of randomly chosen cells from each well can be determined using image analysis software.

but the cells should not overlap or even touch to the extent that it is difficult to discern whether an individual cell is spread or not spread. The number of cells recommended in the protocol in Table II should work for all but the most extreme cases. Another key parameter is the time that cells are allowed to spread. Generally, cells of fibroblastic origin spread quite rapidly and 80–90% cell spreading can often be reached in 30–60 min. Other cell types can take up to several hours. The time chosen for the assay should be long enough to get substantial cell spreading on the extracellular matrix protein but not so long that more than about 2–3% of the cells begin to spread on BSA. If allowed to incubate too long, many cells can modify the matrix-coated surface by using membrane-bound proteases and synthesize and secrete matrix proteins of their own, allowing them to appear to spread on BSA. If long spreading times are necessary, it is possible to use cyclohexamide to inhibit synthesis of new proteins.

Cell spreading assays are usually subjectively quantitated by microscopically inspecting sets of cells through phase contrast optics and deciding whether individual cells are "spread" or "not spread." In order for cell spreading assays to be reproducible, it is critical to maintain a standard, consistent definition of "spread" throughout a series of experiments. Two most useful definitions of "spread" are: (1) the achievement of a normal cell morphology similar to that found in sparse culture or (2) the attainment of visible cytoplasm completely surrounding the nucleus. When analyzing spreading assays, the usual approach is to count a set of 100 cells in a randomly chosen microscope field, scoring cells as "spread" or "not spread." This counting procedure is usually repeated for 3–6 sets of 100 cells per set to yield an average percent cells spread. The practice of giving fractional values to cells that appear to be "partially spread" is strongly discouraged. It should be possible to decide whether or not a cell has cytoplasm completely surrounding its nucleus. A simple "spread" vs "not spread" scoring system will lead to more consistent, reproducible results. Another approach to quantitating cell spreading assays is to use image processing software such as NIH Image to measure a parameter related to cell spreading, such as the percent area of a field covered by cell bodies. This can eliminate subjectivity or investigator bias in the reading of the assay. However, the difference between spread cells and nonspread cells may be quantitatively small because spherical and poorly spread cells that are not responding well to the extracellular matrix protein may often cover a surprisingly large surface area. NIH Image is in the public domain and can be found at http://rsb.info.nih.gov/nih-image/about.html.

V. Direct Binding Assays

Although not precisely in the category of cell adhesion, assays quantitating the direct binding of matrix proteins to cells in suspension can give important insights into the mechanism of cell–matrix interactions. This approach has been used to quantitate the interactions of laminin and fibronectin with tumor cells, platelets, and fibroblasts (Akiyama and Yamada, 1985; Malinoff and Wicha, 1983; Plow and Ginsberg, 1981; Rao *et al.,* 1983). As in most classical direct binding studies, radiolabeled ligand is required. Iodination appears to be the method of choice for laminin using ^{125}I and lactoperoxidase

Table III
Direct Binding Assay

1. Harvest cells and suspend at a concentration of 1–10×10^7/ml in Dulbecco's modified Eagle's medium supplemented with 25 mM Hepes buffer and 2% bovine serum albumin (binding medium).
2. Add [^{3}H]fibronectin to cells in a total volume of 150 μl to yield a final concentration of 0.5–4×10^7 cells/ml.
3. Mix cells and [^{3}H]fibronectin by rotating the tubes end-over-end.
4. To terminate the binding assay, dilute 100 μl of cells in 1 ml binding medium and microfuge 1 min to pellet the cells.
5. Resuspend the cells in 100 μl binding medium, dilute in 1 ml binding medium, and microfuge 1 min to pellet the cells.
6. Resuspend the cells in 100 μl binding medium, dilute in 1 ml binding medium containing 10% BSA, and microfuge 1 min to pellet the cells.
7. Suspend the cells in two aliquots of 200 μl deionized water. Pool both aliquots and count in a liquid scintillation counter.

(Rao *et al.*, 1983; Malinoff and Wicha, 1983). Radiolabeling of fibronectin by reductive methylation with ^{3}H may be required when examining binding to fibroblasts and tumor cells, which bind fibronectin with only moderate affinity, although [^{125}I]fibronectin has been used for assaying binding of fibronectin to platelets or cells of hematopoietic origin on which integrins have been activated (Plow and Ginsberg, 1981; Faull *et al.*, 1993). The method of choice to tritium-label fibronectin is reductive methylation (Tack *et al.*, 1980) using carrier-free sodium [^{3}H]borohydride ($\sim$60 Ci/mmol) and formaldehyde. This approach should yield [^{3}H-methyl]fibronectin at a specific activity of more than 10^9 cpm/mg. A complete description of the preparation, reductive methylation, and handling of human plasma fibronectin is provided (Akiyama and Yamada, 1985).

The method for performing direct fibronectin binding assays (Table III) was originally developed for baby hamster kidney cells adapted for suspension culture (Akiyama and Yamada, 1985) but it has been used for a variety of adherent cells including B16 murine melanoma cells, HT-1080 fibrosarcoma cells, and human foreskin fibroblasts (e.g., see Miyamoto *et al.*, 1995). There are experimental parameters, such as the number of cells, the number of washes, and the incubation time, that need to be determined for each cell type, but as is the case with the cell spreading assay, these do not usually require fine-tuning. Direct binding assays require extensive controls. Binding must be shown to be specific, reversible, time-dependent, and saturable. Specific binding usually ranges between 50% and 90% of total amount of bound ligand. In addition, functional equivalence between radiolabeled and unmodified matrix protein must be demonstrated. (For examples of these controls, see Plow and Ginsberg, 1981, or Akiyama and Yamada, 1985.)

VI. Cell Migration Assays

Migration requires even more complex interactions than does cell spreading, involving interactions among adhesion receptors, cytoskeleton, signaling pathways, and membrane transport. Furthermore, migration is optimal when the cell–extracellular matrix affinity is moderate. As shown by DeMilla *et al.* (1993), cells tend to migrate faster on moderately adhesive subtrates. There are several examples of migration assays including the filter assay, haptotaxis, random walk assay, "wound healing" assay, and the agarose drop assay, all of which measure different aspects of cell migration. For example, the filter assay measure cell migration toward a chemoattractant and the random walk assay measures the rate of random motion of cells sparsely distributed on a matrix-coated surface.

The assays described here are the "agarose drop" assay and the "wound healing" assay. Neither of these assays requires specialized materials or equipment. The agarose drop assay is based on a method initially described by Carpenter (1963) and subsequently rediscovered, modified, and refined (Varani *et al.*, 1978; Akiyama *et al.,* 1989). As described in Table IV, cells are suspended in an agarose droplet, which is then placed on a matrix-coated surface. The cells migrate out of the now-solid droplet over the course of 1–3 days. Migration is quantitated by measuring the diameter of the halo of cells that

Table IV
Agarose Drop Migration Assay

1. Precoat 96-well plates with matrix protein. Usually, Immulon ELISA plates work best for migration assays. Each condition should be replicated at least 3 times.
2. Block if necessary. Blocking of the matrix-coated surface is not always necessary because of the several-day time scale of most migration assays.
3. Melt sterile low-melting-point agarose (2% w/v in deionized water) and equilibrate at 37°C.
4. Prewarm all pipettors, pipette tips, and tubes to 37°C. The easiest way to do this is to place these items in a warm cabinet or incubator for several hours. This method was developed using a 100-μl Hamilton syringe to dispense cells, but a pipettor (Eppendorf or equivalent) set to dispense 1 μl with disposable sterile tips also works.
5. Prealiquot inhibitors or activators, if needed, into 100 μl volumes of medium in tubes and equilibrate at 37°C in an air/CO_2 atmosphere.
6. Aspirate dry the prepared 96-well plates and air dry. Great care must be taken to avoid scratching or otherwise damaging the protein coat.
7. Harvest cells, count, and pellet.
8. Calculate the volume needed to resuspend the cells at a concentration of 3×10^7/ml.
9. Thoroughly aspirate the supernatant off of the cell pellet.
10. Add 87% of the volume calculated in step 8 of fibronectin-depleted serum-containing medium to the cells and resuspend the cells, avoiding bubbles. Transfer to a warm microfuge tube.
11. Add 13% of the volume calculated in step 8 of 2% low-melting-point agarose at 37°C.
12. Place 1 μl droplets of cells in agarose to the center of each coated well. The wells must be dry or the droplets will spread out. Add droplets to at least 3 extra coated wells to serve as zero time points.
13. Incubate 5 min in a refrigerator at 4°C to solidify the droplets.
14. Fill each well with 100 μl medium. Leave the 3 wells corresponding to the zero time point determinations dry to prevent any migration from occurring.
15. Check the wells at 12-h intervals. Terminate the assay in all wells when the cells in any well first begin to migrate as far as the walls of the well.
16. Terminate the assay by aspirating the medium and fixing for 1 h with 0.1 M cacodylate containing 2% methylene blue, 3% formaldehyde, and 3% glutaraldehyde.
17. Wash the plate with tap water until only the pellets are stained.
18. Air dry the plate and quantitate cell migration. The simplest way to do this is to measure the diameter of the cell halo and calculate the area of cell outgrowth after subtracting the area of the original pellet from the zero time control.

migrate out of the droplet, either using image capture equipment and analysis software or, most simply, using a calibrated grid in a microscope eyepiece.

Because of the duration of the assay, serum-containing medium may be required to keep some cell types alive. Fibronectin may stimulate cell migration, so medium prepared with fibronectin-depleted serum should be used. Fibronectin can be removed by passing serum slowly over a gelatin–Sepharose (or equivalent) affinity column. The column can be regenerated by washing with 8 M urea in D-PBS and then washing the column clear of urea with more D-PBS.

Agarose droplets should be placed in at least triplicate wells per condition. At least three drops should be left to dry without covering with medium as controls. When fixed and stained at the end of the assay, these control wells will provide a negative, "zero time" control to allow the experimental determination of the average size of agarose droplets before any outward migration occurs.

There are several critical steps in the agarose drop assay. It is important for all tubes and pipette tips as well as the agarose to be equilibrated at 37°C. If the agarose is too hot, the cells may die. Temperatures below 37°C can cause the agarose to solidify prematurely. Drying the matrix-coated surface prior to the addition of the agarose droplets is key to the success of this assay. Any liquid remaining in the wells can dilute the agarose, causing it to flow across the entire bottom of the well instead of creating a well-formed, almost hemispherical droplet. Finally, placement of the droplet in the center of the wells is crucial. Placing droplets too close to one side will force the termination of the assay early. If the assay is allowed to continue beyond the time the migrating cells reach the well of the wells, quantitation will be impossible.

The wound healing assay for cell migration was described as such by Lipton *et al.* (1971). This assay, described in Table V, involves seeding cells in matrix-coated tissue

Table V
Wound Healing Migration Assay

---

1. Coat 6-well tissue culture plates with 2 ml/well of matrix protein and block with heat-denatured BSA (prepared in CMF-PBS).
2. Remove BSA and wash plate twice with 2 ml/well of CMF-PBS.
3. Add 2 ml/well of cells to plate and incubate (37°C, 5% CO_2) overnight.
4. At 24–48 h, depending on when cells become confluent, wound each well by using a sterile yellow (200 μl) pipette tip to mark a straight line down the center of the plate.
5. Remove the medium and wash each well twice with 2 ml of DPBS.
6. Prepare inhibitors or activators of cell migration in 2 ml serum-free medium per well. Add 2 ml/well of medium or treatments. Photograph 4 fields of each wound.
7. Place cells in a tissue culture incubator.
8. At 24-h intervals, photograph 4 fields on each wound again. Remove medium/treatments and add fresh to each well.
9. Repeat each day as needed until the cells have closed the gap.
10. To fix and stain after the final photograph, add 8% glutaraldehyde in DPBS and fix for 1 h at room temperature.
11. Remove fix and add 2 ml/well of Giemsa stain (Gibco). Stain for 10–15 min at room temperature.
12. Remove stain and wash wells thoroughly but gently with deionized water at least 5 times.

---

culture clusters, culturing the cells 24–48 h to confluence, then forming a "wound" in the monolayer by scraping away a strip of cells and observing cell migration across the blank strip. As with the agarose drop assay, the time required for the wound healing assay may necessitate the addition of fibronectin-depleted serum to ensure cell survival. Various tools have been used to scrape the cell monolayer including sterile razor blades, cell scrapers, and polypropylene disposable (yellow) pipette tips. We have found that Rainin 200 μl (yellow) pipette tips work the best. They come presterilized, fit in almost any size tissue culture wells, and can cleanly remove the cells without damaging the matrix-coated surface.

This assay is most easily quantitated by measuring the rate at which cells close the wound using a calibrated eyepiece. Since many cells can extend long processes from the cell body, it is recommended that the cell nuclei be used to define the position of the cell.

VII. Conclusions

Quantitative analyses of cell–matrix interactions, such as those discussed in this chapter, have been used to obtain critical mechanistic information such as the characterization of the biological functions of matrix proteins, to identify novel cell adhesive sites on matrix proteins and novel cell adhesion receptors, to develop an understanding of the role of adhesion receptors as signaling proteins, and to identify and characterize inhibitors and activators of cell adhesion (Akiyama and Yamada, 1985; Akiyama et al., 1985, 1989; Aota et al., 1991, 1994; Bowditch et al., 1994; Dufour et al., 1988; Graf et al., 1987; Grant et al., 1989; Humphries et al., 1987a,b,c; Malinoff and Wicha, 1983; Mould et al., 1991; Nagai et al., 1991; Obara et al., 1988; Obara and Yoshizato, 1995; Pierschbacher and Ruoslahti, 1984; Rao et al., 1983; Tashiro et al., 1989; Yamada and Kennedy, 1984). These are by no means complete lists of either findings or references. Quantitative assays for cell–matrix interactions should continue to be useful in the future. Many of the cell adhesion assays could potentially be modified for use in functional high-throughput screening processes. These assays can be expected to play critical roles in scientific discovery in such areas as the development of novel therapeutics to treat human disease and increase our understanding of cell adhesive function of novel proteins that are certain to be discovered as analysis of the genome and proteome progresses.

Acknowledgment

The author thanks Drs. Sarah Kennett, Paul Noni, and John Roberts for very valuable comments and suggestions.

References

Akiyama, S. K., and Yamada, K. M. (1985). The interaction of plasma fibronectin with fibroblastic cells in suspension. *J. Biol. Chem.* **260,** 4492–4500.

Akiyama, S. K., Hasegawa, E., Hasegawa, T., and Yamada, K. M. (1985). The interaction of fibronectin fragments with fibroblastic cells. *J. Biol. Chem.* **260**, 13256–13260.

Akiyama, S. K., Yamada, S. S., Chen, W.-T., and Yamada, K. M. (1989). Analysis of fibronectin receptor function with monoclonal antibodies: Roles in cell adhesion, migration, matrix assembly, and cytoskeletal organization. *J. Cell Biol.* **109**, 863–875.

Akiyama, S. K., Aota, S., and Yamada, K. M. (1995a). Function and receptor specificity of a minimal 20 kilodalton cell adhesive fragment of fibronectin. *Cell Adhes. Commun.* **3**, 13–25.

Akiyama, S. K., Olden, K., and Yamada, K. M. (1995b). Fibronectin and integrins in invasion and metastasis. *Cancer Metast. Rev.* **14**, 173–189.

Aota, S., Nagai, T., and Yamada, K. M. (1991). Characterization of regions of fibronectin besides the arginine-glycine-aspartic acid sequence required for adhesive function of the cell-binding domain using site-directed mutagenesis. *J. Biol. Chem.* **266**, 15938–15943.

Aota, S., Nomizu, M., and Yamada, K. M. (1994). The short amino acid sequence Pro-His-Ser-Arg-Asn in human fibronectin enhances cell-adhesive function. *J. Biol. Chem.* **269**, 24756–24761.

Barker, S. L., and LaRocca, P. J. (1994). Method of production and control of a commercial tissue culture surface. *J. Tiss. Cult. Methods* **16**, 151–153.

Bowditch, R. D., Hariharan, M., Tominna, E. F., Smith, J. W., Yamada, K. M., Getzoff, E. D., and Ginsberg, M. H. (1994). Identification of a novel integrin binding site in fibronectin: Differential utilization by $\beta 3$ integrins. *J. Biol. Chem.* **269**, 10856–10863.

Carpenter, R. R. (1963). In vitro studies of cellular hypersensitivity. I. Specific inhibition of migration of cells from adjuvant-immunized animals by purified protein derivative and other protein antigens. *J. Immunol.* **91**, 803–818.

Channavajjala, L., Eidsath, A., and Saxinger, W. C. (1997). A simple method for measurement of cell–substrate attachment forces: application to HIV-1 Tat. *J. Cell Sci.* **110**, 249–255.

DeMilla, P. A., Stone, J. A., Quinn, J. A., Albelda, S. M., and Lauffenberger, D. A. (1993). Maximal migration of human smooth muscle cells on fibronectin and type IV collagen occurs at an intermediate attachment strength. *J. Cell Biol.* **122**, 729–737.

Dufour, S., Duband, J.-L., Humphries, M. J., Obara, M., Yamada, K. M., and Thiery, J. P. (1988). Attachment, spreading and locomotion of avian neural crest cells are mediated by multiple adhesion sites on fibronectin molecules. *EMBO J.* **7**, 2661–2671.

Engvall, E., and Wewer, U. M. (1996). Domains of laminin. *J. Cell. Biochem.* **61**, 493–501.

Faull, R. J., Kovach, N. L., Harlan, J. M., and Ginsberg, M. H. (1993). Affinity modulation of intergrin $\alpha 5\beta 1$: Regulation of the functional response by soluble fibronectin. *J. Cell Biol.* **121**, 155–162.

Garcia, A. J., Huber, F., and Boettiger, D. (1998). Force required to break $\alpha 5\beta 1$ integrin–fibronectin bonds in intact adherant cells is sensitive to integrin activation state. *J. Biol. Chem.* **273**, 10988–10993.

Graf, J., Ogle, R. C., Robey, F. A., Sasaki, M., Martin, G. R., Yamada, Y., and Kleinman, H. K. (1987). *Biochemistry* **26**, 6896–6900.

Grant, D. S., Tashiro, K.-I., Segui-Real, B., Yamada, Y., Martin, G. R., and Kleinman, H. K. (1989). *Cell* **58**, 933–943.

Guan, J.-L., and Hynes, R. O. (1990). Lymphoid cells recognize an alternatively-spliced segment of fibronectin via the receptor alpha$_4$beta$_1$. *Cell* **60**, 53–60.

Howe, A., Aplin, A. E., Alahari, A. K., and Juliano, R. L. (1998). Integrin signaling and cell growth control. *Curr. Opin. Cell Biol.* **10**, 220–231.

Humphries, M. J., Akiyama, S. K., Komoriya, A., Olden, K., and Yamada, K. M. (1987a). Identification of an alternatively spliced site in human plasma fibronectin that mediates cell-type specific adhesion. *J. Cell Biol.* **103**, 2637–2647.

Humphries, M. J., Akiyama, S. A., Komoriya, A., Olden, K., and Yamada, K. M. (1987b). Neurite extension of chicken peripheral nervous system neurons on fibronectin: relative importance of specific adhesion sites in the central cell-binding domain and the alternatively-spliced type III connecting segment. *J. Cell Biol.* **106**, 1289–1297.

Humphries, M. J., Komoriya, A., Akiyama, S. K., Olden, K., and Yamada, K. M. (1987c). Identification of two distinct regions of the type III connecting segment of human plasma fibronectin that promote cell type-specific adhesion. *J. Biol. Chem.* **262**, 6886–6892.

Humphries, M. J., Mould, A. P., and Weston, S. A. (1994). Conjugation of synthetic peptides to carrier proteins for cell adhesion studies. *J. Tiss. Cult. Methods* **16,** 239–242.

Huttenlocher, A., Sandborg, R. R., and Horwitz, A. F. (1995). Adhesion in cell migration. *Curr. Opin. Cell Biol.* **7,** 697–706.

Hynes, R. O. (1990). "Fibronectins." Springer-Verlag, New York.

Hynes, R. O. (1992). Integrins: versatility, modulation, and signaling in cell adhesion. *Cell* **69,** 11–25.

Kanemoto, *et al.* (1990). p. 4.

Kanemoto, T., Reich, R., Royce, L., Greatorex, D., Adler, S. H., Shiriashi, N., Martin, G. R., Yamada, Y., Kibbey, M. C., Grant, D. S., and Kleinman, H. K. (1992). Role of the SIKVAV site of laminin in promotion of angiogenesis and tumor growth: an in vivo Matrigel model. *J. Natl. Cancer Inst.* **84,** 1633–1638.

Lipton, A., Klinger, I, Paul, D., and Holley, R. W. (1971). Migration of mouse 3T3 fibroblasts in response to a serum factor. *Proc. Natl. Acad. Sci. USA* **68,** 2799–2801.

Lotz, M. M., Burdsal, C. A., Erickson, H. P., and McClay, D. R. (1989). Cell adhesion to fibronectin and tenascin: Quantitative measurements of initial binding and subsequent strengthening response. *J. Cell Biol.* **109,** 1795–1805.

Malinoff, H. L., and Wicha, M. S. (1983). Isolation of a cell surface receptor protein for laminin from murine fibrosarcoma cells. *J. Cell Biol.* **96,** 1475–1479.

Martin, G. R., and Timpl, R. (1987). Laminin and other basement membrane components. *Annu. Rev. Cell. Biol.* **3,** 57–85.

Mercurio, A. M. (1990). Laminin: multiple forms, multiple receptors. *Curr. Opin. Cell Biol.* **2,** 845–849.

Mosher, D. F., ed. (1989). "Fibronectin." Academic Press, New York.

Mould, A. P., and Humphries, M. J. (1991). Identification of a novel recognition sequence for the integrin alpha4-beta1 in the COOH-terminal heparin-binding domain of fibronectin. *EMBO J.* **10,** 4089–4095.

Mould, A. P., Welson, L. A., Kormoriya, A., Wayner, E. A., Yamada, K. M., and Humphries, M. J. (1991). Affinity chromatography isolation of the melanoma adhesion receptor for the IIICS region of fibronectin and its identification as the integrin alpha4beta1. *J. Biol. Chem.* **265,** 4020–4024.

Miyamoto, S., Akiyama, S. K., and Yamada, K. M. (1995). Synergistic roles for receptor occupancy and aggregation in integrin transmembrane function. *Science* **267,** 883–885.

Nagai, T., Yamakawa, N., Aota, S., Yamada, S. S., Akiyama, S. K., Olden, K., and Yamada, K. M. (1991). Monoclonal antibody characterization of two distant sites required for function of the central cell-binding domain of fibronectin in cell adhesion, cell migration, and matrix assembly. *J. Cell Biol.* **114,** 1295–1306.

Obara, M., and Yoshizato, K. (1995). Possible involvement of the interaction of the a_5 subunit of the $\alpha_5\beta_1$ integrin with the synergistic region of the central cell-binding domain of fibronectin in cells to fibronectin binding. *Exp. Cell Res.* **216,** 273–276.

Obara, M., Kang, M. S., and Yamada, K. M. (1988). Site-directed mutagenesis of the cell-binding domain of human fibronectin: Separable synergistic sites mediate adhesive function. *Cell* **53,** 649–657.

Pierschbacher, M. D., and Ruoslahti, E. (1984). Cell attachment activity of fibronectin can be duplicated by small synthetic fragments of the molecule. *Nature* **309,** 30–33.

Plow, E., and Ginsberg, M. H. (1981). Specific and saturable binding of plasma fibronectin to thrombin-stimulated human platelets. *J. Biol. Chem.* **256,** 9477–9482.

Preissner, K. T. (1991). Structure and biological role of vitronectin. *Annu. Rev. Cell Biol.* **7,** 275–310.

Prockop, D. J., and Kivirikko, K. I. (1995). Collagens: Molecular biology, diseases, and potentials for therapy. *Annu. Rev. Biochem.* **64,** 403–434.

Rao, N. C., Barskey, S. H., Terranova, V. P., and Liotta, L. A. (1983). Isolation of a tumor cell laminin receptor. *Biochem. Biophys. Res. Commun.* **111,** 804–808.

Ruoslahti, E. (1999). Fibronectin and its integrin receptors in cancer. *Adv. Cancer Res.* **76,** 1–20.

Schwartz, M. A., Schaller, M. D., and Ginsberg, M. H. (1995). Integrins: emerging paradigms of signal transduction. *Annu. Rev. Cell Biol.* **1,** 549–599.

Suzuki, S., Oldberg, A., Hayman, E. G., Pierschbacher, M. D., and Ruoslahti, E. (1985). Complete amino acid sequence of human vitronectin deduced from cDNA. Similarity of cell attachment sites in vitronectin and fibronectin. *EMBO J.* **4,** 2519–2524.

Tack, B. F., Dean, J., Eilat, D., Lorenz, P. E., and Schechter, A. N. (1980). Tritium labeling of proteins to high specific radioactivity by reductive methylation. *J. Biol. Chem.* **255,** 8842–8847.

Tashiro, K., Sephel, G. C., Weeks, B., Sasaki, M., Martin, G. R., Kleinman, H. K., and Yamada, Y. (1989). A synthetic peptide containing the IKVAV sequence from the A chain of laminin mediates cell attachment, migration, and neurite outgrowth. *J. Biol. Chem.* **264,** 16174–16182.

Tomasini, B. R., and Mosher, D. F. (1991). Vitronectin. *Prog. Hemostasis Thromb.* **10,** 269–305.

Varani, J., Orr, W., and Ward, P. A. (1978). A comparison of the migration patterns of normal and malignant cells in two assay systems. *Am. J. Pathol.* **90,** 159–171.

Yamada, K. M., and Kennedy, D. W. (1984). Dualistic nature of adhesive protein function: fibronectin and its biologically active peptide fragments can autoinhibit fibronectin function. *J. Cell Biol.* **99,** 29–36.

CHAPTER 14

Measurements of Glycosaminoglycan-Based Cell Interactions

J. Kevin Langford* and Ralph D. Sanderson†

* Department of Pathology
† Departments of Pathology and Anatomy
Arkansas Cancer Research Center
University of Arkansas for Medical Sciences
Little Rock, Arkansas 72205

I. Introduction

Glycosaminoglycans are ubiquitous—being found within cells, at the cell surface, and throughout the extracellular matrix. Although glycosaminoglycans can be bound directly to cell surfaces via noncovalent interactions with cell surface proteins, they are often present as proteoglycans which consist of glycosaminoglycan chains covalently attached to a core protein. There is substantial variation between cell types as to the amount and type of proteoglycan present (e.g., heparan sulfate proteoglycan, chondroitin sulfate proteoglycan); thus the precise glycosaminoglycan status of each cell type must be established using biochemical techniques (Hascall *et al.,* 1994). Although antibodies do exist that can detect specific glycosaminoglycan chains, their utility in characterizing the glycosaminoglycan status of cells is somewhat limited by the ability of a single antibody to accurately quantify these chains given their intrinsic heterogeneity. In fact, a panel of monoclonal antibodies generated against heparan sulfate clearly shows differential staining of various tissues (van Kuppevelt *et al.,* 1998).

Predominantly because of their negative charge, glycosaminoglycans bind to numerous molecules, including both soluble effector molecules such as growth factors, and insoluble molecules present on cell surfaces or within the matrix (Jackson *et al.,* 1991). Once purified, the interaction of glycosaminoglycans with substrates can be measured using various binding assays (Jackson *et al.,* 1991). One of these assays is affinity co-electrophoresis (ACE), which enables calculation of apparent dissociation constants (K_d) between glycosaminoglycans or proteoglycans and specific ligands (San Antonio and Lander, 2001). In this technique, radiolabeled glycosaminoglycans or proteoglycans are subjected to electrophoresis through agarose lanes that contain protein ligands present at a range of concentrations. Following autoradiography, the K_d is calculated as the protein concentration at which the glycosaminoglycan or proteoglycan is half-shifted from being fully mobile between the high and low protein concentration. The advantages of ACE include its low cost, simplicity, and speed and the fact that it requires relatively small amounts of glycosaminoglycan and ligand.

Interactions between glycosaminoglycans and insoluble molecules can profoundly influence the adhesive and motile behavior of cells. As a model to understand glycosaminoglycan-based cell adhesion, our laboratory has explored the adhesive capacity of heparan sulfate-bearing proteoglycans. Most of our studies have focused on syndecans, a major family of transmembrane proteoglycans responsible for displaying heparan sulfate on the surface of many cell types (Bernfield *et al.,* 1999). This chapter describes in detail assays we have employed to examine the role of heparan sulfate in mediating cell adhesion to extracellular matrix (e.g., collagens, fibronectin) and in regulating the invasion of cells into type I collagen gels. A third assay we have developed to examine the role of heparan sulfate in mediating cell–cell adhesion is described elsewhere (Langford and Sanderson, 2001).

The versatility of the cell adhesion and invasion assays described herein is greatly enhanced by the ability to remove heparan sulfate from the cell surface or to block heparan sulfate function. For example, prior to the assay, cells can be treated with enzymes that degrade heparan sulfate, thereby removing it specifically from the cell surface. These heparan sulfate depleted cells can then be tested for their ability to adhere. In another, more simplistic approach, adhesion assays can be done in the presence of exogenous glycosaminoglycans to determine their ability to inhibit adhesion. Yet another approach is to grow cells in the presence of sodium chlorate, a competitive inhibitor of glycosaminoglycan sulfation (Humphries and Silbert, 1988). This results in the presence of glycosaminoglycan on the cell surface that lacks proper sulfation and generally abolishes the ability of cells to adhere via their glycosaminoglycan chains. The assays described in this chapter can also be employed to examine the role of proteoglycan core proteins in adhesion and motility. For example, we have transfected the ARH-77 B lymphoid cell line (which, in contrast to most cells, lacks heparan sulfate at its cell surface) with wild-type or various mutated cDNA constructs of syndecan-1 core protein. These experiments have shown that mutated core proteins that still bear heparan sulfate can mediate cell–matrix and cell–cell adhesion, but certain mutations in the core do result in the loss of syndecan-1-medited inhibition of cell invasion (Liu *et al.*, 1998; Langford and Sanderson, unpublished observations). Thus the assays described in this chapter are amenable to manipulation and modification that expand their utility to examine a number of issues related to glycosaminoglycan-based cell adhesive interactions.

II. Cell Adhesion Assay

A. Overview

In this assay, first described by Koda *et al.* (1985), cells are introduced into 96-well round-bottom polyvinyl plates that are coated with the extracellular matrix components to be tested. Following a short incubation period to allow cell adhesion, the plate is subjected to a centrifugal force. Weakly attached or unattached cells form a pellet in the bottom of the well, while attached cells remain uniformly distributed along the sides of the wells. Following fixation and staining, plates can be photographed for documentation of assay results.

B. Materials

1. "U" bottom 96 well polyvinyl plates (Thermo Lab Systems, Franklin, MA). We highly recommend this specific brand of plates because we have found they yield consistent, reproducible results in this assay. The "U"-shaped bottom yields a broader pellet than a V-shaped plate, thereby enabling better viewing of the pelleted cells.

2. ECM components. For this discussion, we will use rat tail collagen type I (Collaborative Biomedical Products, Bedford, MA) as our example. The working concentration is 100 μg of collagen diluted to 1 ml with phosphate-buffered saline, pH 7.4. Other extracellular matrix molecules such as fibronectin can also be used to coat wells (Saunders and Bernfield, 1988).

3. Bovine serum albumin (1 mg/ml) diluted with phosphate-buffered saline, pH 7.4.

4. Incubation buffer. 10 mM Hepes, 150 mM NaCl, pH 7.4. In the past, we have used either phosphate-buffered saline, pH 7.4, or Hepes buffer as our incubation buffer in this assay. It has been our experience that Hepes is best suited for our needs and performs better for room-temperature incubations, and a lower centrifugal force is needed to pellet the nonadherent cells. When PBS was used, incubation times were shortened to between 5 and 10 min and the centrifugal force required to pellet the negative controls was, at times, higher than the maximum allowed force for the rotor being used. In addition, it has been reported that phosphate can interfere with some interactions between heparan sulfate and extracellular matrix molecules (Stamatoglou and Keller, 1982).

5. Fixative. 4% glutaraldehyde in phosphate buffered saline, pH 7.4.

6. Stain. 5% (w/v) toluidine blue in phosphate-buffered saline, pH 7.4.

7. Phosphate-buffered saline, pH 7.4.

8. Tabletop centrifuge capable of spinning 96-well plates at 120g.

C. Adhesion Assay Procedure

1. Wash the 96-well plate extensively with deionized water. This is important to remove any residual agents from the manufacturing process that may interfere with adhesion of the matrix molecule to the well.

2. Coating wells with matrix. For each sample to be tested, we prepare triplicate wells coated with either BSA (a nonspecific adhesion negative control) or with the matrix molecule of interest (e.g., collagen). Add 200 μl of either BSA solution or matrix solution to the respective wells. BSA-coated wells are used as a negative and non-specific binding control for the myeloma cells we use. It is important to use some benign protein as a negative for each sample to be tested to directly compare the results from the nonadherent wells (BSA) and the potentially adherent wells (collagen). Occasionally, pellets may have ragged edges or empty centers. The negative controls may mimic some of these variations and therefore provide important interpretive data.

3. Cover the plates and incubate overnight at 4°C. Although an overnight incubation of the matrix protein with the wells during the coating step may be excessive, it has provided us with consistent results. This incubation may be shortened for other molecules incubated at different temperatures; however, this must be determined empirically for each matrix protein to be used.

4. Wash all wells 3 times with phosphate-buffered saline, pH 7.4.

5. Blocking of the wells: add 200 μl of the BSA solution to *all* wells. This will block any available sites on the plate prior to the addition of cells.

6. Incubate at room temperature for 1 h.

7. Wash all wells 3 times with phosphate-buffered saline, pH 7.4. Once wells are coated with matrix, they must not be allowed to dry until the end of the experiment. If additional time is required between the washing and incubation steps or between the washing and addition of cells, leave the final wash covering the wells until the material is ready to be added immediately. Drying of the matrix-coated surface may denature the protein and alter the results of the experiment.

8. The cells to be tested are harvested, pelleted by centrifugation, and resuspended into incubation buffer at a concentration of 4×10^5 cells/ml. In our studies with lympho-cytes, the cells are harvested from suspension culture. For cells growing as monolayers in culture, harvesting should be done using PBS containing 0.5 mM EDTA and cells released by gentle scraping. Use of trypsin to remove cells is not recommended as it may remove proteoglycans from the cell surface. Empty the final wash from the wells of the plate and add 200 μl of the cell suspension (8×10^4 cells) to each well. In some instances, it may be desirable to remove glycosaminoglycan chains from the cell surface prior to performing the assay. In this case, once in suspension and prior to harvesting for the adhesion assay, cells can be treated with specific glycosaminoglycan degrading enzymes such as heparitinase (1 mU/ml, 30 min at 37°C) or chondroitinase (50 mU/ml,

30 min at 37°C) (Seikagaku America, Falmouth, MA) (Saunders and Bernfield, 1988). Determining that glycosaminoglycans are actually removed from the cell surface can be somewhat difficult. We have used two methods. One is to perform Western blots on a portion of the digested cells to determine if the syndecan-1 present has been reduced from a broad smear to a tight band (Liu *et al.,* 1998). Another technique for analysis of heparan sulfate removal is staining of cells with monoclonal antibody 3G10, which recognizes stubs remaining on the core protein of proteoglycans following digestion with heparitinase (David *et al.,* 1992).

9. Incubate the cells at room temperature for 15–30 min. This step is done at room temperature because if performed at 37°C, the background adhesion to the BSA control wells is increased.

10. Centrifuge at 120*g* for 20 min.

11. Without disturbing the cells, carefully add 50 μl of fixative to each well.

12. Cover the plate and incubate at least 1 h at 4°C.

13. Add a drop of stain (approximately 50–100 μl is sufficient) and incubate at room temperature for 30–60 min. Carefully, without touching either the sides or bottom of the wells, remove the solution from each well. We use a 200–1000 μl capacity pipette for removal and subsequent washing steps. Carefully wash each well with approximately 500 μl of phosphate-buffered saline, pH 7.4.

14. Continue carefully washing the wells until little, if any, soluble stain is visible.

15. Air dry at room temperature.

16. Although documentation may be done by routine photography, using either a film based or a digital system, we have found that scanning the plates using a color flatbed scanner yields excellent results. For example, the plate in Fig. 1 was inverted onto the scanner and a piece of white paper used as a background. The scanner was set for color scanning with a resolution of 150 dpi.

Results should be similar to those shown in Fig. 1. Here, wild-type ARH-77 cells (available from the American Type Culture Collection, Rockville, MD) that lack heparan sulfate proteoglycan expression fail to bind to either type I collagen or the BSA control

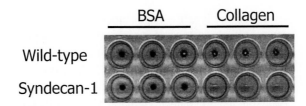

Fig. 1 Image of a completed cell adhesion assay. Triplicate wells were coated with either type I collagen or BSA (as a control) and incubated with either wild-type ARH-77 cells (that are syndecan-1-negative) or with ARH-77 cells transfected with the cDNA for syndecan-1. Following incubation, plates were centrifuged, and cells fixed and then stained. Wells where cell pellets are present indicate a lack of adhesion to the matrix (wild-type cells on collagen); wells not having pellets indicate tight cell binding to the matrix (syndecan-1 transfected cells on collagen). Note that neither cell line binds to the BSA-coated control wells.

and therefore form a tight pellet in the center of the well. In contrast, the cells transfected with the cDNA for the syndecan-1 core protein bind to the collagen-coated wells and are not pelleted by centrifugation. Although this assay is not quantitative, relative strength of attachment between cell types can be assessed by increasing centrifuge speed until differential pelleting is observed. Removal of heparan sulfate chains from the cell surface, addition of exogenous heparin to the assay buffer, or growth of cells in the presence of sodium chlorate all abolish binding of the syndecan-1-expressing cells to the collagen-coated wells (Liebersbach and Sanderson, 1994). An important consideration for each experiment is the inclusion of cells that will act as both positive (adherent to the matrix surface) and negative (nonadherent to the matrix surface) controls. As shown in Fig. 1, the ARH-77 cell line we use is nonadherent on collagen and is always included with each experiment we perform, as are syndecan-1 expressing cells that tightly adhere to collagen and represent the positive control. For different cells or cell lines, appropriate positive and negative controls must be identified.

III. Cell Invasion Assay

A. Overview

To investigate the role of glycosaminoglycans in regulating cell invasion, we employ a hydrated native type I collagen gel. A solution of collagen is prepared in cell culture media lacking fetal bovine serum; the solution is then added to wells of a 24-well plate and the collagen polymerized into a gel at 37°C. Gels are then equilibrated with medium containing fetal bovine serum and cells added to the top of gels and allowed to invade for 48 h. To quantify invasion, the distance or depth of invasion is measured by phase microscopy. The percent of cells invading the gels is measured by proteolytically removing the noninvasive cells from the top of the gel followed by subsequent extensive proteolytic digestion of the gel to remove the invasive cells. After counting cells on a hemocytometer or Coulter counter, a simple calculation determines the percent of invasive cells.

B. Materials

1. Rat tail type I collagen stock. Although type I collagen may be obtained from several sources, we have found that the collagen produced by Collaborative Biomedical Products (Bedford, MA) yields consistent results and reproducible gels.

2. Sodium bicarbonate solution 7.5% (w/v).

3. Complete medium: 1× RPMI solution, 2 mM L-glutamine, 1× antibiotic-antimycotic solution (Mediatech, Herndon, VA), and 5% fetal bovine serum. This is the standard medium used for the B lymphoid cell lines we employ. For other cell lines the type of medium and amount of fetal bovine serum may vary.

4. 10× RPMI solution.

5. Sterile distilled deionized water.

6. 24-well cell culture plate.

7. Trypsin (0.25%) EDTA (0.1%) solution.

8. Collagenase solution (0.5 mg/ml of collagenase type 2 (Worthington Biochemical Corp, Freehold, NJ) in 1× RPMI).

9. PBS + EDTA (0.5 mM) solution.

10. Hemocytometer or Coulter counter.

C. Invasion Assay Procedure

1. Prepare the collagen gel solution. (Because the concentration of collagen in the gel dramatically influences the rate of cell invasion, a major consideration in the preparation of the gels is the collagen concentration best suited to yield cell invasion. If the collagen concentration is too high, cell invasion may not occur; if it is too low, gels may not form properly and cells may passively fall through large pores in the gel. Therefore, the optimal concentration of collagen must be determined empirically for each cell line. We have found that gels with a collagen concentration of 0.5 mg/ml work well with human B lymphoid cells.) Determine the total volume of collagen gel solution required based on using 1 ml of collagen gel solution for each well. First, add cell-culture grade sterile deionized water to approximately one-half of the total collagen gel solution required to an appropriately sized sterile culture tube. Because the total assay time is 48 h, care must be taken during collagen gel preparation to ensure sterility of all solutions. Next add collagen to the final working concentration. Mix the collagen-containing solution by inverting the tube several times and place on ice. Care must be taken throughout the gel preparation to keep all solutions ice-cold to prevent premature polymerization of the collagen. Add 1/10 volume of cold 10× RPMI dropwise while gently shaking the tube, then mix by inverting the tube several times and place on ice.

2. To achieve a pH of approximately 7.2–7.4, add sodium bicarbonate solution drop-wise while gently shaking the tube. Again, mix by inverting the tube several times and place on ice. The pH of the collagen gel is a variable that can greatly influence the integrity of the gel and thus the invasive extent of the cells. The pH is adjusted by varying the amount of sodium bicarbonate added to the collagen gel solution. At slightly more acidic pH (i.e., 6.8–7.0) the gels do not polymerize as well and are unstable. Although this allows cells to invade faster, the gels become fragile, making it difficult to add medium or wash solutions to the gels without damaging the surface. Conversely, a more basic pH (i.e., 7.6–8.0) creates a gel through which cells have difficulty invading. With experience, the optimal pH can be determined by carefully examining the color of the solution. The appropriate pH results in a "salmon pink" color. If close monitoring of pH is necessary, apply a small amount of collagen gel solution to pH test strips. If during the pH adjustment the collagen gel solution becomes too basic, discard the solution.

3. Once the proper pH is reached, degas the solution by allowing it to sit undisturbed on ice for approximately 20 min. Failure to properly degas the solution will result in gels having trapped air bubbles. This compromises the integrity of the gel and provides pockets where cells can invade at an artificially rapid rate.

4. Pipette 1 ml of the collagen gel solution into each well of the 24-well plate. This *must* be done very gently so as not to introduce bubbles.

5. Carefully, place the plate containing the collagen gel solution into a tissue culture incubator at 37°C and 5% CO_2 for at least 1 h. Polymerization can be assessed by slowly tilting the plate 45°. Once polymerized, the gel should not move relative to the side of the wells upon tilting. If the solution does not polymerize, discard and repour the gels.

6. To equilibrate the gel with medium, carefully pipette 1 ml of complete medium to the surface of each gel and place the gel back into the incubator for at least 1 h. Again, pipetting must be done gently so the gel is not damaged or detached from the surface of the well. For convenience, the gels may be equilibrated overnight.

7. To inoculate wells with cells, carefully add cells diluted in 1 ml of complete medium and grow at 37°C in 5% CO_2 for 48 h. The number of cells to be added to each well must be determined empirically for each cell line used. Once seeded on the gels, cells should not have extensive cell–cell contact. Thus, the size and mitotic rate of the cells to be used dictate the density at which they are added to each collagen well. For myeloma cells, we use 5.0×10^4 cells per collagen gel. Another important consideration is the density of cells growing in culture prior to the start of the invasion assay. For myeloma cells, the most consistent results in our invasion assays are obtained from cells growing between 50% and 75% confluency. Myeloma cells do not invade collagen gels well if, prior to seeding the gel, they are growing at low density or are over-confluent.

D. Quantification of Depth of Cell Invasion

The depth of cell invasion is determined by measuring the distance from the top of the gel to the leading front of migrating cells. The leading front is defined as the point at which two of the leading cells within a given field are in the same focal plane under 200× magnification. This measurement is obtained using the calibrated micrometer present on the fine focus dial of an inverted phase microscope.

1. Measurements are taken in five fields within each well; using the center of the well as a landmark and as the first point (center point), the four additional measurements are taken in fields selected by moving the stage to points north, south, east, and west of center. The distance to each field is defined as half the distance from the center point to the edge of the well.

2. Starting in a focal plane near the bottom of the gel, rotate the fine-focus dial so the focal plane moves toward gel surface. When two cells appear within the same focal plane, begin counting revolutions of the fine focus dial until the surface of the gel is reached.

3. Calculate the distance from the gel surface to cells at the leading front by multiplying the number of revolutions counted by the number of μm/revolution. For the Nikon Diaphot used in our lab, one revolution moves the focal plane 100 μm.

E. Quantification of the Percent of Invading Cells

The percent of invading cells is determined by first subjecting the collagen gel to a limited trypsin and collagenase treatment to remove the noninvasive cells (cells attached at or near the gel surface). After collecting cells released by this limited digestion, the remaining gel is subjected to extensive collagenase digestion to destroy the gel and the invading cells are harvested by centrifugation (Fig. 2). After counting the number of invading and non-invading cells, the percentage of invading cells is calculated.

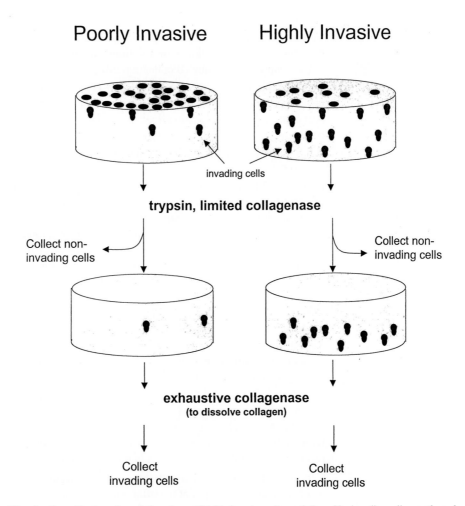

Fig. 2 Quantification of poorly invasive and highly invasive cell populations. Noninvading cells are released from the gel surface by trypsin and then limited collagenase digestion and counted. Next, following complete digestion of the remaining collagen with collagenase, the invading cells are collected and counted. The percent of cell invasion for each gel is then calculated.

1. Collection of the noninvasive cell population.

   a. Aspirate and discard the medium from the surface of each gel. Gently wash each gel with 0.5 ml of PBS + 5 m*M* EDTA and collect by aspiration into an appropriately marked tube. Note that while washing the gels, care must be taken to not damage the gels. If the gel becomes distorted, or a hole appears during the quantification process, accurate quantification cannot be accomplished and the gel must be discarded. When pipetting, slowly allow each drop of liquid to spread over the surface of the gel or liquid surface rather than dropping the liquid from a distance. The impact of a large droplet of liquid on the gel surface can damage the gel.

   b. Repeat this wash/collection step two additional times, placing all solutions (washes, trypsin and collagenase solutions) harvested from each individual well into an appropriately labeled collection tube.

   c. Incubate the gel with 0.5 ml of trypsin/EDTA solution at room temperature for 30 min. Aspirate medium and released cells and place in collection tube. Although this initial digestion does not release many cells, it does prepare the collagen for the collagenase treatment.

   d. Wash gels twice and collect the solution two times as in step b. Incubate gels with 0.5 ml of collagenase solution (0.5 mg/ml of collagenase in 1× RPMI; pre-warmed to 37°C) for approximately 10 min at 37°C. This first collagenase treatment removes the noninvasive cell population from the gel surface and is *the most critical step of the quantification process*. Overdigestion of the gel will remove many of the invasive cells, whereas underdigestion will fail to remove the cells at the surface. To ensure that results are accurate, for each experiment, the extent of digestion should be monitored by examining the digestion occurring in the noninvading control gel. This control serves as a "landmark" that can be used to determine the time required to remove the noninvasive cells from the surface of all gels. This monitoring is accomplished by removing the plate from the incubator 5 min after the collagenase digestion has begun and examining the cells on the surface using an inverted phase microscope. If the majority of cells in the noninvading control are in suspension, proceed to the next step. If many cells remain attached at the gel surface, return the plate to the incubator for 1-min intervals and examine until the noninvading cells are released from the gel surface.

   e. Once the noninvading cells are released from the surface, quickly and carefully remove the collagenase solution and wash the gels three times as in step b and place released cells and all washes in the collection tube.

   f. Centrifuge the tubes containing the noninvasive cell population at 300*g* for 10 min and resuspend the cell pellet in buffer and count the cells.

2. Collection of the invasive cell population.

   a. Add 0.5 ml of collagenase solution to each well and incubate at 37°C until the gels are completely digested (approximately 1.5 h).

b. Collect the solution by aspiration into individual tubes.

c. Wash the wells, collect the solution, and place all solutions collected from each individual well into the same 15 ml culture tube.

d. Centrifuge the cells and count.

3. Calculate the percent of invasive cells as follows:

$$\frac{\text{Invasive cells}}{\text{Noninvasive cells} + \text{invasive cells}} \times 100 = \% \text{ invasive cells}$$

Using this assay we have demonstrated that cells lacking syndecan-1 readily invade collagen gels, whereas cells expressing syndecan-1 do not invade. Addition of heparin or growth of syndecan-1-expressing cells in sodium chlorate prior to placing cells on gels renders the cells highly invasive, indicating that inhibition of invasion requires functional heparan sulfate at the cell surface (Liebersbach and Sanderson, 1994).

Although relatively simple in design, the invasion assay requires close attention to detail. Critical parameters include pH and collagenase digestion times as noted previously. In addition, we have found some variability between lots of fetal bovine serum in their ability to promote cell invasion. Similarly, some lots of type I collagen are not as permissive for invasion as others. Thus, the assay requires close monitoring and quality control. One of the most essential requirements for obtaining interpretable results is the inclusion of positive and negative controls (i.e., invasive and noninvasive cells, respectively). This ensures that the collagen gel is permissive for invasive cells yet still able to prevent noninvasive cells from passively falling through the spaces between collagen fibers. By including these two controls, data can be reported either as raw data (Liebersbach and Sanderson, 1994) or as the percent invasion relative to the controls (Langford *et al.,* 1998). The latter method compensates for interassay variability.

Acknowledgment

R.D.S. was supported by NIH grants CA68494 and CA55819.

References

Bernfield, M., Gotte, M., Park, P. W., Reizes, O., Fitzgerald, M. L., Lincecum, J., and Zako, M. (1999). Functions of cell surface heparan sulfate proteoglycans. *Annu. Rev. Biochem.* **68,** 729–777.

David, G., Bai, X. M., Van der Schueren, B., Cassiman, J. J., and Van den Berghe, H. (1992). Developmental changes in heparan sulfate expression: in situ detection with mAbs. *J. Cell Biol.* **119,** 961–975.

Hascall, V. C., Calabro, A., Midura, R. J., and Yanagishita, M. (1994). Isolation and characterization of proteoglycans. *Methods Enzymol.* **230,** 390–417.

Humphries, D. E., and Silbert, J. E. (1988). Chlorate: a reversible inhibitor of proteoglycan sulfation. *Biochem. Biophys. Res. Commun.* **154,** 365–371.

Jackson, R. L., Busch, S. J., and Cardin, A. D. (1991). Glycosaminoglycans: Molecular properties, protein interactions, and role in physiological processes. *Physiol. Rev.* **71,** 481–539.

Koda, J. E., Rapraeger, A., and Bernfield, M. (1985). Heparan sulfate proteoglycans from mouse mammary epithelial cells. Cell surface proteoglycan as a receptor for interstitial collagens. *J. Biol. Chem.* **260,** 8157–8162.

Langford, J. K., and Sanderson, R. D. (2001). Regulatory roles of syndecans in cell adhesion and invasion. *Methods Mol. Biol.* **171,** 495–503.

Langford, J. K., Stanley, M. J., Cao, D., and Sanderson, R. D. (1998). Multiple heparan sulfate chains are required for optimal syndecan-1 function. *J. Biol. Chem.* **273,** 29965–29971.

Liebersbach, B. F., and Sanderson, R. D. (1994). Expression of syndecan-1 inhibits cell invasion into type I collagen. *J. Biol. Chem.* **269,** 20013–20019.

Liu, W., Litwack, E. D., Stanley, M. J., Langford, J. K., Lander, A. D., and Sanderson, R. D. (1998). Heparan sulfate proteoglycans as adhesive and anti-invasive molecules: Syndecans and glypican have distinct functions. *J. Biol. Chem.* **273,** 22825–22832.

San Antonio, J. D., and Lander, A. D. (2001). Affinity coelectrophoresis of proteoglycan–protein complexes. *Methods Mol. Biol.* **171,** 401–414.

Saunders, S., and Bernfield, M. (1988). Cell surface proteoglycan binds mouse mammary epithelial cells to fibronectin and behaves as a receptor for interstitial matrix. *J. Cell Biol.* **106,** 423–430.

Stamatoglou, S. C., and Keller, J. M. (1982). Interactions of cellular glycosaminoglycans with plasma fibronectin and collagen. *Biochim. Biophys. Acta* **719,** 90–97.

van Kuppevelt, T. H., Dennissen, M. A., Van Ven rooij, W. J., Hoet, R. M., and Veer Kamp, J. H. (1998). Generation and application of type-specific anti-heparan sulfate antibodies using phage display technology. Further evidence for heparan sulfate heterogeneity in the kidney. *J. Biol. Chem.* **273,** 12960–12966.

CHAPTER 15

Applications of Adhesion Molecule Gene Knockout Cell Lines

Jordan A. Kreidberg

Department of Medicine
Division of Nephrology
Children's Hospital
and
Department of Pediatrics
Harvard Medical School
Boston, Massachusetts 02115

I. Introduction

The past 10 years has witnessed the targeted mutation of almost the entire panoply of cell-surface adhesion molecules and their cognate ligands, including the components of the extracellular matrix (ECM). In many instances, these gene targeting experiments have yielded interesting phenotypes that have been revealing of the function of the adhesion molecule or ECM component within the context of the entire organism. However, there is no denying that the greatest focus of study involving adhesion molecules during this time period has occurred within a cell-biological context, particularly focusing on signal transduction events that are triggered by interactions of adhesion receptors and the ECM. This being the case, an obvious goal of gene targeting experiments that mutate cell adhesion or ECM genes is to eventually understand at a cell-biological level why a particular mutation has caused a specific phenotype. These findings may lead to beneficial therapeutic interventions in diseases that result from aberrant adhesive processes. However, this is often a very difficult goal to attain, as the whole organism presents a far more complex system to analyze than a homogeneous population of cells in a culture dish. One means of attempting to bridge the gap between the whole organism and the individual cell has been to derive cell lines from organisms that carry targeted mutations, and then use these cells to carry out investigations at a cell-biological level. As expected, there are great advantages and disadvantages in procuring these types of cell lines. On the one hand, they may provide an inexhaustible source of biological material containing a defined mutation. On the other hand, in many cases transfection experiments using dominant-negative mutants may also provide important insights without taking the time needed to develop these cell lines from primary cell populations. In the end, both approaches may be required to answer important biological questions. This chapter will review how some of the gene-null cell lines have been made and used.

An additional type of experiment that will also be reviewed in this chapter utilizes embryonic stem cells carrying homozygous mutations in cell adhesion molecule genes. If an adhesion receptor performs a vital function during embryonic development that cannot be replaced by a related receptor, embryos that are homozygous for a targeted mutation in the gene for that receptor undergo embryonic lethality at the first point when that receptor is required for vital developmental processes. Yet there may be many functions of that receptor that one might wish to study at later points in development or in the adult animal, in either the same or different cell lineages than the gene was expressed in the point of lethality. One approach to this problem is to use conditional mutations, which allow the inactivation of a targeted gene at a selected location or time point, thus avoiding early embryonic lethality. Conditional mutation in mice has been reviewed in multiple reports, but is beyond the scope of the present review (Sauer, 1998). An additional approach that avoids the problem of embryonic lethality is to derive embryonic stem cell (ES cell) lines that are homozygous for a mutation in a cell adhesion receptor. It is then possible to derive chimeric embryos or mice using the homozygously mutated ES cells, and then determine the ability of these ES cells to contribute to various cell lineages (Fassler and Meyer, 1995; Gustafsson and Fassler, 2000). It can then also be determined whether these different cell types function properly within the organism, when unable to express the particular adhesion molecule that has been mutated. In addition, homozygous-null ES cells can be studied under conditions that allow differentiation *in vitro*.

II. Derivation of Somatic Cell Lines

Cell lines or primary cell populations have been derived from a variety of mouse tissues, including kidney epithelial tubules, keratinocytes, and embryonic fibroblasts. In cases where it is difficult to expand primary cultures to provide sufficient material for cell biological experiments, it was necessary to immortalize a small number of primary cells, with the subsequent outgrowth of an immortalized population. When this approach is used, it is of great importance to verify the extent to which the immortalized population retains characteristics of the primary tissue or organ from which it was derived.

Primary cell populations derived from gene targeted mice can be immortalized by a variety of approaches involving the introduction of oncogenes via DNA transfection, or retroviral infection using retroviral vectors that contain oncogenes such as large T antigen (reviewed in Chou, 1989). An additional favored approach involves transgenic mice carrying a temperature-sensitive large T antigen under control of the H-2K promoter that confers nearly ubiquitous expression (Immortomouse, Charles River Laboratories) (Jat *et al.,* 1991). In this latter approach, the T antigen transgenic mouse is mated with a mouse carrying the targeted mutation. Depending on whether homozygous mutant or only heterozygous mice are viable, they are crossed to finally produce mice or embryos that carry a homozygous mutation of the gene under study and the T antigen transgene. Tissues may then be removed from these mice or embryos and placed in culture under conditions permissive for T antigen function (33°C) (Jat *et al.,* 1991). Because the T antigen in this mouse is temperature-sensitive, cells derived from these mice can be shifted to a higher temperature (37–39°C) that is nonpermissive for T antigen function to observe their behavior in the absence of the transforming oncoprotein. In our experience, we have not observed a difference in growth or differentiation characteristics of our collecting duct epithelial cells (discussed below) after shifting to 39°C. On the other hand, immortalized kidney podocyte lines do differentiate after shifting to the higher temperature (Mundel *et al.,* 1997).

In practice, the foregoing approach is not simple, as most differentiated cell types will not immediately start growing while retaining their differentiated characteristics, simply because they now express the T antigen transgene. The actual process for finally obtaining immortalized cell lines must be tailored and optimized for each cell type. As it is not possible to provide a single protocol that is suited for all cell types from different tissues, I discuss the experience of my lab in isolating α3-integrin-null cell lines and some other case histories.

A. Kidney Epithelial Cell Lines from α3 Integrin–Null Mice

In the following section the approach used in the author's lab to obtain immortalized cells from kidney collecting ducts will be described to provide one example (Wang *et al.,*

1999). Mice carrying a targeted mutation of the $\alpha3$ integrin gene were crossed with the T antigen transgenic mice discussed previously, to obtain double heterozygous mice that could then be mated to obtain newborn mice or embryos homozygous-null for the $\alpha3$ mutation that also carried the temperature-sensitive T-antigen transgene.

In cases where homozygous mutant animals are not viable postnatally or beyond certain embryonic time points, it is necessary to obtain cells from embryos or early newborn mice before the time of death and tissue resorption. Moreover, since certain organs or tissues provide rather small amounts of starting cell material, it is desirable to pool tissues from several embryos. However, it is often not possible to identify homozygous mutant embryos from their external appearance. Therefore it is necessary either to culture cells from each embryo individually until genotyping can be performed, or to save the embryos or organs/tissues until genotyping is completed, thereby allowing pooling of mutant and wild-type control organs or tissues. In our experience with embryonic kidneys, we have found it possible to remove them from E18 embryos and place them in DMEM/10% fetal calf serum overnight or even for 36–48 h if necessary, at 4°C, during which time a PCR-based genotyping procedure is completed, thus allowing us to pool mutant and wild-type kidneys. Indeed, we found that to obtain sufficient quantities of starting cell populations it was necessary to pool 10–20 mutant E18 kidneys. DNA for the genotyping procedure was obtained from a limb or tail removed from the E18 embryo at the time of dissection. A rapid procedure was used to prepare DNA from this tissue sample (Laird et al., 1991). In our experience, if kidneys were removed in the afternoon, the DNA sample could be processed in 24 h by allowing the tissue lysis/proteolysis to proceed overnight at 55°C, precipitating the DNA in the morning, allowing the DNA to dissolve in TE for a few hours at 55°C, and then performing a PCR reaction with relatively short elongation times over a 2- to 3-h period, finally analyzing the PCR genotyping results by mid- to late afternoon. Other PCR-based genotyping approaches are available using very small tissue samples that do not involve preparation of a DNA sample, which might allow even faster genotyping.

In general it is most optimal if the starting tissue contains a homogenous or nearly homogenous population of cells. Otherwise it will be necessary to enrich for the desired cells, or clone them from a mixed population once a growing population is established. To obtain a starting population of collecting duct cells, the section of the embryonic kidney that contained these cells was microdissected away from the remainder of the kidney while in Dulbecco's PBS with Ca^{2+} and Mg^{2+}. In our case this yielded bundles of epithelial tubules that were not completely homogenous but largely composed of the desired cell type, as determined by histological analysis of embryonic kidneys prior to initiating this procedure. These were then placed in 0.2% Type I collagenase at 37°C (Life Technologies) for 45 min, after which they were placed in growth medium and allowed to settle. A defined growth medium was used that had been designed for use with embryonic kidney organ culture (Woolf et al., 1995), although epithelial cells appeared to grow better if 1% fetal calf serum was included. After several days in culture, cobblestoned epithelial patches could be observed spreading out from bunches of tubules. During these manipulations several observations were made that may aid others in attempting to isolate primary cells through a similar approach. First, coating the culture surface with Matrigel (1:300 dilution from stock bottle; Becton Dickinson) greatly aided the ability of epithelial cells to migrate out and survive in culture. Coating with Matrigel may also have preferentially aided proliferation of epithelial cells over contaminating mesenchymal cells. Secondly, cell outgrowth was optimal from small bunches of tubules that were estimated to be the diameter of 5–10 tubules; little outgrowth was observed from the larger bunches. On the other hand, single cells dispersed from the bunches invariably died, presumably because of excessive mechanical disruption.

Obviously, investigators may substitute for the foregoing any of the many procedures that have been designed to obtain primary cell populations from various organs. The relative size of the starting organ and homogeneity of the cell population will determine the ease and success of these procedures. It is also possible that a specific cell type may be isolated from a heterogeneous population if monoclonal antibodies to a surface marker allow FACS sorting or panning, although if these procedures are used, a process must be developed for obtaining suspensions of single cells without greatly impairing cell viability.

Once the primary population is derived, it must be expanded through successive passages. When cells from mice transgenic for the temperature-sensitive T antigen are used, they are cultivated at 33°C. γ-Interferon is added to stimulate expression of T antigen from the H-2K promoter (Jat *et al.,* 1991). In our experience, the first few trypsinizations were particularly difficult, with a great loss of cells at each passage. Indeed, cell loss was so pronounced that cells were replated in the same surface area after the first 3–5 trypsinizations until cell proliferation appeared more robust and cells tolerated expansion to greater surface areas. To minimize cell disruption, we found it helpful to first wash twice with PBS/EDTA, and then leave the cells in PBS/EDTA for an extended wash of several minutes, during which time cells began to detach from each other. Trypsin was then added in equal volume to the PBS/EDTA (final concentration 0.1%), and cells were then placed at 37°C until most cells were observed to round up. In some instances it has been possible to cease adding γ-interferon without an observable effect on cell phenotype.

An obvious concern with using large T antigen to immortalize cells is whether this oncogenically transforms the cells, and thus renders them an imperfect model for the cells from which they are derived. In our experience, epithelial cells derived through the foregoing procedure do not behave as if they are oncogenically transformed, i.e., they grow in uniform monolayers and appear to be contact inhibited. (This is true of wild-type cells, and somewhat less so of $\alpha3\beta1$ integrin-deficient cells.) We presume that this is due to relatively low levels of T antigen expressed in these cells.

It is also crucial to verify cell lineages once an immortalized population is obtained. In the case of the collecting duct epithelial cells, it was demonstrated that they express aquaporin-2, a specific marker of collecting duct cells, as well as a Na/K ATPase (Wang *et al.,* 1999). Finally, whenever possible, transfection of a cDNA expression construct into mutant cells to rescue mutant phenotypes is a crucial control to prove that the phenotypes obtained are due to the absence of the mutated gene's product, and not to artifacts that arise during culture of primary or immortalized cell populations.

B. Conditionally Immortalized Keratinocytes

The laboratory of Michael DiPersio (Albany Medical College) has studied keratinocyte cell lines derived from α3 integrin mutant newborn mice (DiPersio *et al.,* 2000), and compared them with parallel wild-type cell lines. The derivation of these cell lines also utilized temperature-sensitive T antigen transgenic mice and was initiated using previously established techniques for the primary culture of mouse and human keratinocytes (DiPersio *et al.,* 1997; Dlugosz *et al.,* 1995). This involved removing the skin from neonatal mice, floating it on a 0.25% trypsin solution overnight at 4°C, and then separating the dermis from the epidermis. The epidermis was then minced, filtered to remove clumps, and placed in culture dishes coated with rat-tail collagen. A previously established keratinocyte growth medium was used (DiPersio *et al.,* 1997; Dlugosz *et al.,* 1995) to maintain these cells in culture. Unlike the kidney epithelial cells discussed previously, which are cultured in medium that contains normal amounts of calcium, keratinocytes are cultured in a low-calcium medium that minimizes cell–cell adhesion and thus prevents terminal differentiation of these cells.

C. Fibroblasts from Mutant Mice

Fibroblasts are the cell type most commonly obtained from mutant mice, and in general they require less technically demanding approaches for their derivation and cultivation. Except in cases where homozygous mutant embryos undergo embryonic lethality prior to or around the time of implantation, standard approaches can be used to derive mouse embryo fibroblasts. These cells do not represent an immortalized population, but they can be expanded extensively as a primary culture and immortalized using conventional oncogenes or temperature-sensitive oncogenes if required (see references in Chou, 1989). The approach to preparing embryonic fibroblasts from late-stage embryos has been published in detail (Robertson, 1987). Briefly, E13 or 14 embryos are placed in DMEM with 10% FCS, and the internal organs and heads are removed. The remaining connective tissue is minced, placed in trypsin for 30 min–1 h, and then pipetted up and down several times to obtain a dispersed suspension of single cells and small clumps of cells. This suspension is placed in culture and expanded. A more recent report has used this approach to obtain fibroblast-like cells from E9.5 embryos (Yang and Hynes, 1996). When dealing with litters containing mutant embryos, the embryos must first be screened to determine which are homozygous mutant, heterozygote, or wild-type. When screening embryos, it is important to obtain a tissue sample that is entirely of embryonic origin, so that maternal DNA will not complicate the genotyping result. Part of the embryo itself is obviously most optimal, but yolk sac may also be used. The placenta, however, may contain maternal tissue and should not be used. In cases where early embryonic lethality of homozygous mutants precludes obtaining E13 or older embryos to prepare fibroblasts, it has also been possible to obtain them by differentiating homozygous-mutant ES cells (discussed later).

III. ES Cell Lines Carrying Homozygous Mutations of Integrin Adhesion Molecule Genes

Embryonic stem cell lines that are homozygous for targeted mutations in adhesion receptor or ECM component genes constitute a valuable resource for studies on adhesion molecule function. The development of such lines contrasts with standard gene targeting experiments, in which a single allele of an adhesion gene has been mutated in ES cells, which are then used to derive heterozygous mice that are in turn mated to obtain homozygous mutant embryos or mice. Traditional gene targeting will not be discussed, and the reader is referred to several reviews of gene targeting studies of cell adhesion and ECM genes (Gustafsson and Fassler, 2000; Hynes, 1996). Homozygously mutated ES cells may be used in several types of experiments. They can be used to derive chimeric mice that will be partially derived from wild-type cells, and partially derived from the homozygous mutant ES cells. In cases where a homozygous mutation would lead to embryonic lethality, it is often observed that the presence of wild-type cells in the chimeric animal allows the homozygous mutant cells to contribute to different tissues to varying extents. Chimeric animals can therefore be studied to determine to which organs or tissues homozygously mutated cells can contribute, and whether they differentiate and function normally in these locations (Fassler and Meyer, 1995; Yang and Hynes, 1996; Yang et al., 1996). A separate use of homozygously mutated ES cells is in in vitro differentiation experiments. Depending on how they are cultured and pharmacologically treated, ES cells will differentiate toward various lineages (Robertson, 1987). Using homozygously mutated ES cells, it can be determined whether the absence of a particular gene product affects a specific differentiation pathway. Particularly useful for the study of adhesion molecules is the ability of aggregates of ES cells to form embryoid bodies, when cultured in situations that prevent adhesion to an underlying surface (Robertson, 1987). For example, embryoid bodies elaborate a surrounding basement membrane; hence the formation of embryoid bodies by ES cells carrying a homozygous mutation in an adhesion or ECM gene can be used to determine the requirement for that gene product in basement membrane assembly.

Homozygously mutated ES cells should prove a useful tool for in vivo and in vitro experiments aimed at analyzing the function of adhesion receptors and their ligands. They may be helpful in delineating signaling pathways acting downstream of particular receptors, and may serve as model systems for the screening of small molecules that affect signaling pathways. Homozygously mutated cells can be used in mutational analyses of adhesion molecules, especially in situations where the mutant receptor would act recessively instead of as a dominant-negative when the wild-type receptor is present.

A. Derivation of ES Cells Carrying Homozygous Mutations

There are three possible approaches to obtaining ES cells carrying homozygous mutation of a gene. The first follows the original approach to deriving wild-type ES cell lines and requires passing the mutation through the germ line, to first derive heterozygous mice. A traditional gene targeting vector is constructed and used to target a single allele. The targeted ES cell clone is used to derived chimeric mice, which will then pass the mutation through the germ line to produce heterozygous mice. Male and female heterozygous mice are mated to obtain homozygous-mutant preimplantation embryos or blastocyts from which ES cells are derived. This process will be briefly reviewed and is found in detail in a volume by E. J. Robertson (Robertson, 1987). It should be stressed that proper handling of ES cells is essential to their successful experimental use, and investigators planning to use these cells are advised to do so in collaboration with a laboratory that has extensive experience with ES cells and the derivation of gene-targeted mice.

ES cells are more easily obtained from "delayed blastocysts" than from normal blastocysts. To obtain delayed blastocysts, female mice are ovariectomized after mating and treated with a synthetic progesterone analog. This treatment allows the normal *in vivo* removal of zona pellucida from the blastocyst, but prevents implantation of the blastocysts in the uterine wall. These free-floating blastocysts can be flushed from the uterus, after sacrifice of the pregnant female, and placed in culture. The inner cell masses can then be isolated from the cultured blastocysts and developed into an ES cell line that can then be screened to identify those derived from homozygous mutant embryos. It should be noted that this procedure is inefficient, and not all ES cell clones obtained will turn out to contribute to well to chimeras. On the other hand, if a good clone is obtained, it will represent a new ES cell clone, and it is a general experience that newer ES cell clones make more robust chimeras than those that have been extensively cultivated.

The two other procedures are more straightforward and technically less demanding. However, they involve additional manipulation of the original targeted ES cell line, which runs the risk of decreasing its efficiency at making chimeras. The second approach is based on the observation that raising the concentration of G418, the neomycin analog that is used to select for cells expressing the neomycin resistance gene included in gene targeting vectors, often results in the selection of ES cells that have "homozygosed" the targeted locus (Mortensen *et al.,* 1992). In practice, ES cells carrying a targeted mutation of a single allele are placed in culture, and the G418 concentration is raised to at least double the concentration originally used to select for ES cells that incorporated the original targeting vector. The vast majority of ES cells will die, but a few clones will grow up that can be screened to determine if any are now homozygous for the targeted mutation (Mortensen *et al.,* 1992). The actual concentration of G418 used in this selection must be empirically determined, as it is affected by how well the Neo gene is expressed from the original targeted allele. In some instances when the Neo gene is highly expressed, or an enhanced-activity version of the neo cDNA has been used , it has not been possible to use this approach to derive homozygously targeted ES cell clones.

A third approach to obtaining homozygously mutated ES cells is to construct a second gene targeting vector using a different antibiotic resistance gene. For example, if the original vector incorporated the Neo gene, the second vector could use either the Hygromycin or puromycin resistance gene, or any of several others currently available. With this strategy, the original targeted ES cell line is put through an identical process of electroporation and screening, to identify clones where the remaining wild-type allele has now been eliminated. It should be mentioned that sometimes one is very fortunate and the screening of the first round of gene-targeted ES cells reveals a clone that has been targeted in both alleles, this obviating the need for any additional manipulations to obtain homozygously targeted clones.

A crucial aspect of studies using homozygous-mutant ES cells is to place a marker in these cells that will allow identification of cells derived from the injected ES cells. Otherwise it may be difficult or impossible to make firm interpretations of chimeric phenotypes. There are several approaches to marking ES cells. It is possible to construct a gene targeting vector such that the β-galactosidase (LacZ) gene is "knocked-in" to the locus being targeted, so that its expression now comes under control of the regulatory elements of the gene being targeted. This has the advantage of serving as an indicator of whether the gene would be expressed in a specific location, had it not been mutated. Additional approaches to marking ES cells with constitutively expressed markers include transfection of a GFP or LacZ expression vector under control of a broadly expressed promoter such as the CMV early promoter or the β-actin promoter. In the case of GFP, it is possible to use FACS sorting to select highly expressing cells. These latter approaches have the disadvantage that they involve additional *in vitro* manipulations that might decrease the ability of the cells to form chimeras that are highly derived from ES cells. It is conversely possible to inject unmarked ES cells into blastocysts produced by mating males carrying the ROSA 26 locus to wild-type C57Bl/6 female mice. ROSA 26 is a ubiquitously expressed LacZ transgene (Friedrich and Soriano, 1991), and ES cell-derived tissue would be indicated by the lack of LacZ expression. This approach relies on a negative result, i.e., failure to stain for LacZ, and works less well in the experience of the author's laboratory than approaches involving positive marking of ES cells with GFP or LacZ.

IV. Studies with Adhesion Receptor Null Cell Lines

Some studies with cell lines carrying mutations in adhesion receptor or ECM genes are summarized in the following sections to provide examples of how these cell lines may used to extend our understanding of the function of adhesion molecules.

A. β1 Integrin-Null ES Cells

ES cells carrying a homozygous mutation in the β1 integrin gene have been used by R. Fassler and collaborators in a variety of studies. Embryos deficient in all β1 integrins experience a perimplantation lethality, precluding study of whether any organs and tissues can develop in the complete absence of all β1 integrins (Fassler and Meyer, 1995; Stephens *et al.*, 1995). However, when chimeras were derived with ES cells carrying a homozygous mutation in the β1 integrin gene, it was found that β1 integrin-null cells can contribute at very low percentages to a variety of cell types (Fassler and Meyer, 1995). Embryos that contained greater than 25% contribution from ES cells were not viable, but those with a smaller overall contribution showed variable contribution of

β1-null cells to organs including brain, lung, muscle, heart, and gut. In general, higher percentage contributions were observed in embryos than adult tissues, suggesting that β1-null cells were lost through attrition during development. For example, adult liver and spleen showed no contribution from β1 integrin-null cells, while many other organs showed a small percentage ($<2\%$) contribution. The highest contribution was in skeletal muscle, although this was due to fusion of wild-type and null myoblasts, thus obscuring the degree to which β1 integrin-null cells can contribute to extensive patches of skeletal muscle. In contrast, cardiac muscle showed only scattered single cell contribution by β1-null cells. Further study of cardiac muscle using *in vitro* differentiation of null ES cells as well as additional studies in chimeras showed that β1 integrin-null cardiac muscle cells are abnormal and are not observed in chimeras beyond the age of 6 months (Fassler *et al.,* 1996). Studies on the hematopoietic ability of β1 integrin-null cells demonstrate that cells of erythroid and myeloid lineages are found in the yolk sac but not in the fetal liver or adult animals (Hirsch *et al.,* 1996). Interestingly, neural crest-derived tissues showed contribution from β1-null cells, suggesting that neural crest cells can migrate using non-β1 integrin adhesion receptors (Fassler and Meyer, 1995). These results with β1 null cells are difficult to interpret. When only scattered and isolated β1 integrin-null cells are found in an organ, does this absolutely imply that there is not a cell-autonomous requirement for β1 integrins, or do the surrounding cells somehow rescue these few deficient cells?

β1 integrin-null ES cells have also been injected subcutaneously to obtain teratomas, and it has been observed that basement membrane formation is impaired in comparison to basement membrane assembly by teratomas derived with wild-type ES cells (Sasaki *et al.,* 1998). Further studies using β1 integrin-null ES cells *in vitro* demonstrated that although these cells could bind laminin, presumably through the dystroglycan complex, they could not assemble laminin into a more complex matrix without the use of β1 integrins (Henry *et al.,* 2001).

B. β1 Integrin–Null Somatic Cell Lines

β1 integrin-null somatic cell lines have also been prepared from homozygous-null ES cells that have been induced to undergo differentiation. GD25 cells are a fibroblast-like cell line prepared by treating homozygous-null ES cells with DMSO, followed by transformation with large T antigen and cloning (Wennerberg *et al.,* 1996). They have been used to demonstrate that limited fibronectin-matrix assembly could occur in the absence of β1 integrins, with αvβ3 integrin possibly substituting for α5β1 integrin (Wennerberg *et al.,* 1996). When this cell line was rescued by transfection of a cDNA encoding the β1A subunit, expression of α5β1 integrin and robust assembly of a fibronectin matrix was restored (Wennerberg *et al.,* 1996). GD25 cells have also been used to study the function of transfected null forms of β1A integrin in the absence of the wild-type β1 integrin subunit, simplying the interpretation of the results of these experiments (Wennerberg *et al.,* 2000). Thus, these cells have a major advantage over the use of most cell lines, as β1 integrins are widely expressed.

A β1 integrin-null epithelial cell line, GE11, was derived by placing cells derived from minced E10.5 chimeric embryos in culture, transforming with T antigen, and then isolating clonal cell populations by limiting dilution (Gimond *et al.,* 1999). This cell line was then used to demonstrate that when $\beta 1$ integrins are reexpressed in these cells, epithelial cell clusters show a scattering behavior that is associated with decreased expression of cadherin and catenins, and increased activity of Rho and Rac (Gimond *et al.,* 1999). In summary, cell lines lacking all $\beta 1$ integrins have proved to be valuable tools for the study of integrin function *in vivo* and *in vitro*.

In an additional set of experiments, $\beta 1$ integrin-null fibroblastoid lines were prepared and then transformed with Ras and Myc oncogenes to produce highly transformed cell lines that could be used to study the requirement for $\beta 1$ integrins in metastasis. These lines were injected subcutaneously and the ability of subcutaneous tumors to metastasize to lung or liver was examined (Brakebusch *et al.,* 1999). There appeared to be an inverse correlation between levels of $\beta 1$ integrin expression and growth of the primary tumor at sites where the fibroblastoid cells were injected. However, $\beta 1$ integrin-null cells were markedly less metastatic compared with cells from wild-type mice (Brakebusch *et al.,* 1999).

C. $\alpha 3\beta 1$ Integrin–Null Cells

$\alpha 3\beta 1$ integrin functions predominantly as a receptor for certain isoforms of laminin, including laminin-5 and 10/11 (Delwel *et al.,* 1994; Kikkawa *et al.,* 1998). For reasons that are not well understood, many immortalized adherent cell lines express $\alpha 3\beta 1$ integrin as a predominant integrin, even if they are derived from tissues that did not express this integrin *in vivo* (Elices *et al.,* 1991; Hemler *et al.,* 1987). This observation must be taken into account when interpreting results obtained by comparing the behavior of wild-type and $\alpha 3\beta 1$ integrin-null cell lines described in this section. One of the basic observations made with $\alpha 3\beta 1$-null cells was that they failed to organize a cortical cytoskeleton, and instead assembled actin stress fibers (Wang *et al.,* 1999). This phenotype was also observed in primary cell cultures of $\alpha 3\beta 1$ integrin-deficient collecting duct cells prior to immortalization. Importantly, cortical cytoskeletal organization could be rescued by expression of the human $\alpha 3$ cDNA in null cells, providing important confirmation that this phenotype was indeed due to the absence of $\alpha 3\beta 1$ integrin expression, and not a spurious artifact of the immortalization process. This finding ascribed to integrins a crucial role in stimulating organization of the cortical cytoskeleton by cadherin : catenin complexes along lateral membranes of epithelial cells. However, it should be noted that in the kidneys of $\alpha 3\beta 1$ integrin-null embryos or newborn mice, there are many fewer collecting ducts than in wild-type littermate kidneys, but those collecting ducts that are present contain relatively normal appearing epithelial cells (Kreidberg *et al.,* 1996). Thus the *in vitro* phenotype of $\alpha 3\beta 1$-null cells is considerably more drastic than that observed *in vivo,* at least with regard to cytoskeletal organization. Our interpretation of these findings takes into account the differences in integrin expression of these cells *in vitro* and *in vivo*. Collecting duct cells *in vivo* express $\alpha 2\beta 1$, $\alpha 3\beta 1$, and $\alpha 6\beta 1$ integrins (Korhonen *et al.,* 1991), whereas *in vitro* they express $\alpha 3\beta 1$ integrin as the predominant $\beta 1$ integrin (Wang *et al.,* 1999), in common with many adherent cells as noted previously. No other $\beta 1$ integrin appears to be up-regulated *in vitro* to compensate for the loss of $\alpha 3\beta 1$ integrin in the null cell lines. Therefore we suggest that $\alpha 2\beta 1$ and $\alpha 6\beta 1$ integrins provide a partially redundant function *in vivo* in organizing the cytoskeleton, but *in vitro*, the overall reduction in $\beta 1$ integrins leads to a loss of normal epithelial cytoskeletal organization. These results emphasize that experimental observations must take into account overall differences in integrin expression between immortalized cell lines and the *in vivo* cells from which they were derived.

Primary cells and immortalized cell lines from $\alpha 3\beta 1$ integrin-null basal keratinocytes have also (DiPersio *et al.*, 1997, 2000; Hodivala-Dilke *et al.*, 1998) demonstrated a role for $\alpha 3\beta 1$ in modulating cytoskeletal organization. Additionally, these studies have offered the intriguing observation that matrix metalloproteinase-9 (MMP-9) is expressed by immortalized wild-type keratinocytes, but not by those from $\alpha 3\beta 1$-null mice, suggesting that $\alpha 3\beta 1$ integrin may have an important role in regulating MMP expression (DiPersio *et al.*, 2000). MMP-9 expression was restored by expression of the human $\alpha 3$ cDNA, again demonstrating the importance of rescue experiments (DiPersio *et al.*, 2000). However, it is also important to note that this difference in MMP expression was not observed in primary cell populations prior to their immortalization. Therefore it is important to keep in mind that the immortalization process is not entirely benign in conferring additional phenotypes not present in primary cells, yet these novel phenotypes may provide important insights into the interconnected functions of cell adhesion molecules.

D. Integrin-Null ES Cells to Study Hematopoeisis

Chimeras derived with a4 integrin homozygous-null ES cells were used to study the ability of $\alpha 4\beta 1$ integrin-null cells to contribute to hematopoietic lineages (Arroyo *et al.*, 1996). These ES cells were derived using the high G418 procedure referred to earlier (Mortensen *et al.*, 1992). These studies showed that although lymphocyte development in the fetal liver occurred normally, differentiation of T and B cells (but not monocytes and natural killer cells) in the bone marrow required $\alpha 4\beta 1$ integrin (Arroyo *et al.*, 1996). $\alpha 4\beta 1$ integrin was also required for homing of T cells to Peyer's patches, although not to lymph nodes or spleen (Arroyo *et al.*, 1996). These findings were extended in additional studies which demonstrated that in contrast to $\alpha 4\beta 1$ integrin, $\alpha 5\beta 1$- or αv-containing integrins were not required for lymphoid development (Arroyo *et al.*, 2000). This study also further demonstrated a role for $\alpha 4\beta 1$ integrin in mediating lymphoid migration at sites of inflammation (Arroyo *et al.*, 2000).

E. $\alpha 4$ and $\alpha 5$ Integrin-Null Fibroblasts

Embryonic fibroblasts obtained from E9.5 embryos carrying homozygous mutations in either the $\alpha 4$ or $\alpha 5$ integrin genes, or both genes, were used to demonstrate that these cells could migrate and form focal adhesions on fibronectin, despite the absence of these two major fibronectin receptors. This suggested that an additional fibronectin receptor was present and functional in these cells (Yang and Hynes, 1996). Using a function-blocking antibody to αv-containing integrins, it was shown that αv integrins are required by $\alpha 5\beta 1$-null cells to migrate on fibronectin, or to assemble a fibronectin matrix (Yang and Hynes, 1996).

F. α5 Integrin-Null ES Cells

Since $\alpha5\beta1$ integrin-null embryos die too early to evaluate the role of $\alpha5\beta1$ integrin in myogenesis, chimeras were derived with $\alpha5$ homozygous-null ES cells (Taverna *et al.*, 1998). Although $\alpha5\beta1$ integrin-null cells contributed well to muscles, muscle tissue composed of $\alpha5\beta1$-null cells became dystrophic during late developmental and adult stages of chimeric mice. Corroborative results were obtained *in vitro*, where it was shown that $\alpha5$ homozygous-null ES cells or myoblasts differentiated into myotubes, but the survival of these myotubes on fibronectin was decreased compared to wild-type controls.

V. Studies with Extracellular Matrix Component Null Cell Lines

A. Fibronectin and Syndecan-4 Null Fibroblasts

Fibronectin-null fibroblasts were obtained by differentiating fibronectin homozygous-null ES cells, originally derived from delayed blastocysts as described previously (Saoncella *et al.*, 1999). Use of these cells allowed observations to made concerning focal adhesion assembly in response to defined fragments of fibronectin, in the absence of endogeneously produced fibronectin, thus permitting a less complicated experimental format. Fibronectin-null fibroblasts did not spread and form focal adhesions when plated on the cell-binding domain of fibronectin, which is recognized by integrins (Saoncella *et al.*, 1999). However, when an antibody against the ectodomain of syndecan-4 was included in the growth medium, fibronectin-null cells were now able to form focal adhesions, presumably because the antibody served as a bridge between the cells and syndecan-4 and effectively replaced the heparin-binding domain of fibronectin (Saoncella *et al.*, 1999). This experiment served to identify syndecan-4 as a key heparan-sulfate proteoglycan involved in focal adhesion assembly. The role of syndecan-4 in focal adhesion formation was studied further using syndecan-4-null primary embryonic fibroblasts (Ishiguro *et al.*, 2000). (Because of their use in understanding adhesion to fibronectin, syndecan-4 deficient cells are discussed in this section, even though syndecans in this situation are acting as receptors and these cells lines could have been appropriately discussed in the preceding section.) Syndecan-4-null fibroblasts assembled focal adhesions on surfaces coated with intact fibronectin, or the separated heparin and cell-binding domains, but had impaired focal adhesion assembly if the heparin-binding domain was presented in solution. These observations led the authors to conclude that syndecan-4 may have a role in promoting focal adhesion assembly when the heparin-binding domain of fibronectin is presented in solution, as may occur from apical membranes of cells (Ishiguro *et al.*, 2000).

B. Thrombospondin 2-Null Fibroblasts

Skin fibroblasts were prepared from adult thrombospondin 2 (TSP-2) homozygous-null and wild-type control mice to examine the role of TSP-2 in cell attachment and spreading (Yang *et al.*, 2000). TSP-2-null fibroblasts attached and spread less well on plates coated with different ECM proteins including fibronectin, vitronectin, and type I collagen. It was determined that TSP-2-null fibroblasts express increased levels of MMP-2, and that the adhesive defects could be corrected by treating the cells with MMP inhibitors. Rescue with a TSP-2 cDNA restored normal levels of MMP-2 expression and normal adhesive behavior (Yang *et al.,* 2000). Thus, similarly to $\alpha 3 \beta 1$ integrin-null keratinocytes discussed earlier, TSP-2-null cell lines were used to demonstrate that regulation of MMP expression is complex and is related to interactions between ECM glycoproteins and their receptors.

VI. Summary

Cell lines derived from mice carrying targeted mutations in adhesion or ECM genes, and ES cell lines homozygous for mutations in these genes, have proved to be valuable tools to examine the functions of these molecules in development, and at a molecular level by *in vitro* experimentation. In many cases, the development of cell lines has exposed novel phenotypes not apparent from *in vivo* observations, possibly because the *in vitro* system under observation is more dependent on a specific molecule than is an entire organ or organism. These cell lines should continue to provide interesting model systems to study the molecular function of adhesion receptors.

Acknowledgments

The author's laboratory is supported by funding from the NIDDK, the March of Dimes Research Foundation, and the National Kidney Foundation.

References

Arroyo, A. G., Taverna, D., Whittaker, C. A., Strauch, U. G., Bader, B. L., Rayburn, H., Crowley, D., Parker, C. M., and Hynes, R. O. (2000). In vivo roles of integrins during leukocyte development and traffic: insights from the analysis of mice chimeric for alpha 5, alpha v, and alpha 4 integrins. *J. Immunol.* **165,** 4667–4675.

Arroyo, A. G., Yang, J. T., Rayburn, H., and Hynes, R. O. (1996). Differential requirements for alpha4 integrins during fetal and adult hematopoiesis. *Cell* **85,** 997–1008.

Brakebusch, C., Wennerberg, K., Krell, H. W., Weidle, U. H., Sallmyr, A., Johansson, S., and Fassler, R. (1999). Beta1 integrin promotes but is not essential for metastasis of ras-myc transformed fibroblasts. *Oncogene* **18,** 3852–3861.

Chou, J. Y. (1989). Differentiated mammalian cell lines immortalized by temperature-sensitive tumor viruses. *Mol. Endocrinol.* **3,** 1511–1514.

Delwel, G. O., de, M. A., Hogervorst, F., Jaspars, L. H., Fles, D. L., Kuikman, I., Lindblom, A., Paulsson, M., Timpl, R., and Sonnenberg, A. (1994). Distinct and overlapping ligand specificities of the alpha 3A beta 1 and alpha 6A beta 1 integrins: recognition of laminin isoforms. *Mol. Biol. Cell* **5,** 203–215.

DiPersio, C. M., Hodivala, D. K., Jaenisch, R., Kreidberg, J. A., and Hynes, R. O. (1997). Alpha3beta1 Integrin is required for normal development of the epidermal basement membrane. *J. Cell Biol.* **137,** 729–742.

DiPersio, C. M., Shao, M., Di Costanzo, L., Kreidberg, J. A., and Hynes, R. O. (2000). Mouse keratinocytes immortalized with large T antigen acquire alpha3beta1 integrin-dependent secretion of MMP-9/gelatinase B. *J. Cell Sci.* **113,** 2909–2921.

Dlugosz, A. A., Glick, A. B., and Tennebaum, T. (1995). Isolation and utilization of epidermal keratinocytes for oncogene research. *Methods Enzymol.* **254,** 3–20.

Elices, M. J., Urry, L. A., and Hemler, M. E. (1991). Receptor functions for the integrin VLA-3: fibronectin, collagen, and laminin binding are differentially influenced by Arg-Gly-Asp peptide and by divalent cations. *J. Cell Biol.* **112,** 169–181.

Fassler, R., and Meyer, M. (1995). Consequences of lack of beta 1 integrin gene expression in mice. *Genes Dev.* **9,** 1896–1908.

Fassler, R., Rohwedel, J., Maltsev, V., Bloch, W., Lentini, S., Guan, K., Gullberg, D., Hescheler, J., Addicks, K., and Wobus, A. M. (1996). Differentiation and integrity of cardiac muscle cells are impaired in the absence of beta 1 integrin. *J. Cell Sci.* **109,** 2989–2999.

Friedrich, G., and Soriano, P. (1991). Promoter traps in embryonic stem cells: a genetic screen to identify and mutate developmental genes in mice. *Genes Dev.* **5,** 1513–1523.

Gimond, C., van Der Flier, A., van Delft, S., Brakebusch, C., Kuikman, I., Collard, J. G., Fassler, R., and Sonnenberg, A. (1999). Induction of cell scattering by expression of beta1 integrins in beta1-deficient epithelial cells requires activation of members of the rho family of GTPases and downregulation of cadherin and catenin function. *J. Cell Biol.* **147,** 1325–1340.

Gustafsson, E., and Fassler, R. (2000). Insights into extracellular matrix functions from mutant mouse models. *Exp. Cell Res.* **261,** 52–68.

Hemler, M. E., Huang, C., and Schwarz, L. (1987). The VLA protein family. Characterization of five distinct cell surface heterodimers each with a common 130,000 molecular weight beta subunit. *J. Biol. Chem.* **262,** 3300–3309.

Henry, M. D., Satz, J. S., Brakebusch, C., Costell, M., Gustafsson, E., Fassler, R., and Campbell, K. P. (2001). Distinct roles for dystroglycan, beta1 integrin and perlecan in cell surface laminin organization. *J. Cell Sci.* **114,** 1137–1144.

Hirsch, E., Iglesias, A., Potocnik, A. J., Hartmann, U., and Fassler, R. (1996). Impaired migration but not differentiation of haematopoietic stem cells in the absence of beta1 integrins. *Nature* **380,** 171–175.

Hodivala-Dilke, K. M., DiPersio, C. M., Kreidberg, J. A., and Hynes, R. O. (1998). Novel roles for alpha3beta1 integrin as a regulator of cytoskeletal assembly and as a trans-dominant inhibitor of integrin receptor function in mouse keratinocytes. *J. Cell Biol.* **142,** 1357–1369.

Hynes, R. O. (1996). Targeted mutations in cell adhesion genes: what have we learned from them? *Dev. Biol.* **180,** 402–412.

Ishiguro, K., Kadomatsu, K., Kojima, T., Muramatsu, H., Tsuzuki, S., Nakamura, E., Kusugami, K., Saito, H., and Muramatsu, T. (2000). Syndecan-4 deficiency impairs focal adhesion formation only under restricted conditions. *J. Biol. Chem.* **275,** 5249–5252.

Jat, P. S., Noble, M. D., Ataliotis, P., Tanaka, Y., Yannoutsos, N., Larsen, L., and Kioussis, D. (1991). Direct derivation of conditionally immortal cell lines from an H-2Kb-tsA58 transgenic mouse. *Proc. Natl. Acad. Sci. USA* **88,** 5096–5100.

Kikkawa, Y., Sanzen, N., and Sekiguchi, K. (1998). Isolation and characterization of laminin-10/11 secreted by human lung carcinoma cells. Laminin-10/11 mediates cell adhesion through integrin alpha3 beta1. *J. Biol. Chem.* **273,** 15854–15859.

Korhonen, M., Ylanne, J., Laitinen, L., Cooper, H. M., Quaranta, V., and Virtanen, I. (1991). Distribution of the alpha 1-alpha 6 integrin subunits in human developing and term placenta. *Lab. Invest.* **65,** 347–356.

Kreidberg, J. A., Donovan, M. J., Goldstein, S. L., Rennke, H., Shepherd, K., Jones, R. C., and Jaenisch R. (1996). Alpha 3 beta 1 integrin has a crucial role in kidney and lung organogenesis. *Development* **122,** 3537–3547.

Laird, P. W., Zijderveld, A., Linders, K., Rudnicki, M. A., Jaenisch, R., and Berns, A. (1991). Simplified mammalian DNA isolation procedure. *Nucleic Acids Res.* **19,** 4293.

Mortensen, R. M., Conner, D. A., Chao, S., Geisterfer-Lowrance, A. A., and Seidman, J. G. (1992). Production of homozygous mutant ES cells with a single targeting construct. *Mol. Cell Biol.* **12,** 2391–2395.

Mundel, P., Reiser, J., Borja, A., Pavenstadt, H., Davidson, G. R., Kriz, W., and Zeller, R. (1997). Rearrangements of the cytoskeleton and cell contacts induce process formation during differentiation of conditionally immortalized mouse podocyte cell lines. *Exp. Cell Res.* **236,** 248–258.

Robertson, E. J. (1987). Isolation of embryonic stem cells. *In* "Teratocarcinomas and Embryonic Stem Cells: A Practical Approach" (E. J. Robertson, ed.). IRL Press, Oxford.

Saoncella, S., Echtermeyer, F., Denhez, F., Nowlen, J. K., Mosher, D. F., Robinson, S. D., Hynes, R. O., and Goetinck, P. F. (1999). Syndecan-4 signals cooperatively with integrins in a Rho-dependent manner in the assembly of focal adhesions and actin stress fibers. *Proc. Natl. Acad. Sci. USA* **96,** 2805–2810.

Sasaki, T., Forsberg, E., Bloch, W., Addicks, K., Fassler, R., and Timpl, R. (1998). Deficiency of beta 1 integrins in teratoma interferes with basement membrane assembly and laminin-1 expression. *Exp. Cell Res.* **238,** 70–81.

Sauer, B. (1998). Inducible gene targeting in mice using the Cre/lox system. *Methods* **14,** 381–392.

Stephens, L. E., Sutherland, A. E., Klimanskaya, I. V., Andrieux, A., Meneses, J., Pedersen, R. A., and Damsky, C. H. (1995). Deletion of beta 1 integrins in mice results in inner cell mass failure and peri-implantation lethality. *Genes Dev.* **9,** 1883–1895.

Taverna, D., Disatnik, M. H., Rayburn, H., Bronson, R. T., Yang, J., Rando, T. A., and Hynes, R. O. (1998). Dystrophic muscle in mice chimeric for expression of alpha5 integrin. *J. Cell Biol.* **143,** 849–859.

Wang, Z., Symons, J., Goldstein, S., McDonald, A., Miner, J., and Kreidberg, J. A. (1999). $\alpha 3\beta 1$ integrin regulates epithelial cytoskeletal organization. *J. Cell Sci.* **112,** 2925–2935.

Wennerberg, K., Armulik, A., Sakai, T., Karlsson, M., Fassler, R., Schaefer, E. M., Mosher, D. F., and Johansson, S. (2000). The cytoplasmic tyrosines of integrin subunit beta1 are involved in focal adhesion kinase activation. *Mol. Cell Biol.* **20,** 5758–5765.

Wennerberg, K., Lohikangas, L., Gullberg, D., Pfaff, M., Johansson, S., and Fassler, R. (1996). Beta 1 integrin-dependent and -independent polymerization of fibronectin. *J. Cell Biol.* **132,** 227–238.

Woolf, A. S., Kolatsi, J. M., Hardman, P., Andermarcher, E., Moorby, C., Fine, L. G., Jat, P. S., Noble, M. D., and Gherardi, E. (1995). Roles of hepatocyte growth factor/scatter factor and the met receptor in the early development of the metanephros. *J. Cell Biol.* **128,** 171–184.

Yang, J. T., and Hynes, R. O. (1996). Fibronectin receptor functions in embryonic cells deficient in alpha 5 beta 1 integrin can be replaced by alpha V integrins. *Mol. Biol. Cell* **7,** 1737–1748.

Yang, J. T., Rando, T. A., Mohler, W. A., Rayburn, H., Blau, H. M., and Hynes, R. O. (1996). Genetic analysis of alpha 4 integrin functions in the development of mouse skeletal muscle. *J. Cell Biol.* **135,** 829–835.

Yang, Z., Kyriakides, T. R., and Bornstein, P. (2000). Matricellular proteins as modulators of cell–matrix interactions: adhesive defect in thrombospondin 2-null fibroblasts is a consequence of increased levels of matrix metalloproteinase-2. *Mol. Biol. Cell* **11,** 3353–3364.

CHAPTER 16

Flexible Polyacrylamide Substrata for the Analysis of Mechanical Interactions at Cell–Substratum Adhesions

Karen A. Beningo,* Chun-Min Lo,† and Yu-Li Wang*

*Department of Physiology
University of Massachusetts Medical School
Worcester, Massachusetts 01605

†Department of Physics
Cleveland State University
Cleveland, Ohio 44115

METHODS IN CELL BIOLOGY, VOL. 69
Copyright 2002, Elsevier Science (USA). All rights reserved.
0091-679X/02 $35.00

I. Introduction

Cultured cells undergo complex mechanical interactions with the environment through focal adhesions and other adhesive structures. These interactions form the physical basis for cell migration and provide an important means of communication during embryonic development, tissue formation, and wound healing (Martin, 1997; Bray, 2001). However, despite significant advances in understanding the chemical interactions between cells and the environment, only limited progress has been made in the characterization of such mechanical interactions. To gain a thorough understanding of this process, it is important both to measure the forces exerted by a cell, and to characterize cellular responses to external physical forces. Elastic substrata provide an effective means for achieving both purposes.

A number of methods have been developed for the preparation of elastic substrata. Harris and colleagues first introduced the wrinkling elastic substrata (1980), prepared by polymerizing a thin film on the surface of fluid silicone elastomers. Compressive forces introduced by the cell create wrinkles, which provide a limited, qualitative indication of the forces exerted on the substratum. This approach was later improved to allow a more precise control of the elastic property of the substratum (Burton and Taylor, 1997). In addition, quantitative measurements were made with nonwrinkling silicone film embedded with microbeads (Oliver *et al.,* 1998), which move and recoil upon the exertion and relaxation of forces. Molds with etched micropatterns have been used to cast a matrix of dots or a grid on the silicone elastomer films, which allows precise measurements of mechanical forces based on the distortion of the matrix pattern (Balaban *et al.,* 2001). In addition, flexible cantilevers constructed on a microchip have been introduced as an alternative means for force detection (Galbraith and Sheetz, 1997). However, many of these methods are quite involved and costly in their preparation and do not easily allow the modification of their chemical and/or physical properties.

We have developed flexible sheets of polyacrylamide as substrata for studying the mechanical interactions between the cell and the substratum. This approach has several appealing properties. The material is easily prepared with common materials. The polyacrylamide surface is inert and interactions with the cell are mediated by extracellular matrix proteins covalently linked to the surface. The optical properties allow for high-resolution imaging, and the porous nature of the gel provides a more physiological environment than do glass or plastic surfaces. In addition, the material shows nearly ideal elasticity over a range of applied forces, and the flexibility is easily controlled by varying the concentrations of acrylamide/bisacrylamide without changing the chemical properties. By embedding fluorescent beads, the substratum can be used to measure traction forces with a resolution close to 2 microns. When combined with simple micromanipulation procedures, the substratum allows the application of mechanical forces to cultured cells. This chapter provides an updated account of the method, which has been modified and expanded considerably since the previous description (Wang and Pelham, 1998).

II. Preparation of the Polyacrylamide Substratum

A. Preparation of Glass Surfaces

Thin sheets of polyacrylamide are cast by sandwiching a drop of polymerizing acrylamide solution between two coverslips, or a glass slide and a coverslip. Glass slides are used for supporting the gel for upright microscopes, and large No. 1 coverslips (Fisher Scientific) are typically used for inverted microscopes. Thicker coverslips, although easier to handle, should be avoided because of potential problems with focusing and optical aberration with an inverted microscope. The sheet must adhere tightly to the supporting coverslip or glass slide, to prevent the substratum from floating off during experiments. However it should adhere minimally to the other coverslip, to allow easy removal and exposure of the surface for cell attachment. The former is achieved by chemically activating the glass surface for the covalent attachment of polyacrylamide. The latter is facilitated by siliconizing the glass to make the surface hydrophobic. The method of Alpin and Hughes (1981) for chemical activation, based on glutaraldehyde (detailed later), provides consistent bonding that lasts for 3–5 days in a 37°C incubator. The association lasts for at least several weeks when stored at 4°C.

The clean glass surface to be activated is marked (e.g., with a diamond-tipped pen) and passed quickly through the inner flame of a Bunsen burner to make it hydrophilic. The surface is then smeared with a small volume of 0.1 N NaOH using the side of a Pasteur pipette and let dry in air. 3-Aminopropyltrimethoxysilane (Sigma) is smeared over the dried NaOH and the glass incubated for 5 min in a hood. The coated glass is then rinsed extensively with distilled water until the treated surface is clear. Next, 0.5% glutaraldehyde in phosphate-buffered saline (PBS) is pooled onto the treated glass surface and incubated at room temperature in the hood for 30 min. The surface is washed thoroughly with distilled water on a shaker and allowed to dry in air. Activated glass can be stored in a desiccator at room temperature for up to a month.

B. Preparation of the Polyacrylamide Gel

If the glass is to be used in a culture chamber, it may be more convenient to mount it into the chamber before casting the gel (Fig. 1). In our simple design, a chamber is formed by sealing the activated side of the coverslip to a piece of acrylic glass with a hole drilled at the center, using high-vacuum grease (Dow Chemical; McKenna and Wang, 1989). We use stock solutions of acrylamide (40% w/v) and N,N'-methylene bisacrylamide (BIS; 2% w/v) from Bio-Rad Inc, which are mixed with distilled water and 1 M Hepes, pH 8.5 (final concentration 10 mM) to obtain a desirable concentration (see later discussion). If the substratum is to be used for measuring traction forces, fluorescent latex beads (0.2-μm Fluospheres, Molecular Probes, Eugene, OR) are sonicated and added to the mixture, typically at a dilution of 1:30. The mixture is then degassed for a consistent period of time, e.g., 20 min for 5 ml, to ensure reliable polymerization. Polymerization is induced by adding 1/200 volume of 10% ammonium persulfate (Bio-Rad) and 1/2000

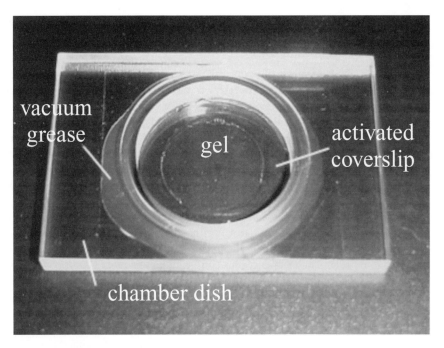

Fig. 1 Cell culture chamber. An image of a 22-μm diameter, 75-μm thick polyacrylamide substratum adhered to a coverglass and mounted with vacuum grease onto a Plexiglas chamber dish. The dish has an overall dimension of $70 \times 48 \times 6$ mm and a hole of 35 mm diameter at the center. An optional groove is machined around the hole for accommodating the lid of a 35 mm petri dish. This simple design allows easy access to the cells and the substratum for micromanipulations.

volume of TEMED (N,N,N,N'-tetramethylethylenediamine; Bio-Rad) to the degassed solution. The solution is mixed gently, and a drop of defined volume (discussed later) pipetted immediately onto the activated glass surface. A second coverslip is then placed onto the droplet. The assembly is inverted during polymerization, with the side for cell culture facing down, to encourage settling of the beads to the surface of the substratum. Properly degassed acrylamide should polymerize within 30 min at room temperature. Following polymerization, the chamber assembly is turned right-side-up and the top nonactivated coverslip is removed after flooding the gel with 50 mM Hepes, pH 8.5, to reduce the surface tension. We use two pairs of fine-tipped forceps, one for bracing an edge and the second for lifting the opposite edge. The exposed polyacrylamide surface is washed thoroughly with 50 mM Hepes, pH 8.5, on a shaker and stored in PBS at 4°C.

C. Helpful Suggestions for Gel Preparation

To obtain clean images, the solutions of Hepes and PBS used for gel preparation should be filtered with 0.22 μm filters to remove particles. The thickness of the gel is determined by the volume used for casting. Typically 15–20 μl is used for a surface

22 mm in diameter and 75–100 μm in thickness. Too small a thickness ($<$50 μm) would reduce the deformability of the surface, because of the constraint imposed by the bonding between polyacrylamide and the supporting glass surface. Too large a thickness ($>$100 μm) would cause serious degradation of the image quality.

A wide range of flexibility of the substratum can be obtained by adjusting the concentrations of acrylamide/BIS. The optimal flexibility depends on the cell type and the scientific questions being addressed. For example, to measure the traction forces under an NIH3T3 fibroblast we use 5% (w/v) acrylamide and 0.1% (w/v) BIS. For cells with weaker traction forces, such as neutrophils and neurons, we have used concentrations as low as 5% acrylamide/0.03% BIS. Although it is possible to increase the flexibility by lowering the concentration of either acrylamide or BIS, too low a concentration of BIS would lead to cracking surfaces and eventually failure of polymerization. Thus surface cracking can usually be eliminated by lowering the acrylamide concentration while increasing the BIS concentration.

It is important to avoid excessively flexible substrata, since cell morphology and growth can be adversely affected (Pelham and Wang, 1997; Wang *et al.*, 2000). However, too stiff a substratum would make any displacement difficult to measure. Through systematic trial and error, an optimal composition of acrylamide/BIS (cell dependent) can usually be identified, to yield a displacement of approximately about 10 pixels at the experimental magnification without significantly affecting the cell behavior. If the purpose is to determine the response to different flexibility, then a range of stiffnesses can be obtained by varying the concentration of BIS cross-linker while maintaining a constant total concentration of acrylamide.

D. Conjugation of Matrix Proteins to the Polyacrylamide Substratum

Since tissue cultured cells adhere poorly to naked polyacrylamide, the surface must be coated with extracellular matrix proteins. We have used two types of reagents for covalent conjugation, a photoactivatable heterobifunctional reagent or a classical carbodiimide such as EDC (1-ethyl-3-(3-dimethylaminopropyl)carbodiimide-HCl). The former is quick and effective; however, the latter has worked more consistently for certain proteins such as BSA.

Sulfosuccinimidyl-6-(4′-azido-2′-nitrophenylamino)hexanoate (sulfo-SANPAH; Pierce Chemical) is a heterobifunctional reagent, one end of which contains a phenylazide group that reacts nonspecifically with polyacrylamide upon photoactivation. The other end contains a sulfosuccinimidyl group, which reacts constitutively with primary amines. A solution of 1 mM sulfo-SANPAH in 50 mM Hepes, pH 8.5, and 0.5% DMSO is prepared immediately before use (although a more acidic pH may promote the stability of the sulfosuccinimidyl group, it also reduces the solubility of the reagent). The reagent is best solubilized by adding DMSO to the powder followed by the addition of room-temperature Hepes while vortexing. Sufficient solution is added immediately to cover the gel surface, typically 200 μl for a surface of 22 mm diameter. The substratum is then exposed to UV light generated by 302-nm bulbs (VWR #21476-010) at a distance of 2.5 in. for 5–8 min. The photoactivation should cause the sulfo-SANPAH solution to darken considerably. The solution is then removed and the process repeated with fresh sulfo-SANPAH to ensure activation of the entire surface. The substratum is rinsed with 50 m*M* Hepes, pH 8.5, two to three times to remove any excess reagent. Because of the aqueous instability of the sulfosuccinimidyl group it is important to add the matrix proteins immediately after activation. We typically layer either 0.2 mg/ml type I collagen in PBS or 15 μg/ml fibronectin in Hepes onto the substratum and allow the proteins to react on a shaker at 4°C overnight, although 1–4 h at room temperature should suffice. Other proteins used include fibrinogen, polylysine, and the ECL matrix mixture (Upstate Biotechnology). Laminin is sensitive to chemical modifications and is best coated secondarily on top of a surface of polylysine. Coated substratum is rinsed with several changes of PBS on a shaker and may be stored for up to 6 weeks in PBS at 4°C.

The carbodiimide method relies on the reaction of EDC (Pierce Chemicals) with a free carboxyl group, forming an amine-reactive intermediate for protein conjugation (Grabarek and Gergely, 1990). To supply a free carboxyl group, acrylic acid is included in a solution of 8% acrylamide/0.05–0.2% BIS solution to a final concentration of 0.2%. This concentration of acrylic acid should be adjusted in proportion to the concentration of total acrylamide, to avoid inhibition of polymerization. After polymerization, the substratum is washed thoroughly in 0.1 *M* MES (2-[*N*-morpholino]ethanesulfonic acid; Sigma), pH 4.9. A solution of EDC at 26 mg/ml in 0.1 *M* MES is pooled onto the polymerized substratum and incubated at room temperature for 2 h on a shaker. The substratum is then rinsed three times with the MES buffer and the protein of interest (diluted in MES) is pooled immediately onto the substratum and incubated overnight at 4°C on a shaker. After the reaction, the substratum is rinsed with PBS and stored a 4°C. The amine-reactive intermediate of EDC is highly unstable in aqueous solutions. This may be compensated by using a high concentration of both EDC and protein as in the present procedure, although NHS (*N*-hydroxysuccinimide; Pierce) may be included in the solution for an increased stability (Staros *et al.*, 1986). It should be noted that the buffer used to solubilize proteins can affect the coupling reaction with either method. In particular, Tris buffer should be removed by dialysis and the pH should not deviate substantially from neutrality.

III. Characterization of the Gel

A. Characterization of Protein Coating

The density of coated protein is determined by standard indirect immunofluorescence procedure, using fluorescent secondary antibodies with a different color from that of the fluorescent beads in the substratum. The intensity of fluorescence is measured with a cooled slow-scan CCD camera attached to the microscope. This method provides a simple means for determining the relative density of coating. However, it does not take into account the possibility that some proteins may be able to penetrate into the substratum, and may not be available for interactions with the cell. To determine the density of proteins on the surface, we use secondary antibodies conjugated to the surface of fluorescent particles (1 μm diameter; Polysciences), in a manner identical to the application of standard secondary antibodies (Lo *et al.,* 2000). Since large beads may create steric hindrance for the binding, the smallest available size should be used for this purpose.

B. Characterization of Mechanical Properties

Young's modulus is a measurement of elasticity. It is required not only for the calculation of traction forces but also for the qualitative comparison of cell behavior as a function of substratum flexibility. We have used three methods to measure the Young's modulus of polyacrylamide substratum. The gel strip method, as previously described by Pelham and Wang (1997), measures the elongation of strips of gels upon the application of a defined weight. Young's modulus is calculated by the equation

$$Y = (F\perp/A)/(\Delta l/l)$$

where l is the original length of the gel strip, Δl is the change in length, A is the cross-sectional area, and $F\perp$ is the applied force (1 g of weight applies 980 dynes of force). The microneedle method measures the horizontal deformation of the gel when lateral forces are applied through a bending microneedle (Lee *et al.,* 1994). The calculation is based on the deformation of the gel and the corresponding deformation of the needle, whose bending behavior is calibrated with known submilligram weights. Although conceptually straightforward, these methods become very difficult as the flexibility of substrata increases and the gels become increasingly susceptible to tear.

An alternative method, suitable for a wide range of stiffness, measures the depth of surface depression caused by the weight of a steel microball (0.64 mm in diameter, 7.2 g/cm^3, Microball Company, Peterborough, NH; Lo *et al.,* 2000). The theoretical basis is similar to that used for reological measurements in atomic force microscopy (Radmacher *et al.,* 1992). A microball is dropped with a pair of fine-tipped forceps onto the surface of the polyacrylamide substratum containing fluorescent beads, placed on the stage of a leveled inverted microscope. The microscope is first focused on the fluorescent beads immediately underneath the center of the microball. The microball is then removed with a magnet, and the upward movement of the substratum surface determined with the focusing mechanism of the microscope. Young's modulus is calculated as

$$Y = 3(1 - v^2)f/4d^{3/2}r^{1/2}$$

where f is the force exerted on the gel (calculated based on the volume and density of the microball and corrected for buoyancy), d is the indentation, r is the radius of the steel ball, and v is the Poisson ratio. This method can be most easily performed with a motorized focusing mechanism.

When measuring Young's modulus with any method, it is critical that the substratum be equilibrated with the medium and at the temperature for the actual experiment, because of the shrinking and swelling of the gel in response to osmolarity and temperature. For the same reason, the thickness of the gel cannot be simply calculated from the volume of the solution for polymerization. The measured thickness at steady state is typically 3–4 times that of the calculated value.

Calculation of traction forces also requires an estimate of the Poisson ratio, which measures the compressibility of the material. Accurate measurement of Poisson ratio is difficult; however, the value does not vary dramatically with the acrylamide concentration (Li *et al.,* 1993). A value of ~0.3 was reported in a published study (Li *et al.,* 1993).

IV. Data Collection for the Analysis of Traction Forces

A. Substratum Preparation

A variety of cell types have been studied on the polyacrylamide substratum including fibroblasts, epithelial cells, endothelial cells, smooth muscle cells, cardiomyocytes, macrophages, neutrophils, and neurons. To prepare the coated substratum for seeding, the substrata are exposed for 10 min to the irradiation from the germicidal (UV) light of a tissue culture hood. This "quasi-sterilization," in combination with a mixture of antibiotics (50 units/ml penicillin, 50 μg/ml streptomycin), keeps contamination under control for at least 2–3 days. The substratum is then incubated for 1 h in growth medium in an incubator, to ensure equilibration of the space within the gel with the medium. For the study of traction forces, it is important that cells be plated at a low density and that neighboring cells be separated by at least one cell's length, to minimize the superimposition of deformation by forces from multiple cells. We typically seed fibroblasts at ~5000 cells per substratum of 22 mm diameter. If the density is too high, a microneedle mounted on a micromanipulator may be used to clear unwanted cells immediately before the experiment.

B. The Microscope Setup

To maintain healthy cells during data acquisition, it is essential that temperature, humidity, and CO_2 be controlled and monitored carefully. Various aspects of cell culture on the microscope stage have been described in detail previously (McKenna and Wang, 1989). We use custom-designed, Plexiglas incubators that completely enclose the stage of an inverted microscope and the space surrounding the objective lens (Carl Zeiss, Inc). To minimize evaporation of the medium, mineral oil (Sigma #400-5) is layered onto the surface of the medium (however, this also prevents the subsequent replacement of the medium or addition of drugs). We prefer a simple chamber with easy access to the cells (Fig. 1). Although a number of microscope incubators/chambers are commercially available, many of them confine cells within a sealed space and limit the accessibility for cell manipulation, which is required for a number of purposes as discussed later.

Besides suboptimal culture conditions, excessive light used for imaging fluorescent beads also causes cell damage. To minimize the light exposure we use a 12V–100W quartz halogen lamp connected to a variable power supply, and set the input voltage at a level just sufficient for imaging. Heat from the lamp is removed with a heat-absorbing filter (BG-38) and/or heat-reflecting mirror. To further reduce the radiation damage, the light is controlled with an automated shutter. Although fluorescent beads are brightly labeled, it is recommended that a cooled, high-sensitivity CCD camera be used to reduce the level of excitation light.

A microscope equipped with a 40× phase objective is typically used for the measurement of substratum deformation. The strong signal from fluorescent microspheres allows nonimmersion lenses to be used for most purposes. However, if additional fluorescence images such as GFP signals are to be collected, a high-N.A. immersion lens may be necessary. Because of the thickness of the substratum, image quality may be seriously degraded by spherical aberration when using an inverted microscope. The degree of aberration varies with the lens design, and some objectives have proven to be surprisingly tolerant (the Zeiss 40× Plan-Neofluor, phase 2, N.A. 0.75, works well). An ideal, although costly, solution to this problem is to use a long-working-distance, water-immersion lens designed for deep focusing in confocal microscopy.

In addition to fluorescent beads in the substratum, the cell itself may be labeled with fluorescent probes such as microinjected analogs or GFP-tagged proteins, to allow simultaneous observations of dynamic changes in focal adhesions or cytoskeletal components (Beningo et al., 2001; Kaverina et al., 2000). Microinjection of cells on flexible substrata is possible, although considerably more difficult than the injection of cells on coverslips, because of the movement of the substratum upon needle penetration. In addition, the injection can cause profound, reversible effects on traction forces; therefore the cell should be allowed to recover for a period of 30–60 min before observation. On the other hand, GFP-tagged proteins, expressed through transient and stable transfection, can be readily observed in conjunction with red fluorescent beads (Beningo et al., 2001), for determining the relationship between traction forces and focal adhesions.

C. Image Collection

Generally a target cell is selected with a dry 40× objective lens based on several criteria. First, the cell must have a normal healthy morphology. Second, the beads under the cell should be evenly dispersed and as dense as possible while allowing individual beads to be resolved. Third, the distance of the cell from the edge of the substratum (which can be marked before observation with a Magic Marker) and from neighboring cells should be at least 200 μm. For each time point, a phase image of the cell and a fluorescent image of the beads are collected. Some find it helpful to also collect an image with simultaneous phase and fluorescence illumination. These images should be recorded quickly within seconds of each other.

Since the displacement of beads decreases rapidly as a function of the depth into the substratum, it is important to focus the microscope on the layer immediately underneath the cell. The most challenging aspect of data collection is to maintain a consistent level of focusing. Because the beads and the cell do not lie on the same focal plane, refocusing is mandatory when taking each set of phase and fluorescence images. Some practice in manual focusing is required to acquire a set of bead images where the corresponding beads are focused consistently throughout the sequence. Generally it is easier to focus on a unique pattern of beads that lie at some distance from the cell. Alternatively, automatic, motorized focusing mechanisms on some new microscopes may be used to great advantage for this purpose.

After the desired number of images is collected, a bead image without the cell is collected, which shows the position of beads without forces. This "null-force" image is compared with each previous image to determine the direction and distance of displacements. The simplest approach to remove cells is to treat the dish with Triton or trypsin. Unfortunately this also sacrifices the rest of the dish. Alternatively, cytochalasin B or D may be used to inhibit fibroblast traction forces (Pelham and Wang, 1999), although it requires several hours after extensive washes before the dish can be used again. A more selective method is to simply "pluck" the cell off the substratum with a microneedle and a micromanipulator, or to microinject a substance such as Gc-globulin to relax the forces.

V. Analysis of Traction Forces

The images of beads may be analyzed qualitatively or quantitatively to extract the information on traction forces. Here we will describe the analysis in qualitative terms. A more rigorous discussion of the algorithm involved will be described in a separate article (Marganski, Dembo and Wang, in press).

A. Generation of Deformation Data

Before analysis, it is essential to align all of the bead images to correct any movements due to the drift of the stage or sample. An area far away from the cell, with clearly defined, presumably stationary beads, is selected as the reference region. A computer algorithm based on pattern recognition is then used to search for the corresponding pattern throughout the set of bead images, and the images are translated accordingly to make the reference region stationary. Alternatively, manual inspection can be used to register a pair of images, by displaying the images alternatively on the screen and "panning" the images relative to each other until no wobbling of the reference region is observed.

Once the images are registered, the second step is to map the deformation of the substratum. Although manual comparisons may be performed to determine relative movements of individual beads in a pair of images (Kaverina *et al.,* 2000), this is tedious and time-consuming. We have developed an automated approach based on computer pattern recognition. The program divides the image into many small regions, each with a unique pattern of beads. Positions containing the corresponding patterns are identified in the null-force image and the image with cell-exerted forces, and a map of vectors is generated accordingly.

B. Analysis of Substratum Deformation

Qualitative analysis of traction forces may be performed based on the deformation vectors. Regions with many large vectors generally correspond to regions with strong tractions, and the direction of these large vectors is generally close to the direction of exerted forces. However, it is important to note that the deformation propagates and superimposes across the substratum surface, and the magnitude of deformation does not simply correspond to the magnitude of traction forces at the corresponding position. Therefore a precise value of traction force cannot be obtained simply by multiplying the deformation by the Young's modulus.

Quantitative analysis of traction forces can only be obtained by a "deconvolution" process, which mathematically identifies a set of forces within the cell boundary that best recreates the observed deformation (Dembo and Wang, 1999). The cell boundary provides an important condition for the calculation and can be defined interactively with the phase image using a drawing program. The calculation generally yields a map of traction stress, i.e., forces per unit area.

C. Visualization of Traction Forces

The distribution of traction stress vectors can be visualized as a map of arrows, as standard practice for studying vectorial fields (Fig. 2A). However, an equally useful approach is to render the magnitude of traction stress vectors as color images, with "hot" color corresponding to strong forces and "cool" color corresponding to weak forces (Fig. 2C). Vector plots have the advantage of displaying both the directional and magnitude information. Color rendering, on the other hand, proves to be superior for visualizing dynamics in time-lapse studies (Munevar *et al.,* 2001).

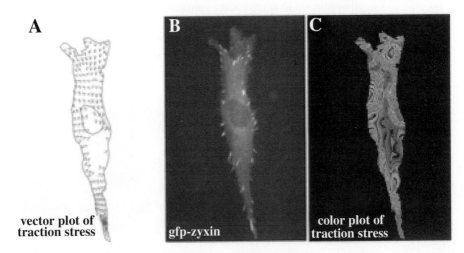

Fig. 2 Visualization of traction stress. (A) A vector map of traction stress shows the direction and magnitude of forces per unit area generated by a fish fin fibroblast. (B) GFP-zyxin identifies the focal adhesions. (C) Color rendering of the magnitude of traction stress, where "hot" colors (red) indicate strong forces and "cool" colors weak forces. (See Color Plate.)

VI. Other Applications using Polyacrylamide Substrata

The chemical characteristics (flexibility, inertness) and optical clarity of polyacrylamide have made it useful for addressing a number of biological questions of a mechanical nature. In addition to the measurement of traction forces, we have used the polyacrylamide substrata for the application of mechanical stimulation, for observing cellular responses to a gradient of substratum flexibility, and for testing the effects of target flexibility on phagocytosis.

A. Application of Mechanical Stimulations

This technique involves pushing and pulling the substratum with a blunted microneedle that grips the substratum near a cell (Fig. 3A). Forces are transmitted across the substratum and through the adhesion sites into the cell, eliciting interesting responses (Lo *et al.*, 2000). Unlike many other methods of mechanical stimulation, this approach allows a highly localized and defined mechanical input to be exerted and the immediate responses of the cell to be observed.

A relatively stiff polyacrylamide substratum of 5–8% acrylamide/0.1% bisacrylamide is typically used, to minimize tearing of the gels. Blunted microneedles are prepared by first pulling glass capillary tubing in a micropipette puller, then shaping the tip with a microforge (Narashige). The latter is done by placing the tip near the heating element of the microforge until the tip melts into a tiny sphere. The shank of the microneedle is then softened by heating, to create a kink several hundred microns behind the tip. This allows the tip to approach the substratum surface at a steep angle to minimize slippage. The blunted microneedle is then mounted on a micromanipulator and lowered onto the substratum within 50–100 microns of the cell. The needle is then either pushed or pulled to create the desired local deformation, as determined by the displacement of fluorescent

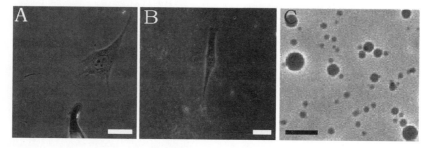

Fig. 3 Other applications using polyacrylamide substrata. (A) Localized mechanical stimulation is exerted on the cell by pushing or pulling on the substratum with a blunted microneedle (Scale bar = 40 μm). (B) Cellular responses to changes in substratum rigidity are tested with a gradient substratum. Since only the soft side contains fluorescent beads, changes in substratum rigidity is visualized as changes in the bead density (Scale bar = 40 μm). (C) Polyacrylamide microbeads of 1–6 μm generated in a microemulsion system (Scale bar = 10 μm).

beads underneath the tip. The magnitude of exerted forces can be estimated based on the displacement of fluorescent beads under the tip of the needle (and the Young's modulus of the material). The actual force exerted on the cell is strongly dependent on the distance from the needle, such that significant forces are generally limited to the region proximal to the microneedle.

B. Preparation of Substrata with a Gradient of Stiffness

Substrata with a gradient of stiffness have been used to test the hypothesis that cells are able to read local mechanical cues for the guidance of their migration (Lo *et al.,* 2000). The substratum is created by placing two-10 μl droplets of activated acrylamide/BIS acrylamide mixtures at different concentrations next to each other on activated glass, followed by careful placement of a coverslip to flatten and merge the two droplets. In this procedure, only the concentration of BIS is varied while the total acrylamide concentration, (acrylamide plus BIS), is maintained at a constant level. The boundary between hard and soft substrata is identified by adding fluorescent beads to one of the mixtures (Fig. 3B). The transition area between high and low rigidity is typically 50–100 μm in width, as judged from the distribution of beads. It is important to perform experiments on such gradient substrata at a low cell density, as both direct cell–cell interactions and forces generated by neighboring cells and transmitted across the substratum can obscure the effect of the change in substratum flexibility.

C. Preparation of Polyacrylamide Beads

Microbeads of polyacrylamide (Fig. 3C) have been used to test the sensitivity of phagocytosis to the flexibility of the target (Beningo and Wang, 2002). The beads are prepared in a microemulsion system, using bis(2-ethylhexyl) sulfosuccinate (AOT) (Fluka Chemicals) in toluene as the emulsifier. This agent is capable of forming inverse micelles (i.e., micelles of aqueous solution in an organic solvent) of relatively uniform size. The size of the beads is determined by several factors, including the AOT concentration and the relative volume of the aqueous solution to the organic solvent (Kunioka and Ando, 1996; Candau and Leong, 1985).

For phagocytosis studies we have generated beads 1–6 μm in diameter. The solution of acrylamide is prepared as described earlier, except that, instead of fluorescent beads, a high molecular weight FITC dextran (464 kDa, Molecular Probes) is added at a concentration of 20 mg/ml. This becomes is trapped within the polyacrylamide beads and serves as a label. AOT is dissolved in toluene in a fume hood at a concentration of 10.2 mg in 1 ml. While stirring under a stream of nitrogen, degassed acrylamide mixture is added to the AOT solution. The beads are allowed to polymerize for 1 h with constant stirring under nitrogen. Beads are recovered by centrifugation at 23g for 5 min and washed repeatedly (3–5 times) in methanol to remove the emulsifier. Following multiple washes in PBS, the beads can be stored at 4°C for up to 8 months. Conjugation of proteins to the beads is carried out as for polyacrylamide substrata, although the method with EDC has yielded more consistent results. Because the beads swell and shrink in response to changes in osmolarity and temperature, it is important to acclimate them to the experimental condition before use. Protein-conjugated beads are used within 2 days.

VII. Summary

We have described a powerful tool for the study of mechanical interactions between cells and their physical environment. Although the approach has already been used in a variety of ways to measure traction forces and to characterize active and passive responses of cultured cells to mechanical stimulation, it can be extended easily and combined with other microscopic approaches, including fluorescent analog imaging (Beningo *et al.*, 2001), photobleaching, calcium imaging, micromanipulation, and electrophysiology. This method will be particularly useful for studying the functions of various components at focal adhesions, and the effects of mechanical forces on focal adhesion-mediated signal transduction. In addition, the method can be extended to a 3D setting, e.g., by sandwiching cultured cells between two layers of polyacrylamide to create an environment mimicking that in the tissue of a multicellular organism.

Whereas chemical interactions between cells and the environment have been investigated extensively, many important questions remain as to the role of physical forces in cellular functions and the interplay between chemical and physical mechanisms of communication. The present approach, as well as other approaches capable of probing physical interactions, should fill in this important gap in the near future.

Acknowledgment

The authors thank Dr. Paul Jamney for sharing the information on EDC activation, and members of the Wang laboratory for the discussions and joint efforts in developing various methods. The project was supported by research grants from NIH GM-32476 and NASA NAG2-1197. K.A.B. is an NIH NRSA fellow supported by grant GM-20578.

References

Alpin, J. D., and Hughes, R. C. (1981). Protein-derivatised glass coverslips for the study of cell-to-substratum adhesions. *Anal. Biochem.* **113,** 144.

Balaban, N. Q., Scharwz, U. S., Riveline, D., Goichberg, P., Tzur, G., Sabanay, I., Mahalu, D., Safran, S., Bershadsky, A., Addadi, L., and Geiger, B. (2001). Force and focal adhesion assembly: a close relationship studied using elastic micropatterned substrates. *Nat. Cell Biol.* **3,** 466–472.

Beningo, K. A., Dembo, M., Kaverina, I., Small, J. V., and Wang, Y.-L. (2001). Nascent focal adhesions are responsible for the generation of strong propulsive forces in migrating fibroblasts. *J. Cell Biol.* **153,** 881–887.

Beningo, K. A., and Wang, Y.-L. (2002). Fc-receptor mediated phagocytosis is regulated by mechanical properties of the target. *J. Cell Sci.* **115,** 849–856.

Bray, D. (2001). "Cell Movement: From Molecule to Motility." Garland Publishing, New York.

Burton, K., and Taylor, D. L. (1997). Traction forces of cytokinesis measured with optically modified substrata. *Nature* **385,** 450–454.

Candau, F., and Leong, Y. S. (1985). Kinetic studies of the polymerization of acrylamide in inverse microemulsion. *J. Polymer Sci.* **23,** 193–214.

Dembo, M., and Wang, Y.-L. (1999). Stresses at the cell-to-substrate interface during locomotion of fibroblasts. *Biophys. J.* **76,** 2307–2316.

Galbraith, C. B., and Sheetz, M. P. (1997). A micromachined device provides a new bend on fibroblast traction forces. *Proc. Natl. Acad. Sci. USA* **94,** 9114–9118.

Grabarek, Z., and Gergely, J. (1990). Zero-length crosslinking procedure with the use of active esters. *Anal. Biochem.* **185,** 131–135.

Harris, A. K., Wild, P., and Stopak, D. (1980). Silicone rubber substrata: a new wrinkle in the study of cell locomotion. *Science* **208,** 177–179.

Kaverina, I., Krylyshkina, O., Gimona, M., Beningo, K., Wang, Y.-L., and Small, J. V. (2000). Enforced polarization and locomotion of fibroblasts lacking microtubules. *Curr. Biol.* **10,** 739–742.

Kunioka, Y., and Ando, T. (1996). Innocous labeling of the subfragment-2 region of skeletal muscle heavy meromyosin with a fluorescent polyacrylamide nanobead and visualization of heavy meromyosin molecules. *J. Biochem.* **119,** 1024–1032.

Lee, J., Leonard, M., Oliver, T., Ishihara, A., and Jacobson, K. (1994). Traction forces generated by locomoting keratocytes. *J. Cell Biol.* **127(6 Pt 2),** 1957–1964.

Li, Y., Hu, Z., and Li, C. (1993). New method for measuring Poisson's ratio in polymer gels. *J. Appl. Polym. Sci.* **50,** 1107–1111.

Lo, C.-M., Wang, H.-B., Dembo, M., and Wang, Y.-L. (2000). Cell movement is guided by the rigidity of the substrate *Biophys. J.* **79,** 144–152.

Marganski, W. A., Dembo, M., and Wang, Y.-L. (2002). Measurements of cell-generated deformations on flexible substrata using correlation-based optical flow. *Methods Enzymol.,* in press.

Martin, P. (1997). Wound healing: aiming for perfect skin regeneration. *Science* **276,** 75–81.

McKenna, N. M., and Wang, Y.-L. (1989). Culturing cells on the microscope stage. *In* "Methods in Cell Biology" (Y.-L. Wang and L. D. Taylor, eds.), Vol. 29, pp. 295–305. Academic Press, San Diego.

Munevar, S., Wang, Y.-L., and Dembo, M. (2001). Traction force microscopy of migrating normal and H-ras transformed 3T3 fibroblasts. *Biophys. J.* **80,** 1744–1757.

Oliver, T., Jacobson, K., and Dembo, M. (1998). Design and use of substrata to measure traction forces exerted by cultured cells. *Methods Enzymol.* **298,** 497–488.

Pelham, R. J., and Wang, Y.-L. (1997). Cell locomotion and focal adhesions are regulated by substrate flexibility. *Proc. Natl. Acad. Sci. USA* **94,** 13661–13665.

Pelham, R. J., and Wang, Y.-L. (1999). High resolution detection of mechanical forces exerted by locomoting fibroblasts on the substrate. *Mol. Biol. Cell* **10,** 935–945.

Radmacher, M. R., Tillmann, W., Fritz, M., and Gaub, H. E. (1992). From molecules to cells: imaging soft samples with the atomic force microscope. *Science* **257,** 1900–1905.

Staros, J. V., Wright, R. W., and Swingle, D. M. (1986). Enhancement by N-hydroxysulfosuccinimide of water-soluble carbodiimide-mediated coupling reactions. *Anal. Biochem.* **156,** 220–222.

Wang, H.-B., Dembo, M., and Wang, Y.-L. (2000). Substrate flexibility regulates growth and apoptosis of normal but not transformed cells. *Am. J. Physiol.* **279,** C1345–C1350.

Wang, Y.-L., and Pelham, R. J. (1998). Preparation of a flexible, porous polyacrylamide substrate for mechanical studies of cultured cells. *Methods Enzymol.* **298,** 489–496.

CHAPTER 17

Cell Migration in Slice Cultures

Donna J. Webb, Hannelore Asmussen, Shin-ichi Murase, and Alan F. Horwitz

Department of Cell Biology
University of Virginia
Charlottesville, Virginia 22908

I. Introduction

Cell migration plays a pivotal role in diverse biological processes including embryo-genesis, inflammation, wound repair, and tumor metastasis (Lauffenburger and Horwitz,

1996). Much of the insight into the molecular basis of cell migration has been obtained using cells growing on tissue culture plates in the presence of a purified growth factor or serum. However, these cell culture systems do not adequately duplicate the complex *in vivo* environment. For example, cells *in vivo* are exposed to gradients of multiple motogenic and chemotactic factors from surrounding tissue, whereas cultured cells are typically subjected to a single bolus of a soluble growth factor. In addition, cells migrating *in vivo* encounter a heterogeneous mixture of matrix molecules at varying concentrations in three dimensions. By contrast, cells in culture migrate on a fixed concentration of immobilized extracellular matrix molecules. The ECM and growth factor environment can have a profound influence on cell behavior including migration. Therefore, it is not known whether the mechanisms that regulate migration *in vitro* contribute similarly to migration *in vivo*.

Clearly, it is necessary to develop systems for studying cell migration *in vivo*. However, imaging migration *in vivo* is a formidable challenge. The problems include light scattering by the tissue, immobilization of the animal for long-term observations, limited depth of focus at high magnifications, phototoxicity and photobleaching of fluorescently labeled cells, direction of migration with respect to the microscope stage, and labeling of the cells and select molecules within the cells. To circumvent these problems, we developed systems for examining cell migration *in situ*, which closely mimics the *in vivo* environment.

In this chapter, we describe techniques for imaging 200–300-μm slices as one approach to observing the dynamics of migrating cells using three different model systems. These include somitic cell migration to the forelimb, neuronal precursors in the rostral migratory stream, and motor axon outgrowth to the limb bud. In all three systems, we visualize cells by labeling with the fluorescent marker DiI. This allowed us to image the migration of individual cells using time-lapse microscopy. In order to examine the molecular mechanisms that regulate migration, we developed a technique for expressing exogenous GFP-tagged proteins in the slices using a cationic reagent, Transfast. This represents an alternative to previous methods of transfecting slices, including electroporation and retroviruses, which are time-consuming, costly, and toxic to cells. The combination of these methodologies allowed us to study the regulation of *in vivo* migration.

II. Somitic Migration

A. Overview

Myogenic precursor cells migrate from the lateral part of the somites into the forelimb where they form muscles of the limb (Chevallier *et al.*, 1977; Christ *et al.*, 1977; Ordahl and Le Douarin, 1992; Williams and Ordahl, 1994). In the chick, somitic cell migration begins at stage 15 and is completed within approximately 10 h. Since the beginning and endpoints of this migratory path (from the somites to the limb bud) are known and migration occurs over a relatively short, defined period of time, we chose this system to begin to study the molecular mechanisms that regulate migration *in vivo*.

This section provides the detailed methodology that we developed for visualizing the migration of muscle precursors from the somites to the limb bud. Detailed protocols are provided later and depicted schematically in Fig. 1. Somitic cells are labeled with either DiI or GFP-tagged proteins. With the first method, a DiI solution or cDNAs encoding fusions with GFP are microinjected into the somitocoele of stage 13–14 chick embryos and the embryos are incubated at 37.5°C. When the embryo reaches stage 16–17, an explant containing somites 16–21 is harvested. The tissue is embedded in agarose, sliced in 200–300 μm thick sections and transferred to a Millipore insert in a microscopy dish. With the second method, a small DiI crystal is placed on the dermomyotome region of the slice after transferring to the insert. In some experiments, instead of transferring to inserts, the slices are placed on glass coverslips in dishes prepared for microscopic observation. The slices are layered with agarose to maintain the morphology of the tissue. Migration of individual, labeled cells is observed by time-lapse microscopy.

B. Preparation of Embryo Media

BSS dissecting solution:

1× HBSS (Gibco Cat. #14180-061 for 10× HBSS)
10 mM Hepes (Gibco Cat. #15630-080)
0.5% Fungizone (optional)
4 mM glutamine
1 mM sodium pyruvate
100 units/ml penicillin
100 units/ml streptomycin

Embryo medium (without phenol red) for imaging:

0.5% Fungizone (optional)
4 mM glutamine
1 mM sodium pyruvate
50 U/ml penicillin
8 mg/ml streptomycin
Ham's/F-12 (HyClone Cat. #sh30026.01)

C. Embryo Staging

1. Incubate white leghorn chicken eggs (Charles River) at 37.5°C until they reach stage 13–14.

2. Spray the eggs with 70% ethanol and dry on paper towels under a dissecting hood. Open the eggs in an aseptic environment and place the embryos in 100 × 20 mm petri dishes.

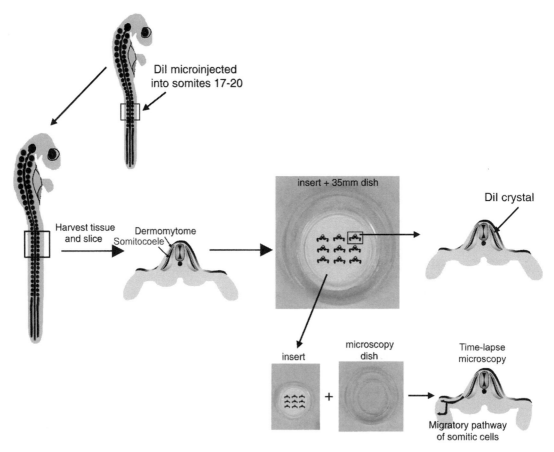

Fig. 1 Schematic illustration depicting the procedure for visualizing somitic migration in chick embryo slices. DiI solution was microinjected into somites 17–20 of stage 13–14 embryos and incubated at 37.5°C until they reached stage 16–17. An explant containing somites 16–21 was excised and sliced into 200– 300 μm thick sections. The slices were transferred to a Millipore insert in a 35 mm dish. In subsequent experiments, an alternative method for DiI labeling the cells was used. Tissue was harvested as described above from stage 16–17 embryos that were not previously microinjected with DiI. A DiI crystal was then gently placed on the dermomyotome region of the slice. The Millipore insert with the slices was transferred to a microscopy dish and somitic cell migration was visualized by time-lapse microscopy. The curved arrows indicate the migratory pathway of the somitic cells.

3. Inject 10% India ink in PBS under the embryo using a tuberculin syringe with a 30-gauge needle to visualize the embryos.

4. Maintain the embryos at 37.5°C with 2.5% CO_2.

5. Stage embryos as previously described (Hamburger and Hamilton, 1992).

6. When the embryos have 19–22 somites, microinject Wesson oil into somites 16 and 21 to indicate the position for cutting.

7. Microinject somites 17–20 with DiI or the transfection solution.

D. Microinjections

Pull the needles with a Sutter Instruments Flaming/Browning Micropipette puller (Model P-87) using glass capillary tubes (Warner Instruments Corp. Cat. #GC100TF-10). When pulling the needles, we observed some variability in the glass capillary tubes. For DiI microinjections, we use the following settings for pulling the needles:

| First cycle: | Heat (550) | Pull (5) | Velocity (35) | Time (50) |
| Second cycle: | Heat (550) | Pull (50) | Velocity (150) | Time (180) |

Prepare a stock solution of DiI (0.25% w/v; Molecular Probes Cat. #C-7000) in 100% ethanol and then dilute the stock solution 1:20 in 0.3 M sucrose. For transfections, resuspend Transfast (Promega Cat. #E2431) in water (at a concentration of 1 mg/ml), freeze, thaw, and then incubate at a 1:1 ratio with cDNA in 0.3 M sucrose for 10 min at 25°C. Dispense 10 μl of DiI or the transfection solution into a glass microneedle with a Hamilton syringe. Inject approximately 50 nl into the somitocoele of somites 17–20 using a Narashige Microinjector (IM 300) at 10 psi for 30 ms. Incubate the embryos at 37.5°C in 2.5% CO_2 for 12–16 h until they have 27–30 somites.

E. Embryo Slicing and Embedding

1. Pack the reservoir of the Vibratome with ice for approximately 1 h before use.

2. Decapitate embryos. Using sharpened micro-forceps (Fine Scientific Tools, Cat. #11252-20; Dumont #5), isolate somites 16–21 and transfer to a 60 mm petri dish with BSS dissection medium.

3. Remove the connective tissue and viscera with sharpened microforceps.

4. Transfer the explants to a petri dish containing BSS dissecting medium.

5. Slowly add 0.25g ultralow-melt agarose (Sigma Type IX, Cat. #A-5030) to 5 ml of cold BSS in a 50 ml tube while vortexing. To dissolve, microwave the agar for 6 s, mix, and repeat 1–2 additional times. Briefly centrifuge to remove the bubble and then fill a histology mold (Curtin Matheson Scientific, Inc.; dimensions 24 × 24 × 5 mm, Cat. #038218) with the hot agarose and allow to cool to approximately 37°C (the agarose hardens quickly).

6. Place 3 limb buds in the agarose approximately 2 mm from one end of the mold. Arrange so that the tissue is approximately 1 mm apart and perpendicular to the long axis of the mold. Place the mold (level surface) on ice for 5–10 min to allow agarose to harden. Cover the mold with the top of a 100 mm petri dish and embed it into the ice for 5–10 min.

7. Remove the ice from the Vibratome until the level is below the carrier stage. Take the agarose block away from the mold and transfer to a Kimwipe. Trim the block on both sides to fit the cover glass (Thomas, Cat. #5972M10) attached to the Vibratome stage. With a thin layer of Super Glue, secure the agarose onto the cover glass. Make a straight cut about 2 mm behind the tissue. Place the chuck portion of the stage in the Vibratome carrier. Have the blade about 2 mm from the agarose, so the blade will be away from the slice you need to collect after each cutting. Slice 250–300 μm thick sections. Discard the first and last section containing tissue slices. Do not touch the tissue.

8. Place a Millipore insert (Millipore, Cat. #PICM ORG 50) into a 35 mm petri dish with 1 ml of embryo medium. Gently transfer the slices with surrounding agarose to the insert. As an alternative method for labeling the somitic cells, place a small DiI crystal on the dermomyotome region of the slice. Before imaging, transfer the insert to a microscopy dish with 1 ml of embryo medium (without phenol red).

9. Alternatively, instead of transferring to inserts, place the slices without agarose on glass coverslips in dishes prepared for microscopic observation. To secure the slice to the glass, place a drop of 1% warm agarose (FMC BioProducts Cat. #50081) on each slice and allow the agarose to harden on ice. Then, layer 1% agarose over all the slices. After the agarose hardens, put 3 ml of embryo medium over the slices. Prior to imaging gently overlay with mineral oil (Sigma, Cat. #M-8410).

F. Preparation of Microscopy Dishes

1. Punch a hole in the bottom of a 35 mm petri dish with a heated #18 cork borer.

2. Wash glass coverslips (22 mm^2, #1; VWR Cat. #48380 080) with 20% H_2SO_4 for 1 h and rinse 3 times with deionized water.

3. To neutralize the acid, wash coverslips with 0.1 M NaOH for 30 min, rinse 3 times with deionized water, and then dry thoroughly.

4. Attach the coverslip to the dish with Norland Optical Adhesive (Norland Products, Inc., Cat. #6801).

5. Cure the glue and sterilize by exposing the dishes to UV light for 16 h on each side.

G. Imaging

The microscopy dish with nine embryo slices is placed in a heated stage maintained at 37°C. Fields from the slices that have characteristic morphology with fluorescently labeled cells in the dorsolateral aspect of the somite are selected for observation. The slices are illuminated with a halogen lamp and images are recorded every 5–10 min using the 40× (Nikon, N.A. 0.75) or the 20× (Nikon, N.A. 0.50) objectives. The tissue is exposed to the excitation light for only a short period of time (0.05–0.20 s) and bright fluorescence is attenuated with neutral density filters. EGFP is visualized with an endow GFP filter cube (ex HQ470/40, emHQ525/50, Q495LP dichroic mirror) (Chroma, Brattleboro, VT) and DiI using a rhodamine/TRITC cube (ex BP520-550, barrier filter BA580IF, dichroic mirror DM565). In a typical experiment, migration of myogenic cells to the limb bud is observed in several slices. For the other slices, either the cells do not migrate or they migrate to the lateral flank.

Images are obtained from a cooled CCD camera (Hammatsu OrcaII) attached to a Nikon TE-300 inverted microscope. Image acquisition is controlled with the Inovision ISee software program interfaced to a Ludl modular automation controller (Inovision, Raleigh, NC). Phase and fluorescence illumination are regulated by electronic shutters.

H. Results and Discussion

The embryo slices must maintain the physical and molecular characteristics of the intact embryo, over the time period of observation, for the slice cultures to be a valid model. Both morphologic changes and expression of molecular markers in the embryo slices were similar to those seen in intact embryos. In the slice cultures, wing bud growth and expansion as well as somite elongation were similar to those in the stage-matched controls from the intact embryos for at least 24 h. In addition, neural tube expansion, neurite outgrowth, and kidney maturation were also morphologically comparable in the embryo slices and the stage-matched controls. Thus, the morphogenesis of the slice cultures and the intact embryo were comparable during our experimental interval.

We further tested the validity of the slice cultures as a model system by determining whether the somitic cells express the molecular markers for myogenic precursor cells at the appropriate times and locations. N-cadherin is expressed by myogenic cells and can be used as a molecular marker to distinguish the myoblast precursors cells from other limb mesenchymal cells (Hayashi and Ozawa, 1995). The monoclonal antibodies L4 and 6b3 are specific for N-cadherin and thus should label the muscle precursor cells, but not other cells such as fibroblasts (George-Weinstein et al., 1988, 1997). Their expression patterns were similar to those of stage-matched controls.

The next level of validation was to show that the somitic cells migrated to their appropriate locations (Fig. 2). The muscle precursor cells migrating out of the somite were visualized by labeling with DiI. In initial experiments, DiI was injected into the somitocoele of stage 13–14 embryos. The embryos were allowed to reach stage 16–17 and then migration was examined using time-lapse microscopy. In subsequent experiments, an alternative method for DiI labeling the somitic cells was used. Stage 16–17 embryos were harvested and sliced, and a DiI crystal was placed on the dermomyotome region of the slice. Migration was then observed. The latter method provided robust labeling.

With both DiI labeling methods, somitic cells were not observed to migrate along a closely restricted path. Rather, rapid bursts of directed migration were followed by

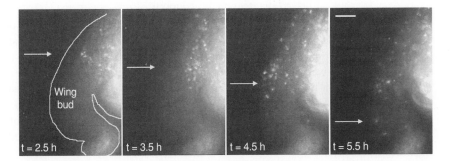

Fig. 2 Migration of DiI-labeled somitic cells to the wing bud. Tissue from stage 16 embryos was harvested and sliced as described in Fig. 1. A DiI crystal was placed on the slice and then migration was visualized. A series of time-lapse images are shown illustrating the migration of somitic cells into the wing bud (arrows). The wing bud is outlined in the far left panel. Scale bar = 50 μm.

periods in which the cells strayed from the defined pathway. During this wandering period, perhaps the cells were searching for guidance cues to direct their migration. This behavior is analogous to outgrowing axons in which growth cones are constantly sampling the surrounding environment for cues to guide them to their targets (Stoeckli and Landmesser, 1998). The extracellular environment in the embryo slices provides multiple signals both attractive and repulsive to navigate axons, and probably other cells as well, along the appropriate pathway to their targets. The importance of this surrounding environment was apparent in our study since directional migration was no longer observed when the somitic cells moved out of the slice.

The basic migratory cycle observed in cultured cells, which includes extension of protrusions, formation of stable attachments at the leading edge, translocation of the cell body forward, and detachment and release at the cell rear, was recapitulated by the muscle precursor cells. However, the exaggerated length, persistence, and polarity of protrusions in the muscle precursor cells are not typically observed in migrating fibroblasts *in vitro*. The somitic cells formed large, persistent protrusions in the direction of migration prior to movement of the cells. The protrusions extended up to 50 μm in length and persisted on the average for approximately 1 h. Before forming stable attachments, the protrusions moved laterally. The lateral movement of protrusions may be similar to that of growth cones searching the environment for directional cues. However, once the protrusions stabilized, the cells translocated forward. Cells that were not translocating also extended and retracted protrusions, often for long periods of time. This suggested that although protrusive activity was necessary, it was not sufficient to initiate migration.

The cellular environment of the slice also influenced protrusive activity. In somitic cells that migrated out of the slice, the large, persistent protrusions were not seen. Rather, the protrusions were short and they extended and retracted more rapidly than the persistent protrusions observed with somitic cells in the slice. Taken together, our results in the embryo slices suggest that environmental cues play a role in directing the migration of cells *in vivo*. The complex cellular environment in the embryo slices may not be recapitulated in culture systems *in vitro,* and thus, this could at least partially account for the differences that we observed in the migratory processes of the slices as compared with cultured cells (Knight *et al.,* 2000).

In cultured cells *in vitro,* the small GTPase Rac is a key regulator of protrusive activity (Ridley and Hall, 1992). We wanted to determine whether Rac had a similar effect on protrusive behavior in the embryo slices. To accomplish this, wild-type Rac, dominant-negative Rac (N17Rac), and constitutively active Rac (L61Rac) were expressed as GFP fusion proteins in the slices. The large, persistent protrusions were not observed in somitic cells expressing GFP-N17Rac and migration was blocked. In cells expressing GFP-wild-type Rac, the protrusions were shorter and less persistent, but they were still polarized. In cells expressing GFP-L61Rac, the protrusions extended and retracted more frequently as compared with wild-type Rac. The protrusions were no longer polarized in the GFP-L61Rac expressing cells, and random, but not directional, migration was observed. Interestingly, GFP-L61Rac was observed preferentially at the plasma membrane, suggesting that local activation of Rac might regulate polarization of the protrusions. Our results suggest that Rac may modulate protrusive activity in the somitic cells.

III. Migration from the Subventricular Zone to the Olfactory Bulb

A. Overview

In the rodent brain, olfactory interneuron precursors migrate from the subventricular zone to their targets in the center of the olfactory bulb. Migration proceeds along a restricted pathway known as the rostral migratory stream (RMS). This migration is first observed in embryonic mice and continues in the adult. The mechanisms that regulate this migratory process are not well understood. However, studies from knockout mice (Tomasiewicz *et al.*, 1993; Cremer *et al.*, 1994) suggest that the polysialylated form of the neural adhesion molecule (PSA-N-CAM) plays a role in regulating this migration. PSA may exert its effect by reducing the adhesiveness of N-CAM (Hoffman and Edelman, 1983; Sadoul *et al.*, 1983), which allows the cells to migrate in the RMS.

In this section, we develop slice technology to visualize the dynamics and molecular mechanisms of migration of the neuronal precursors as they move to the olfactory bulb using function blocking antibodies; we identify integrins and DCC/netrins as key players in this directed migration. Our procedure, which is a modification of the protocols outlined previously, is described in this section and shown schematically in Fig. 3. Briefly, the brains from pre- and postnatal mice are collected and embedded in agarose. The brains are then sliced into 200 μm parasagittal sections and placed on a Millipore insert. The neuronal precursors are labeled by placing small DiI crystals along various points of the migratory pathway. Migration is observed using time-lapse microscopy. In some experiments, slices that showed migrating DiI-labeled cells are cultured in CCM1 with function-blocking antibodies.

B. Preparation of Brain Slices

1. Prepare slices from embryonic day 18 or postnatal day 0–16 mice. Anesthetize, decapitate, and remove the brains of the postnates. For the fetal brains, anesthetize the mother and remove the fetuses from the abdominal cavity. Decapitate the fetuses and remove the brains.

2. Place the brains into ice-cold CCM1 (Hyclone Laboratories) medium.

3. Transfer the brains to CCM1 or F-12 medium with 5% heat-inactivated horse serum and penicillin/streptomycin.

4. Embed the brains in 8% agarose.

5. Slice into 200 μm parasagittal sections with a Vibratome.

6. Transfer the slices to the Millipore inserts.

7. Select only the slices that contain the entire migratory stream for further studies.

To visualize the migratory cells, small DiI crystals are placed on the slice with a microneedle at various points along the migratory path between the subventricular zone and the olfactory bulb. Some slices that showed DiI-labeled neuronal precursor migration are cultured in CCM1 with function-blocking antibodies.

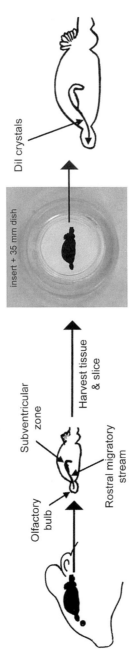

Fig. 3 Experimental design for observing migration of neuronal precursor cells in the rostral migratory stream in mice. Brains from embryonic day 18 or postnatal day 0–16 mice were isolated and sliced into 200-μm parasagittal sections. Slices containing the entire migration stream were placed on a Millipore insert in CCM1 medium with 5% horse serum. DiI crystals were placed at various points along the RMS to visualize migrating cells using time-lapse microscopy.

C. Results and Discussion

The movement of the neuronal precursor cells from the subventricular zone to the olfactory bulb provided a different system for studying migration in the slice cultures. The viability of the cells and direction of migration provided the criteria for validating the system. Under our experimental conditions, the neuronal precursor cells in the slices were viable as determined by trypan blue exclusion and migrated directionally along the RMS to the olfactory bulb. Actively migrating cells were labeled by placing small DiI crystals along various points of the RMS. Comparable results were observed irrespective of where the crystal was placed along the pathway.

Labeled neuronal precursors were not observed to migrate into regions surrounding the RMS, showing that the cells move along a restricted pathway. The migrating cells were highly polarized with a single, long, persistent protrusion in the direction of migration (Fig. 4). Rapid bursts of migration were followed by periods of wandering or rest, which is similar to what we observed with the somitic cells. However, unlike the somitic cells, the movement of the neuronal precursor cells was highly directed with cells migrating strictly along the defined path.

The factors that contribute to the directional migration of the neuronal precursors are not known. Previous studies suggest that chemorepulsive factors, such as the Slit proteins, from the caudal region of the septum or choroids plexus may play a role by repelling migrating cells (Hu and Rutishauser, 1996; Hu, 1999). However, the Slit proteins are not expressed in the tissue surrounding the RMS, which suggests that other factors may contribute to the directional migration of the neuronal precursors. It is tempting to speculate that chemoattractive factors in the olfactory bulb are involved in navigating the neuronal precursor cells to their target. As with axonal outgrowth, the directional migration is probably a delicate balance between chemoattractive and chemorepulsive factors.

Molecules other than PSA-N-CAM most likely contribute to RMS migration. To address this possibility, we examined integrin expression and function during various developmental stages. We found that six integrin subunits were expressed in a stage-specific manner during development (our unpublished observation). As determined using function-blocking antibodies, $\alpha 1$, $\beta 1$, and αv were necessary for RMS migration in the stage of development in which they were expressed. Unfortunately, we could not test the role that the other three integrins play in RMS migration because function-blocking antibodies are not available. These results suggested that integrins are necessary for neuronal precursor migration in the RMS.

Since our results indicated that the neuronal precursor migration toward the olfactory bulb was highly directed, we wanted to determine which molecules contributed to the guided movement. Because receptors for netrin are known to function in neuronal guidance, we examined the involvement of netrin, netrin receptors, neogenin, and Deleted in Colorectal Carcinoma (DCC) using function blocking antibodies. When the slices were treated with a DCC antibody, the neuronal precursors extended single, long protrusions in all directions. This nondirected extension of protrusions was not observed when the slices were incubated with the integrin function-blocking antibodies.

Fig. 4 Migration of neuronal precursor cells along the RMS. Brain slices from a P9 mouse were labeled with small DiI crystals and migration was visualized. A series of time-lapse images are shown depicting migration of cells from the subventricular zone (right side of panel) to the olfactory bulb (left side of panel). The arrows depict a cell extending a single, long, persistent protrusion. Note the distance that this cell migrated toward the olfactory bulb. Scale bar = 50 μm.

The DCC antibody significantly reduced the migration rates of the neuronal precursors and the cells no longer migrated directionally. These results suggested that DCC and related proteins are involved in the guidance of the neuronal precursors to the olfactory bulb.

IV. Axonal Outgrowth

A. Overview

In the developing limb bud, growing axons respond to cues that direct their movement to specific targets (Lance-Jones and Landmesser, 1981; Ferns and Hollyday, 1993; Hollyday, 1995). For example, axons from motor and sensory neurons extend from the neural tube along specific pathways to their targets in the developing limb bud (Lance-Jones and Landmesser, 1981; Tosney and Landmesser, 1985; Hollyday, 1995). Once at the target, the motor neurons innervate muscles in the limb. Since the time course and pathway for axonal outgrowth to the limb have been described and the nerve bundle emanating from the neural tube is highly visible, we used this system for observing migration mediated via growth cone extension.

The embryo slice cultures provide an environment for studying neurite outgrowth that is comparable to that observed in the intact embryo (Hotary *et al.,* 1996). Our protocol is a combination of the procedures that we used for the other migration systems outlined previously and that previously described for studying neuronal guidance (Landmesser, 1988; Hotary *et al.,* 1996). In brief, tissue containing the wing bud is harvested from stage 22–24 embryos and sliced as described above. The slices are then transferred to Millipore inserts. In some of our experiments, the neurites are labeled by microinjecting a DiI solution into the neural tube after slicing (Fig. 5). Alternatively, after harvesting and slicing, DiI crystals are placed on the tissue (Fig. 5). Neurite outgrowth is then observed by time-lapse microscopy. The detailed procedure follows.

B. Preparation of Slices

1. Incubate white leghorn chicken eggs at 37.5°C for 3.5–4.5 days.

2. Spray eggs with 70% ethanol, open the eggs under a dissecting hood, and put into a 100 × 20 mm petri dish.

3. Decapitate the embryos (stage 22–24).

4. Transfer the embryos to a 60 mm petri dish with BSS dissecting media.

5. Isolate tissue containing the wing bud with sharpened microforceps. Transfer to a 60 mm dish with BSS dissecting media.

6. Remove connective tissue and viscera with sharpened microforceps.

7. Slice the tissue as described above.

8. Transfer the slices with the surrounding agarose to a Millipore insert.

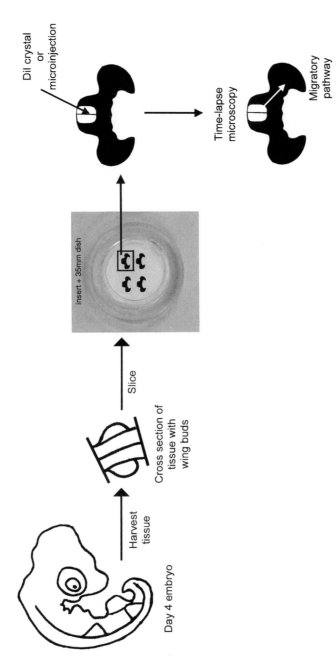

Fig. 5 Schematic representation of the experimental design for studying axonal outgrowth in chick embryo slices. Tissue from stage 22–24 embryos containing the wing bud was harvested, sliced into 250–300 μm thick sections, and transferred to a Millipore insert. Motor axons were visualized either by microinjecting DiI solution into the neural tube or by gently placing a DiI crystal on the neural tube region of the slice. Axonal outgrowth was then observed by time-lapse microscopy. The white arrow indicates the migratory pathway of the axons from the neural tube to the wing bud.

C. DiI Microinjections

Pull the needles on a Sutter Instruments Flaming/Browning Micropipette puller (Model P-87) using glass capillary tubes. We pull the needles with the following settings:

First cycle:	Heat (516)	Pull (0)	Velocity (35)	Time (50)
Second cycle:	Heat (516)	Pull (0)	Velocity (50)	Time (50)

Dilute the DiI stock (0.25% w/v) solution 1:10 in 0.3 M sucrose. Dispense 10 μl of the DiI solution into a glass microneedle. Inject about 50 nl into the neural tube of the slice using a Narashige Microinjector at 10 psi for 30 ms. Repeat the microinjections 2–4 times. Alternatively, label the neurites by gently placing a DiI crystal on the neural tube of the slices. Visualize neurite outgrowth using time-lapse microscopy.

D. Results and Discussion

In the embryo, axons from motor neurons extend from the neural tube into the limb bud where they innervate muscle. Navigational cues from the surrounding tissues guide the axons to their appropriate targets. Axonal outgrowth in the embryo slices provided a third model system for our studies. Axonal extension was visualized by labeling the neurites with DiI. Initially, the tissue was sliced and DiI solution was microinjected into the neural tube. In subsequent experiments, an alternative method for DiI labeling was used in which a DiI crystal was gently placed on the neural tube region of the slice. The slices were then imaged by time-lapse microscopy. Comparable results were obtained with both DiI labeling methods.

In the embryo slices, DiI labeled neurites extended from the neural tube to the plexus region of the limb bud (Fig. 6). As the neurites extended toward their targets, the growth

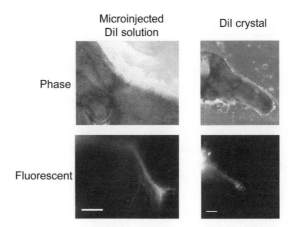

Fig. 6 Outgrowth of DiI-labeled motor axons to the wing bud. Tissue from stage 24 embryos was harvested, sliced, and labeled with DiI. The two labeling methods are depicted in Fig. 5 and described in the text. Fluorescent and phase images from both are shown. A bundle of motor axons extended from the neural tube to the plexus region of the wing bud. Scale bar = 50 μm.

cones were highly dynamic and their movement was not continuous. Like the somitic and neuronal precursor cells, rapid advancement of the growth cones followed periods of resting. During this time of resting, the growth cones were observed to pause, and sometimes retract, before eventually resuming movement toward their target.

V. Conclusions and Perspectives

This chapter describes technologies for observing the dynamics of migrating cells in slice preparations. The slice cultures closely emulate the *in vivo* environment, making this system ideal for studying migration. The three model systems that we studied include myogenic precursors migrating from the somite to the limb, neuronal precursors migrating along the rostral migratory stream, and motor axon extension from the neural tube to the limb. The cells are visualized by labeling with the fluorescent probe DiI or by transfection with GFP fusion proteins. Migration is then observed using time-lapse microscopy.

The basic migratory processes observed for cultured cells *in vitro* are recapitulated in the slice cultures; however, these processes are highly polarized and exaggerated when compared with cultured fibroblasts. The somitic and neuronal precursor cells extend a single, long, and persistent protrusion, which is not typically observed in cultured cells. Interestingly, these protrusions persist after the cells reach their targets, but they are no longer directional and do not result in movement of the cell body.

The protrusive activity in the somitic cells, like that in cultured fibroblasts, is regulated by Rac. Although the same Rac pathway appears to be activated *in vitro* and *in vivo,* the duration of activation may be differentially regulated *in vivo* and thus account for the altered persistence of the protrusions. It also seems possible that the activation of Rac is highly localized, perhaps by chemoattractive agents, thus producing the single protrusion that we observed. Since the effectors of Rac in our slice cultures have not been identified, it is unclear whether the downstream signaling pathway is the same as described in cultured cells. The directed nature of the protrusions, the timing of the activation, and the signaling molecules that regulate protrusive activity in the slice cultures are all important subjects for future studies.

Although rapids burst of migration followed by periods of resting are observed in all three systems, the migratory pathways in the slice cultures showed clear differences. The neuronal cells migrated directionally along a highly defined and restricted route. In contrast, the somitic cells migrated along a far less restricted and defined pathway; these cells often strayed from their path before reaching their target. At this time, the factors that account for these differences in the migratory patterns are not known. ECM components, such as laminin and tenascin C, that line the RMS may provide a pathway on which the cells could migrate; however, these substrata would not account for the highly directional migration that we observe. Instead it is more likely that a delicate balance between chemoattractive and chemorepulsive factors contributes to the migration patterns of both the somitic and neuronal precursor cells. The environmental cues, including the attractive and repulsive gradients provided by the surrounding tissue, that navigate the cells to their appropriate targets are not present in cell-culture systems. Another avenue for future studies is to identify the guidance molecules that direct migration of the neurites.

Although the slice technology has already been highly useful, it should readily extrapolate to a number of other systems in both embryonic and adult tissues. Transgenic mice provide another useful tool for studying migration *in vivo*. Mice with select genes deleted or mutated should provide important insights into the function of various genes involved in migration. In addition, tissue specific reexpression of GFP fused proteins, which have been deleted in knockout mice, would be highly useful. Finally, it may also be possible to use photomanipulation technology in the slices to locally activate and inactivate various molecules.

Acknowledgments

D. J. Webb was supported by National Institutes of Health postdoctoral training grant HD07528-01. The authors thank Karen Donais for help with the manuscript and figures.

References

Chevallier, A., Kieny, M., and Mauger, A. (1977). Limb–somite relationship:origin of the limb musculature. *J. Embryol. Exp. Morphol.* **41**, 245–258.

Christ, B., Jacob, H. J., and Jacob, M. (1977). Experimental analysis of the origin of the wing musculature in avian embryos. *Anat. Embryol.* **150**, 171–186.

Cremer, H., Lange, R., Christoph, A., Plomann, M., Vopper, G., Roes, J., Brown, R., Baldwin, S., Kraemer, P., Scheff, S., Barthels, D., Rajewsky, K., and Wille, W. (1994). Inactivation of the N-CAM gene in mice results in size reduction of the olfactory bulb and deficits in spatial learning. *Nature* **367**, 455–459.

Ferns, M. J., and Hollyday, M. (1993). Motor innervation of dorso-ventrally reversed wings in chick/quail chimaeric embryos. *J. Neurosci.* **13**, 2463–2476.

George-Weinstein, M., Decker, C., and Horwitz, A. (1988). Combinations of monoclonal antibodies distinguish mesenchymal, myogenic, and chondrogenic precursors of the developing chick embryo. *Dev. Biol.* **125**, 34–50.

George-Weinstein, M., Gerhart, J., Blitz, J., Simak, E., and Knudsen, K. A. (1997). N-cadherin promotes the commitment and differentiation of skeletal muscle precursor cells. *Dev. Biol.* **185**, 14–24.

Hamburger, V., and Hamilton, H. L. (1992). A series of normal stages in the development of the chick embryo. *Dev. Dyn.* **195**, 231–272.

Hayashi, K., and Ozawa, E. (1995). Myogenic cell migration from somites is induced by tissue contact with medial region of the presumptive limb mesoderm in chick embryos. *Development* **121**, 661–669.

Hoffman, S., and Edelman, G. M. (1983). Kinetics of homophilic binding by embryonic and adult forms of the neural cell adhesion molecule. *Proc. Natl. Acad. Sci. USA* **80**, 5762–5766.

Hollyday, M. (1995). Chick wing innervation. I. Time course of innervation and early differentiation of the peripheral nerve pattern. *J. Comp. Neurol.* **357**, 242–253.

Hotary, K. B., Landmesser, L., and Tosney, K. W. (1996). Embryo slices. *In* "Methods in Avian Embryology" (M. Bronner-Fraser, eds.), pp. 109–124. Academic Press, San Diego.

Hu, H. (1999). Chemorepulsion of neuronal migration by Slit2 in the developing mammalian forebrain. *Neuron* **23**, 703–711.

Hu, H., and Rutishauser, U. (1996). A septum-derived chemorepulsive factor for migrating olfactory interneuron precursors. *Neuron* **16**, 933–940.

Knight, B., Laukaitis, C., Akhtar, N., Hotchin, N., Edlund, M., and Horwitz, A. (2000). Visualizing muscle cell migration in situ. *Curr. Biol.* **10**, 576–585.

Lance-Jones, C., and Landmesser, L. (1981). Pathway selection by chick lumbosacral motoneurons during normal development. *Proc. R. Soc. Lond. [Biol.]* **214,** 1–18.

Landmesser, L. (1988). Peripheral guidance cues and the formation of specific motor projections in the chick. *In* "From Message to Mind" (S. E. Easter, Jr., K. F. Barald, and B. M. Carlson, eds.), pp. 121–133. Sinauer Assoc. Sunderland, MA.

Lauffenburger, D., and Horwitz, A. (1996). Cell migration: a physically integrated molecular process. *Cell* **84,** 359–369.

Ordahl, C. P., and Le Douarin, N. M. (1992). Two myogenic lineages within the developing somite. *Development* **114,** 339–353.

Ridley, A. J., and Hall, A. (1992). The small GTP-binding protein rho regulates the assembly of focal adhesions and actin stress fibers in response to growth factors. *Cell* **70,** 389–399.

Sadoul, R., Hirn, M., Deagostini-Bazin, H., Rougon, G., and Goridis, C. (1983). Adult and embryonic mouse neural cell adhesion molecules have different binding properties. *Nature* **304,** 347–349.

Stoeckli, E. T., and Landmesser, L. (1998). Axon guidance at choice points. *Curr. Opin. Neurobiol.* **8,** 73–79.

Tomasiewicz, H., Ono, K., Yee, D., Thompson, C., Goridis, C., Rutishauser, U., and Magnuson, T. (1993). Genetic deletion of a neural cell adhesion molecule variant (N-CAM-180) produces distinct defects in the central nervous system. *Neuron* **11,** 1163–1174.

Tosney, K. W., and Landmesser, L. (1985). Specificity of early motoneuron growth cone outgrowth in the chick embryo. *J. Neurosci.* **5,** 2336–2344.

Williams, B. A., and Ordahl, C. P. (1994). Pax-3 expression in segmental mesoderm marks early stages in myogenic cell specification. *Development* **120,** 785–796.

CHAPTER 18

Application of Cell Adhesion to Study Signaling Networks

Cindy K. Miranti

Laboratory of Integrin Signaling and Tumorigenesis
Van Andel Research Institute
Grand Rapids, Michigan 49503

I. Introduction

Integrins are a family of heterodimeric transmembrane proteins that mediate adhesion of cells to extracellular matrix proteins such as fibronectin, vitronectin, laminin, and collagen (Howe *et al.,* 1998). In addition, some integrin family members mediate cell–cell interactions. Specific pairing of different α- and β-integrin chains determines which extracellular matrix protein binds to which heterodimer. Each cell type expresses a different combination of these heterodimers and the sum of their expression determines which extracellular matrices the cell will bind to. Some integrins are very cell type specific, such as $\alpha IIb\beta3$, which is only expressed on platelets, or $\alpha6\beta4$, which is found in hemidesmosomes of epithelial cells. That these integrins are important for their respective cell functions is demonstrated by the profound effect on their respective tissues when the integrin is missing. Loss of $\beta3$ function in mice and humans causes a bleeding disorder and loss of $\beta4$ in mice results in abnormal epithelial structure and blistering of the skin (Clemetson and Clemetson, 1994; van der Neut *et al.,* 1996). Loss of other integrin subunits often results in severe developmental defects in mice (Fassler *et al.,* 1996).

Integrins are an integral part of many cell functions, including development, differentiation, cell cycle progression, cell survival, cell migration and gene regulation (Berman and Kozlova, 2000; De Arcangelis and Georges-Labouesse, 2000; Howe *et al.,* 1998). Through their ability to bind cytoskeletal regulatory proteins at the plasma membrane, integrins regulate the assembly of specific actin-containing cytoskeletal structures inside the cell, namely focal adhesions, clusters of integrins that define the points of attachment of the cell to the extracellular matrix, filapodia, and lamellipodia (Keely *et al.,* 1998; Petit and Thiery, 2000).

In addition to their role in adhesion, integrins are capable of transmitting and receiving signals through a series of different intracellular signaling pathways. Integrin engagement with various different extracellular matrices stimulates the activation (and inhibition) of an extensive array of signaling molecules (Aplin *et al.,* 1998; Petit and Thiery, 2000). The signaling pathways and mechanisms of activation by integrins are an intense area of study. Many interesting relationships and paradigms have begun to emerge. One of the more recent findings is the interdependence or "crosstalk" between signals initiated by integrins and those initiated by other receptor–ligand interactions. One example of this is the dependence for normal cells on integrin engagement for cell cycle progression in response to growth factors. Cells that are removed from their matrix and placed in suspension cannot progress through the cell cycle even in the presence of growth factor (Assoian, 1997). Many tumor cells have overcome this block and are able to grow in suspension. The exact mechanisms by which these signals are regulated by integrins are still under investigation. Identification of the molecules involved in coordinating signals between integrins and growth factor receptors will be important for generating targets that can be used to inhibit tumor growth.

It has also become apparent that integrins do not act alone in conveying signals from the extracellular matrix into the cell. A number of integrin-associated proteins and coregulators have been identified and their role in mediating integrin functions is currently being characterized (Hemler, 1998; Petit and Thiery, 2000). There are three general categories of integrin regulatory proteins: intracellular proteins that interact directly with integrin cytoplasmic domains; proteins that span the membrane and interact with integrins through their extracellular domains, referred to as integrin-associated proteins; and proteoglycan-like molecules that interact with specific domains on extracellular matrix molecules and assist or augment adhesion and signaling by the integrin, the syndecans being the most well characterized (Woods and Couchman, 1998). Specific integrin-associated proteins appear to interact with specific subsets of integrin subunits to affect the function of those integrins. For instance the transmembrane protein CD47 appears to primarily regulate $\alpha v\beta3$ and $\alpha IIb\beta3$ integrins by enhancing their adhesive and signaling functions, whereas CD151 and CD81 preferentially interact with $\alpha 3\beta 1$ or $\alpha 6\beta 1$ to regulate cell motility and neurite outgrowth (Hemler, 1998; Stipp and Hemler, 2000).

Syndecan-1, -2, and -3 are expressed in epithelial, fibroblast, and neuronal cells, respectively. Syndecan-4, on the other hand, is more widely expressed, but at lower levels. The effects of syndecans on integrin function depends on the cell type and type of matrix or biological response being measured (Woods and Couchman, 1998). Integrin-associated proteins and syndecans are also capable of regulating a subset of the signaling events that are attributable to integrin engagement. For instance, both syndecan-4 and several integrin-associated proteins, CD151, CD81, and CD82, associate with and regulate integrin activation of the PKC family (Zhang *et al.,* 2001). Other signaling molecules regulated by integrin-associated proteins include PI-4K, Src, and Rho (Woods and Couchman, 1998). The presence of these "cofactors" adds to the complexity of integrin signaling and it is therefore important to consider what their contributions to overall integrin function might be.

II. Applications

When trying to understand the nature of different biological processes it is often necessary to understand the underlying signaling pathways and their role in regulating cell functions. Since integrins are an integral part of many cell functions, cell matrix adhesion assays have increasingly become necessary for deciphering the contributions of integrin-mediated signaling events to the overall function and physiology of the cell. The specific integrins that are expressed and engaged in any given situation contribute to the overall response of a cell to a specific stimulus. For example, cells expressing $\alpha v\beta 3$ integrin plated on a vitronectin matrix will migrate, but if they express $\alpha v\beta 5$ integrin and are plated on the same matrix, they will not migrate unless the PKC pathway is simultaneously stimulated (Lewis *et al.,* 1996). In some cell types, plating on vitronectin induces a different set of signaling events than those induced by plating on fibronectin (Felsenfeld *et al.,* 1999). Similarly, stimulation of cells plated on vitronectin with a growth factor can induce cell migration, but if the cells are plated on fibronectin the same growth factor will have no effect (Borges *et al.,* 2000). To determine how these different biological effects are mediated requires an understanding of the underlying signaling events.

The techniques for studying signaling transduction pathways were first developed using standard adherent cell cultures that were treated with different soluable stimuli, such as hormones, growth factors, neuropeptides, and cytokines. Many of those same techniques are applicable to studying integrin-mediated signaling. This chapter will outline some of the general approaches to studying signal transduction and how they can be applied to studying integrin-mediated signaling in particular. The most important steps involve preparation of the cells and matrix-coated surfaces and the selection of proper controls. Most techniques require only some minor changes to accommodate the differences in studying signaling events mediated by cell adhesion vs those mediated by soluble factors.

===== ## III. Methods

A. Matrix-Coated Plates

The most common extracellular matrix materials utilized for integrin-mediated adhesion assays are collagen I, collagen IV, fibronectin, laminin, vitronectin, and fibrinogen. Other proteins that can be used include VCAM, ICAM, tenascin, and osteopontin. For most extracellular matrix molecules the concentration used for coating plates can vary from 50 μg/ml for fibrinogen to 5 μg/ml for fibronectin. Depending on the cell type this may need to be empirically determined, but for most cells in culture 10 μg/ml is usually adequate if the cells express normal levels of the target integrin. As with most receptor–ligand interactions, integrin–matrix interactions are concentration dependent. Changes in matrix concentration can affect the speed and affinity of cell adhesion. Table I lists the integrin subunits, their respective ligands, and expected expression in various cell types. The choice of cell type and matrix protein will determine which specific integrin will be stimulated by adhesion. A simple adhesion assay using several different matrix proteins can provide a quick screen for determining which integrins might be expressed on a particular cell. However, because of the ability of several integrins to bind the same matrix protein, it is important to confirm the presence of a specific integrin either by immunoblotting or by detection of cell surface expression by FACS.

Table I
Integrins and Ligands

Subunits	Ligands	Cell types
$\alpha1\beta1$	Collagen/laminin	Endothelial, smooth muscle, hepatocytes
$\alpha2\beta1$	Collagen/laminin	Endothelial, epithelial
$\alpha3\beta1$	Laminin	Endothelial, epithelial
$\alpha4\beta1$	Fibronectin/VCAM-1	Neural crest, lymphocytes
$\alpha5\beta1$	Fibronectin	Fibroblasts, endothelial
$\alpha6\beta1$	Laminin	Epithelial, endothelial
$\alpha7\beta1$	Laminin	Skeletal muscle, melanoma
$\alpha8\beta1$	Fibronectin	Epithelial, brain
$\alpha9\beta1$	Tenascin	Epithelial, smooth/skeletal muscle
$\alpha_v\beta1$	Vitronectin/fibronectin	Epithelial, osteoblasts
$\alpha_L\beta2$	ICAM-1,2,3	Lymphocytes
$\alpha_M\beta2$	iC3b/fibrinogen	Macrophage
$\alpha_X\beta2$	Fibrinogen/iC3b	Macrophage, lymphocytes
$\alpha_{IIb}\beta3$	Fibrinogen/FN/VN	Platelets
$\alpha_V\beta3$	Vitronectin/FN	Endothelial, fibroblasts
$\alpha6\beta4$	Laminins	Epithelial
$\alpha_V\beta5$	Vitronectin	Epithelial, fibroblasts
$\alpha_V\beta6$	Fibronectin/TN.III3	Epithelial
$\alpha4\beta7$	Fibronectin/V-CAM	Lymphocyte
$\alpha_E\beta7$	E-cadherin	Lymphocytes
$\alpha_V\beta8$	Fibronectin/laminin	Kidney, brain

Most extracellular matrix components can be purchased commercially from Gibco or Becton-Dickinson. Resuspend lyophilized components at 0.5 mg/ml in 1× PBS lacking Ca^{2+}/Mg^{2+} (Gibco). Fibrinogen should be made up at 5 mg/ml. It may take several minutes to dissolve and there may be some fraction that remains insoluble. Working size aliquots should be made and stored at $-20°C$ or $-80°C$ according to the manufacturer. Repeated freeze–thawing is not recommended. Any size tissue culture plate, from 24-well multiwell plates to 150-mm plates, can be coated with extracellular matrix materials. Glass coverslips or chamber slides can also be coated. For the biochemical analyses required for signaling transduction assays, a 100-mm plate works best. Depending on the cell type and density, a 100-mm plate can yield 500–1000 μg of cell extract, enough material for one to two immunoprecipitations.

For coating 100-mm plates, dilute the matrix solution to 5–50 μg/ml into 5 ml of 1× PBS per plate and ensure that the entire plate is covered. When increasing or decreasing plate sizes, adjust the volume relative to the surface area. Plates should be placed on a level surface to prevent uneven coating. Plates can be incubated overnight at 4°C or for 1–2 h at 37°C. Because the amount of matrix protein that binds to the plate is dependent on concentration, time, and temperature it is important to pick one method rather than switch back and forth so that all experiments remain consistent.

At least one hour prior to the time at which cells will be plated on the newly coated plates, suction off the PBS and wash the coated plates two times with 10 ml PBS each. Add 1% BSA in 5 ml PBS to each plate and incubate at 37°C for at least 1 h. The BSA acts to block any unbound sites on the plate and as a surface tension reducer during cell plating. This incubation can be done while the cells are being prepared.

B. Cell Preparation

In order to be sure that the biological and biochemical events being monitored are due solely to signals mediated by integrins, it is important to remove as many extracellular and exogenous stimulatory factors as possible. How this is accomplished may vary with the cell type under study. Cells should be serum starved at least 16 h prior to use. For some cell types this can be accomplished by simply omitting the serum from the growth medium; for others this may require a low concentration of serum in the medium, such as 0.1%. It is not necessary for the cells to completely arrest their growth, but only to reduce the serum growth factor levels and background signaling events. Plates that are nearly confluent, but not overgrown, work well. The number of cells required depends on the number of samples being generated, but a good rule of thumb is that for each 100-mm matrix-coated plate you will need two nearly confluent 100-mm plates of cells. This is because some cells are lost during processing and the percentage of cells that readhere is usually around 70%. The number of cells required may vary with cell type. Generally the more spread and less confluent the cells, the more plates will be needed.

After serum starvation, remove the medium and wash the cells 1 time with 10 ml 1× Ca^{2+}/Mg^{2+}-free PBS. It is very important at this stage to remove all sources of calcium, since calcium promotes integrin and cadherin-dependent cell adhesion. Remove the PBS and add 1 ml of PBS containing 0.01% trypsin and 5 mM EDTA to each plate.

Place the plates back into the incubator for approximately 5–10 min—until the cells are dislodged from the plate. Add 3 ml of PBS containing 5 mM EDTA and 1 mg/ml soybean trypsin inhibitor to each plate. Pipette up and down to break up clumps of cells until a single-cell suspension is achieved. If cells have a tendency to clump, a second wash with PBS prior to trypsinization and a longer incubation with trypsin might help. Starting with subconfluent plates is also helpful. Pool the cells into 50-ml centrifuge tubes. Rinse the plates with 2 ml more of the PBS/EDTA/STI solution to recover any remaining cells and pool with the others. Pellet the cells for 4 min at 130g. Remove the liquid and wash the cells with 50 ml PBS containing 5 mM EDTA. Do this by first resuspending the pellet in 1 ml until the pellet is loose and all cells appear to be in suspension. Then add 4 ml and remix to be sure all cells are resuspended, then dilute to 50 ml. Pellet again. Resuspend the cells in serum-free phenol red-free DMEM using the 1 ml, then 5 ml resuspension method. No phenol red should be present in the medium since it interferes with the protein assay. It is also important to use serum-free medium; amino acid supplements and antibiotics are fine. The final volume for resuspension will depend on the number of controls and matrix-coated plates. Use 3 ml of suspended cells for each matrix-coated plate and 2 ml for each suspension sample. Loosen the caps on the test tubes and place in the incubator for at least 30–60 min. Most cells can be incubated in suspension for up to 2 h without serious consequences.

Some cell lines, particularly primary cell lines, have a tendency to adhere to the walls of the centrifuge tube during this incubation in spite of the fact there is no serum or matrix coating. To avoid this, first coat the tubes with 1% BSA in PBS for 1 h at 37°C. Rinse them out with PBS before adding the cells. Some cell lines are also susceptible to apoptosis when they are removed from matrix (anoikis) and no serum is present (Frisch and Ruoslahti, 1997). It is important that these cell lines not be left in suspension for more than 1 h.

C. Adhesion Assay

Prior to plating the cells, remove the 1% BSA mixture from the matrix-coated plates and wash 2 times with 10 ml PBS each. Tilt the plates to remove any residual liquid, but do not "dry" the plates. Gently pipette 3 ml of cell suspension on the center of the plate and allow the cell suspension to slowly spread out on the plate. Let sit 1–2 min and then gently, without disturbing the cells, place the plates in the incubator. The small volume promotes quick settling of the cells onto the surface of the plate and helps to "synchronize" adhesion of the population. For quick time courses it is possible to set up the experiment in 6-well multiwell panels and centrifuge the cells at 100g onto the matrix to obtain better synchrony.

If the cells express the integrin required for adhesion to the matrix chosen, then cell adhesion can be observed within 10 min after plating; however, 15 min is usually more optimal. These times may vary by cell type and matrix and will have to be determined for each one. After adhering, many cells spread out, lose their refractility, extend processes or lamellipodia, and over the course of an hour dramatically change their morphology. Highly motile cells can also be seen migrating. For signal transduction assays most experiments are terminated within 1 h of plating. For cells that have lower levels of specific integrins it may take longer for cells to adhere and spread. Increasing the concentration of matrix protein for coating can sometimes shorten the time course. If cells do not express the integrin that recognizes the specific matrix they will not adhere and can easily be dislodged by gently shaking the plate.

It is important to not let the assay go for too long, no more than a couple of hours. Many cells synthesize and begin secreting their own matrix material several hours after plating, so any effects that might be due to the matrix originally selected will now be masked by the effects due to the new matrix. If the experimental design requires longer term analysis, synthesis and secretion of new matrix can be inhibited by pretreatment of the cells for 2 h prior to plating with 50 μg/ml cyclohexamide (Burridge *et al.,* 1992).

D. Controls

One of the critical aspects of adhesion assays is the selection of proper controls. For signal transduction assays, there is always a control in which similarly treated cells are not subjected to the primary signal-inducing event. For serum-starved, growth factor-stimulated cells this is usually a plate of serum-starved cells that do not receive the growth factor. For adhesion assays this kind of control is not acceptable for two reasons. Cells that are serum starved, but still remain adherent to the tissue culture plates, still have their integrins engaged with matrix that was deposited on the plate when the cells were growing. Good controls require that you subject the control cells to the same conditions as the samples, i.e., trypsinization, washing, and incubation in suspension.

Three types of controls have been routinely used in adhesion assays. The first is the use of plates coated with a non-integrin ligand, such as polylysine or polyhema. These are positively charged substances that promote cell adhesion through mechanisms involving proteoglycans, which were initially thought to be independent of integrins. However, more recent evidence suggests that in some cell types certain proteoglycans can enhance and cooperate with integrins in mediating cell adhesion, spreading, and signaling, making this a less optimal control (Woods and Couchman, 1998).

A second control that is sometimes used is to place cells onto a plate that has been coated only with 1% BSA. Most cells will not adhere under these conditions, generating suspended cells in a culture dish. This is a good control, except that in order to harvest the cells to make lysates, the cells have be recovered and centrifuged to remove the large volume of medium. Unfortunately this adds more manipulations to the control cells that the matrix-plated cells do not see. For some cell types the physical process of pipetting and centrifugation can be sufficient to initiate some cell-signaling events.

The most common type of control is to simply incubate the control cells in suspension in a test tube. This can be accomplished in several ways. The control cells can be left in the original 2-ml volume and incubation continued for the same amount of time as the matrix-plated cells. The cells will tend to settle to the bottom of the tube, which is all right as long as they do not adhere to the tube or to each other. To avoid nonspecific adhesion, the tubes can be precoated with 1% BSA, or the tubes can be rotated to keep the cells suspended. However, prior to generation of the cell lysates, the cells will have to be centrifuged to remove the excess liquid. Another method is to pellet the 2 ml of control cells after the 30–60 min preincubation period and resuspend them in a small volume, 0.1–0.3 ml, of the serum-free phenol red-free DMEM. Then incubate these for the same time as the matrix-plated cells.

It is advisable to test each cell type for which control works best. The aim is to reduce nonspecific signaling events that are not mediated by integrins so that integrin-specific events can be monitored. Another way in which to determine if the signaling events being monitored are integrin-specific is to pretreat the cells with integrin-blocking antibodies (Chemicon) prior to plating on matrix. Blocking antibodies are generally not 100% effective, but will reduce adhesion and signaling if the integrin being tested is truly involved. This method is particularly useful if two integrins can mediate cell adhesion to the same matrix protein and it needs to be determined which one is involved in generating a particular signal.

E. Cell Lysates

1. Introduction

Once good cell adhesion and/or cell spreading has been achieved, cell extracts need to be generated in order to analyze the biochemical signaling events that are stimulated by integrin engagement. There are numerous kinds of cell lysis buffers that can be used to generate cell extracts, the choice of which depends on what signaling events are being monitored and the antibodies being employed. For a rapid, extensive extraction of protein from both the cytoplasm and the nucleus, RIPA lysis buffer is the best. Because RIPA contains three detergents and moderate levels of salt it is effective at lysing all membranes and quickly inactivates proteases and phosphatases, enzymes that if not controlled can degrade the signals generated by adhesion. RIPA is also effective at extracting proteins from Triton-X "insoluble" structures such as focal adhesions and actin structures, commonly formed following cell adhesion. Many immunoprecipitating antibodies are also compatible with RIPA buffer, making it the buffer of preference. However, RIPA buffer will inactivate most enzymes and can break apart some protein complexes. If the activity of a particular signaling molecule is going to be measured or if preservation of protein–protein interactions is important, then less harsh lysis conditions employing the nonionic detergents NP-40 or Triton X-100 should be employed. Buffers containing Triton X-100 are also generally useful if the signaling molecules of interest are easily solubilized and you wish to enrich the extract for them. The total amount of protein extracted will be less, but 500 μg of Triton-X extract will contain more of the soluble signaling molecules than 500 μg of RIPA extract. A general-purpose Triton-X lysis buffer that is good for preserving enzymatic activity is the one used for analyzing MAPK kinase, MAPK lysis buffer. It should be noted that while MAPK lysis buffer may work for most enzymes, careful attention should be paid to the lysis buffer if enzymatic activity needs to be preserved. Many kinases and phosphatases require specific extraction conditions to maximally preserve activity. The literature should be consulted and the buffers used therein should be employed.

Table II
RIPA Lysis Buffer

Components	100 ml 2× stock
10 mM Tris pH 7.2	2 ml 1 M Tris pH 7.2
158 mM NaCl	6.3 ml 5 M NaCl
1 mM EDTA[a]	0.4 ml 0.5 M EDTA
0.1% SDS	2 ml 10% SDS
1% sodium deoxycholate	20 ml 10% NaDOC
1% Triton X-100	20 ml 10% Triton X-100
ddH$_2$O	49.3 ml ddH$_2$O

[a] Na$_3$VO$_4$ concentration cannot exceed EDTA concentration.

2. Lysis Buffers

All lysis buffers should be made up as 2× stocks and can usually be generated by diluting standard lab stocks for each of the components. In addition to salts and detergents, lysis buffers are also supplemented with several protease and phosphatase inhibitors. Some of these can be added to the 2× solution at the time the stock is made, but others, which are more labile, need to be added just prior to lysis. The components for RIPA and MAPK lysis buffers and their 2× stocks are shown in Tables II and III.

Just prior to lysis, 10 ml of the 2× lysis buffer stock should be supplemented with protease and phosphatase inhibitors as listed in Table IV.

Most of the inhibitor stocks are made up in water, except for pepstatin, which is made in DMSO : acetic acid (9:1), and PMSF, which is made up in methanol. All stock components for the buffers are usually stored at room temperature, except for the protease inhibitors, leupeptin, pepstatin, and aprotinin, which are aliquotted and stored at −20°C.

An effective preparation of sodium orthovanadate requires polymerization, which enhances its ability to prevent tyrosine dephosphorylation by tyrosine phosphatases and is achieved by boiling. Dissolve 9.2 g of sodium orthovanadate in 100 ml water. Bring

Table III
MAPK Lysis Buffer

Components	100 ml 2× MAPK stock
50 mM Tris, pH 7.5	10 ml 1 M Tris, pH 7.5
0.5 mM EDTA	0.2 ml 0.5 M EDTA
50 mM NaF	10 ml 1 M NaF
100 mM NaCl	4 ml 5 M NaCl
50 mM B-glycerolPO	10 ml 1 M B-glyceroPO
5 mM NaPyroPO	10 ml 0.1 M NaPyroPO
1% Triton X-100	20 ml 10% Triton X-100
ddH$_2$O	35.8 ml ddH$_2$O

Table IV
Lysis Buffer Inhibitors

Final 2× lysis buffer	Per 10 ml
2× lysis buffer stock	10 ml
1 mM Na$_3$VO$_4$[a]	40 μl 0.5 mM Na$_3$VO$_4$
1 mM PMSF	40 μl 0.5 mM PMSF
5 μg/ml leupeptin	20 μl 5 mg/ml leupeptin
5 μg/ml pepstatin	20 μl 5 mg/ml pepstatin
10 μg/ml aprotinin	20 μl 10 mg/ml aprotinin
1 mM benzamidine	20 μl 1 M benzamidine

[a] Na$_3$VO$_4$ concentration cannot exceed EDTA concentration.

the pH to 10 with HCl. This will turn the solution yellow. Boil the solution until it is clear and then bring back up to the original 100 ml volume. The solution will appear cloudy and will settle out, but this is normal. Be sure to resuspend the particulate matter prior to use. When diluted into the lysis buffer it will dissolve. The use of a concentrated sodium orthovanadate solution prevents unnecessary dilution of the lysis buffer. When adding sodium orthovanadate to different types of lysis buffers make sure that the level of EDTA in the buffer does not exceed 1 mM, since EDTA is inhibitory to sodium orthovanadate action.

3. Lysis Protocol

a. Matrix-Adherent Cells

Gently rock the plates containing adherent cells, to dislodge any cells that have not adhered to the plate, and suction off the nonadherent cells and liquid. Tilt the plates to let any residual liquid flow to one side and gently suction it off. Then lyse the adherent cells on a 100 mm plate in 500 to 1000 μl of ice-cold 1× lysis buffer, prepared by diluting the final 2× buffer 1:1 with water. The volume used for lysis will depend on the number of adherent cells and the desired protein concentration of the cell extract, usually around 1 mg/ml. Immediately place the plates on ice or at 4°C.

RIPA lysis extracts are usually collected after a 15-min incubation at 4°C. For Triton-X buffers, lysis should be done for 30 min, to be sure of maximum solubilization. Using a cell lifter or plate scraper, collect the lysates into a microfuge tube and place on ice. RIPA extracts will be viscous because of lysis of the nucleus and release of DNA into the extract. RIPA extracts have to be passed through a 25-gauge needle on a 1–2 cc syringe 2–3× to break up the DNA. This will generate some foaming but is unavoidable. Place all samples into a cold microfuge and spin at 13,000g for 10 min to remove unsolubilized components.

At this stage all the samples can be quick frozen in liquid nitrogen and stored at −80°C until needed or they can be used immediately for biochemical analyses. However, if kinase assays are going to be performed (see Section III.F.3.b), it is better to not freeze the lysate, but to continue with the assay.

b. Suspension Cells

Remove the suspension cells from the incubator and place on ice. To lyse the suspension cells that were placed in a small volume (0.1–0.3 ml) of DMEM prior to incubation add an equal volume of 2× lysis buffer to the cells and then add twice the volume of 1× lysis buffer. It is necessary to dilute the DMEM 4× in order to prevent interference in the protein assays.

If cells were placed on BSA-coated plates, gently collect the cells by rocking the plates and collecting with a pipette. Place the cells in a centrifuge tube on ice. If a large volume of DMEM was used for cells in suspension, place those tubes on ice. Incubate all tubes for 5 min, to chill the cells and stop any signaling events. Pellet 5 min at 4°C and 130g and then suction off the excess liquid. Lyse cells in 500–1000 μl of 1× lysis buffer and transfer to microfuge tubes. Process samples as described in Section III.E.3.a.

c. Alternative Lysis Methods

Occasionally a lysis protocol may require a hypotonic lysis step, especially if cell fraction experiments are done to determine the localization or relative abundance of a protein within specific cell compartments. Because the cells are in culture medium that has a normal isotonicity it will be necessary to remove the medium and wash the cells at least once with PBS—preferably with Ca^{2+}/Mg^{2+}-containing PBS to prevent disruption of cell adhesion. Similarly suspension cells will have to be pelleted and washed once with PBS to remove the medium. A lysis volume of 1 ml or more is highly recommended to make sure the salt concentration is low enough to allow swelling of the cells. Cells can then be broken up by Dounce homogenization or repeated freeze–thawing.

4. Protein Assay

Once cell extracts have been generated, it is necessary to determine the protein concentration. There are two protein assay reagents that work well, the Bradford assay (BioRad) and the Bicinchoninic acid assay (Pierce). The Bradford Assay is based on Coomassie blue dye binding to protein. The Bicinchoninic acid assay is based on the reduction of Cu^{2+} to Cu^+ when proteins are present in an alkaline solution. The Bicinchoninic acid assay is the preferred method for RIPA extracts since SDS in the lysis buffer can interfere with the Bradford assay. The disadvantage for the Bicinchoninic acid assay is that it requires a 30-min incubation, but it is otherwise a more linear assay over a broader range of concentrations.

a. Bradford Assay

Instructions on how to set up the Bradford assay are supplied with the reagent. The expected range of concentration in the cell extracts should be between 1 and 2 mg/ml. The "micro" assay works the best using 0, 2, 4, 6, 8, 10, 12, and 15 μg BSA standards and 5 μl of cell extract in 800 μl of water and 200 μl of reagent. A few minutes after mixing read the OD in a spectrophotometer set at 595 nm.

b. Bicinchoninic Acid Assay

For this assay 0, 5, 10, 20, 30, and 40 μg BSA standards and 10 μl of cell extract in 50 μl of water and 1 ml of reagent mixture are adequate. Incubate the reaction at 37°C for 30 min. It is necessary to generate a standard curve each time, since slight changes in incubation temperature and time can shift the curve. After the 30-min incubation read the OD at 562 nm.

Generate a standard curve and calculate the protein concentration of each sample taking into account the dilution factor and volumes.

F. Biochemical Analysis

1. Immunoprecipitation

Immunoprecipitation is used to isolate a particular protein of interest from the rest of the cell extract in order to study its properties. Some of the properties of signaling molecules that are often examined include tyrosine phosphorylation state (an indicator of activation), actual activity, other signaling molecules or proteins that associate with it, or a change in cellular localization. These events are usually compared to control cells to determine which events are due solely to the stimulus, i.e., cell adhesion.

Depending on the abundance of the molecule of interest, or the affinity of the antibody being used, 500–1000 μg of cell extract is usually sufficient for most applications. Which property of the molecule is being studied will determine the conditions for lysis, but once the lysate is generated, the immunoprecipitation protocol is the same. When doing immunoprecipitations it is important that the concentration of protein in the lysate not be too dilute or too concentrated; 1 mg/ml is generally recommended. There are some rare exceptions where inhibitory activities in an extract make diluting it to 0.5 μg/ml better or where the expression of the molecule of interest is so low that 2 mg/ml is required. The cell lysates should be diluted to 1 mg/ml with the same lysis buffer in which it was generated, being careful to include all the proper protease and phosphatase inhibitors. All manipulations should be carried out on ice or at 4°C since any warming of the sample can activate proteases and phosphatases, even in the presence of inhibitors.

The most commonly measured integrin-induced signaling event is the level of increase in tyrosine phosphorylation of a particular signaling molecule or its downstream effectors. Integrin engagement leads to the activation of many molecules by tyrosine phosphorylation, including FAK, paxillin, Cas, Shc, PKCδ, tensin, cortactin, talin, and vinculin. The ideal approach for studying tyrosine phosphorylation of these molecules is to select an antibody generated in rabbits that is good for immunoprecipitation (Santa Cruz). This can be in the form of rabbit sera or purified antibody, the latter being better since it reduces the background. Sometimes a rabbit antibody is not always available and some mouse monoclonal antibodies are good for immunoprecipitation (Becton Dickinson). One μg of purified rabbit or monoclonal antibody or 1 μl of rabbit antiserum per sample is usually adequate. Some manufacturers will provide recommendations, while others require that you determine the optimal amount.

When testing out a new antibody for the first couple of times it is important to include a couple of antibody controls. One control is to add either preimmune sera or nonimmune immunoglobulin to one sample. This will control for any proteins that might nonspecifically adhere to immunoglobulin or protein A. Another control is to do a competition by preincubating some of the antibody with a peptide or protein to which the antibody was generated. Some antibody suppliers will offer the peptide to which the antibody was made. This should block specific binding of the antibody to protein in the cell lysate and help to confirm that the protein seen on the Western immunoblot is the correct one.

The principle behind immunoprecipitation is that protein A, a cell surface protein found on a particular strain of bacteria, is used to recover the protein–antibody complexes because of its high affinity for the Fc domain of some immunoglobulins. All rabbit immunoglobulins bind to protein A efficiently; however, most mouse immunoglobulins do not. There are two ways to get around this problem. One is to add a secondary antibody made in rabbit directed against mouse immunoglobulin to the immunoprecipitation. If 1 μg of mouse antibody is used for immunoprecipitation, then adding 1 μg of purified rabbit anti-mouse antibody to the cell lysate is usually sufficient. The second approach is to use protein G instead of protein A, which binds mouse immunoglobulin better than protein A. If the molecular weight of the protein molecule being examined falls within the range of the molecular weight of the two immunoglobulin chains, $\sim$60 kDa or $\sim$20 kDa, and a mouse antibody is being used, then protein G would be the method of choice since it will reduce the interference by immunoglobulins during visualization of the molecule upon Western immunoblotting. Another way to reduce this latter problem is to use a different species of antibody for immunoprecipitation than for Western immunoblotting or to use the primary immunoprecipitating antibody already conjugated to beads. Some manufacturers will provide antibodies in this form.

Protein A can be purchased in two forms, either on the surface of formalin-fixed bacteria, or conjugated on agarose or sepharose beads (Pierce). The bacteria are more difficult to resuspend and wash and can produce higher backgrounds, but are much cheaper than the conjugated beads. Protein G is available conjugated to beads (Pierce). Usually 30 μl of a 50% slurry/suspension is used for each immunoprecipitation sample. The protein A or G suspension needs to be washed 3 times with 1$\times$ lysis buffer to remove preservatives before it can be added to the cell lysate. In order to ensure uniform distribution of the beads into the cell lysates, cut the narrow tip off the end of the pipette tip so that the beads are easier to pipette. Keep the beads suspended in the tube by vortexing between each sample when adding beads to consecutive samples. Once the antibodies and protein A or protein G are added to the cell lysates, the tubes are incubated at 4°C on a rotation device to keep everything suspended to aid in capture of the complexes. Incubation times can vary from 1 h to overnight. However, long-term incubations increase the risk of protein cleavage and phosphate loss due to proteases and phosphatases. A 3-h incubation is usually sufficient to capture most of the protein of interest without too much risk of losing the protein or its modifications.

At the end of the incubation, the protein A complexes should be pelleted, about 1 min in a cold microfuge, and washed 3 times with 0.5–1.0 ml of lysis buffer to remove all nonbound proteins. After the last wash, all the liquid is removed from the pellet (use a gel loading tip for this to prevent accidental loss of beads) and the pellet is resuspended in 30 μl of 2× SDS–PAGE sample buffer and boiled for 5–10 min to release the protein/antibody complexes from protein A. If the protein of interest has a molecular weight around 50–60 kDa, then it is preferable not to boil the sample, so as to not reduce the antibody into its two components, a 50-kDa and a 20-kDa form that will interfere with visualization of the protein during Western immunoblotting. Instead the samples should be incubated at room temperature for 15–30 min to allow removal of the complexes from the beads. The samples can now be loaded onto SDS polyacrylamide gels for separation and Western immunoblotting.

2. Western Immunoblotting

After running the immunoprecipitates out on an appropriate percentage SDS polyacrylamide gel using standard SDS–PAGE procedures, the proteins on the gel are transferred to immunoblotting paper. Two types of blotting membrane are commonly used, nitrocellulose or PVDF. PVDF is superior because the background is lower, but is more expensive and requires a methanol prewetting step. There are several different conditions that can be used for gel transfer, but the standard is Tris/glycine (25 mM Tris, 192 mM glycine) with 20% methanol (no SDS), which suits most applications. Follow the manufacturers directions for the particular apparatus being used to transfer. Those models with solid plate electrodes as opposed to wire electrodes are far superior in their efficiency and speed of transfer.

The single most important step in Western immunoblotting is blocking the membrane to prevent nonspecific binding of the antibodies. It is really difficult to "overblock" a blot, so if in doubt go for longer and more if possible. Two different blocking agents are commonly used, dry milk and BSA. Dry milk is cheap and very effective, especially if used at 5%. However, dry milk contains phosphatases and will cleave off the phosphates that have been so painstakingly preserved. So when immunoblotting for tyrosine or serine/threonine phosphorylation it is imperative to use BSA. The best blocking is achieved with 5% BSA, though to conserve on an expensive reagent some blocking is done with 1–3% BSA.

The blocking agent is typically made up in TBST, a Tris-buffered saline containing Tween-20 (10 mM Tris, pH 7.5, 150 mM NaCl, 0.5% Tween-20). BSA, 5 g per 100 ml, should be allowed to dissolve slowly into the buffer and not be stirred or mixed to prevent denaturation. Once dissolved sterilize the solution by filtering through a PES membrane filter and store at 4°C. This removes any nondissolved particles and prevents bacterial growth. The 5% BSA solution used for blocking can be collected and reused during the secondary antibody incubation step to help conserve on reagent. For dry milk, dissolve in TBST buffer just prior to use and discard after use since it is not stable or filterable.

Incubate the blots in blocking agent for a minimum of 1 h at room temperature. Two hours is better. Alternatively, blocking can be done overnight at 4°C. Incubation requires a tilting platform or rotation device to ensure equal coverage over the blot. The use of sealing bags is recommended for all incubations to minimize the volume needed and to prevent waste. Roughly a minimum of 1 ml of blocking agent is required for each 10 cm^2 of blot if using a sealing bag. Larger volumes will be necessary if using open trays.

The primary antibody used for immunoblotting will depend on what is being analyzed. To detect tyrosine phosphorylation a good monoclonal antibody that recognizes most phosphotyrosine residues is 4G10 (Upstate Biochemical). Others include PY20 and PY99 (Santa Cruz Biotechnology) or PY69 (Becton Dickinson Transduction Labs). For immunoblotting, most purified antibodies can be diluted to a final concentration of 1–0.1 μg/ml. This will vary with different antibodies and most suppliers will provide a suggested dilution. When in doubt, a 1:1000 dilution is a good place to start. For antisera a 1:2000 dilution or greater is recommend to reduce the background which is inherent when using antisera. The primary antibody solution is made up in TBST containing 5% of the blocking agent; for 4G10 use BSA. Again a minimum of 1 ml per 10 cm^2 of blot is required if using sealing bags. If BSA is used to make up the primary antibody and 0.02% sodium azide is added, it can be collected afterwards and reused many times. It can also be refiltered if it does become contaminated. This saves on BSA and antibody and is highly recommended. Incubate the blot with primary antibody for 2 h at room temperature. Longer incubations typically lead to an increase in background. Overnight incubations at 4°C can also be done. After removing the primary antibody wash the blots 4 times over the course of an hour. Western wash buffer (50 mM Tris, pH 7.5, 150 mM NaCl, 0.2% NP-40) is recommended.

The choice of secondary antibody depends on what the primary antibody was and what detection system is being used. The secondary antibody is usually an antibody that was generated to either mouse or rabbit immunoglobulin that is conjugated to an enzyme, such as horseradish peroxidase (HRP) or alkaline phosphatase (AP). On rare occasions an enzyme-linked primary antibody can be employed, but sensitivity is increased when using a two-step procedure. The conjugate of choice is HRP, because of the superiority of the chemiluminescence detection reagent, ECL. Cleavage of H$_2$O$_2$ by peroxidase releases free radicals that excite a chemofluor to generate light emissions from antibody bound to the blot that can then be captured on X-ray film. The AP reaction generates a dark precipitate on the surface of the blot where the antibody is bound.

The secondary antibody, like the primary antibody, is diluted in TBST containing 5% BSA. The standard dilution is usually 1:5000 to 1:10,000 depending on the supplier. This antibody solution can only be used one time and cannot be preserved with sodium azide because it will inhibit the peroxidase reaction. The blot is incubated with secondary antibody for no more than 1 h. After removal of the secondary antibody the blot is again washed 4× over an hour with Western wash buffer.

The ECL reagent is available commercially but is very easy and cheap to make and is usually more potent. To make solution 1 add 500 μl of 250 mM luminol (Sigma #A-8511) and 220 μl of 90 mM p-coumaric acid (Sigma #C-9008) to 50 ml 100 mM Tris, pH 8.6. The luminol stock and solution 1 are light sensitive and should be placed in a foil-covered or amber bottle. To make solution 2 add 30 μl of fresh 30% H$_2$O$_2$ to 50 ml of 100 mM Tris, pH 8.6. Solutions 1 and 2 are stable at 4°C for at least a month. For optimal results it is best to replace the 30% H$_2$O$_2$ stock at least once a year.

To initiate the ECL reaction, mix equal volumes of solution 1 and solution 2 in a flat-bottom tray; 3–5 ml of each is usually sufficient. For blots that are likely to produce a strong signal it may be necessary to dilute the ECL reaction further by adding an equal volume of water. Immediately place the blot into the ECL solution and incubate for 1 min, tilting the tray throughout and removing air bubbles to ensure even coverage. Place the blot on Saran wrap and cover, being careful to remove air bubbles. Place in an X-ray film cassette and expose to X-ray film for 25 s, 1 min, and 4 min. If the 4-min exposure is too light, a 10-min exposure can be taken. The ECL reagent is light sensitive and will start to degrade after 30 min. Blots can be reincubated in fresh reagent and exposed for up to 30 min if necessary, but usually if the antibodies are working a signal should be visible in 10–15 min. It is important to generate several exposures, since overexposure can saturate the film and produce a nonlinear result.

After the desired exposures are obtained wash the blot for 10 min in wash buffer to wash off excess ECL reagent. The blot can be stored in wash buffer for short-term storage (2–3 days), but should be stored in a very small amount of buffer in sealed bags for long-term storage. The 4G10 blot should be stripped and reprobed for the total levels of protein that were present in the immunoprecipitation. The antibody that was originally used for the immunoprecipitation can be used if it works on blots. This is where the use of two different species of antibodies works to an advantage. If the tyrosine antibody is made to one species (mouse), then when the blot is reprobed with a different species of antibody (rabbit), any mouse primary antibody that might not have been efficiently stripped off the blot will not be seen when probed with the rabbit secondary antibody, increasing the probability that what is being seen is due to the second antibody.

There are several methods that can be used to strip Western blots. They involve using low pH, high pH, high temperature and/or detergent to remove the first and secondary antibodies from the blot. Invariably a small amount of the protein is also lost. Repeated stripping is not advised. Incubate the blot in 50–100 ml of stripping buffer (67.5 mM Tris, pH 6.8, 2% SDS, 100 μM 2-mercaptoethanol) at 65°C with rocking for 1 h in a covered container. Caution should be taken when using this much 2-mercaptoethanol. Rinse the blot a couple of times with Western wash buffer and wash 2–3× for 10 min each to remove the stripping buffer. There is usually no need to reblock the blot. Proceed by incubating with primary antibody and following the standard protocol from there on. Another method is to incubate the blot in water for 5 min, then add 0.2 N NaOH for 2–3 min, immediately followed by two 5-min washes in water. This may turn the blot light brown, but does not appear to affect subsequent blotting steps.

3. Other Assays

a. Phospho-Specific Antibodies

The recent development of phospho-specific antibodies has simplified the detection of signaling-induced phosphorylation events. It is now possible to use the total cell lysates generated from the adhesion assay and run them out on a gel without first immunoprecipitating the molecule of interest. Depending on the size of the gels being used, 30 μg for a mini-gel up to 100 μg of cell extract on a large gel is sufficient for detection of phosphorylation events on specific molecules. When running out cell lysates on a gel, first dilute the amount of protein desired into an equal volume of 2× SDS–PAGE sample buffer. If running several samples at the same time that are of different concentrations it is best to dilute them to the same concentration with 1× lysis buffer first before adding the 2× sample buffer. That way all the samples are of the same volume and any differences in band intensity can be attributed to differences in protein phosphorylation or concentration and not to differences in loading and running condition artifacts that can be generated when different volumes of samples are loaded on the same gel.

Once the gel is transferred to a membrane, the same immunoblotting procedures used for the 4G10 blots should be followed, only substituting the phospho-specific antibody of interest for 4G10. A blot for total levels of the protein should also be done after probing for specific phosphorylation, just to make sure the level of protein in the samples did not change. Integrin-stimulated signaling molecules for which phospho-specific antibodies are commercially available include MAPK, Akt, Jnk, Src, FAK, paxillin, MEK, and PLCγ (Biosource International; New England BioLabs). As more phosphorylation sites are mapped on other molecules and their role in regulating signaling activity is defined, more antibodies will become available.

b. Kinase Assays

Another way in which to measure the activation of some signaling molecules, particularly those that are kinases, is to directly measure their activity in an in vitro kinase assay. The exact condition used for each kinase is usually quite specific and the literature should be consulted for the exact details. The basic principle is that the kinase is immunoprecipitated from cell lysates using the same procedure outlined in Section III.F.1. For some kinases it is necessary to shorten the time of immunoprecipitation to 1–2 h instead of 3. After the last wash with lysis buffer the pellets are then washed 2× with kinase buffer, which usually contains no detergent and low or no salt and Mg^{2+}. The pellet is resuspended in a small volume, 30–50 μl, of kinase buffer. Some ATP and a small amount of radioactive ^{32}P-γ-ATP are added to the pellet along with a few micrograms of appropriate substrate. The reaction is usually incubated at 30°C for 10–30 min and then 2× SDS sample buffer is added to quench the reaction. It can then be boiled and loaded on an SDS–PAGE gel. The gel can be transferred to blotting paper and exposed to film to detect radioactive incorporation of ATP in the substrate and then subsequently blotted with antibody to the kinase to determine if the amount of enzyme in each reaction is the same. For some assays the reactions are quenched with EDTA and then spotted onto charged filter paper; the paper is washed to remove unincorporated radioactivity and then counted in a scintillation counter (Reuter *et al.*, 1995). This only works if the substrate is highly charged and has the disadvantage that it cannot determine the levels of enzyme in each reaction.

c. Coimmunoprecipitation

Immunoprecipitation can also be used to determine if integrin-induced signaling stimulates the formation of specific complexes between different types of molecules, referred to as coimmunoprecipitation. By immunoprecipitating one molecule with one antibody and then probing the Western immunoblot with antibody to another molecule, the ability of these two molecules to associate with each other can be monitored. Several factors influence the success of this approach. The type of lysis buffer used is important. Some protein complexes are very stable to buffers as harsh as RIPA, whereas others are more easily captured if lower amounts of detergents and different types of detergents are used. The antibody being used is also important. If the antibody has been generated by using peptides that fall within the domain of the protein that interacts with another, then the antibody may disrupt their association. For interactions that have not previously been demonstrated it is important to be sure that the interaction is specific. One way to do this is to reverse the procedure: immunoprecipitate with the second antibody and probe the immunoblot with the first. If the interaction is still observed, then this is a good indication these two molecules may specifically interact *in vivo*. Another important control is to do the competition experiment (Section III.F.1) using a peptide that blocks the immuno-precipitating antibody and show that when you lose binding of the first molecule to the antibody, the second molecule is also blocked and thus its appearance is not due to a nonspecific association with immunoglobulin or protein A. It should also be emphasized that just because two molecules coimmunoprecipitate does not necessarily mean that these molecules actually physically interact with each other. It simply means they can be found in a complex together.

d. Drug Inhibitors

Identifying which molecules are activated following integrin-mediated adhesion is only a part of understanding the signaling process. The more interesting aspect is to determine how the signals are transmitted and what their targets are. The first step is to take a broad approach and screen for the involvement of specific known signaling pathways. The recent availability of more specific inhibitors for a number of signaling molecules has made this approach more useful (Calbiochem, Sigma). Table V lists some integrin-regulated molecules, their specific inhibitors, and effective concentrations.

For cell adhesion assays the timing of drug treatment can be complicated by the need to put cells into suspension prior to plating on matrix. This can be solved by extending the time of preincubation in suspension or pretreating with the inhibitor prior to removing the cells from the plate. Many inhibitors are cell permeable and act fairly quickly to inactivate their targets. For these inhibitors simply add them to the cells after their 30-min incubation in suspension (Section III.B) and incubate for an additional 15 min. The inhibitors remain in the cell mixture during plating on the matrix-coated plates to block activation of their targets. Some inhibitors, such as cyclohexamide, actinomycin, and phorbol esters, require several-hour to overnight incubations to effectively remove their targets from the cells. In this case add the drug to the serum-starved cells and incubate for the appropriate time. After the cells are harvested and placed in suspension another dose of the drug should be added prior to incubation to ensure that the inhibitor is still effective.

The advantage of this approach is that it can be used to determine what signaling molecules or pathways might be important for regulating other molecules or biochemical events as well as any of the biological events associated with cell adhesion, such as cell spreading, migration, gene expression, cell cycle, or survival. To analyze the

Table V
Commonly Used Signal Transduction Inhibitors

Drug	Target	IC$_{50}$ *in vitro*	*In vivo* dose	Other targets
LY294002[a]	PI-3K	1.4 μM	10 μM	
Wortmannin[b]	PI-3K	5 nM	50 nM	MLCK at 100×
PP2[c]	src family	5 nM	5 μM	PDGFR
SU6656[d]	src, fyn, yes, lyn, abl	0.3 μM	2 μM	
PD98059[e]	MEK		10 μM	
U0126[f]	MEK 1, 2	72 nM	2 μM	p38 at 500×
AG1478[g]	EGFR	3 nM	0.5 μM	
PD168393[h]	EGFR	700 pM	2 μM	
PD169316[i]	p38	89 nM	10 μM	
SB202190[j]	p38	8 nM	10 μM	
Bisindolylmaleimide[k]	PKC$^{\alpha,\beta,\gamma,\delta,\varepsilon}$	10 nM	5 μM	PKA at 200×
Chelerythrine Chloride[l]	PKC	660 nM	4 μM	
Y27632[m]	Rho kinase	140 nM	10 μM	PKC$^{\zeta}$ at 50×
KT5770[n]	PKA	60 nM	10 μM	
ML-7[o]	MLCK	300 nM	5 μM	PKA, PKC at 100×
KN62[p]	CaMKII	900 nM	2 μM	
Okadaic acid[q]	PP1, PP2A	10,0.1 nM	1 μM	PP2B at 500×
Calyculin A[r]	PP1, PP2A	2 nM, 1 nM	0.3 μM	
3,4-Dephostin[s]	PTPase	18 μM	1 μg/ul	

[a] Vlahos, C. J., *et al.* (1994). *J. Biol. Chem.* **269**, 5241–5248.
[b] Yamamoto-Honda, R., *et al.* (1995). *J. Biol. Chem.* **270**, 2729–2734.
[c] Hanke, J. H., *et al.* (1996). *J. Biol. Chem.* **271**, 695–701.
[d] Blake, R. A., *et al.* (1999). *Mol. Cell. Biol.* **20**, 9018–9027.
[e] Pang, L., *et al.* (1995). *J. Biol. Chem.* **270**, 13585–13588.
[f] Favata, M. F., *et al.* (1998). *J. Biol. Chem.* **273**, 18623–18632.
[g] Eguchi, S., *et al.* (1998). *J. Biol. Chem.* **273**, 8890–8896.
[h] Fry, D. W., *et al.* (1998). *Proc. Natl. Acad. Sci. USA* **95**, 12022–12027.
[i] Kummer, J. L., *et al.* (1997). *J. Biol. Chem.* **272**, 20490–20494.
[j] Kramer, R. M., *et al.* (1996). *J. Biol. Chem.* **271**, 27723–27729.
[k] Toullec, D., *et al.* (1991). *J. Biol. Chem.* **266**, 15771–15781.
[l] Herbert, J. M., *et al.* (1990). *Biochem. Biophys. Res. Commun.* **172**, 993–999.
[m] Hirose, M., *et al.* (1998). *J. Cell. Biol.* **141**, 1625–1636.
[n] Gadbois, D. M., *et al.* (1992). *Proc. Natl. Acad. Sci. USA* **89**, 8626–8630.
[o] Saitoh, M., *et al.* (1987). *J. Biol. Chem.* **262**, 7796–7801.
[p] Tokumitsu, H., *et al.* (1990). *J. Biol. Chem.* **265**, 4315–4320.
[q] Gjertsen, B. T., *et al.* (1994). *J. Cell Sci.* **107**, 3363–3377.
[r] Bennecib, M., *et al.* (2001). *Biochem. Biophys. Res. Commun.* **280**, 1107–1115.
[s] Fujiwara, S., *et al.* (1997). *Biochem. Biophys. Res. Commun.* **238**, 213–217.

relationships between different signaling events, treat the cells with an inhibitor of one signaling molecule and then monitor the effect on the ability of another molecule to be activated, using the immunoprecipitation and immunoblotting procedures outlined in Sections III.F.1 and III.F.2. Using this approach the relationships between different signaling molecules can be ascertained and signaling pathways characterized.

It should be cautioned that the use of inhibitors, although effective for screening for relationships between signaling events and biological events, should not be the final test. While most inhibitors exhibit a good deal of specificity, there always exists the possibility that the drug is acting on an unknown molecule or on a molecule whose response to the drug is unknown. This can be overcome in some cases by the use of an alternative inhibitor, especially if it inhibits by a different mechanism. However, the best approach is to use molecular genetic techniques to more fully characterize the proposed pathways.

e. GTPases

GTPases are small molecular weight signaling proteins that when activated bind GTP and have intrinsic GTPase activity that cleaves GTP to GDP and inactivates them (Takai *et al.*, 2001). Members of the Ras and Rho family of GTPases are regulated by integrin-mediated adhesion. These molecules are not typically phosphorylated, so their activation is monitored by directly measuring GTP binding. Because of their intrinsic GTPase activity, the half-life of GTP binding is short and binding is not generally stable under normal immunoprecipitation conditions. One exception is Ras, where a Ras blocking antibody (H-Ras 259) is available that will "trap" the GTP on Ras long enough to permit immunoprecipitation (Santa Cruz, sc-35). However, measuring the GTP bound to Ras requires *in vivo* labeling of the cells with high levels of radioactive inorganic ^{32}P.

Recently a new approach, a GTP Binding Assay, has been developed that is simpler and safer, but does require some extra preparation steps. The principle behind the GTP Binding Assay is that most GTPases when bound to GTP, but not GDP, interact with a downstream signaling molecule via a specific GTP-binding domain. For the Ras and Rho family members these downstream effector molecules and domains have been identified. These binding domains are expressed as GST fusions in bacteria and then purified from bacteria by using glutathione-conjugated agarose beads, which bind specifically to the GST portion of the protein. This GST-binding domain/bead slurry is then used to "immunoprecipitate" the GTP-bound forms of Ras or Rho family members from cell extracts. The captured material is run out on a gel and immunoblotted with antibody for the respective GTPase. For Ras the binding domain is a segment from Raf (Taylor and Shalloway, 1996), for Rho it is a segment from Rhotekin (Ren *et al.*, 1999), and for Rac and cdc42 it is a segment of PAK (Bagrodia *et al.*, 1998).

The only modification that is needed for cell adhesion assays is to use cell extracts at a concentration of 2 mg/ml. The relative abundance of molecules to which GTP is actually bound is quite low, even in growth factor-treated cells, and is usually even lower after integrin stimulation. Other important pointers are to include high levels of Mg^{2+} in the lysis buffer to stabilize the GTP-bound form and to not incubate much longer than 1 h. Make sure that the GST-fusion protein is in large excess in order to favor the reaction. Be sure to include a GST-only control and a positive control extract—cell extract from cells transfected with an active form of the GTPase of interest. Use a high-affinity antibody for the Western blot such as Ras clone10 (Upstate Biotechnology).

G. Molecular Genetics

1. Introduction

One of the most useful tools for elucidating integrin-mediated signaling pathways and networks is molecular genetics. From simple overexpression of one molecule and monitoring the effect on downstream events, to more complicated mutant analyses, this approach can be used to determine what the signaling events and pathways are, and give detailed information about how those signaling events occur and, most importantly, how they regulate the final biological outcome. When this is combined with studies in cells derived from mice deficient in a specific signaling molecule, this approach can be very powerful.

2. Dominant Inhibitory Mutants

Molecular genetics involves the use of molecular biology techniques to clone and genetically modify specific genes of interest. One way to assess the potential role of a particular signaling molecule in regulating a specific event in the cell is to introduce the gene that encodes that molecule into the cell and determine what effect that has on other signaling molecules or biological processes following adhesion to extracellular matrices. Expression of a particular kinase may cause the cells to extend large lamellipodia and simultaneously induce the phosphorylation or activation of another signaling molecule. Such a result would indicate that this kinase specifically targets that molecule and suggests that both might be involved in regulating lamellipodia formation.

To confirm the relationships between these events it is necessary to generate mutants of the kinase as well as the target molecule that, when expressed, would block activation of the downstream events following adhesion: generally referred to as a dominant inhibitory mutant. For some kinases, this may involve generating a kinase inactive mutant, or a truncation that cleaves off the kinase domain. For other types of molecules it could be expression of a regulatory domain alone. When the mutant kinase gene is expressed in the cell, stimulation of the target molecule by cell adhesion should be blocked or expression of the mutant target should block lamellipodia formation in response to either cell adhesion or overexpression of the active kinase. Generation of an "activated" form of the target should be able to rescue inhibition by the mutant kinase if these two molecules do lie within the same signal transduction pathway.

3. Transfection

Molecular genetic experiments can be extremely powerful, but also require advanced planning. When planning this kind of approach the first issue to be addressed is how the genes will be introduced into the cell. There are three methods that are commonly used: transient transfection, stable transfection, and viral infection. Viral infection is the most effective method because all the cells in a given population will contain the introduced gene and express it. The disadvantage is that generating virus for every single different gene and mutant studied can be tedious. Additionally, some retroviral systems are limited by the size of the gene that can be introduced and require that the cells be mitotic (Dunn *et al.,* 2001; Miller *et al.,* 1993; Onishi *et al.,* 1996). Adenoviral vectors are more versatile, but are more hazardous to the user (Yeh and Perricaudet, 1997). For these reasons transfection protocols are more commonly used.

For stable and transient transfections, genes are introduced into cells by the use of a transfection reagent, usually a lipid-based product (Gibco, Panvera, Invitrogen). Other ways to get genes into cells include generating a $CaPO_4$/DNA precipitate, or electroporation. The disadvantage of all these methods is that the efficiency is low and varies a great deal between cell types and with the method used. In a given population, only 1–20% of the cells may actually contain the gene. This makes analysis of a particular event difficult. Two approaches are commonly used to overcome this. One is to introduce a drug resistance marker into the cells at the same time as the gene. The cells are then treated with the drug, and those cells not receiving the marker (and the gene) are killed. Those that do express the marker (and the gene) survive, and a whole population of cells is generated that stably express the gene and will continue to do so for many generations. There are several disadvantages: the length of time it takes to generate these cells, the fact that putting cells through such a selective process can cause other genetic changes not anticipated, and the possible inability to generate these cells if expression of the gene is detrimental to the cell when expressed. The last problem can be overcome by the use of an inducible expression system that permits the gene to be stably integrated, but not express active protein until an external stimulus induces it. Such inducible systems include the tetracycline-inducible/repressor system or the use of an estrogen receptor (ER) epitope that inactivates the protein until estrogen is applied to the cell (Chambard and Pognonec, 1998; Samuels *et al.,* 1993). Selecting several different clones can compensate for any nonspecific genetic changes that might occur during the selection process and testing each to make sure the differences observed are due to the gene introduced.

The most rapid approach is to use a transient transfection method. To compensate for the low efficiency of most transfection methods, a gene encoding the target molecule is introduced along with the gene that it is thought to regulate it, the regulator. The target gene is usually designed so that it expresses an epitope tag, such as HA, myc, GST, or flag. Since both genes will end up in the same cells in the population, the effect of the regulatory gene on the target molecule can be monitored by immunoprecipitating the target molecule with antibody that recognizes the epitope. If antibody to the target molecule is used, all the target molecules, even those in the untransfected cells (which also will not express the regulator), will be analyzed, and because only a small number of those molecules came from transfected cells, any effects due to expression of the regulator may not be seen.

A similar problem also exists if monitoring the effect of expression of the regulatory molecule on cell functions, such as lamellipodia formation. Only a small number of cells in the population will express the regulator and thus be affected. To determine which cells actually received the gene, a gene encoding a fluorescent protein, such as GFP, can be introduced into the cells at the same time. It is then possible, under a UV light, to determine which cells received the regulatory gene and to observe the morphology of those cells and compare it to the cells that did not receive the gene.

There are really no major changes in the normal protocols used to carry out transfections or viral infections in cells that will be used in adhesion assays since introduction of the genes into the cells occurs 1–2 days (transient transfection), 3–5 days (viral infection), or 1–2 weeks (stable transfection) prior to the adhesion assay. If an inducible expression system is employed then induction is usually done 16–24 h prior to the cell adhesion assay also. In some situations it might be necessary to include the inducing agent in the adhesion assay to help maintain expression levels.

4. Antisense RNA

Another approach that can be used to inhibit the activity of a particular molecule is to express an antisense mRNA transcript or oligonucleotide in the cell. This can be achieved by transfecting the cells with either a plasmid construct designed to generate a full reverse transcript of the gene of interest or a short oligonucleotide from a reverse segment of the gene. Alternatively, an oligonucleotide can be introduced directly into cells by microinjection or standard transfection procedures. In all cases it is important to include several control oligonucleotides or plasmid constructs that would be predicted not to have an impact on the molecule of interest. It is also important to monitor the cells for loss of the protein and/or mRNA of interest to verify that the antisense is actually working. As with all other molecular genetic approaches, the efficiency of delivery must be considered. Additionally, the use of an inducible system maybe necessary if deleting the molecule will be a detriment to the cell.

A relatively new approach in antisense technology, the use of RNAi, is proving to be a much more efficient and effective method than standard antisense RNA for inhibiting protein expression in cells (Caplen *et al.,* 2001). This technique was originally used for inhibiting protein expression in *Caenorhabditis elegans*. By a mechanism that is still not completely understood, expression of a short 21-base-pair double-stranded RNA stimulates endogenous machinery to stimulate degradation of target mRNAs encoding the same 21-base sequence.

5. Introduction of Proteins

In some cases it can be useful to introduce the protein of interest directly into cells rather than attempting to express it from a cDNA construct. This is particularly true if expression of the protein over long periods of time might be detrimental to the cell or if analysis of short-term effects on cell signaling or function may be masked by long-term expression. This method could also be used in place of an inducible system. This approach can also be used to introduce antibodies or small peptides that might inhibit or activate a specific molecule in the signaling pathway of interest.

The limitation of this approach is the normal impermeability of the membrane to proteins and peptides. Two techniques have been used to overcome this. One is microinjection of the protein of interest and the other is to fuse a membrane-permeable peptide to the protein of interest. Microinjection requires less preparation of reagents, but does require expensive equipment and development of the necessary skills. It is the preferred method for introduction of antibodies. It is useful for looking at small numbers of cells and requires a visual assay for determining the effects. This can be accomplished by using either GFP-labeled proteins to monitor changes in localization or tagged proteins that can be stained with tag antibodies. Introduction of some proteins may induce specific expression of cellular proteins or induce changes in architecture or cell morphology. These events can be monitored by direct immunofluorescent staining or by cell imaging. This method is not conducive to direct biochemical analysis.

The use of a membrane-permeable peptide fused to the end of the protein of interest has the distinct advantage of being 100% efficient, allowing all cells in the population to be affected, and is good for biochemical analysis. This technique is based on the properties of membrane-permeable peptides, the most common ones being those derived from the TAT protein of HIV virus or the Antennapedia protein from *Drosophila* (Derossi *et al.,* 1998; Schwarze *et al.,* 2000). These peptides share the same structural motif, a highly basic helix that has a high affinity for cell membranes—possibly through binding to heparin. Attachment of this peptide to a protein will induce rapid internalization of that protein into all cells. Although this technique is very efficient and occurs very rapidly, within 20 min, it does require generation of an appropriate cDNA construct and expression and purification of the fusion protein. This method will often work in cells that are not normally susceptible to standard transfection or viral infection techniques. As for microinjection, this technique is useful when expression of the protein of interest for long periods of time might be detrimental to the cell. A precaution should be noted here. Because of the high efficiency of entry of TAT-fusion proteins into cells, there is a small risk that the user's own cells could become targets of this protein if proper attire and safety precautions are not observed. This method can also be used to transfer oligonucleotides or drug compounds into cells.

References

Aplin, A. E., Howe, A., Alahari, S. K., and Juliano, R. L. (1998). Signal transduction and signal modulation by cell adhesion receptors: the role of integrins, cadherins, immunoglobulin-cell adhesion molecules, and selectins. *Pharmacol. Rev.* **50,** 197–263.

Assoian, R. K. (1997). Anchorage-dependent cell cycle progression. *J. Cell Biol.* **136,** 1–4.

Bagrodia, S., Taylor, S. J., Jordon, K. A., Van Aelst, L., and Cerione, R. A. (1998). A novel regulator of p21-activated kinases. *J. Biol. Chem.* **273,** 23633–23636.

Berman, A. E., and Kozlova, N. I. (2000). Integrins: structure and functions. *Membr. Cell Biol.* **13,** 207–244.

Borges, E., Jan, Y., and Ruoslahti, E. (2000). Platelet-derived growth factor receptor β and vascular endothelial growth factor receptor 2 bind to the $\beta 3$ integrin through its extracellular domain. *J. Biol. Chem.* **275,** 39867–39873.

Burridge, K., Turner, C. E., and Romer, L. H. (1992). Tyrosine phosphorylation of paxillin and pp125FAK accompanies cell adhesion to extracellular matrix: a role in cytoskeletal assembly. *J. Cell Biol.* **119,** 893–903.

Caplen, N. J., Parrish, S., Imani, F., Fire, A., and Morgan, R. A. (2001). Specific inhibition of gene expression by small double-stranded RNAs in invertebrate and vertebrate systems. *Proc. Natl. Acad. Sci. USA* **98,** 9742–9747.

Chambard, J. C., and Pognonec, P. (1998). A reliable way of obtaining stable inducible clones. *Nucleic Acids Res.* **26,** 3443–3444.

Clemetson, K. J., and Clemetson, J. M. (1994). Molecular abnormalities in Glanzmann's thrombasthenia, Bernard–Soulier syndrome, and platelet-type von Willebrand's disease. *Curr. Opin. Hematol.* **1,** 388–393.

De Arcangelis, A., and Georges-Labouesse, E. (2000). Integrin and ECM functions: roles in vertebrate development. *Trends Genet.* **16,** 389–395.

Derossi, D., Chassaing, G., and Prochiantz, A. (1998). Trojan peptides: the penetratin system for intracellular delivery. *Trends Cell Biol.* **8,** 84–87.

Dunn, K. J., Incao, A., Watkins-Chow, D., Li, Y., and Pavan, W. J. (2001). In utero complementation of a neural crest-derived melanocyte defect using cell directed gene transfer. *Genesis* **30,** 70–76.

Fassler, R., Georges-Labouesse, E., and Hirsch, E. (1996). Genetic analyses of integrin function in mice. *Curr. Opin. Cell Biol.* **8,** 641–646.

Felsenfeld, D. P., Schwartzberg, P. L., Venegas, A., Tse, R., and Sheetz, M. P. (1999). Selective regulation of integrin–cytoskeleton interactions by the tyrosine kinase Src. *Nat. Cell Biol.* **1,** 200–206.

Frisch, S. M., and Ruoslahti, E. (1997). Integrins and anoikis. *Curr. Opin. Cell Biol.* **9,** 701–706.

Hemler, M. E. (1998). Integrin associated proteins. *Curr. Opin. Cell Biol.* **10,** 578–585.

Howe, A., Aplin, A. E., Alahari, S. K., and Juliano, R. L. (1998). Integrin signaling and cell growth control. *Curr. Opin. Cell Biol.* **10,** 220–231.

Keely, P., Parise, L., and Juliano, R. (1998). Integrins and GTPases in tumour cell growth, motility and invasion. *Trends Cell Biol.* **8,** 101–106.

Lewis, J. M., Cheresh, D. A., and Schwartz, M. A. (1996). Protein kinase C regulates $\alpha v \beta 5$-dependent cytoskeletal associations and focal adhesion kinase phosphorylation. *J. Cell Biol.* **134,** 1323–1332.

Miller, A. D., Miller, D. G., Garcia, J. V., and Lynch, C. M. (1993). Use of retroviral vectors for gene transfer and expression. *Methods Enzymol.* **217,** 581–599.

Onishi, M., Kinoshita, S., Morikawa, Y., Shibuya, A., Phillips, J., Lanier, L. L., Gorman, D. M., Nolan, G. P., Miyajima, A., and Kitamura, T. (1996). Applications of retrovirus-mediated expression cloning. *Exp. Hematol.* **24,** 324–329.

Petit, V., and Thiery, J. P. (2000). Focal adhesions: structure and dynamics. *Biol. Cell* **92,** 477–494.

Ren, X. D., Kiosses, W. B., and Schwartz, M. A. (1999). Regulation of the small GTP-binding protein Rho by cell adhesion and the cytoskeleton. *EMBO J.* **18,** 578–585.

Reuter, C. W., Catling, A. D., and Weber, M. J. (1995). Immune complex kinase assays for mitogen-activated protein kinase and MEK. *Methods Enzymol.* **255,** 245–256.

Samuels, M. L., Weber, M. J., Bishop, J. M., and McMahon, M. (1993). Conditional transformation of cells and rapid activation of the mitogen-activated protein kinase cascade by an estradiol-dependent human raf-1 protein kinase. *Mol. Cell. Biol.* **13,** 6241–6252.

Schwarze, S. R., Hruska, K. A., and Dowdy, S. F. (2000). Protein transduction: unrestricted delivery into all cells? *Trends Cell Biol.* **10,** 290–295.

Stipp, C. S., and Hemler, M. E. (2000). Transmembrane-4-superfamily proteins CD151 and CD81 associate with alpha 3 beta 1 integrin, and selectively contribute to alpha 3 beta 1-dependent neurite outgrowth. *J. Cell Sci.* **113,** 1871–1882.

Takai, Y., Sasaki, T., and Matozaki, T. (2001). Small GTP-binding proteins. *Physiol. Rev.* **81,** 153–208.

Taylor, S. J., and Shalloway, D. (1996). Cell cycle-dependent activation of Ras. *Curr. Biol.* **6,** 1621–1627.

van der Neut, R., Krimpenfort, P., Calafat, J., Niessen, C. M., and Sonnenberg, A. (1996). Epithelial detachment due to absence of hemidesmosomes in integrin beta 4 null mice. *Nat. Genet.* **13,** 366–369.

Woods, A., and Couchman, J. R. (1998). Syndecans: synergistic activators of cell adhesion. *Trends Cell Biol.* **8,** 189–192.

Yeh, P., and Perricaudet, M. (1997). Advances in adenoviral vectors: from genetic engineering to their biology. *FASEB J.* **11,** 615–623.

Zhang, X. A., Bontrager, A. L., and Hemler, M. E. (2001). Transmembrane-4 superfamily proteins associate with activated protein kinase C (PKC) and link PKC to specific beta(1) integrins. *J. Biol. Chem.* **276,** 25005–25013.

CHAPTER 19

Use of Micropatterned Adhesive Surfaces for Control of Cell Behavior

**Philip LeDuc,* Emanuele Ostuni,†,‡ George Whitesides,†
and Donald Ingber***

* Departments of Pathology and Surgery
Children's Hospital and Harvard Medical School
Boston, Massachusetts 02115

† Departments of Chemistry and Chemical Biology
Harvard University
Cambridge, Massachusetts 02138

I. Introduction

Extracellular matrix (ECM) plays a central role in cell regulation, both *in vivo* during tissue development and in cell culture (Ingber, 1997). Analysis of the molecular basis of cell regulation by ECM has led to the identification of specific transmembrane receptors,

‡ Current address: Surface Logix, Inc., 50 Soldiers Field Place, Brighton, Massachusetts 02135.

METHODS IN CELL BIOLOGY, VOL. 69

known as integrins, that can activate many of the same intracellular signaling cascades that are induced when soluble growth factors bind to their own receptors (Clark and Brugge, 1995; Giancotti and Ruoslahti, 1999; Calderwood *et al.*, 2000). However, other studies suggest that the ECM may provide different regulatory signals to the cell depending on the substrate's ability to resist cell tractional (tensional) forces and thereby modulate cell shape (Ingber and Folkman, 1989a; Ingber, 1991). In support of this possibility, many studies have demonstrated a direct correlation between cell spreading and growth (Folkman and Moscona, 1978; O'Neill *et al.*, 1986; Ingber, 1990) and between retraction or rounding and differentiation (Glowacki and Lian, 1987; Watt *et al.*, 1988; Ingber and Folkman, 1989b; Mooney *et al.*, 1992).

Analysis of this structural form of cell regulation has been limited by the availability of defined culture systems. In some of the past *in vitro* studies, cell shape was modulated by overlaying conventional tissue culture dishes with varying amounts or distributions of a nonadhesive blocking polymer (e.g., poly[hydroxyethyl methacrylate]) and then plating the cells in serum-containing medium: in this method, cell rounding is induced as the thickness of the blocking layer is increased (Folkman and Moscona, 1978; O'Neill *et al.*, 1986; Glowacki and Lian, 1987; Watt *et al.*, 1980; Ingber, 1990). In other experiments, cells were cultured in chemically defined medium on otherwise nonadhesive surfaces that were precoated with different densities of ECM molecules, causing cell spreading to increase as the coating density was raised (Ingber and Folkman, 1989b; Ingber, 1990; Mooney *et al.*, 1992). Because serum contains high amounts of ECM proteins, in particular vitronectin and fibronectin, in neither type of experiment was it possible to distinguish signals conveyed by cell distortion from those elicited by differences in integrin (ECM receptor) binding. For this reason, we set out to develop methods to control structural interactions between cells and ECM independently of growth factors or changes in integrin binding, and thereby to make cell distortion an independent variable.

Our approach to cell shape control involved microfabrication of adhesive islands of defined size, shape, and position on the micron scale, surrounded by nonadhesive boundary regions. The surfaces were coated with a saturating density of an ECM molecule and cells were then plated on these surfaces in chemically defined medium containing saturating amounts of recombinant growth factor. Thus, the only variable was the size and shape of the island and hence the degree to which the cell could physically distend. The formation of the islands was accomplished by creating patterns of self-assembled monolayers (SAMs) of alkanethiolates on gold that either support or prevent surface interactions with proteins (Prime and Whitesides, 1991, 1993; Mrksich and Whitesides, 1996). Hydrophobic SAMs rapidly and irreversibly adsorb ECM proteins, and hence promote cell adhesion; SAMs that present polyethylene glycol (PEG) moieties effectively resist protein adsorption, and thus prevent cell adhesion (Prime and Whitesides, 1991, 1993). Our approach was to pattern these two SAMs on a surface with micronscale resolution to define the pattern of ECM that adsorbs onto the surface, and hence the pattern of cell adhesion and spreading.

Micropatterned culture surfaces (e.g., containing square islands with edge lengths of 5, 10, 20, 30, or 40 μm separated by defined nonadhesive regions; Figs. 1A, 1B) were created using a soft-lithography-based, microcontact printing technique (Singhvi

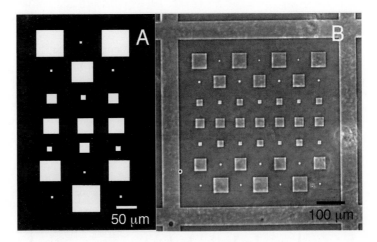

Fig. 1 Creation of a micropatterned adhesive surface using microcontact printing with self-assembled monolayers (SAMs) of alkanethiolates. (A) Drawing of the design for a photomask that includes squares with edge lengths of 5, 10, 20, 30, and 40 μm, separated by varying nonadhesive regions. (B) A phase contrast micrograph of a microfabricated surface containing the pattern shown in A and coated with fibronectin.

et al., 1994; Chen *et al.*, 1997). In this method, the desired pattern is created using computer design software, printed to a mask, and then transferred to a thin film of photoactive polymer (photoresist) on a silicon wafer; standard photolithographic techniques are used in transferring the corresponding pattern onto the layer of photoresist overlaying the wafer, thus creating a "master." A "rubber stamp" containing the imprint (negative form) of the topographical surface on the silicon wafer is then generated by pouring and polymerizing polydimethylsiloxane (PDMS) against the master (Figs. 2A–2D) (Wilbur *et al.*, 1995; Xia *et al.*, 1996; Takayama *et al.*, 2001). The dimensions of these stamps can be as low as 200 nm if necessary and can be designed to exhibit almost any geometric pattern. The stamp is peeled off the master, "inked" with hydrophobic alkanethiols, and then tightly apposed to the surface of a gold-coated cover glass. This results in the transfer of the hydrophobic alkanethiol molecules to the glass surface, which self-assemble into an almost crystalline molecular monolayer that is limited to the regions of the geometric pattern of the islands created on the original master (Figs. 2E–2I) (e.g., 30 μm square island that appeared as an elevated 30 μm square region on the surface of the rubber stamp). Next, a solution containing nonadhesive PEG alkanethiolate, which contains terminal tri(ethylene glycol) groups such as $HS(CH_2)_{11}O(CH_2CH_2O)_2CH_2CH_2OH$, is added to the patterned surface. This alkanethiol self-assembles in the remaining uncoated surfaces between the hydrophobic-SAM covered islands, thereby creating a continuous SAM covering the entire gold-coated area. While the hydrophobic islands support protein adsorption, these intervening PEG-covered barrier regions remain nonadhesive. Thus, when saturating amounts of ECM molecules, such as fibronectin, are added to these SAM-covered surfaces, they adsorb only to the surfaces of the adhesive islands coated with the hydrophobic SAM (Fig. 3).

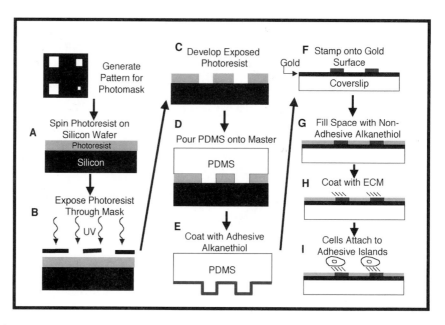

Fig. 2 Schematic of the soft lithography-based microcontact printing method. First, the design for the micropattern is drawn to scale using a commercial drawing package and saved as an electronic file. This file is used to fabricate a photomask with the identical features. (A) A silicon wafer is coated with a thin layer of photoresist. (B) The photomask is overlaid on the wafer and exposed to ultraviolet light thus protecting underlying regions of the photoresist from light exposure. (C) The photoresist is chemically developed resulting in the dissolution of photoresist from the light-exposed regions; the remaining bas-relief pattern retains the same pattern as the photomask. This pattern is exposed to vapors of (tridecafluoro-1,1,2,2,-tetrahydrooctyl)-1-trichlorosilane to reduce its adhesivity to the stamp. (D) Polydimethylsiloxane (PDMS) elastomer is poured over the pattern and cured for at least 2 h at 60°C. When the stamp is cut out and peeled off the original surface, it retains the complementary features of the master silicon wafer. (E) The molded surface of the PDMS stamp is "inked" with the adhesive alkanethiol, which promotes protein adsorption and cell adhesion. (F) The stamp is brought into contact with a gold-coated cover glass for 15 s to transfer the adhesive alkanethiol, which self-assembles into a monolayer in the regions where the stamp contacts the surface. (G) The stamp is removed and the entire surface is covered with a nonadhesive alkanethiol, which coats the remaining exposed regions of the gold surface and resists protein adsorption. (H) When extracellular matrix (ECM) molecules are added to this surface, they only adsorb to the adhesive islands. (I) Cells are seeded onto the cover glass where they attach and spread on these geometrically defined, ECM-coated islands.

Importantly, living cells only adhere to the ECM-coated adhesive islands when they are cultured on these surfaces. We found that when cells adhered to the ECM-coated islands, they spread over the fixed ECM anchors to cover the surface of the island, yet spreading ceased when the cell periphery reached the nonadhesive boundary (Singhvi *et al.,* 1994; Chen *et al.,* 1997, 1998; Huang and Ingber, 1999, 2000). Thus, as a result of the cells exerting tractional forces on their ECM anchors, they changed their morphology and literally "took on the shape of their container" (i.e., the geometric form of their micropatterned island); cells exhibited 90° corners when cultured on square islands and appeared round on circular islands (Figs. 4A–4D). In this manner, the effects

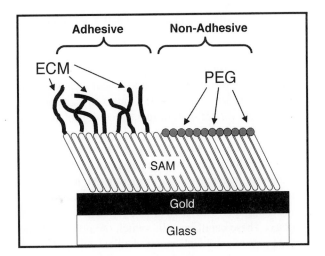

Fig. 3 Schematic representation of the local distribution of adhesive and nonadhesive SAMs of alkanethiolates at the edge of an adhesive island. Molecules such as ECM adsorb to the adhesive SAM (hexadecanethiol) while the regions coated with the nonadhesive SAM polyethylene glycol resist protein attachment.

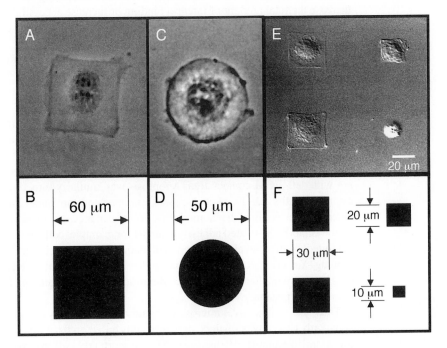

Fig. 4 Microscopic images (A,C,E), and corresponding adhesive island designs (B,D,F), of capillary endothelial cells whose shape and size were controlled using micropatterned adhesive surfaces. Cells cultured on square (A,B,E,F) or circular (C,D) adhesive islands coated with fibronectin. Note that cells may be held in different shapes when adherent to the same ECM density and cultured within the same medium containing saturating amounts of growth factor using this method, as shown in E,F. Scale bar = 20 μm.

of varying cell size and shape on cell functions (e.g., growth, apoptosis, differentiation) could be analyzed independently of both ECM density and soluble growth factors (Figs. 4E, 4F).

When we studied hepatocyte function using this method, we found that cell growth increased in direct proportion as island area was increased and cell spreading was promoted (Singhvi *et al.,* 1994). Conversely, as spreading was prevented and growth was turned off, differentiation (e.g., secretion of liver-specific products, such as albumin) was switched on. Similarly, when we studied capillary endothelial cells, we observed a similar link between spreading and growth (Chen *et al.,* 1997). In these cells, restricting cell spreading to a very high degree (island diameter less than 20 μm) switched on the apoptosis (cellular suicide) program. More recently, we created linear micropatterned adhesive surfaces that restrict capillary cells to a moderate degree of spreading that is not consistent with growth or apoptosis and that permit cell–cell interactions to take place. These capillary cells switch on a differentiation program and undergo capillary tube formation *in vitro* under identical culture conditions (Dike *et al.,* 1999).

One possible caveat in these studies was that the cells contacted more area of ECM (and hence more ECM molecules) on large vs small islands, even though the local ECM molecular coating density was saturating (which should promote optimal integrin clustering and intracellular signaling). However, increasing ECM contact area could increase the total amount of integrin binding or increase accessibility of cells to matrix-bound growth factors, thereby influencing cell behavior through chemical rather than structural means.

To explore this possibility more fully, micropatterned adhesive surfaces were created containing closely spaced adhesive circular islands of either 3 or 5 μm in diameter, to approximate the size of individual focal adhesions. When capillary cells were cultured on these surfaces, individual cell bodies spread across the intervening nonadhesive areas of the surfaces by stretching processes from one small adhesive island to another (Fig. 5). By changing the spacing between adhesive islands, cell spreading could be increased more than 10-fold without significantly altering the total cell-ECM contact area (Chen *et al.,* 1997). On these surfaces, DNA synthesis scaled directly with projected cell area and not with cell–ECM contact area. Apoptosis was similarly turned off by cell spreading, even though the cell–ECM contact area remained constant under these conditions. More recent studies confirm that some forms of integrin receptor signaling are fully activated on the small ECM-coated adhesive islands: for example, cells on small and large islands exhibited identical levels of activation of matrix metalloproteinase-2 in response to matrix binding (Yan *et al.,* 2000). Additional studies have shown that the effects of cell distortion on growth are mediated by specific alterations in the cytoskeleton, including changes in cytoskeletal tension.

Thus, use of this novel microscale patterning technology has revealed that the ECM apparently can convey distinct signals to cells based on its ability to resist cell tractional forces and to promote cell distortion. More importantly, cell shape per se appears to be the critical determinant that switches growth factor-stimulated cells between life and death, and also between proliferation and quiescence. This is critical because local changes in cell–ECM interactions appear to be responsible for establishment of the local growth

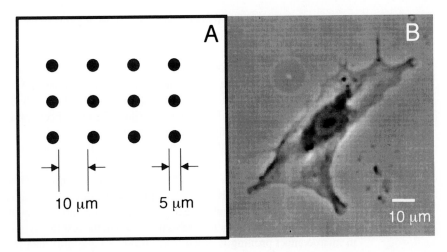

Fig. 5   A cell spread on a micropatterned surface that was used to investigate the effects of cell size and shape independently of cell–ECM contact area. (A) Schematic of the micropattern containing circular adhesive islands that are 5 μm in diameter and separated by 10-μm nonadhesive regions. (B) Phase contrast photograph of a cell that spreads over multiple focal adhesion-sized fibronectin-coated islands as shown in A. Scale bar = 10 μm.

differentials that drive pattern formation during morphogenesis *in vivo* (Banerjee *et al.,* 1977; Huang and Ingber, 1999).

Surfaces with geometrically defined chemistry can be fabricated with techniques other than microcontact printing with SAMs of alkanethiolates, such as vapor deposition and photolithography. In vapor deposition, adhesive metals (e.g., palladium) are deposited onto a surface of nonadhesive polyhydroxyethyl methacrylate, in specific patterns, by delivering them through a removable occlusive mask (e.g., a commercially available electron microscope grid or copper foil with defined pore size, shape, and position) (O'Neill *et al.,* 1986, 1990). Ultraviolet light can also be used with photoreactive chemicals to spatially pattern adhesive and nonadhesive regions. Specifically, photochemical changes in organosilanes are used to create adhesive and nonadhesive regions through deep UV (193 nm) irradiation (Stenger *et al.,* 1992). Photoimmobilization has been used after conjugating oligopeptides with the Arg-Gly-Asp cell-adhesion sequence on a surface of oligo(ethylene glycol) alkanethiolate. The surface was exposed to a UV light or laser beam which modified the molecular density and thus created spatially patterned adhesive and nonadhesive regions for cell attachment (Herbert *et al.,* 1997). Photoactive polymers including phenyl azido-derivatized polymers also have been coated onto an adhesive polymer surface (tissue culture dish) and irradiated through an overlying photomask, and the unreacted components removed to create spatially patterned surfaces for cell attachment (Matsuda and Sugawara, 1995). Another method involved covering photosensitive poly(ethylene terephthalate) films with poly(benzyl *N,N*-diethyldithiocarbamate-*co*-styrene) and irradiating through a photomask; this approach generated polymer stripes with distinct ionic characteristics that inhibit cell adhesion, thus creating spatially defined regions of varying adhesivity (DeFife *et al.,* 1999). Finally, other forms of microcontact printing can be used to spatially pattern adhesive and nonadhesive regions to investigate cell–ECM interactions. Specifically, microcontact printing is used to spatially pattern avidin molecules onto a polymeric surface such as poly(ethylene terephthalate) through high-affinity biotin–avidin molecular binding, to form an adhesive region for functionally active biotinylated molecules such as biotin-(G)(11)-GRGDS (Patel *et al.,* 2000; Yang and Chilkoti, 2000). Microcontact printing has also been used in conjunction with the self-assembly of fluid lipid bilayers to spatially pattern regions for cell attachment by printing adhesive regions of fibronectin on glass and then immersing the glass in a lipid bilayer solution: this creates a spatially patterned surface for cell–ECM attachment by producing minimally adhesive regions between the fibronectin areas (Kam and Boxer, 2001).

Although these methods are viable approaches for creating patterns of spatially distributed adhesive and nonadhesive regions for cell attachment through cell–ECM interactions, our method has numerous advantages. First, the resolution of molecular patterning in microcontact printing is much lower (down to 200 nm) than for some of the vapor deposition and photoreactive chemical techniques, which are often limited to 5 micrometers or more. One stamp in microcontact printing can be used for at least one hundred repetitions, in contrast to the photoreactive chemical or vapor deposition methods where photomasks or complex irradiation techniques are required for the fabrication of each surface. The nonadhesive regions of our technique use an "inert" region composed of PEG-SAM, instead of utilization of a coating of passive molecules (such as BSA) which degrade over a period of several days. Finally, the time and material costs of fabricating multiple surfaces by our technique are low relative to the alternate techniques.

In summary, our studies suggest that this micropattern-based method for spatial control of cellular attachment by ECM may have important implications for future work on the structural basis of morphogenetic regulation. This system also can be used to investigate a variety of questions involving complex cell behaviors, including cell motility and pattern generation (Brangwynne *et al.,* 2000) as well as differentiation, apoptosis, and growth. Studying these problems is important in the fields of cell biology, bioengineering, drug discovery, and tissue engineering. In this chapter, we provide a detailed description of the methods used for the preparation and use of micropatterned surfaces, coated with the ECM protein fibronectin, for analysis of cell behavior using bovine capillary endothelial cells. This approach may be easily adapted for use with other adherent molecules and cell types by adopting appropriate culture medium.

II. Materials and Instrumentation

A. Creation of the Photomask, Photolithographic Master, and PDMS Stamps

AutoCAD software from Designers' CADD Company, Inc. (Cambridge, MA) was used to create the geometric features of the micropattern. The photomask was custom made by Advance Reproductions Corporation (North Andover, MA) with chrome deposited on quartz for the pattern transfer. Hexamethyldisilazane was from Sigma-Aldrich

(Cat. No. 44109-1) and Shipley 1813 positive photoresist and 351-photoresist developer were from Microchem Corporation (Newton, MA). Silicon wafers in 3- and 5-inch diameters are from Silicon Sense, Incorporated (Nashua, NH). PDMS (Sylgard 184 silicone elastomer kit) was from Dow Chemical Company. All pattern creation was in a class 100 clean room with a photoresist spinner from Headway Research, Inc. (Garland, TX), a Karl Suss (Waterbury Center, VT) mask aligner, and hot plates. An electron-beam evaporator was used to deposit the metals on the glass surface. Desiccators, distilled water, argon gas, and nitrogen gas were purchased from standard scientific distributors.

B. Synthesis of Alkanethiols and Fabrication of Gold–Coated Surfaces

The chemicals necessary for the synthesis of adhesive and nonadhesive alkanethiol SAMs included (tridecafluor-1,1,2,2,-tetrahydrooctyl)-1-trichlorosilane (Cat. No. T2492; United Chemical Technologies), tetrahydrofuran (Cat. No. 43921-5), deuterated chloroform (Cat. No. 15185-8), 11-bromoundec-1-ene (Cat. No. 46764-2), calcium hydride (Cat. No. 21332-2), hexadecanethiol (Cat. No. H7637), sodium (Cat. No. 48374-5), hexanes (Cat. No. 43918-5), sodium hydroxide (Cat. No. 22146-5), benzophenone (Cat. No. 42755-1), dichloromethane (Cat. No. 43922-3), sodium sulfate (Cat. No. 23931-3), recrystallized 2,2'-azobisisobutyronitrile (Cat. No. 44109-0), thiolacetic acid (Cat. No. T30805), sodium methoxide (Cat. No. 16499-2), tri(ethylene glycol) (Cat. No. T5945-5), and DL-camphor-10-sulfonic acid (Cat. No. 14792-3) (all from Sigma-Aldrich). Merck 0.25-mm silica gel plates (Cat. No. 5554/7) were from EM Science, and silica gel with a 60–200 mesh (Cat. No. 6551) from Mallinckrodt. Other common basic laboratory equipment and materials (e.g., thin-layer chromatography, column chromatography, nuclear magnetic resonance, 450-W medium-pressure mercury lamp, ace glass, rotary evaporator) commonly found in chemistry labs (Zubrick, 1992) were also necessary for these studies. Trichloroethylene (Cat. No. 25642-0), methanol (Cat. No. 43919-3), and acetone (Cat. No. 27072-5) were from Sigma-Aldrich. Microscope cover glasses (#2), petri dishes, cotton swabs, and razor blades were from Fisher Scientific. Titanium (Cat. No. 43367-5) and gold (Cat. No. 37316-8) for metal deposition were from Sigma-Aldrich.

C. Pattern Stamping, ECM Coating, and Cell Plating

Phosphate-buffered saline (PBS, Cat. No. BM-220) was from Boston Bioproducts. Fibronectin (Cat. No. 40008A), was from Collaborative Biomedical Products. Bovine serum albumin (Cat. No. 3220-75) was from Intergen. Nalgene Filter systems (0.22 μm) were from Fisher Scientific (Cat. No. 09-761-111). Dulbecco's modified Eagle's medium (DMEM, Cat. No. 11885-084), penicillin, streptomycin, glutamine (Cat. No. 10378-016), and trypsin (25300-054) were from Gibco BRL. Hepes (Cat. No. 50-20577), human high-density lipoprotein (Cat. No. RP-035), bovine calf serum (Cat. No. SH30072.03), and transferrin (Cat. No. T-1283) were from JRH Biosciences, Intracell, Hyclone Laboratories, and Sigma, respectively.

III. Procedures

A. Creation of the Photomask, Photolithographic Master, and PDMS Stamps

The size, shape, and position of the islands and intervening nonadhesive regions within the micropatterns are empirically determined through preliminary studies with each cell type. A starting point is to design one pattern with multiple islands of different shape and size, separated by different-length nonadhesive regions (e.g., square and circular islands from 5 to 100 μm on each side and diameter, respectively, separated by 10- to 200-μm spaces). The geometric design of the pattern is created using the AutoCAD software drawing package and sent electronically to a commercial photomask printing company that fabricates the photomask. This mask is used to create a master by overlaying the photomask on the photoresist-coated silicon wafer. Through exposure to ultraviolet light and developing the photoresist, the pattern of the photomask is transferred to the photoresist. The resulting features of the photoresist can be used as a mold to fabricate the flexible PDMS stamp by pouring and curing the polymer over the surface of the wafer.

1. Creation of the Photomask

a. Steps

i. Select micropattern with desired island size, shape, and position making sure that the total pattern area does not exceed the area of the silicon wafer (either 3- or 5-inch diameter).

ii. Create a file containing the pattern design using AutoCAD; draw pattern in white with black as the background color.

iii. Scale each drawing object to proper dimensions by adjusting the object size (draw a square and then modify the dimensions to 30 μm on each side).

iv. Save as a dxf or dwg file in electronic format for fabrication of the photomask and transfer to a professional high-precision photomasking company (Advance Reproductions Corporation).

v. Obtain mask and store in a clean environment until needed (preferably the clean room).

2. Creation of Photolithographic Master

a. Steps

i. Clean silicon wafers in sonicating bath for 5 min each in trichloroethylene, acetone, and finally methanol.

ii. Bake wafers at 180°C for 10 min.

iii. Spin coat the wafers with 1–2 ml hexamethyldisilazane for 40 s at 4000 rotations per minute (rpm) followed by Shipley 1813 photoresist for 40 s at 4000 rpm for a thickness of 1.3 μm.

iv. Bake the photoresist for 2.5 min at 105°C.

v. Place the mask on top of the wafer in a mask aligner and expose wafer and photo-mask to UV light for 5.5 s at 10 mW/cm^2.

vi. Place wafers in Shipley 351 developer for 45 s and rinse with distilled water and dry wafer with nitrogen.

vii. Desiccate wafers under vacuum with 1–2 ml of (tridecafluoro-1,1,2,2,-tetrahydro-octyl)-1-trichlorosilane for 2 h.

3. Creation of PDMS Stamps

a. Steps

i. Mix PDMS using a 10:1 monomer to initiator ratio and degas mixture under vacuum for 1 h.

ii. Cover wafer with prepolymer in a petri dish and cure for a minimum of 2 h at 60°C.

iii. Remove PDMS by cutting around outside of petri dish and peeling the stamp from the wafer.

iv. Cut the stamps to the desired sizes with a razor blade preferably to 1.5 cm by 1.5 cm.

B. Synthesis of Alkanethiols and Fabrication of Gold-Coated Surfaces

Before the flexible PDMS stamp can be used, SAMs must be synthesized and the glass cover glass must be coated with a layer of gold. The nonadhesive alkanethiolate, (1-mercaptoundec-11-yl)tri(ethylene glycol), is created by first combining (undec-1-en-1-yl)tri(ethyleneglycol) with sodium hydroxide, tri(ethylene glycol), and 11-bromoundec-1-ene. This product is purified and then used to synthesize {1-[(methylcarbonyl)thio]undec-11yl}tri(ethylene glycol), and eventually the final nonadhesive product. The adhesive alkanethiol, hexadecanethiol, is purchased from a commercial source and purified before use. Finally, a cover glass must be coated sequentially with titanium and gold for the alkanethiolates to attach to the surface.

1. Synthesis of Alkanethiolate SAMs

a. (Undec-1-en-1-yl)tri(ethylene glycol)
Solutions:

i. Tetrahydrofuran (THF): distilled freshly with benzophenone at 1 g/liter and sodium at 1 g/liter.

ii. Dichloromethane: distilled freshly with calcium hydride at 1 g/liter.

Steps:

i. Mix 0.34 ml of 50% aqueous sodium hydroxide with 3.2 g of tri(ethylene glycol).

ii. Stir for 30 min in oil bath at 100°C.

iii. Add 1 g 11-bromoundec-1-ene, stir at 100°C for 24 h under argon, and allow to cool.

iv. Extract six times with 50–100 ml aliquots of hexanes and dry with sodium sulfate.

v. Combine the hexane aliquots, concentrate them, and purify the resulting yellow oil.

vi. The resulting R_f value of the product (undec-1-en-1-yl)tri(ethylene glycol) is 0.3 with 70% yield and NMR at 250 MHz values of δ 1.2 (broad singlet, 12H), 1.55 (quintet, $J = 7$ Hz), 2.0 (quartet, 2H, $J = 7$ Hz), 2.7 (broad singlet, 1H), 3.45 (triplet, 2H, $J = 7$ Hz), 3.5–3.8 (multiplet, 12H), 4.9–5.05 (multiplet, 2H), 5.75–5.85 (multiplet, 1H).

b. {1-[(Methylcarbonyl)thio]undec-11yl}tri(ethylene glycol)
Steps:

i. Dissolve 0.6 g of (undec-1-en-1-yl)tri(ethylene glycol) into 20 ml of tetrahydrofuran.

ii. Add 10 mg of recrystallized 2,2'-azobisisobutyronitrile along with 1.4 ml of thiol-acetic acid.

iii. Irradiate for 6–8 h with 450-W medium pressure mercury lamp.

iv. Remove a 0.1 ml aliquot for NMR, and then purify. The R_f of the resulting {1-[(methylcarbonyl)thio]undec-11yl}tri(ethylene glycol) is 0.3 with a yield of approximately 80%.

v. NMR at 250 MHz is δ 1.2 (broad singlet, 14H), 1.6 (multiplet, 4H), 2.3 (singlet, 3H), 2.85 (triplet, 2H, $J = 7$ Hz), 3.45 (triplet, 2H, $J = 7$ Hz), 3.5–3.75 (multiplet, 12H).

c. (1-Mercaptoundec-11-yl)tri(ethylene glycol)
Steps:

i. Degas 8 ml of methanol under argon or nitrogen for 30 min.

ii. Dissolve 0.4 g of {1-[(methylcarbonyl)thio]undec-11yl}tri(ethylene glycol) in 2 ml of freshly distilled dichloromethane and 8 ml of degassed methanol.

iii. Combine 0.9 ml or 1.3 M sodium methoxide in the methanol solution.

iv. Use DL-camphor-10-sulfonic acid to neutralize pH in the reaction solution after waiting 45 min.

v. Concentrate and purify (1-mercaptoundec-11-yl)tri(ethylene glycol) to $R_f = 0.25$ with a yield of 50%.

vi. NMR at 250 MHz δ 1.1 (broad singlet, 14H), 1.2 (triplet, 1H, $J = 7$ Hz), 1.5 (multiplet, 4H), 2.3 (singlet, 3H), 2.5 (quartet, 2H, $J = 7$ Hz), 3.0 (broad singlet, 1H), 3.4 (triplet, 2H, $J = 7$ Hz), 3.5–3.75 (multiplet, 12H). The nonadhesive alkanethiols should be stored under an inert gas at 2–8°C, protected from light, and used within 1 year.

d. Hexadecanethiol
Steps:

i. Hexadecanethiol, which is purchased from commercial sources, is purified through flash chromatography using hexanes or distillation at reduced pressure. The R_f of the product is approximately 0.4 with the typical NMR spectrum peaks at δ 1.25 (broad singlet, 29H), 1.6 (quintet, 2H), 2.5 (quartet, 2H). The adhesive alkanethiols should be stored at room temperature, protected from light, and used within 1 year.

2. Fabrication of Gold–Coated Surfaces

a. Steps

i. Rinse cover glasses with hexane and ethanol, then dry with nitrogen.

ii. Load cover glasses onto the carousel of an electron-beam evaporator, and reduce pressure in evaporator to less than 1×10^{-6} torr.

iii. Standardize evaporation rates in chamber to 1 Å/s and allow 200–300 Å of metal to evaporate before opening the shutters.

iv. First, evaporate 15 Å of titanium, and then 115 Å of gold onto the cover glasses.

C. Pattern Formation, ECM Coating, and Cell Plating

The gold-coated surfaces can now be manually stamped with the adhesive alkanethiols. After stamping, the intervening spaces are filled with nonadhesive alkanethiols. Next, the ECM is coated onto the adhesive regions. Cells are then plated onto the micropatterned surfaces to study the ECM-dependent control of cell shape and function.

1. Pattern Stamping

a. Steps

i. Rinse both sides of the cover glass with ethanol and blow-dry with nitrogen.

ii. Place the cover glass with the gold-coated side face up on a clean, flat surface and rinse the stamps with ethanol; dry completely with nitrogen.

iii. Immerse the tip of a cotton swab into a 2 mM ethanolic solution of hexadecanethiol, then completely wet and cover the surface of the stamp by swabbing the entire surface at least two times; blow dry stamp with nitrogen.

iv. Manually place the patterned side of the stamp down onto the gold-coated side of the cover glass.

v. Gently place firm pressure from one end of stamp to the opposite end to avoid trapping air bubbles under the stamp and allow the stamp to sit on the gold-coated glass for 15 s.

vi. Hold down cover glass by placing tweezers on an exposed edge of the cover glass and then remove stamp by manually peeling the stamp from this edge to the opposite.

vii. Cover the surface of the cover glass with a 2 mM ethanolic solution of (1-mercaptoundec-11-yl)tri(ethylene glycol) drop by drop with a Pasteur pipette and incubate with the solution for 45 minutes.

viii. Rinse the cover glass with ethanol on both sides, completely dry with nitrogen, and store the cover glasses with the patterned side face up under nitrogen in 2–8°C; these should be used within 2 weeks.

ix. Rinse the stamp with ethanol, dry with nitrogen, and store with the PDMS stamp face up in a covered petri dish.

2. ECM Coating

a. Solutions

i. 1 *M* sterile phosphate buffered saline (PBS): Dilute 10× PBS in sterile distilled water and pass through a 0.22 μm filter.

ii. ECM solution: Dilute ECM protein in sterile PBS to a concentration that will produce optimal surface coating (e.g., 50 μg/ml).

iii. 1% Bovine serum albumin (BSA) in PBS (1% BSA/PBS) solution: Dissolve 1% w/v BSA into sterile PBS.

b. Steps

i. Pipette a 250 μl droplet of ECM solution onto a sterile bacteriological petri dish.

ii. After flaming tweezers for sterilization purposes, place cover glass with the patterned side face down onto top of the droplet; this will cause the cover glass to float.

iii. Allow cover glass to sit in a humidified chamber at room temperature for 2 h and then pipette 500 μl of PBS under cover glass to raise level of cover glass for easier removal.

iv. After flaming tweezers again, lift cover glass and place face up in a petri dish containing a 1%BSA/PBS solution for 30 min.

v. Rinse cover glass three times in PBS and then add culture medium to cover the coated surface.

3. Cell Plating

a. Solutions

i. Defined medium: Dulbecco's modified Eagle's medium containing 2 m*M* glutamine, 100 μg/ml streptomycin, 100 μg/ml penicillin, 1 ng/ml basic fibroblast growth factor, 10 m*M* Hepes, and 1% BSA.

b. Steps

i. Aspirate medium from adherent monolayers of bovine capillary endothelial cells or other cell types cultured in sterile tissue culture dishes.

ii. Add 2 ml of PBS with calcium to the monolayer to cover the surface of the cells.

iii. Aspirate PBS and add 0.3 ml of trypsin in a 35 mm dish for 2 min, or until visual confirmation of rounding and detachment from dish.

iv. Using 10 ml of medium containing 1% BSA, pipette cells off surface of dish and collect; centrifuge solution for 5 min at 1000 rpm, aspirate supernatant, and resuspend cells in 5 ml of defined medium with 1% BSA (trypsin inhibitors can be added).

v. After determining cell density, add 10,000 cells per cover glass and then allow cells to attach and spread.

IV. Comments

The SAM and microcontact printing technique presented here creates adhesive islands surrounded by nonadhesive regions and leads to the adsorption of proteins onto the gold-coated surface in geometrically defined patterns. The use of the microcontact printing technique obviates the need for a dust-controlled laboratory environment after fabrication of the master. This technique also reduces the cost significantly compared to reproducing patterned surfaces for each cover glass using standard photolithographic techniques. It also permits larger scale production of the surfaces, because the photolithographic processing step is only used once during the fabrication process, for the initial fabrication of the master.

When drawing the original pattern designs for the mask, a 5-inch silicon wafer should have a total design area only 4.5 inches in diameter, a 3-inch wafer should not use more than 2.75 inches. The use of number 2 cover glass vs number 1 reduces the chance of brittle fracture of the coated surface. Evaporation of metals onto these cover glasses is accomplished using an electron beam evaporator, because sputtering causes inconsistencies in the layer of deposited metal. Each of the synthesized chemical products is concentrated through rotary evaporation at reduced pressure. Thin layer chromatography with 0.25-mm silica gel plates is then used to examine the progress of the reactions: the column chromatography is carried out using silica gel with 60–200 mesh under air, nitrogen, or argon conditions.

V. Pitfalls

A. Surfaces that contain adhesive islands separated by large distances can be compromised because the intermediate areas of the PDMS stamp sag and come into contact with the cover glass. This will force the stamp to print in regions intended to be inert, as the adhesive alkanethiols become deposited in these barrier regions.

B. Stamping of the surfaces is accomplished not just by placement of the stamps on the surface, but also by the application of firm pressure on the stamp. When applying this pressure, work from one corner of the stamp to the opposite corner to avoid trapping air bubbles between the stamp and the surface.

C. After coating the surfaces with ECM, be careful to avoid contact of the printed surface with other surfaces, or drying of the ECM, which can diminish the stability of the printed surface.

D. The conditions for cell culture should be optimized to minimize the presence of ECM proteins in the medium (preferably by removing or minimizing serum), which could alter the cell–matrix interactions. When seeding cells on the coated surfaces we use a density of 10,000 cells/cm^2 for our bovine capillary endothelial cells, although every cell type should be optimized for plating density. High densities can cause monolayers to form that extend from island to island over the entire surface, whereas low densities will cover printed regions sparsely.

E. Keep the gold-coated cover glass in enclosed containers after the initial electron beam deposition and clean with ethanol to prevent dust and other contaminants from adhering to the surfaces. These impurities will cause problems in stamping and in nonspecific attachment of cells to the final surfaces.

F. For deeper feature sizes, SU-8 negative photoresist from Shipley can be used as an alternative to positive photoresist, although a negative of the initial mask must then be utilized.

References

Banerjee, S. D., Cohn, R. H., and Bernfield, M. R. (1977). Basal lamina of embryonic salivary epithelia—production by epithelium and role in maintaining lobular morphology. *J. Cell Biol.* **73,** 445–463.

Brangwynne, C., Huang, S., Parker, K. K., and Ingber, D. E. (2000). Symmetry breaking in cultured mammalian cells. *In Vitro Cell. Dev. Biol. Animal* **36,** 563–565.

Calderwood, D. A., Shattil, S. J., and Ginsberg, M. H. (2000). Integrins and actin filaments: reciprocal regulation of cell adhesion and signaling. *J. Biol. Chem.* **275,** 22607–22610.

Chen, C. S., Mrksich, M., Huang, S., Whitesides, G. M., and Ingber, D. E. (1997). Geometric control of cell life and death. *Science* **276,** 1425–1428.

Chen, C. S., Mrksich, M., Huang, S., Whitesides, G. M., and Ingber, D. E. (1998). Micropatterned surfaces for control of cell shape, position, and function. *Biotechnol. Progr.* **14,** 356–363.

Clark, E. A., and Brugge, J. S. (1995). Integrins and signal transduction pathways: the road taken. *Science* **268,** 233–239.

DeFife, K. M., Colton, E., Nakayama, Y., Matsuda, T., and Anderson, J. M. (1999). Spatial regulation and surface chemistry control of monocyte/macrophage adhesion and foreign body giant cell formation by photochemically micropatterned surfaces. *J. Biomed. Mater. Res.* **45,** 148–154.

Dike, L. E., Chen, C. S., Mrksich, M., Tien, J., Whitesides, G. M., and Ingber, D. E. (1999). Geometric control of switching between growth, apoptosis, and differentiation during angiogenesis using micropatterned substrates. *In Vitro Cell. Dev. Biol. Animal* **35,** 441–448.

Folkman, J., and Moscona, A. (1978). Role of cell-shape in growth-control. *Nature* **273,** 345–349.

Giancotti, F. G., and Ruoslahti, E. (1999). Transduction—integrin signaling. *Science* **285,** 1028–1032.

Glowacki, J., and Lian, J. B. (1987). Impaired recruitment and differentiation of osteoclast progenitors by osteocalcin-depleted bone implants. *Cell Differ. Dev.* **21,** 247–254.

Herbert, C. B., McLernon, T. L., Hypolite, C. L., Adams, D. N., Pikus, L., Huang, C. C., Fields, G. B., Letourneau, P. C., Distefano, M. D., and Hu, W. S. (1997). Micropatterning gradients and controlling surface densities of photoactivatable biomolecules on self-assembled monolayers of oligo(ethylene glycol) alkanethiolates. *Chem. Biol.* **4,** 731–737.

Huang, S., and Ingber, D. E. (1999). The structural and mechanical complexity of cell-growth control. *Nat. Cell Biol.* **1,** E131–E138.

Huang, S., and Ingber, D. E. (2000). Shape-dependent control of cell growth, differentiation, and apoptosis: switching between attractors in cell regulatory networks. *Exp. Cell Res.* **261,** 91–103.

Ingber, D. E. (1990). Fibronectin controls capillary endothelial-cell growth by modulating cell-shape. *Proc. Natl. Acad. Sci. USA* **87,** 3579–3583.

Ingber, D. E. (1991). Integrins as mechanochemical transducers. *Curr. Opin. Cell Biol.* **3,** 841–848.

Ingber, D. E. (1997). Extracellular matrix: a solid-state regulator of cell form, function, and tissue development. *In* "Handbook of Cell Physiology" (J. D. Jamieson and J. F. Hoffman, eds.), pp. 541–556. Oxford University Press, New York.

Ingber, D. E., and Folkman, J. (1989a). How does extracellular matrix control capillary morphogenesis? *Cell* **58,** 803–805.

Ingber, D. E., and Folkman, J. (1989b). Mechanochemical switching between growth and differentiation during fibroblast growth factor-stimulated angiogenesis in vitro—role of extracellular-matrix. *J. Cell Biol.* **109,** 317–330.

Kam, L., and Boxer, S. G. (2001). Cell adhesion to protein-micropatterned-supported lipid bilayer membranes. *J. Biomed. Mater. Res.* **55,** 487–495.

Matsuda, T., and Sugawara, T. (1995). Development of surface photochemical modification method for micropatterning of cultured cells. *J. Biomed. Mater. Res* **29,** 749–756.

Mooney, D., Hansen, L., Vacanti, J., Langer, R., Farmer, S., and Ingber, D. (1992). Switching from differentiation to growth in hepatocytes—control by extracellular-matrix. *J. Cell. Physiol.* **151,** 497–505.

Mrksich, M., Chen, C. S., Xia, Y., Dike, L. E., Ingber, D. E., and Whitesides, G. M. (1996). Controlling cell attachment on contoured surfaces with self-assembled monolayers of alkanethiolates on gold. *Proc. Natl. Acad. Sci. USA* **93,** 10775–10778.

O'Neill, C., Jordan, P., and Ireland, G. (1986). Evidence for two distinct mechanisms of anchorage stimulation in freshly explanted and 3T3 Swiss mouse fibroblasts. *Cell* **44,** 489–496.

O'Neill, C., Jordan, P., Riddle, P., and Ireland, G. (1990). Narrow linear strips of adhesive substratum are powerful inducers of both growth and total focal contact area. *J. Cell Sci.* **95,** 577–586.

Patel, N., Bhandari, R., Shakesheff, K. M., Cannizzaro, S. M., Davies, M. C., Langer, R., Roberts, C. J., Tendler, S. J. B., and Williams, P. M. (2000). Printing patterns of biospecifically-adsorbed protein. *J. Biomater. Sci. Polymer Ed.* **11,** 319–331.

Prime, K. L., and Whitesides, G. M. (1991). Self-assembled organic monolayers: model systems for studying adsorption of proteins at surfaces. *Science* **252,** 1164–1167.

Prime, K. L., and Whitesides, G. M. (1993). Adsorption of proteins onto surfaces containing end-attached oligo(ethylene oxide)—a model system using self-assembled monolayers. *J. Am. Chem. Soc.* **115,** 10714–10721.

Singhvi, R., Kumar, A., Lopez, G. P., Stephanopoulos, G. N., Wang, D. I., Whitesides, G. M., and Ingber, D. E. (1994). Engineering cell shape and function. *Science* **264,** 696–698.

Stenger, D. A., Georger, J. H., Dulcey, C. S., Hickman, J. J., Rudolph, A. S., Nielsen, T. B., Mccort, S. M., and Calvert, J. M. (1992). Coplanar molecular assemblies of aminoalkylsilane and perfluorinated alkylsilane—characterization and geometric definition of mammalian-cell adhesion and growth. *J. Am. Chem. Soc.* **114,** 8435–8442.

Takayama, S., Ostuni, E., Qian, X. P., McDonald, J. C., Jiang, X. Y., LeDuc, P., Wu, M. H., Ingber, D. E., and Whitesides, G. M. (2001). Topographical micropatterning of poly(dimethylsiloxane) using laminar flows of liquids in capillaries. *Advanced Mater.* **13,** 570–580.

Watt, F. M., Jordan, P. W., and O'Neill, C. H. (1988). Cell-shape controls terminal differentiation of human epidermal-keratinocytes. *Proc. Natl. Acad. Sci. USA* **85,** 5576–5580.

Wilbur, J. L., Kim, E., Xin, Y. N., and Whitesides, G. M. (1995). Lithographic molding—a convenient route to structures with submicrometer dimensions. *Advanced Mater.* **7,** 649–652.

Xia, Y., Kim, E., Zhao, X. M., Rogers, J. A., Prentiss, M., and Whitesides, G. M. (1996). Complex optical surfaces formed by replica molding against elastomeric masters. *Science* **273,** 347–349.

Yan, L., Moses, M. A., Huang, S., and Ingber, D. E. (2000). Adhesion-dependent control of matrix metalloproteinase-2 activation in human capillary endothelial cells. *J. Cell Sci.* **113,** 3979–3987.

Yang, Z. P., and Chilkoti, A. (2000). Microstamping of a biological ligand onto an activated polymer surface. *Advanced Mater.* **12,** 413–418.

Zubrick, J. W. (1992). "The Organic Chem Lab Survival Manual," 3rd ed. Wiley, New York.

CHAPTER 20

Adenoviral–Mediated Gene Transfer in Two-Dimensional and Three-Dimensional Cultures of Mammary Epithelial Cells

Harriet Watkin and Charles H. Streuli

School of Biological Sciences
University of Manchester
Manchester M13 9PT, United Kingdom

I. Introduction

The majority of cells within tissues are in contact with adjacent cells and an extracellular matrix (ECM). The ECM is a specialized, complex network of secreted extracellular macromolecules and can broadly be divided into basement membrane—a thin laminate of laminin, collagen IV, entactin, and perlecan that subtends all parenchymal cells—and a stromal matrix that lies distal to the basement membrane that contains fibrillar collagen, fibronectin, tenascin, and many others. ECM has an invaluable role not only in promoting cell anchorage and migration, but also in preventing apoptosis and regulating both cell cycle and cellular differentiation. Indeed, the regulation of cell phenotype in complex tissues is determined both by soluble factors and by interactions between cells and their ECM (Streuli, 1999; Assoian and Schwartz, 2001).

Adhesion to the ECM influences differentiation in a number of epithelial cell systems (Stoker *et al.*, 1990; Hay, 1993) including intestinal (Stutzmann *et al.*, 2000) and epidermal (Adams and Watt, 1989, 1990; Fuchs *et al.*, 1997) epithelium. However, one of the most tractable systems for analysis has been the mammary gland. The mammary gland is an ideal organ in which to study the mechanisms of cell differentiation, apoptosis, and epithelial tissue morphogenesis, as it is one of few tissues that develops largely postnatally. In pubertal animals, extensive ductal morphogenesis occurs. Following the onset of pregnancy the mammary gland undergoes dramatic morphological and functional changes when epithelial cells proliferate and differentiate into milk-secreting cells. During weaning, the lactational epithelial cells die by apoptosis and the gland is remodeled, by a process known as involution, and returns to a resting state.

In order to dissect the mechanisms controlling ductal morphogenesis, cell differentiation and survival, culture methods to study and manipulate the changes in cell morphology and function are required. Mammary epithelial cells are valuable because primary cultures retain their potential to differentiate in culture providing they are given appropriate extracellular cues, and large numbers of cells are readily available *ex vivo*. A range of markers have been developed to detect differentiation endpoints, signaling pathways for transcription of tissue-specific genes, and apoptosis commitment points. The cells form multicellular structures that resemble lactating alveoli *in vivo*, express similar levels of milk protein mRNA to those found *in vivo* (Aggeler *et al.*, 1991), and secrete milk proteins vectorially (Barcellos-Hoff *et al.*, 1989). Moreover, cell culture conditions can be manipulated to provide three-dimensional environments that facilitate the morphogenesis of multicellular ducts and alveoli (Runswick *et al.*, 2001) and that either promote or suppress mammary epithelial apoptosis (Pullan *et al.*, 1996; Gilmore and Streuli, 1998).

Many of the phenotypes associated with mammary cell function *in vivo* do not occur efficiently in cell lines. Lines that have been established over a period of months or years are refractory to apoptosis, or at least their apoptosis control mechanisms have been mutated to permit cellular survival. In addition, very few mammary cell lines retain their capacity to form organized multicellular structures that resemble either bona fide ducts or alveoli *in vivo*, or to differentiate into lactational cells. The cell lines that do retain differentiation capabilities often do so poorly and express extremely low levels of milk proteins. This has necessitated the use of primary cultures of cells isolated from mammary tissue and cultured either for one or two passages only. Unfortunately gene transfer into primary cultures of mammary cells has proved to be extremely problematic, thus precluding modern molecular analysis of signal transduction and cell phenotype.

To overcome problems associated with gene transfer into primary cultures, we have developed a technique using adenovirus that allows us to express heterologous genes with 75–95% efficiency in primary mammary epithelial cells cultured in a three-dimensional matrix (see Fig. 6). This provides a new source of molecular tools to dissect signaling and phenotypic regulation.

In this chapter we will describe how to isolate primary mammary epithelial cells, how to culture them in three-dimensional matrices, and how to obtain high levels of heterologous gene expression.

II. Using Collagen I as a Three-Dimensional Matrix for Cell Culture

A. Background

Collagen I had been used as a substratum for culturing mammary epithelial cells since the pioneering studies of Pitelka and Emerman (Pitelka *et al.,* 1973; Emerman *et al.,* 1977; Emerman and Pitelka, 1977). Thick gels of collagen provide a stromal-type environment that permits mammary cells to undergo morphogenetic changes in cell shape and in multicellular clustering. Cells plated onto the top of thick collagen gels polarize and deposit basement membrane provided that the gels are not restricted by attachment to the culture dish (Streuli and Bissell, 1990). This provides an environment for cellular differentiation and the expression of tissue-specific genes (Lee *et al.,* 1985). Mammary cells plated within the three-dimensional matrix of a collagen gel form multicellular structures that can be induced to undergo integrin-dependent sprouting in the presence of HGF and/or TGFβ (Soriano *et al.,* 1995; Klinowska *et al.,* 1999). Although sprouting is rather primitive and restricted in most cases, this system has been used frequently as a model for ductal morphogenesis. In some cell lines long epithelial cell tubes can be induced.

One problem with attempting to recapitulate mammary cell phenotype by plating purified epithelial cells within collagen I gels is that the cellular composition of mammary gland is much more than just a single cell system. Adipocytes, stromal cells within the stromal matrix (largely consisting of collagen I *in vivo*), and myoepithelial and other cell types contribute to mammary tissue architecture (Silberstein, 2001). Fortunately, collagen I can provide a versatile matrix for developing more sophisticated three-dimensional environments that resemble mammary tissue more closely, even though such multicomponent systems have not been perfected yet. For example, including 3T3 fibroblasts within the collagen gel, as has been done for advanced keratinocyte cultures that differentiate to form squames (Kopan *et al.,* 1987), may improve mammary epithelial morphogenesis and differentiation in culture. The use of mammary fibroblasts, and the inclusion of myoepithelial cells (Lakhani and O'Hare; 2001), is likely to be even better.

In addition to its use in three-dimensional cultures, collagen I coated onto tissue culture plastic can provide an optimal environment for the formation of mammary epithelial monolayers. Mammary cells adhere well to fibronectin and vitronectin, ECM components that are also found in serum, but cell adhesion and spreading are more efficient on a collagen substratum. Our laboratory routinely plates mammary organoids isolated directly from tissue on thin layers of collagen I in order to promote cell outgrowth from organoids and the formation of confluent monolayers. The use of collagen I as a matrix for culturing mammary epithelial cells has also been described in detail by others (Pullan and Streuli, 1996; Imagawa *et al.,* 2000; Medina and Kittrell, 2000).

B. Materials

Rat-tail collagen I, 1–3 mg/ml in 0.1% acetic acid. Collagen is made according to previously published protocols (Pullan *et al.,* 1996) or can be purchased (Sigma C7661). It is important that native, not denatured, collagen I be used.

Sterile phosphate-buffered saline (PBS): 172 mM NaCl, 3.3 mM KCl, 10 mM Na$_2$HPO$_4$, 17.6 mM KH$_2$PO$_4$, pH 7.4

10× concentrated culture medium, e.g., F12 or DMEM

0.1% NaOH

C. Procedure for Coating Culture Dishes with a Thin Layer of Collagen I

Dilute rat-tail collagen I in ice-cold, sterile PBS to give a concentration of 80 μg/ml. Coat the dish with 100 μl per cm^2 dish area; this results in a coating density of 8 μg/cm^2. Allow the collagen to adhere to the tissue culture dish at 4°C overnight, or at 37°C for 1 h. Wash dishes three times in cold, sterile PBS. Mammary cells can be plated directly onto prepared dishes.

D. Procedure for the Use of Thick Collagen Gels

Add 1 volume of 10× medium to 8 volumes of ice-cold collagen I. Mix well and place on ice. Add 1 volume ice-cold 0.1% NaOH and mix well. Aliquot the required amount of neutralized collagen solution to a culture dish and allow to set at 37°C in an incubator. Neutralized collagen should be used within half an hour; otherwise it will gel in the tube.

Thick collagen gels can be made with 0.5 ml collagen per cm^2 tissue culture dish, to create gels that are approximately 5 mm thick. Before plating cells, the gels should be washed carefully with 3 changes of fresh 1× medium.

Once cells have formed a monolayer, after culture for 2–3 days, the gels can be released from the tissue culture plastic by scoring around the gel, between it and the dish, with a sterile spatula or blue Gilson tip. Releasing the gel allows the cells to contract it, and thereby to become polarised (Streuli and Bissell, 1990). Often the apparent surface area of the gel is reduced to one-sixth of the original area by cell polarization and consequent gel contraction.

Cells can also be placed within collagen gels (Streuli *et al.,* 1991). To minimize excessive salt imbalance, dialyze a solution of collagen on ice in 0.1% acetic acid against

3 changes of sterile, ice-cold, water for 30 min each. Remove the dialyzed collagen into sterile tubes, prechilled on ice, and add a tenth volume of $10\times$ medium. To embed cells, either single cells or organoids prepared from tissue, first pellet the cells (800–1200 rpm), remove medium, and add the required volume of prepared collagen solution. Triturate very carefully, to avoid the formation of bubbles, and plate onto dry tissue culture dishes or coverslips at room temperature. Allow the gels to set at $37°C$, then flood with medium.

Note that for all these procedures, it is important that the cells be well washed before plating on or in collagen. Residual trypsin, and in particular residual collagenase used to prepare organoids directly from tissue, can degrade the collagen gel.

III. Using Reconstituted Basement Membrane as a Three-Dimensional Matrix for Cell Culture

A. Background

A reliable mammary epithelial cell culture model that permitted lactation and *in vivo* morphological characteristics of differentiated cells came with the development of EHS matrix, a reconstituted basement membrane isolated from the Engelbreth–Holm–Swarm tumor (Kleinman *et al.,* 1986). This matrix is commercially referred to as Matrigel, and it contains laminin-1, collagen IV, nidogen, and perlecan, as well as some minor ECM proteins such as fibronection. Primary mouse mammary epithelial cells, isolated as whole alveoli from 14- to 17-day pregnant mice, cultured on the Matrigel basement membrane matrix under the influence of lactogenic hormones, express high levels of β-casein gene transcription and gene expression and retain their alveolar morphology (Li *et al.,* 1987; Barcellos-Hoff *et al.,* 1989; Schmidhauser *et al.,* 1990; Aggeler *et al.,* 1991). Even single mammary cells, trypsinized from plastic culture dishes, aggregate and undergo dramatic morphogenesis to form functional alveoli when cultured on Matrigel (Streuli *et al.,* 1991). More recently Matrigel has been used as a substratum to promote efficient survival of mammary cells and thereby suppress apoptosis that occurs in cells cultured over long time frames in monolayer (Pullan *et al.,* 1996; Farrelly *et al.,* 1999). In addition, mammary cells cultured on Matrigel can be induced to undergo branching and tubular morphogenesis, when stimulated with appropriate soluble factors such as HGF (Niemann *et al.,* 1998).

Matrigel has also been used in other sophisticated assays to examine highly specific aspects of mammary cell behavior (Ip and Darcy, 1996). For example, cells can be cultured as single cells within Matrigel to examine aspects of cell behavior that are not dependent on cell–cell interactions (Streuli *et al.,* 1991; Farrelly *et al.,* 1999). Moreover, Matrigel has been used as a diluted ECM preparation and supplied in the culture medium to mammary cell monolayers to promote both differentiation and survival (Roskelley *et al.,* 1994; Streuli *et al.,* 1995; Farrelly *et al.,* 1999). It can also be used as a matrix to support differentiation of mammary cell clusters containing more than on cell type, e.g., myoepithelial and luminal cells (Smalley *et al.,* 1999).

B. Materials

    1. Matrigel: Becton Dickinson Labware (40234), Sigma (E 1270)

C. Procedure for Coating Dishes with Matrigel

Matrigel is kept at $-80°C$ for long-term storage. The first time that Matrigel is thawed after purchase, it is advisable to aliquot it into volumes that will be used in single experiments. Prior to use it is thawed on ice as it forms a gel at room temperature. This takes a few hours and is best done overnight in a Dewar flask containing ice and kept in a refrigerator. If necessary Matrigel can be diluted using two parts Matrigel to one part ice-cold medium. Culture dishes are prechilled on ice and coated with 20 μl Matrigel per cm^2 dish area, resulting in a thin layer of Matrigel matrix. The Matrigel is spread evenly over the surface of the dish using the wide end of a sterile 1 ml Gilson tip. As the dishes are cold it is important to have the lid off the dish for as short a time as possible to prevent water from condensing on the dish and further diluting the Matrigel. The dishes should be as level as possible to prevent the Matrigel from accumulating at one side of the dish. Take care to ensure that any air bubbles are removed; the narrow end of a 1 ml Gilson tip may be used for this. Allow the Matrigel to set, with the dishes on a level platform in a $37°C$ incubator, for a minimum of 1 h before plating cells.

D. Procedure for Culturing Single Cells within Matrigel

Coat tissue culture dishes with Matrigel as described earlier. A single cell suspension of mammary epithelial cells is obtained by trypsinizing a monolayer of cells cultured on a collagen I coated tissue culture dish. Pellet the cells by centrifugation (1200 rpm, 4 min) and resuspend in growth medium before straining through a 70 μm membrane (Becton Dickinson Labware 2350). A viable cell count is performed using the trypan blue exclusion method.

Centrifuge the required number of cells at 1200 rpm for 4 min, remove the medium, then add 150 μl of ice-cold Matrigel. Resuspend the cell pellet by triturating very carefully, to avoid the formation of bubbles, and place drops of 50–70 μl onto the Matrigel-coated dishes. Allow the gel drops to set at $37°C$ for 1.5 h, then flood with medium. The quality of the cell preparation can be determined by light microscopy using long-distance phase-contrast objectives. The cultures should appear as single, rounded cells rather than clumps or clusters of cells.

E. Procedure for Overlaying Cells with Matrigel

For cells that are already cultured in monolayer dilute Matrigel 1:50 to 1:100 in ice-cold medium and warm gently to $37°C$. Place the Matrigel-containing medium over the cells. Culture the cells in a $37°C$ incubator. Protein networks formed by the Matrigel that precipitate over the cell layer can often be seen by phase contrast microscopy (Streuli *et al.,* 1995).

IV. Preparing Mouse Mammary Epithelial Cell Cultures

A. Background

Mammary glands are located between the skin and the peritoneum. There are five pairs of glands in the mouse, and their quantity and ease of location make them an excellent source of a large number of primary mammary epithelial cells for functional analysis. The morphology of the tissue and the cell composition of the gland change dramatically during the different stages of development, and the choice of when to harvest mice depends on the major cell type required. For example, most of the cells isolated from late pregnant animals are luminal epithelial cells. In contrast, nonpregnant animals are a valuable source of fibroblasts or adipocytes and the ratio of myoepithelial : luminal cells is higher, thus providing a better source of the former. Primary cultures from nonpregnant animals are also useful for genetic manipulation of mammary cells followed by transplantation into mammary fat pads, as these cell populations contain a higher proportion of ductal stem cells (Edwards *et al.,* 1996). Methods for isolating and culturing mammary epithelial cells have previously been described in detail (Pullan and Streuli, 1996; Darcy *et al.,* 2000; Imagawa *et al.,* 2000).

V. Preparing Recombinant Adenovirus for Infecting Mammary Epithelial Cells

A. Background

Recombinant adenoviruses have become a very attractive and versatile vector system for the transfer of genes into mammalian cells and are currently used for a variety of purposes, including gene transfer *in vitro,* vaccination *in vivo* and gene therapy (Rosenfeld *et al.,* 1991; Haddada *et al.,* 1995; Giannoukakis *et al.,* 2000). The versatility of recombinant adenoviruses is due to several features of their biology. For example, adenoviruses are able to infect and transfer their genes to virtually all cell types, with the exception of some lymphoid cells, which are more resistant to adenovirus infection (Chu *et al.,* 1992). Unlike retrovirus vectors, which require cycling target cells for vector integration and efficient gene expression, adenovirus infection and subsequent gene expression can occur in quiescent cells, making it therefore ideal for gene transfer into postmitotic cells (Dewey *et al.,* 1999; Brunner *et al.,* 2000). Adenovirus is a useful gene transfer vehicle for differentiating mammary epithelial cells that have largely exited the cell cycle in culture medium not containing serum growth factors or EGF. Adenovirus is primarily maintained in the nuclei of infected cells, in association with the nuclear matrix, without integration. This precludes insertional mutagenesis of the host genome and renders virally encoded genes refractory to silencing as a consequence of insertion adjacent to negative regulatory elements.

We have used a new system, the pAdEasy system (He *et al.,* 1998), to generate recombinant adenoviruses. The adenoviral particles produced are infectious; however, they are replication-deficient and incapable of producing more infectous viral particles in target cells. The system uses the major immediate early human cytomegalovirus (CMV) promoter to drive expression of the transgene, which has been found to induce extremely high-levels of expression (Wilkinson and Akrigg, 1992) and is an effective promoter in differentiated mammary epithelial cells (Schmidhauser *et al.*, 1994). The AdEasy system greatly simplifies traditional approaches for generating recombinant adenoviruses by eliminating the need for technically challenging direct ligations of the transgene into the adenoviral genome. In addition, repeated rounds of plaque purification are not required, thus significantly reducing the time required to generate usable viruses.

The gene of interest is first cloned into a shuttle vector and the resultant plasmid is linearized by digestion with the restriction endonuclease Pme1 (Fig. 1A). The linearized plasmid in then cotransformed into *E. coli* BJ5183 cells with a supercoiled adenoviral backbone plasmid, pAdEasy. Recombination takes place between the homologous sequences within the two plasmids such that the recombinant gene cassette and kanamycin gene are incorporated into the adenoviral plasmid (Fig. 1B). The recombinant adenoviral plasmid is then linearized by restriction endonuclease digestion with Pac1 and transfected into HEK293 E1 cells. These cells express the adenoviral proteins necessary for packaging infectious adenoviral particles (Graham *et al.*, 1977). Recombinant adenoviruses are typically generated within 7 to 12 days. Here we describe the AdEasy system including several helpful modifications.

B. Protocols for Generating Recombinant Adenovirus Using the pAdEasy System

1. Generation of recombinant adenoviral plasmids in *E. coli*
2. Screening for positive recombinants
3. Virus production in HEK293 E1 cells
4. Titration of virus
5. Preparation and purification of high-titer viral stocks

1. Generation of Recombinant Adenoviral Plasmids in *E. coli*

Materials:

Vectors: pShuttle, pShuttle-CMV, pAdTrack or pAdTrack-CMV, Supercoiled pAd-Easy-1 or pAdEasy-2: Q-BIOgene (AES1000A)

cDNA encoding your gene of interest

Standard cloning bacteria, e.g., Competent DH5α cells. (Gibco BRL, 18265-017)

LB-Agar plates

Kanamycin: Sigma (K-0879)

LB broth

Mini and Maxi Prep Kits: Qiagen (27104, 12162)

Buffer-saturated phenol: Gibco BRL (15513-039)

Chloroform

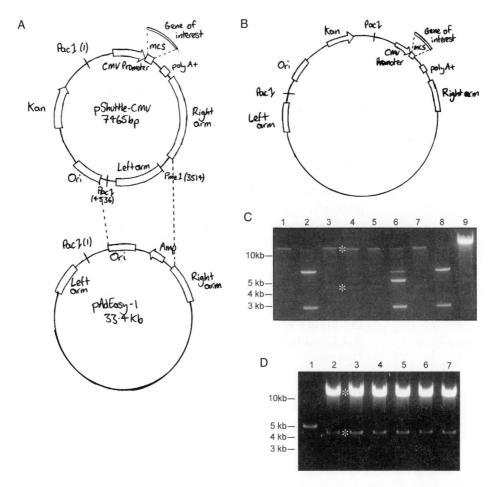

Fig. 1 Summary of the plasmid construction scheme for preparing adenoviral DNA. (A) The pAd Easy system for generating recombinant adenoviral DNA ready to transfect into 293 cells involves: (i) cloning the gene of interest into the multiple cloning site of the shuttle vector; (ii) linearizing the shuttle vector with PmeI; (iii) inducing recombination with the pAdEasy adenoviral backbone vector by cotransformation into BJ5183 *E. coli*. The only plasmids that are permissive for growth on kanamycin plates are ones that contain the shuttle vector. These can be religated intact shuttle vector alone, shuttle vector/pAdEasy recombinants where recombination occurs between the Left and Right arm sequences, or shuttle vector/pAdEasy recombinants where recombination occurs between the Ori sequence and Right arm sequence. Original maps of the plasmids are available on www.coloncancer.org/adeasy.htm. (B) The recombinant plasmid is then linearized with Pac 1 and the large 33 kb fragment containing the gene of interest plus adenoviral backbone is transfected into HEK293 El cells for virus production. (C) A Pac1 restriction digest is used to distinguish diagnostically between religated shuttle vectors and recombinants. Lanes 2, 6, and 8 contain shuttle vector alone, whereas lanes 1, 3, 4, 5, and 7 yield fragments of 4.5 and 33 kb, indicating recombination has occurred between the Ori and Right arm sequences. PacI digested pAdEasy in lane 9 as a control. (D) The plasmid tested in lane 3 of (C) was amplified in DH5α cells. Plasmids obtained were rescreened and a positive recombinant was used for transfection into HEK293 El cells. Pac1 digested DNA from amplified positive recombinants seen in lanes 2, 3, 4, 5, 6, and 7. Adapted with permission from www.coloncancer.org.

Ethanol

3 *M* NaAc pH 5.2

Sterile dH$_2$O

Bio-Rad Gene Pulser

Electrocompetent BJ5183 *Escherichia coli* cells, genotype *endA sbcBC recBC galK met thi-1 bioT hasdR (Strr):* Q-BIOgene (AES1005)

*Pme*I restriction enzyme: New England BioLabs (R0560S)

*Pac*I restriction enzyme: New England BioLabs (R0547S)

- Ensure that the gene of interest does not contain the sites for the restriction enzymes *Pme*1 or *Pac*1, then clone the gene of interest into a shuttle vector, e.g., pShuttle, pAdTrack, or pAdTrack-CMV.
- Linearize the shuttle plasmid containing your gene of interest by digesting 1 μg of Qiagen Maxi prepped DNA with *Pme*1. Extract DNA using phenol–chloroform, ethanol precipitate, and resuspend in 6 μl sterile dH$_2$O.
- Combine the 1 μg of Pme1-digested shuttle plasmid with 0.1 μg of supercoiled pAdEasy-1 or pAdEasy-2 and cotransform into 30 μl of electrocompetent *E. coli* BJ5183 cells by electroporation in 2.0 mm cuvettes at 2500V, 200 ohms, and 25 μFD in a Bio-Rad Gene Pulser electroporator. Immediately transfer the cells to a 1.5 ml tube containing 500 μl of LB-broth and grow at 37°C for 20 min. Plate onto 4 LB-agar plates containing 50 μg/ml kanamycin and incubate at 37°C overnight.
- The next day colonies can be picked and grown for screening. The level of homologous recombination was found to be very low in our hands, and we have therefore adapted the method to ensure that only recombinants are selected.

2. Screening for Positive Recombinants

Materials:

DNA oligonucleotide: 5'-GCG CAA ACC CCA CGC ACG AGA-3'

[γ-^{32}P]ATP: Amersham Pharmacia Biotech (AA0018)

Polynucleotide kinase: Boethringer Mannheim (84365637)

Micro Bio-Spin 6 Chromatography Columns: Bio Rad (732-6221)

Nylon membrane, Bioband 82-mm disk: Sigma (N-4156)

Microlance 3, 21 gauge needle: Becton Dickinson UK Ltd. (304432)

Denaturing solution: 1.5 *M* NaCl, 0.5 *M* NaOH

Neutralising solution: 1.5 *M* NaCl, 0.5 *M* Tris pH 7.2, 1 m*M* EDTA

Amersham Ultraviolet Crosslinker

Prehybridization Solution: 6× Salt sodium chloride/sodium citrate (SSC), 5×

Denhardt's reagent, 0.5% SDS, 100 μg/ml denatured, fragmented calf thymus DNA (Sigma D-8661) boiled for 5 min prior to use.

50× Denhardt's reagent: 5g Ficoll, 5g polyvinylpyrrolidone, 5g BSA, to 500 ml with dH$_2$O

Wash solution: 2× SSC, 0.1% SDS

- Design an 18- to 21-mer oligodexynucleotide probe, which will bind to a region within the adenoviral DNA of the pAdEasy plasmid. See above for the sequence we use.

- End-label the probe with [γ-^{32}P] ATP in a 10 μl reaction containing 50 ng oligonucleotide, 1 μl polynucleotide kinase enzyme, 1 μl polynucleotide kinase buffer, 10 μCi [γ-^{32}P]ATP, sterile dH$_2$O to 10 μl

- Incubate at 37°C for 30 min.

- Drain a Micro Bio-Spin 6 chromatography column by removing the bottom and lid; centrifuge at 1000g for 2 min to ensure remaining buffer is removed.

- Add 60 μl sterile dH$_2$O to the kinase reaction and place onto the column drop by drop.

- Centrifuge the labeled oligodeoxynucleotide into a sterile 1.5 ml tube, 1000g, 4 min. The probe is now ready for use.

- Instead of picking single colonies the day after the cotransformation of BJ5183 cells, lift entire plate of colonies onto Bioband membrane. Place the dry membrane onto the dish: there is no need to apply any force as the colonies will stick to the dry membrane. While the membrane is on the dish, position holes asymmetrically in the membrane and agar below using a needle. This will enable orientation of the membrane and subsequent identification of recombinants.

- Carefully lift the membrane and float it colony side up on Whatman filter paper presoaked with denaturing solution for 7 min, then on filter paper soaked in neutralizing solution for 3 min. Wash the membrane by floating on a bath of 2× SSC and leave to dry. Cross-link the DNA to the membrane in an Ultraviolet Crosslinker 100 mJ/cm^2 for 1 min.

- Only the top of the colonies will have been removed from the plates, and to be able to pick recombinants leave plates at room temperature overnight to allow the colonies to regrow.

- Place the cross-linked membranes in roller bottles and add 30 ml prehybridization solution prewarmed to 42°C. Prehybidize at 42°C in rotary hybridization oven for 30 min, then add 30 μl labeled probe and hybridize overnight at 42°C.

- Wash the membranes twice with wash solution 15 min each at 42°C.

- Positive colonies are detected by exposing the membranes covered in Saran wrap onto autoradiograph film, with an intensifying screen, overnight at −70°C.

- Mark the asymmetric holes on the film by lining them up over the filters. Identify positive recombinants by lining the film up with the dishes. Select 5–10 positive colonies and add each one to 100 μl LB broth, vortex, then streak out onto LB-agar plates containing 50 μg/ml kanamycin, 4–5 colonies per plate. Grow at 37°C overnight, then repeat the colony lift procedure to ensure that single colonies are identified.

- Select 10 positive single clones and grow overnight at 37°C in 5 ml LB broth containing 50 μg/ml kanamycin. Prepare DNA from each clone using a Qiagen mini-prep kit and resuspend in 50 μl dH$_2$O. Endonuclease restriction digest 10 μl of the mini-prepped DNA with Pac1 and separate on an agarose gel. Recombinant clones yield a large fragment, near 33 kb, plus a smaller fragment of 3.0 kb or 4.5 kb (Fig. 1C). The 3.0 kb fragment results from homologous recombination through the adenovirus left and right arm sequences in both plasmids and the 4.5 kb fragment results from homologous recombination between the plasmid Ori sequences shared between the shuttle and AdEasy vectors.

- The yield of plasmid DNA from BJ5183 cells is generally low; therefore it is necessary to amplify recombinant plasmids in DH5α cells to give good plasmid preps. Transform DH5α cells using recombinant plasmids and carry out larger scale plasmid DNA preps (Fig. 1D).

3. Virus Production in HEK293 E1 Cells

Materials:

HEK293 E1 cells
Growth medium: MEM
Filter cap tissue culture flasks
Lipofectamine: Gibco BRL (18324-012)
Sterile PBS
Dry ice/alcohol bath
1,1,2-Trichlorotrifluorethane (Arklone): Sigma (T-5271)

- The recombinant plasmid DNA is transfected into HEK293 E1 cells for virus production. Linearize 8 μg of the recombinant adenovirus plasmid by endonuclease restriction digest with Pac1, ethanol precipitate, and resuspend in 40 μl sterile water.

- Seed HEK293 E1 cells in two T-25 flasks at 2×10^6 cells per flask 24 h prior to transfection. The confluency should be about 50% to 70% at the time of transfection. Perform a standard lipofectamine transfection according to the manufacturer's manual using 4 μg DNA (20 μl) for each flask. Viral production kills HEK293 cells and can be monitored by the cytopathic effect (CPE). Adenovirus CPE consists of rounding and grapelike clustering of the swollen infected cells and will cause patches to appear in the HEK293 E1 monolayer (Fig. 2).

- When 60% to 70% of the cells show CPE, 7 to 10 days post transfection, the virus particles can be harvested. Adenovirus is not shed into the supernatant but

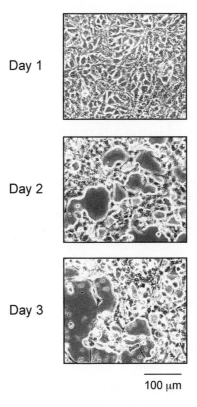

Day 1

Day 2

Day 3

100 μm

Fig. 2 Morphology of 293 cells after adenoviral infection. Confluent HEK 293 El cells were infected with recombinant adenovirus at an MOI of 10; 24 h post infection the cells remain healthy in appearance. The following day a cytopathic effect is observed as the cells round up and patches appear in the monolayer. This becomes more acute on the third day after infection. Scale bar = 100 μm.

remains cell-associated, so recombinant virus particles need to be extracted from the HEK293 E1 cells. Give the flasks a strong tap (this will make the cells float off into the medium), then transfer the medium to tubes for centrifugation. Pellet the cells in a benchtop centrifuge, 1500 rpm, 5 min and resuspend in 500 μl sterile PBS.

• To lyse the cells freeze them in a dry ice/alcohol bath and thaw rapidly in a 37°C water bath, then vortex vigorously. Repeat the freeze, thaw, vortex cycle a further two times, taking care not to let the virus supernatant warm up, and then transfer to 1.5 ml tubes and centrifuge at 14,000 rpm for 5 min to pellet the cell debris. Transfer the supernatant containing released infectious recombinant adenovirus particles to fresh 1.5 ml tubes and store at −70°C. The number of virus particles produced at this stage is not sufficient for them to be purified and used in experiments; further growth and packaging in HEK293 E1 cells is required.

- Infect two 50% to 70% confluent T-75 flasks with half of the viral supernatant harvested from the transfected HEK293 E1 cells each. Add the viral supernatant to fresh growth medium in the flasks. CPE should become evident at 2 to 3 days post infection.
- Harvest the virus from the cells as described earlier 3 to 5 days post infection; as the pellet of HEK293 E1 cells is bigger, resuspend in 800 μl sterile PBS.
- Transfer the viral lysate to 1.5 ml tubes, as before, but before centrifugation add an equal volume of Arklone and mix by shaking. Centrifuge (14,000 rpm, 5 min), transfer the top layer containing the virus particles to fresh tubes, and store at $-70°$C.
- At this point the viral titer should be at least 10^7 infectious particles per ml and is enough to use for gene transfer experiments in cultured cells; however, to prepare and purify a high-titer viral stock a further round of amplification in HEK293 E1 cells is required. Before this final amplification step the titer of the viral stock harvested from the T-75 flasks must be obtained.

4. Titration of Virus

Materials:

HEK293 E1 cells
96-well tissue culture plates
Growth medium

- A number of different viral titration methods exist; we use the Tissue Culture Infectious Dose 50 (TCID$_{50}$) method. This method is based on the development of CPE in HEK293 E1 cells using endpoint dilutions to estimate the titer.
- Seed HEK293 E1 cells in two 96-well tissue culture plates with 1×10^4 cells per well and allow to attach in a 37°C incubator overnight.
- Serial dilutions of the viral stock are prepared, in duplicate, as follows in 5 ml sterile tubes.
- In duplicate place 0.9 ml growth medium in the first tube and 1.8 ml in all others. Add 0.1 ml of the viral stock to the first tube. Pipette up and down five times to mix and change the tip after each dilution.
- Take 0.2 ml of 10^{-1} dilution and transfer to the second tube.
- Repeat dilutions up to the highest dilution required.
- For each row of the 96-well plate, add 0.1 ml of diluted virus to the wells *1 to *10 for the 8 highest dilutions (Fig. 3). Starting with the highest dilution in the top row.
- Use columns *11 and *12 for the negative control to test cell viability and add 0.1 ml growth medium to these wells.
- Leave the plates in a 37°C incubator for 10 days.
- After 10 days observe the wells and count observable CPE per row using an inverted microscope. A well is counted as positive even if only a small spot of a few cells

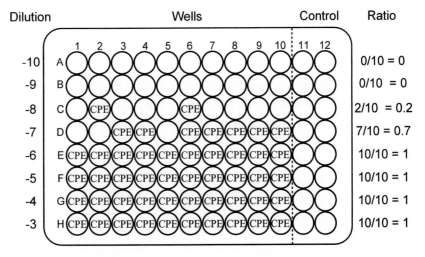

Fig. 3 Typical results of TCID$_{50}$. In this hypothetical experiment no infection is detected at dilutions of -9 or -10. Two wells with cytopathic effect (CPE) are present at -8 dilution and seven wells show CPE at -7 dilution. Higher concentrations of virus (-6 to -3 dilutions) result in all the wells showing CPE.

show CPE. If in doubt between CPE and dead cells, compare with the negative control.

- The test is valid if the negative controls do not show any CPE or cell growth problems and the lowest dilution shows 100% infection (10/10) while the highest dilution shows 0% infection (0/10).
- Calculate the ratio of positive wells per row (Fig. 3).
- Use the KÅRBER statistical method to determine the titer (T), using the following equation:

$$T = 10^{1+d(S-0.5)}$$

d is log$_{10}$ of the dilution ($=1$ for a 10 fold dilution); S is the sum of ratios (always starting from the first 10^{-1} dilution). If some of the lowest dilutions are omitted, like 10^{-1} and 10^{-2} in Fig. 3, they still have to be included in the calculation as ratios of 1.

- For the example shown in Figure 3:

$$d = 1$$
$$S = 1 + 1 + 1 + 1 + 1 + 1 + 0.7 + 0.2 + 0 + 0 = 6.9$$
$$T = 10^{1+1(6.9-0.5)} = 10^{7.4} \text{ for } 100 \ \mu\text{l of virus}$$
$$T = 10^{8.4}/\text{ml}$$
$$\text{TCID}_{50} = 2.5 \times 10^{8} \text{ per ml virus stock}$$

- Between duplicates, the difference in titres should be less than 0.7 log.

5. Preparation and Purification of High–Titer Viral Stocks

Materials:

As for 3

Ultrapure cesium chloride: Gibco BRL (15507-015)

CsCl buffer: 5 mM Tris, 1 mM EDTA, pH 7.8

Beckman Ultra-Clear 14 ml centrifuge tubes: Beckman (344060)

Mineral oil: Sigma (M-5904)

Long Pasteur pipette

Slide-A-Lyzer dialysis cassettes: Pierce (66332)

Buffer A: 10 mM Tris pH 7.5, 1 mM MgCl$_2$, 135 mM NaCl

Buffer B: Buffer A plus 10% glycerol

- Seed HEK293 E1 cells in T-175 flasks and grow until 90% confluent at time of infection. Usually 10-15 T-175 flasks are sufficient to make a high-titer stock.
- Infect cells with virus at a multiplicity of infection (MOI) of 1 to 5, i.e., 1 to 5 virus particles per cell.
- When all the cells have rounded up and 50% show CPE, usually 3 to 4 days post infection, harvest the virus from the flasks. Resuspend the cell pellet from each flask in 800 μl sterile PBS and combine together. Freeze thaw the cells as described in Section V.B.3.
- Add an equal volume of Arklone and shake before centrifuging (14,000 rpm, 5 min).
- The virus is purified by CsCl banding using Beckman 14 ml centrifuge tubes that fit the swing-out SW40 Ti rotor as follows.
- Prepare CsCl at a density of 1.33 g/ml by dissolving 8.349 g in 16 ml CsCl buffer and at a density of 1.45 g/ml by dissolving 8.349 g CsCl in 11.4 ml CsCl buffer.
- Create a gradient by first placing 2.5 ml of 1.33 g/ml CsCl in a centrifuge tube and then carefully placing 1.5 ml of 1.45 g/ml CsCl underneath using a Gilson pipette. Layer the virus prep, harvested from the T-175 dishes, onto the top of the CsCl gradient and place mineral oil over until 2 mm from the top to prevent the tube from collapsing (Fig. 4A).
- Prepare a balance tube and centrifuge in an ultracentrifuge using an SW40 Ti rotor at 22,500 rpm for 2 h at 4°C.
- The virus band is the lowest of three bands that should be visible (Fig. 4B) and is carefully removed using a long glass Pasteur pipette.
- Dilute the virus band with a half-volume of CsCl buffer and layer onto a second CsCl gradient consisting of 1.5 ml of 1.33 g/ml CsCl and 1 ml of 1.45 g/ml CsCl. Again fill the rest of the tube with mineral oil and prepare a balance tube.
- Centrifuge in an ultracentrifuge using an SW40 Ti rotor at 23,800 rpm for 18 h at 4°C. Remove the virus band as before and dialyze against 1 liter buffer A for 1 h

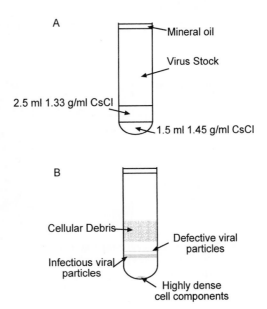

Fig. 4   A diagram showing the CsCl gradient (A) before centrifugation; (B) following the first centrifugation step.

and against 1 liter buffer B for 2 h to remove the CsCl. Use Slide-A-Lyzer dialysis cassettes (Pierce) according to the manufacturer's instructions. Store dialyzed virus in 10- to 50-μl aliquots at $-70°C$.

- Titer the purified viral stock using the method described earlier. The final yields should be in the region of 1×10^{11} to 1×10^{12} virus particles per ml.

VI. Infecting Mammary Epithelial Cell Cultures with Adenovirus

A. Background

The standard method for infecting cells in culture is to place recombinant adenovirus in the medium over cells in a culture dish. The cells are left at 37°C for 1–3 h and, then the medium containing virus particles is replaced with fresh medium. Primary mammary epithelial cells plated on collagen I form a monolayer; however, organoids are still present, and when these cells are infected *in situ* the adenovirus is unable to infect tightly packed clusters of cells (Fig. 5). Moreover, transgene expression in mammary epithelial cells cultured on Matrigel is less than 15% following adenoviral infection. To increase the efficiency of infection, we have adapted the method by infecting mammary epithelial cells as a single cell suspension for up to 1 h prior to plating on an appropriate

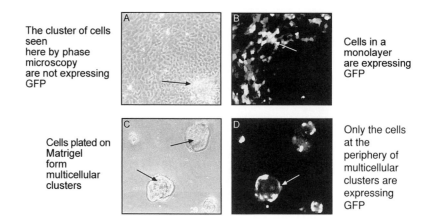

Fig. 5 Primary mammary epithelial cells infected *in situ* with an adenovirus expressing GFP. (A, B) Cells plated on collagen I grow out of organoids (black arrow) to form a monolayer. (C, D) On Matrigel the cells form multicellular clusters (black arrows) representative of alveoli *in vivo*. (B, D) *In situ* infection results in relatively few of the cells becoming infected, shown by the GFP fluorescence (white arrows).

ECM, either collagen I or Matrigel. The cells plated on Matrigel are able to reaggregate and form multicellular clusters, and virtually all of the cells now express the transgene (unpublished data). Infected cells can also be cultured using any one of the techniques described above, for example on thick gels of collagen or as single cells within collagen or basement membrane gels.

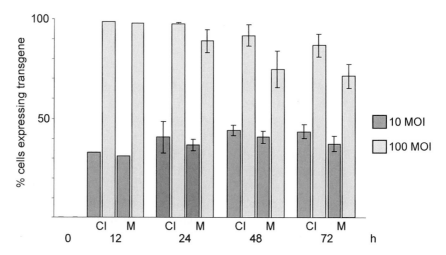

Fig. 6 Highly efficient transgene expression in primary mammary epithelial cells infected in suspension with recombinant adenovirus. Cells were infected with an MOI of 10 or 100 in suspension for 1 h, then replated onto either collagen I (CI) or Matrigel (M). Expression levels over time were monitored, $N = 3$, except for the 12-h time point where $N = 1$ and a minimum of 1000 cells were counted.

B. Materials

Culture medium

70 μm nylon cell strainer: Becton Dickinson Labware (2350)

Recombinant adenovirus

C. Procedure for Infecting Mammary Epithelial Cells

Culture primary mammary epithelial cells on collagen I in order to promote outgrowth from organoids and the formation of a confluent monolayer. Following 3–4 days of growth detach the cells from the culture dish by trypsinization. Resuspend the trypsinized cells in growth medium containing fetal calf serum and strain through a 70 μm nylon cell strainer. It is important to obtain an accurate cell count prior to infection in order to calculate the number of virus particles required for the infection. Perform a viable cell count using the trypan blue exclusion method.

Place the required number of cells into a tube or bottle with 5 to 10 ml growth medium per 2×10^6 cells and add the recombinant adenovirus particles at the desired multiplicity of infection (MOI). Incubate for 1 h at 37°C and agitate the cells every 5 to 10 min. Pellet the cells by centrifugation (1200 rpm, 4 min), resuspend in medium, and plate onto collagen I or Matrigel coated dishes.

In control studies to establish the method and efficiency of infection, cells can be infected with Lac Z or GFP-expressing adenovirus. Under these conditions we have found that up to 95% of cells are infected if they are cultured either as a monolayer or as three D alveolar structures on Matrigel (Fig. 6; unpublished data).

Acknowledgments

Harriet Watkin is a Wellcome Trust Prize Student and Charles H. Streuli is a Wellcome Trust Senior Fellow in Basic Biomedical Science. Our thanks to Nasreen Akhtar for reading the manuscript.

References

Adams, J. C., and Watt, F. M. (1989). Fibronectin inhibits the terminal differentiation of human keratinocytes. *Nature* **340**, 307–309.

Adams, J. C., and Watt, F. M. (1990). Changes in keratinocyte adhesion during terminal differentiation: reduction in fibronectin binding precedes alpha 5 beta 1 integrin loss from the cell surface. *Cell* **63**, 425–435.

Aggeler, J., Ward, J., Blackie, L. M., Barcellos-Hoff, M. H., Streuli, C. H., and Bissell, M. J. (1991). Cytodifferentiation of mouse mammary epithelial-cells cultured on a reconstituted basement-membrane reveals striking similarities to development in vivo. *J. Cell Sci.* **99**, 407–417.

Assoian, R. K., and Schwartz, M. A. (2001). Coordinate signaling by integrins and receptor tyrosine kinases in the regulation of G1 phase cell-cycle progression. *Curr. Opin. Gen. Dev.* **11**, 48–53.

Barcellos-Hoff, M. H., Aggeler, J., Ram, T. G., and Bissell, M. J. (1989). Functional differentiation and alveolar morphogenesis of primary mammary cultures on reconstituted basement membrane. *Development* **105**, 223–235.

Brunner, S., Sauer, T., Carotta, S., Cotten, M., Saltik, M., and Wagner, E. (2000). Cell cycle dependence of gene transfer by lipoplex, polyplex and recombinant adenovirus. *Gene Ther.* **7**, 401–407.

Chu, Y., Sperber, K., Mayer, L., and Hsu, M. T. (1992). Persistent infection of human adenovirus type 5 in human monocyte cell lines. *Virology* **188,** 793–800.

Darcy, K. M., Zangani, D., Lee, P. H., and Ip, M. M. (2000). Isolation and culture of normal rat mammary epithelial cells. In "Methods in Mammary Gland Biology and Breast Cancer Research" (M. M. Ip and B. B. Asch, eds.), pp. 163–175. Kluwer Academic/Plenum Publishers, New York.

Dewey, R. A., Morrissey, G., Cowsill, C. M., Stone, D., Bolognani, F., Dodd, N. J., Southgate, T. D., Klatzmann, D., Lassmann, H., Castro, M. G., and Lowenstein, P. R. (1999). Chronic brain inflammation and persistent herpes simplex virus 1 thymidine kinase expression in survivors of syngeneic glioma treated by adenovirus-mediated gene therapy: implications for clinical trials. *Nat. Med.* **5,** 1256–1263.

Edwards, P. A., Abram, C. L., and Bradbury, J. M. (1996). Genetic manipulation of mammary epithelium by transplantation. *J. Mammary Gland Biol. Neoplasia* **1,** 75–89.

Emerman, J. T., Enami, J., Pitelka, D. R., and Nandi, S. (1977). Hormonal effects on intracellular and secreted casein in cultures of mouse mammary epithelial cells on floating collagen membranes. *Proc. Nat. Acad. Sci. USA* **74,** 4466–4470.

Emerman, J. T., and Pitelka, D. R. (1977). Maintenance and induction of morphological differentiation in dissociated mammary epithelium on floating collagen membranes. *In Vitro Dev. Biol.* **13,** 316–328.

Farrelly, N., Lee, Y. J., Oliver, J., Dive, C., and Streuli, C. H. (1999). Extracellular matrix regulates apoptosis in mammary epithelium through a control on insulin signaling. *J. Cell Biol.* **144,** 1337–1347.

Fuchs, E., Dowling, J., Segre, J., Lo, S. H., and Yu, Q. C. (1997). Integrators of epidermal growth and differentiation: distinct functions for beta 1 and beta 4 integrins. *Curr. Opin. Genet. Dev.* **7,** 672–682.

Giannoukakis, N., Mi, Z., Rudert, W. A., Gambotto, A., Trucco, M., and Robbins, P. (2000). Prevention of beta cell dysfunction and apoptosis activation in human islets by adenoviral gene transfer of the insulin-like growth factor I. *Gene Ther.* **7,** 2015–2022.

Gilmore, A. P., and Streuli, C. H. (1998). Cell adhesion to ECM regulates intracellular localisation of Bax and apoptosis. *Mol. Biol. Cell* **9,** 381a.

Graham, F. L., Smiley, J., Russell, W. C., and Nairn, R. (1977). Characteristics of a human cell line transformed by DNA from human adenovirus type 5. *J. Gen. Virol.* **36,** 59–74.

Haddada, H., Cordier, L., and Perricaudet, M. (1995). Gene therapy using adenovirus vectors. *Curr. Top. Microbiol. Immunol.* **199,** 297–306.

Hay, E. D. (1993). Extracellular matrix alters epithelial differentiation. *Curr. Opin. Cell Biol.* **5,** 1029–1035.

He, T. C., Zhou, S., da Costa, L. T., Yu, J., Kinzler, K. W., and Vogelstein, B. (1998). A simplified system for generating recombinant adenoviruses. *Proc. Natl. Acad. Sci. USA* **95,** 2509–2514.

Imagawa, W., Young, J., Guzman, R. C., and Nandi, S. (2000). Collagen gel method for the primary culture of mouse mammary epithelium. *In* "Methods in Mammary Gland Biology and Breast Cancer Research" (M. M. Ip and B. B. Asch eds.), pp. 111–123. Kluwer Academic/Plenum Publishers, New York.

Ip, M. M., and Darcy, K. M. (1996). Three-dimensional mammary primary culture model systems. *J. Mammary Gland Biol. Neoplasia* **1,** 91–110.

Kleinman, H. K., McGarvey, M. L., Hassell, J. R., Star, V. L., Cannon, F. B., Laurie, G. W., and Martin, G. R. (1986). Basement membrane complexes with biological activity. *Biochemistry* **25,** 312–318.

Klinowska, T. C. M., Soriano, J. V., Edwards, G. M., Oliver, J. M., Valentijn, A. J., Montesano, R., and Streuli, C. H. (1999). Laminin and beta 1 integrins are crucial for normal mammary gland development in the mouse. *Dev. Biol.* **215,** 13–32.

Kopan, R., Traska, G., and Fuchs, E. (1987). Retinoids as important regulators of terminal differentiation: examining keratin expression in individual epidermal cells at various stages of keratinization. *J. Cell Biol.* **105,** 427–440.

Lakhani, S. R., and O'Hare, M. J. (2001). The mammary myoepithelial cell—Cinderella or ugly sister? *Breast Cancer Res.* **3,** 1–4.

Lee, E. Y., Lee, W. H., Kaetzel, C. S., Parry, G., and Bissell, M. J. (1985). Interaction of mouse mammary epithelial cells with collagen substrata: regulation of casein gene expression and secretion. *Proc. Natl. Acad. Sci. USA* **82,** 1419–1423.

Li, M. L., Aggeler, J., Farson, D. A., Hatier, C., Hassell, J., and Bissell, M. J. (1987). Influence of a reconstituted basement membrane and its components on casein gene expression and secretion in mouse mammary epithelial cells. *Proc. Natl. Acad. Sci. USA* **84,** 136–140.

Medina, D., and Kittrell, F. (2000). Establishment of mouse mammary cell lines. *In* "Methods in Mammary Gland Biology and Breast Cancer Research" (M. M. Ip and B. B. Asch eds.) pp. 137–145. Kluwer Academic/Plenum Publishers, New York.

Niemann, C., Brinkmann, V., Spitzer, E., Hartmann, G., Sachs, M., Naundorf, H., and Birchmeier, W. (1998). Reconstitution of mammary gland development in vitro: requirement of c-met and c-erbB2 signaling for branching and alveolar morphogenesis. *J. Cell Biol.* **143,** 533–545.

Pitelka, D. R., Hamamoto, S. T., Duafala, J. G., and Nemanic, M. K. (1973). Cell contacts in the mouse mammary gland. I., Normal gland in postnatal development and the secretory cycle. *J. Cell Biol.* **56,** 797–818.

Pullan, S., and Streuli, C. H. (1996). The mammary gland epithelial cell. *In* "Epithelial Cell Culture" (A. Harris ed.), pp. 97–121. Cambridge University Press, Cambridge, UK.

Pullan, S., Wilson, J., Metcalfe, A., Edwards, G. M., Goberdhan, N., Tilly, J., Hickman, J. A., Dive, C., and Streuli, C. H. (1996). Requirement of basement membrane for the suppression of programmed cell death in mammary epithelium. *J. Cell Sci.* **109,** 631–642.

Rosenfeld, M. A., Siegfried, W., Yoshimura, K., Yoneyama, K., Fukayama, M., Stier, L. E., Paakko, P. K., Gilardi, P., Stratford-Perricaudet, L. D., Perricaudet, M., Jallet, S., Pavirani, A., Lecocq, J.-P., and Crystal, R. G. (1991). Adenovirus-mediated transfer of a recombinant alpha 1-antitrypsin gene to the lung epithelium in vivo. *Science* **252,** 431–434.

Roskelley, C. D., Desprez, P. Y., and Bissell, M. J. (1994). Extracellular matrix-dependent tissue-specific gene expression in mammary epithelial cells requires both physical and biochemical signal transduction. *Proc. Natl. Acad. Sci. USA* **91,** 12378–12382.

Runswick, S. K., O'Hare, M. S., Jones, L., Streuli, C. H., and Garrod, D. R. (2001). Desmosomal adhesion regulates epithelial morphogenesis and cell positioning. *Nat. Cell Biol.* **3,** 823–830.

Schmidhauser, C., Bissell, M. J., Myers, C. A., and Casperson, G. F. (1990). Extracellular-matrix and hormones transcriptionally regulate bovine beta-casein 5′ sequences in stably transfected mouse mammary cells. *Proc. Nat. Acad. Sci. USA* **87,** 9118–9122.

Schmidhauser, C., Casperson, G. F., and Bissell, M. J. (1994). Transcriptional activation by viral enhancers—critical dependence on extracellular-matrix cell-interactions in mammary epithelial-cells. *Mol. Carcinogenesis* **10,** 66–71.

Silberstein, G. B. (2001). Role of the stroma in mammary development. *Breast Cancer Res.* **3,** 218–223.

Smalley, M. J., Titley, J., Paterson, H., Perusinghe, N., Clarke, C., and O'Hare, M. J. (1999). Differentiation of separated mouse mammary luminal epithelial and myoepithelial cells cultured on EHS matrix analyzed by indirect immunofluorescence of cytoskeletal antigens. *J. Histochem. Cytochem.* **47,** 1513–1524.

Soriano, J. V., Pepper, M. S., Nakamura, T., Orci, L., and Montesano, R. (1995). Hepatocyte growth factor stimulates extensive development of branching duct-like structures by cloned mammary gland epithelial cells. *J. Cell Sci.* **108,** 413–430.

Stoker, A. W., Streuli, C. H., Martins Green, M., and Bissell, M. J. (1990). Designer microenvironments for the analysis of cell and tissue function. *Curr. Opin. Cell Biol.* **2,** 864–874.

Streuli, C. H. (1999). Extracellular matrix remodelling, and cellular differentiation. *Curr. Opin. Cell Biol.* **11,** 634–640.

Streuli, C. H., and Bissell, M. J. (1990). Expression of extracellular matrix components is regulated by substratum. *J. Cell. Biol.* **110,** 1405–1415.

Streuli, C. H., Bailey, N., and Bissell, M. J. (1991). Control of mammary epithelial differentiation—basement membrane induces tissue-specific gene expression in the absence of cell cell interaction and morphological polarity. *J. Cell Biol.* **115,** 1383–1395.

Streuli, C. H., Schmidhauser, C., Bailey, N., Yurchenco, P., Skubitz, A. P., Roskelley, C., and Bissell, M. J. (1995). Laminin mediates tissue-specific gene expression in mammary epithelia. *J. Cell Biol.* **129,** 591–603.

Stutzmann, J., Bellissent-Waydelich, A., Fontao, L., Launay, J. F., and Simon-Assmann, P. (2000). Adhesion complexes implicated in intestinal epithelial cell–matrix interactions. *Microsc. Res. Tech.* **51,** 179–190.

Wilkinson, G. W., and Akrigg, A. (1992). Constitutive and enhanced expression from the CMV major IE promoter in a defective adenovirus vector. *Nucleic Acids Res.* **20,** 2233–2239.

PART V

General Information

APPENDIX A

List of Suppliers

Advance Reproductions Corporation	www.advancerepro.com
Alpha Laboratories	www.alphalabs.co.uk
American Type Culture Collection	www.atcc.org
Amersham Pharmacia Biotech	www.apbiotech.com
Anachem	www.anachem.co.uk
Appligene Oncor	www.progen.com
Asylum Research	www.asylumresearch.com
Autogen Bioclear	www.autogen-bioclear.co.uk
Bangs Laboratories	www.bangslabs.com
Bender Med Systems	www.bendermedsystems.com
BIAcore	www.biocore.com
Becton Dickinson Biosciences	www.bdbiosciences.com
Biodesign International	www.biodesign.com
Bioline	www.bioline.com
Biomol Research Laboratories	www.biomol.com
Bio-Rad	www.discover.bio-rad.com
Biosource	www.biosource.com
Boehringer Mannheim	http://biochem.boehringer-mannheim.com
Calbiochem	www.calbiochem.com
Caltag Laboratories	www.caltag.com
Cambio	www.cambio.co.uk
Cambrex Biosciences	www.cambrex.com
Cambridge Bioscience	www.bioscience.co.uk
Camlab	www.camlab.co.uk
Cappel/Helix Diagnostics	Now part of Bio-Rad
Cel-Line	www.cel-line.com
Charles River Laboratories	www.criver.com
Chemicon	www.chemicon.com
Chroma	www.chroma.com
Clontech	www.clontech.com
Collaborative Biomedical Products	Now linked to Becton Dickinson
Corning	www.corning.com
Cor Therapeutics	www.millenium.com

Covance BAbCO	www.covance.com
	www.babco.com
Cytoskeleton	www.cytoskeleton.com
Daigger	www.daigger.com
Dako	www.dako.co.uk
Digital Instruments	www.di.com
D-NAmix Biotechnology	www.d-namix.com
Dow Chemical	www.dow.com
Dynal Biotech	www.dynalbiotech.com
EM Science	www.emscience.com
Endogen	www.endogen.com
Erie Scientific	www.eriesci.com
Exalpha Biologicals	www.exalpha.com
Fine Science Tools	www.finescience.com
Fisher Scientific	www.fisher.co.uk
Flowgen	www.flowgen.co.uk
Fluka	www.fluka.com
Glycorex	www.glycorex.se
GRI	www.gri.co.uk
Hamilton	www.hamiltoncompany.com
Harlan Sera Lab	www.harlanseralab.co.uk
Hyclone	www.hyclone.com
ICN	www.icnbiomed.com
Immunetics	www.immunetics.com
Incyte Genomics	www.incyte.com
Integra Biosciences	www.integra-biosciences.com
Intergen	www.intergenco.com
Intracell Therapeutics	http://informagen.com/Resource_ Informagen/Full/63.html
Invitrogen Life Technologies	www.invitrogen.com
Jencons-PLS	www.jencons.co.uk
JRH Biosciences	www.jrhbio.com
Kirkegaard & Perry Labs	www.kpl.com
Lab Vision	www.labvision.com
Mediatech	www.cellgro.com
Mallinckrodt	www.mallinckrodt.com
Medical & Biological Laboratories	www.mbl.co.jp
Microchem Corporation	www.microchem.com
Millipore	www.millipore.com
Miltenyi Biotech	www.miltenyibiotech.com
Molecular Probes	www.probes.com
Nalge Nunc International	http://nunc.nalgenunc.com
National Diagnostics	www.nationaldiagnostics.com
Neoparin Inc.	www.heparinoids.com
New England Biolabs	www.neb.com

Nikon	www.nikonusa.com
Norland Products	www.norlandprod.com
Novabiochem	www.novabiochem.com
Novagen	www.novagen.com
Novocastra Laboratories	www.novocastra.co.uk
Oncogene Research Products	www.apoptosis.com
Origene Technologies	www.origene.com
Q Biogene	www.qbiogene.com
QIAGEN	www.qiagen.com
Panvera	www.panvera.com
Peprotech	www.peprotech.com
Perkin Elmer Life Sciences	http://lifesciences.perkinelmer.com
Pharmacia	www.pnu.com
Pharmingen	www.pharmingen.com
Pierce Chemical	www.piercenet.com
Polysciences	www.polysciences.com
Promega	www.promega.com
Promo Cell	www.promocell.com
Rainin Instruments	www.rainin.com
R&D systems	www.rndsystems.com
Roche	http://biochem.roche.com
Santa Cruz	www.scbt.co.kr
Scientific Laboratories	www.scientific-labs.com
Seikagaku America	www.seikagaku.com
Serotec	www.serotec.co.uk
Serva	www.serva.de
Sigma-Aldrich	www.sigma-aldrich.com
Stratagene	www.stratagene.com
Stratech	www.stratech.co.uk
Southern Biotechnology Associates	www.southernbiotech.com
Synthecon Inc.	www.synthecon.com
ThermoHybaid	www.thermohybaid.com
Thermo Lab Systems	www.labsystems.com
Titertek	www.titertek.com
Thermo Life Sciences	www.thermols.com
United Chemical Technologies	www.unitedchem.com
Upstate Biotechnology	www.upstatebiotech.com
Vector Laboratories	www.vectorlabs.com
Versene Chelating Agents	www.dow.com/versene
VWR Scientific Products	www.vwrsp.com
Warner Instruments Corp.	www.warnerinstrument.com
Worthington Biochemical Corporation	www.worthington-biochem.com
Zymed Laboratories Inc.	www.zymed.com
Zeiss Inc.	www.zeiss.com

APPENDIX B

Relevant Microarray Dataset Experiments

Responses to Matrix Adhesion

Brenner, V., Lindauer, K., Parker, A., Fordham, J., Hayes, I., Stow, M., Gama, R., Pollock, K., and Jupp, R. (2001). Analysis of cellular adhesion by microarray expression profiling. *J. Immunol. Methods* **250,** 15–28.

Calaluce, R., Kunkel, M. W., Watts, G. S., Schmelz, M., Hao, J., Barrera, J., Gleason-Guzman, M., Isett, R., Fitchmun, M., Bowden, G. T., Cress, A. E., Futscher, B. W., and Nagle, R. B. (2001). Laminin-5-mediated gene expression in human prostate carcinoma cells. *Mol. Carcinog.* **30,** 119–129.

de Fougerolles, A. R., Chi-Rosso, G., Bajardi, A., Gotwals, P., Green, C. D., and Koteliansky, V. E. (2000). Global expression analysis of extracellular matrix-integrin interactions in monocytes. *Immunity* **13,** 749–758.

Falsey, J. R., Renil, M., Park, S., Li, S., and Lam, K. S. (2001). Peptide and small molecule microarray for high throughput cell adhesion and functional assays. *Bioconjug. Chem.* **12,** 346–353.

Kessler, D., Dethlefsen, S., Haase, I., Plomann, M., Hirche, F., Krieg, T., and Eckes, B. (2001). Fibroblasts in mechanically stressed collagen lattices assume a "synthetic" phenotype. *J. Biol. Chem.* **276,** 36575–36585.

Medico, E., Gentile, A., Lo Celso, C., Williams, T. A., Gambarotta, G., Trusolino, L., and Comoglio, P. M. (2001). Osteopontin is an autocrine mediator of hepatocyte growth factor-induced invasive growth. *Cancer Res.* **61,** 5861–5868.

Adhesion Molecules in Vascular Biology/Angiogenesis

Feng, Y., Yang, J. H., Huang, H., Kennedy, S. P., Turi, T. G., Thompson, J. F., Libby, P., and Lee, R. T (1999). Transcriptional profile of mechanically induced genes in human vascular smooth muscle cells. *Circ. Res.* **85,** 1118–1123.

Hendrix, M. J., Seftor, E. A., Meltzer, P. S., Gardner, L. M., Hess, A. R., Kirschmann, D. A., Schatteman, G. C., and Seftor, R. E. (2001). Expression and functional significance of VE-cadherin in aggressive human melanoma cells: role in vasculogenic mimicry. *Proc. Natl. Acad. Sci. USA* **98,** 8018–8023.

Lee, R. T., Yamamoto, C., Feng, Y., Potter-Perigo, S., Briggs, W. H., Landschulz, K. T., Turi, T. G., Thompson, J. F., Libby, P., and Wight, T. N. (2001). Mechanical strain induces specific changes in the synthesis and organization of proteoglycans by vascular smooth muscle cells. *J. Biol. Chem.* **276,** 13847–13851.

McCormick, S. M., Eskin, S. G., McIntire, L. V., Teng, C. L., Lu, C. M., Russell, C. G., and Chittur, K. K. (2001). DNA microarray reveals changes in gene expression of shear stressed human umbilical vein endothelial cells. *Proc. Natl. Acad. Sci. USA* **98,** 8955–8960.

Maniotis, A. J., Folberg, R., Hess, A., Seftor, E. A., Gardner, L. M., Pe'er, J., Trent, J. M., Meltzer, P. S., and Hendrix, M. J. (1999). Vascular channel formation by human melanoma cells in vivo and in vitro: vasculogenic mimicry. *Am. J. Pathol.* **155,** 739–752.

Pendurthi, U. R., Allen, K. E., Ezban, M., and Rao, L. V. (2000). Factor VIIa and thrombin induce the expression of Cyr61 and connective tissue growth factor, extracellular matrix signaling proteins that could act as possible downstream mediators in factor VIIa x tissue factor-induced signal transduction. *J. Biol. Chem.* **275**, 14632–14641.

Riewald, M., Kravchenko, V. V., Petrovan, R. J., O'Brien, P. J., Brass, L. F., Ulevitch, R. J., and Ruf, W. (2001). Gene induction by coagulation factor Xa is mediated by activation of protease-activated receptor 1. *Blood* **97**, 3109–3116.

Seftor, R. E., Seftor, E. A., Koshikawa, N., Meltzer, P. S., Gardner, L. M., Bilban, M., Stetler-Stevenson, W. G., Quaranta, V., and Hendrix, M. J. (2001). Cooperative interactions of laminin 5 gamma2 chain, matrix metalloproteinase-2, and membrane type-1-matrix/metalloproteinase are required for mimicry of embryonic vasculogenesis by aggressive melanoma. *Cancer Res.* **61**, 6322–6327.

Physiological Changes Involving Alterations in Matrix–Adhesion Molecules

Baker, T. K., Carfagna, M. A., Gao, H., Dow, E. R., Li, Q., Searfoss, G. H., and Ryan, T. P. (2001). Temporal gene expression analysis of monolayer cultured rat hepatocytes. *Chem. Res. Toxicol.* **14**, 1218–1231.

Bilban, M., Head, S., Desoye, G., and Quaranta, V. (2000). DNA microarrays: a novel approach to investigate genomics in trophoblast invasion—a review. *Placenta* **21**, Suppl A; S99–105.

Chen, J., Zhong, O., Wang, J., Cameron, R. S., Borke, J. L., Isales, C. M., and Bollag, R. J. (2001). Microarray analysis of Tbx2-directed gene expression: a possible role in osteogenesis. *Mol. Cell. Endocrinol.* **177**, 43–54.

Cirelli, C., and Tononi, G. (2000). Gene expression in the brain across the sleep-waking cycle. *Brain Res.* **885**, 303–321.

de Veer, M. J., Holko, M., Frevel, M., Walker, E., Der, S., and Paranjape, J. M. (2001). Functional classification of interferon-stimulated genes identified using microarrays. *J. Leukoc. Biol.* **69**, 912–920, (Data: www.lerner.ccf.org/labs/williams).

Diehn, M., Eisen, M. B., Botstein, D., and Brown, P. O. (2000). Large-scale identification of secreted and membrane-associated gene products using DNA microarrays. *Nat. Genet.* **25**, 58–62 (protocols: http://cmgm.stanford.edu/pbrown. Data and images: http://genome-www.stanford.edu/mbp).

Dietz, A. B., Bulur, P. A., Knutson, G. J., Matasic, R., and Vuk-Pavlovic, S. (2000). Maturation of human monocyte-derived dendritic cells studied by microarray hybridization. *Biochem. Biophys. Res. Commun.* **275**, 731–738.

Funk, W. D., Wang, C. K., Shelton, D. N., Harley, C. B., Pagon, G. D., and Hoeffler, W. K. (2000). Telomerase expression restores dermal integrity to in vitro-aged fibroblasts in a reconstituted skin model. *Exp. Cell Res.* **258**, 270–278.

Hishikawa, K., Oemar, B. S., and Nakaki, T. (2001). Static pressure regulates connective tissue growth factor expression in human mesangial cells. *J. Biol. Chem.* **276**, 16797–16803.

Kelly, D. L., and Rizzino, A. (2000). DNA microarray analyses of genes regulated during the differentiation of embryonic stem cells. *Mol. Reprod. Dev.* **56**, 113–123.

Le Naour, F., Hohenkirk, L., Grolleau, A., Misek, D. E., Lescure, P., Geiger, J. D., Hanash, S., and Beretta, L. (2001). Profiling changes in gene expression during differentiation and maturation of monocyte-derived dendritic cells using both oligonucleotide microarrays and proteomics. *J. Biol. Chem.* **276**, 17920–17931.

Rockett, J. C., Mapp, F. L., Garges, J. B., Luft, J. C., Mori, C., and Dix, D. J. (2001). Effects of hyperthermia on spermatogenesis, apoptosis, gene expression, and fertility in adult male mice. *Biol. Reprod.* **65**, 229–239.

Tsou, R., Cole, J. K., Nathens, A. B., Isik, F. F. M., Heimbach, D. M., Engrav, L. H., and Gibran, N. S. (2000). Analysis of hypertrophic and normal scar gene expression with cDNA microarrays. *J. Burn Care Rehabil.* **21**, 541–550.

Verrechia, F., Rossert, J., and Mauviel, A. (2001). Blocking sp1 transcription factor broadly inhibits extracellular matrix gene expression in vitro and in vivo implications for the treatment of tissue fibrosis. *J. Invest. Dermatol.* **116,** 755–763. (Full dataset in Table I in the supplementary material section of the PNAS Web site: www.pnas.org).

Verrecchia, F., Chu, M. L., and Mauviel, A. (2001). Identification of novel TGF-beta/Smad gene targets in dermal fibroblasts using a combined cDNA microarray/promoter transactivation approach. *J. Biol. Chem.* **276,** 17058–17062.

Yarwood, S. J., and Woodgett, J. R. (2001). Extracellular matrix composition determines the transcriptional response to epidermal growth factor receptor activation. *Proc. Natl. Acad. Sci. USA* **98,** 4472–4477. (Full dataset of experiments in the supplementary material section of the PNAS Web site: www.pnas.org).

Ziauddin, J., and Sabatini, D. M. (2001). Microarrays of cells expressing defined cDNAs. *Nature* **411,** 107–110. (Protocols: http://staffa.wi.mit.edu/sabatini_public/reverse_transfection.htm).

Changes in Adhesion Molecules in Cancer

Akutsu, N., Lin, R., Bastien, Y., Bestawros, A., Enepekides, D. J., Black, M. J., and White, J. H. (2001). Regulation of gene expression by 1alpha,2,5-dihydroxyvitamin D3 and its analog EB1089 under growth-inhibitory conditions in squamous carcinoma cells. *Mol. Endocrinol.* **15,** 1127–1139.

Brem, R., Hildebrandt, T., Jarsch, M., Van Muijen, G. N., and Weidle, U. H. (2001). Identification of metastasis-associated genes by transcriptional profiling of a metastasizing versus a non-metastasizing human melanoma cell line. *Anticancer Res.* **21,** 1731–1740.

Brem, R., Certa, U., Neeb, M., Nair, A. P., and Moroni, C. (2001). Global analysis of differential gene expression after transformation with the v-H-ras oncogene in a murine tumor model. *Oncogene* **20,** 2854–2858.

Clark, E. A., Golub, T. R., Lander, E. S., and Hynes, R. O. (2000). Genomic analysis of metastasis reveals an essential role for RhoC. *Nature* **406,** 532–535.

Coller, H. A., Grandori, C., Tamavo, P., Colbert, T., Lander, E. S., Eisenman, R. N., and Golub, T. R. (2000). Expression analysis with oligonucleotide microarrays reveals that MYC regulates genes involved in growth, cell cycle, signaling, and adhesion. *Proc. Natl. Acad. Sci. USA* **97,** 3260–3265. (Complete datasets and protocols: www.genome.wi.mit.edu/MPR).

Hong, T. M., Yang, P. C., Peck, K., Chen, J. J., Yang, S. C., Chen, Y. C., and Wu, C. W. (2000). Profiling the downstream genes of tumor suppressor PTEN in lung cancer cells by complementary DNA microarray. *Am. J. Respir. Cell. Mol. Biol.* **23,** 355–363.

Kannan, K., Amariglio, N., Rechavi, G., Jakob-Hirsch, J., Kela, I., Kaminski, N., Getz, G., Domany, E., and Givol, D. (2001). DNA microarrays identification of primary and secondary target genes regulated by p53. *Oncogene* **20,** 2225–2234. (Table of genes upregulated by p53: www.weizmann.ac.il/home/ligivol/primary.html).

Khanna, C., Khan, J., Nguyen, P., Prehn, J., Caylor, J., Yeung, C., Trepel, J., Meltzer, P., and Helman, L. (2001). Metastasis-associated differences in gene expression in a murine model of osteosarcoma. *Cancer Res.* **61,** 3750–3759.

Klus, G. T., Rokaeus, N., Bittner, M. L., Chen, Y., Korz, D. M., Sukumar, S., Schick, A., and Szallasi, Z. (2001). Down-regulation of the desmosomal cadherin desmocollin 3 in human breast cancer. *Int. J. Oncol.* **19,** 169–174.

Ljubimova, J. Y., Lakhter, A. J., Loksh, A., Yong, W. H., Riedinger, M. S., Miner, J. H., Sorokin, L. M., Ljubimov, A. V., and Black, K. L. (2001). Overexpression of alpha4 chain-containing laminins in human glial tumors identified by gene microarray analysis. *Cancer Res.* **61,** 5601–5610.

Medico, E., Gentile, A., Lo Celso, C., Williams, T. A., Gambarotta, G., Trusolino, L., and Comoglio, P. M. (2001). Osteopontin is an autocrine mediator of hepatocyte growth factor-induced invasive growth. *Cancer Res.* **61,** 5861–5868.

Ng, P. W., Iha, H., Iwanaga, Y., Bittner, M., Chen, Y., Jiang, Y., Gooden, G., Trent, J. M., Meltzer, P., Jeang, K. T., and Zeichner, S. L. (2001). Genome-wide expression changes induced by HTLV-1 Tax: evidence for MLK-3 mixed lineage kinase involvement Tax-mediated NF-kappaB activation. *Oncogene* **20,** 4484–4496.

Rubin, M. A., Mucci, N. R., Figurski, J., Fecko, A., Pienta, K. J., and Day, M. L. (2001). E-cadherin expression in prostate cancer: a broad survey using high-density tissue microarray technology. *Hum. Pathol.* **32,** 690–697.

Stratowa, C., Loffler, G., Lichter, P., Stilgenbauer, S., Haberl, P., Schweifer, N., Dohner, H., and Wilgenbus, K. K. (2001). cDNA microarray gene expression analysis of B-cell chronic lymphocytic leukemia proposes potential new prognostic markers involved in lymphocyte trafficking. *Int. J. Cancer* **91,** 474–480.

Virtaneva, K., Wright, F. A., Tanner, S. M., Yuan, B., Lemon, W. J., Caligiuri, M. A., Bloomfield, C. D., de La Chapelle, A., and Krahe, R. (2001). Expression profiling reveals fundamental biological differences in acute myeloid leukemia with isolated trisomy 8 and normal cytogenetics. *Proc. Natl. Acad. Sci. USA* **98,** 1124–1129. (Detailed protocols and datasets: http://cancergenetics.med.ohio-state.edu/microarray. Also data in the supplementary material section of the PNAS Web site: www.pnas.org).

Zhao, R., Gish, K., Murphy, M., Yin, Y., Notterman, D., Hoffman, W. H., Tom, E., Mack, D. H., and Levine, A. J. (2000). Analysis of p53-regulated gene expression patterns using oligonucleotide arrays. *Genes Dev.* **14,** 981–993. (List of induced genes: http://microarray.princeton.edu/oncology).

Other Diseases

Robbins, P. B., Sheu, S. M., Goodnough, J. B., and Khavari, P. A. (2001). Impact of laminin 5 beta3 gene versus protein replacement on gene expression patterns in junctional epidermolysis bullosa. *Hum. Gene Ther.* **12,** 1443–1448.

Purcell, A. E., Rocco, M. M., Lenhart, J. A., Hyder, K., Zimmerman, A. W., and Pevsner, J. (2001). Assessment of neural cell adhesion molecule (N-CAM) in autistic serum and postmortem brain. *J. Autism Dev. Disord.* **31,** 183–194.

APPENDIX C

Web Site Resources of Interest

Extracellular Matrix and Cell Adhesion

• Escape to Adhesion (www.med.virginia.edu/medicine/basic-sci/cellbio/integrins/index1.html): This site has links to various journals related to cell–matrix research. It also has links to companies and information about how to search for antibodies to cell–matrix receptors or related molecules. It also has many databases such as Protein Structure Databases and Two Dimensional Gel Databases, as well as Transgenics Sites.

• The Integrin Page (www.geocities.com/CapeCanaveral/9629/): This site has information about integrins.

• Fibronectin Page (www.gwumc.edu/biochem/ingham/fnpage.htm): This site has detailed information about fibronectin, including its interactions with other matrix components and transmembrane proteins.

• Fibronectin, an Extracellular Adhesion Molecule (www.callutheran.edu/BioDev/omm/fibro/fibro.htm): This site has information about the structure of fibronectin.

• The Horwitz Connection (www.people.virginia.edu/~afh2n/home.html): This site has many images related to cell–matrix interaction.

• Glycosaminoglycans and Proteoglycans (http://we.indstate.edu/theme/mwking/glycans.html): This site has basic information on the structure and clinical significance of glycosaminoglycans and proteoglycans.

• Collagen (www.le.ac.uk/genetics/collagen): A database of human type I and type III collagen mutations.

• Tissue locations of collagens and links to mutational information (www.zygote.swarthmore.edu/cell6.html)

• ADAMTS proteases (www.lerner.ccf.org/bme/staff/apte/adamts/): The ADAMTS proteases homepage.

• Wound Healing Society (www.woundheal.org): This site has links to journals and companies, as well as links to cell–matrix research in wound healing.

• Angiogenesis (www.angio.org/understanding/understanding.html): Information on angiogenesis at the Web site of the Angiogenesis Organisation.

• Cell Migration (www.cellmigration.org): Home page of the Cell Migration Consortium.

Cell Adhesion and Cell Biology

• The WWW Virtual Library of Cell Biology (http://vlib.org/Science/Cell_Biology/): This site contains links to Web sites related to cell biology, including cell adhesion, the extracellular matrix, the cytoskeleton, and cell motility.

• Cell and Molecular Biology Online (www.cellbio.com): An informational resource for cellular and molecular biology, including lists of useful links, online journals, and protocols.

• Biocarta (www.biocarta.com): This site has interactive graphic models of cellular and molecular pathways including adhesion pathways, as well as a gene search tool using multiple online databases.

• Medscape Today (www.medscape.com/): The latest information about preclinical and clinical studies related to various fields including cell–matrix research.

Anatomical and Histological Information

• μMRI Atlas of Mouse Development (www.mouseatlas.caltech.edu): This digital atlas has images and films of developing mouse embryos obtained by magnetic resonance imaging.

• The Mouse Atlas and Gene Expression Database Project (http://genex.hgu.mrc.ac.uk): This site contains a digital atlas of mouse development and a gene expression database.

• BrainInfo (www.braininfo.rprc.washington.edu): This site contains information about structures in the brain.

• Comparative Mammalian Brain Collections (www.brainmuseum.org): This site provides images of brains of a large number of mammals.

• Axeldb (www.dkfz-heidelberg.de/abt0135/axeldb.htm): *Xenopus* gene expression patterns.

• C. elegansparts (www.elegans.swumed.edu): Information on all cells and lineages in *C. elegans*.

Genome Databases and Organism-Related Web Sites

• The Human Genome Project (www.sanger.ac.uk/HGP): The Human Genomic Research Program.

• The Genome Database (www.gdb.org/gdb): An international collaboration in support of the Human Genome Project.

• The TIGR Human Gene Index (www.tigr.org/tdb/hgi): The TIGR Human Gene Index integrates information from international human gene research projects.

• Mouse Genome Informatics (www.informatics.jax.org): The mouse genome.

• Rat Genome Database (http://rgd.mcw.com)

- Ratmap (http://ratmap.gen.gu.se)
- Cow database (www.cri.bbsrc.ac.uk/bovmap)
- Chicken database (www.ri.bbsrc.ac.uk/chickmap): Gene mapping information and general database at the Roslin Institute.
- US poultry genome-mapping project (http://poultry.mph.msu.edu)
- Zebrafish (http://zfin.org): The zebrafish information network.
- Fishbase (www.fishbase.org): Global information system on fish.
- Ichthyology Web Resources (www.biology.ualberta.ca/jackson.hp/IWR/index.php): This site provides a directory of icthyology-related sites.
- Xenopus (www.xenbase.org): Cell and developmental biology of *Xenopus*.
- Berkeley Drosophila Genome Project (www.fruitfly.org): The *Drosophila* genome.
- Flybase (http://flybase.bio.indiana.edu:82/): A database of *Drosophila* biology.
- AceDB (www.hgmp.mrc.ac.uk): *C. elegans* genome information.
- Saccharomyces Genome Database (http://genome-www.stanford.edu/Saccharomyces): The *Saccharomyces cerevisiae* genome.
- MIPS Yeast Genome Database (www.mips.biochem.mpg.de/proj/yeast): The *Saccharomyces cerevisiae* genome.
- *Schizosaccharomyces pombe* Gene DB (www.genedb.org/genedb/pombe): Database of *S. pombe* gene products.
- *S. pombe* genome (www.sanger.ac.uk/Projects/S_pombe.)
- Animal Genome Size Database (www.genomesize.com): This site contains a list of animal genome size data.
- euGenes-Genomic Information for Eukaryotic Organisms (http://iubio.bio.indiana.edu:8089/): This site has information on genes from eukaryotic organism databases.
- Human Genome Central (www.ensembl.org/genome/central): This site was compiled by the International Human Genome Sequencing Consortium and contains a list of genome sites.
- Cybergenome (www.cybergenome.com): Links to information on the major model organisms.
- Genome Web (www.hgmp.mrc.ac.uk/GenomeWeb): This site contains a list of useful genomic sites.
- Frontiers in Biosciences (www.bioscience.org/urllists/genodba.htm): A compilation of genome databases.

Knockout and Transgenic Mice Web Sites

- Gene Knockouts (www.bioscience.org/knockout/alphabet.htm): This site has information about most of the gene knockouts to date.

• The Transgenic and Targeted Mutant Animal Database (www.ornl.gov/Tech-Resources/Trans/hmepg.html)

• Compendium of Published Wound Healing Studies on Genetically Modified Mice (www1.cell.biol.ethz.ch/members/grose/woundtransgenic/home.html): This site contains a list of papers on the use of transgenic mice in wound healing studies.

Disease and Mutation Databases

• Online Mendelian Inheritance in Man—OMIM (www.ncbi.nlm.nih.gov/Omim/): Catalog of human genes and genetic disorders.

• Online Mendelian Inheritance in Animals—OMIA (www.angis.su.oz.au/Databases/BIRX/omia): Catalog of animal genes and genetic disorders.

• OMIM Locus-specific Mutation Database Links (www.ncbi.nlm.nih.gov/Omim/Index/mutation.html): Links to a number of locus-specific mutation databases.

• Positionally Cloned Human Disease Genes (www.ncbi.nlm.nih.gov/Disease_Genes/m97hdt0.html)

• Atlas of Genetics and Cytogenetics in Oncology and Haematology (www.infobiogen.fr/services/chromcancer): This site contains information about genes involved in cancer, including the mutation that leads to cancer, the altered protein made by the mutant gene, and the type of tumor in which it is found.

• GeneCards (http://bioinfo/weizmann.ac.il/cards): A database of human genes, their products, and their involvement in disease.

• The Cancer Genome Anatomy Project (www.ncbi.nlm.nih.gov/ncicgap): Database of the genes responsible for cancer.

• HotMolecBase (http://bioinformatics.weizmann.ac.il/hotmolecbase/hotmolec.htm): This resource contains extended information about selected biological factors that may play crucial roles in diseases.

Protein Web Sites

• Protein Data Bank (www.rcsb.org/pbd): Primary archive of all 3D structures of macromolecules (proteins, DNA, RNA and various complexes).

• Protein Information Resource (www-nbrf.georgetown.edu): This is a protein sequence database that also contains information on the properties and important regions of each protein.

• PIR-International (www.mips.biochem.mpg.de/proj/protseqdb/): A protein database that in addition to sequences provides a variety of biological information, such as protein function and homology information.

• SWISS-PROT (www.expasy.ch/sprot/sprot-top.html): A protein database that provides a description of the function of a protein, its domain structure, post-translational modifications, etc.

- OWL (www.bioinf.man.ac.uk/dbbrowser/OWL): A composite protein sequence database that contains entries from SWISS_PROT, PIR (1-3), GenBank (translation), and NRL-3D.

- Non-Redundant DataBases (www.ncbi.nlm.nih.gov/database/index.html): The Protein sequences database contains sequence data compiled from GenBank, EMBL, DDBJ, PIR, SWISSPROT, PRF, PDB.

- Protein Families Database of Alignments and HMMs (www.sanger.ac.uk/Pfam/): Pfam is a large collection of multiple sequence alignments and profile hidden Markov models (HMMs) covering many common protein domains.

- PROSITE (www.expasy.ch/prosite): PROSITE is a database of residue patterns and profiles that characterize biologically significant sites in proteins and can help identify to which known protein family a new sequence belongs.

- PSORT (http://psort.nibb.ac.jp): A sequence search program for protein localization motifs.

- PRINTS—Protein Fingerprint Database (http://bioinf.mcc.ac.uk/dbbrowser/PRINTS/PRINTS.html): Prints is a compendium of protein fingerprints (conserved motifs used to characterize a protein family).

- SMART (http://smart.embl-heidelberg.de): Domain search tool for protein modular architecture.

- Multiple sequence alignments (http://searchlauncher.bcm.tmc.edu/multialign); http://genomatrix.gsf.de/dalign): Collections of alignment programs for DNA and protein.

- HUGE protein database (www.kazusa.or.jp/huge/index.html): A Database of Human Unidentified Gene-Encoded Large Proteins; contains large (>4kb) human cDNAs.

- Highlighting homologies in multiple sequence alignments: Boxshade at (http://www.ch.embnet.org)

Directories of Online Links

- CMS Molecular Biology Resource (www.sdsc.edu/ResTools/cmshp.html): A compendium of tools and resources for molecular biology and proteomics.

- Biological Databases (www.tdi.es/database.htm): A collection of sites useful to molecular and cellular biologists.

- The EST machine (www.tigem.it/ESTmachine.html): Links to resources for expressed sequence tags analysis.

- Molecular Biology Shortcuts (www.mbshortcuts.com/index.shtml): Links to molecular and cellular biology methods and databases.

- Virtual Lab (www.geocities.com/virtualab/): This site contains many useful links.

- Bio-World (http://search.ebi.ac.uk:8888/compass): An index of resources in the fields of bioinformatics and molecular biology.

- The Free Medical Journals Site (www.freemedicaljournals.com): This site has links to the journals that are available on line for free.
- The Antibody Resource Page (www.antibodyresource.com): A guide to antibody research and suppliers.
- Abcam (www.abcam.com): Gateway and search resource for antibodies.

Other Useful Web Sites

- AcroMed (www.medstract.org): A database of biomedical acronyms.
- Argus (http://vessels.bwh.harvard.edu/software/Argus): A database resource system for the analysis of microarray datasets.
- Biophysics Textbook Online (www.biophysics.org/biophys/society/btol): This is an online biophysics textbook that contains chapters covering topics ranging from membranes, ion channels, and receptors, to NMR and spectroscopy.
- Cell Signalling Networks Database (http://geo.nihs.go.jp/csndb): This database contains information about signaling networks in human cells.
- Genomic Glossary (www.genomicglossaries.com): This site contains a very extended list of glossaries covering areas ranging from cell biology and gene expression to bioinformatics and microarrays.
- Medical Dictionary (www.graylab.ac.uk): Medical dictionary resource of the Gray Cancer Institute.
- Signal Transduction Knowledge Environment (http://stke.sciencemag.org/): This site has information related to signal transduction research.

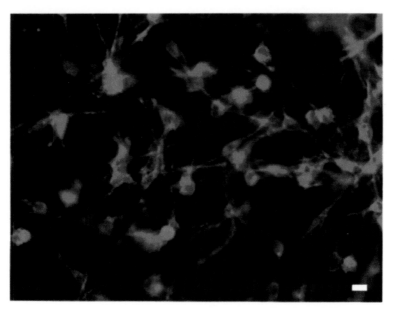

Fig. 6.2 Immunofluorescent staining of cells within a three-dimensional matrix. Cells were cultured for 18 h in a three-dimensional matrix consisting of fibrin and FN. After this time, cells were fixed and the actin cytoskeleton visualized using rhodamine-labeled phalloidin. Scale bar = 20 μm.

A

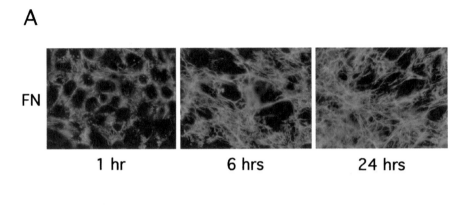

FN

1 hr   6 hrs   24 hrs

B

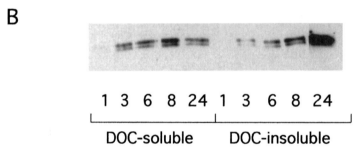

1 3 6 8 24   1 3 6 8 24

DOC-soluble   DOC-insoluble

Fig. 6.3   Time course of FN fibril formation. (A) Analysis by immunofluorescence staining. CHOα5 cells were incubated with 50 μg/ml of rat pFN. At indicated times, the cells were fixed, stained with rat-specific monoclonal antibody IC3, and visualized with fluorescently labeled secondary antibody. (B) Analysis of DOC-soluble and DOC-insoluble material. Cells were lysed with DOC lysis buffer. Fractions (3 μg/lane) were separated by 5% SDS–PAGE and FN was detected with rat-specific monoclonal antibody IC3 and ECL reagents.

Atomic Force Microscope

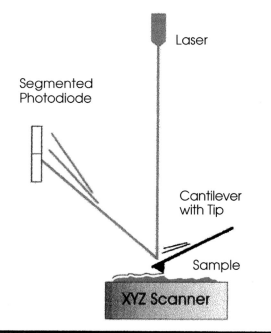

Laser

Segmented
Photodiode

Cantilever
with Tip

Sample

XYZ Scanner

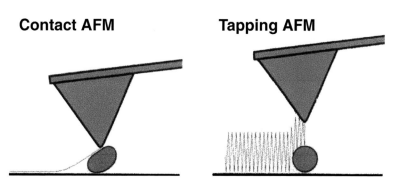

Contact AFM   **Tapping AFM**

Fig. 7.1   Probe microscopy. *(Top)* The AFM (also known as the scanning force microscope, SFM) images samples by raster-scanning a small tip back and forth over the sample surface. When the tip encounters features on the sample surface, the cantilever deflects. This deflection is sensed with an optical lever: a laser beam reflecting off the end of the cantilever onto a segmented pho-todiode magnifies small cantilever deflections into large changes in the relative intensity of the laser light on the two segments of the photodiode. *(Bottom)* Tapping AFM often induces less distortion and movement of soft biomolecules than Contact AFM. In tapping AFM, the cantilever oscillates as it scans the sample surface, which reduces lateral forces on the biomolecules, as illus-trated in these cartoons. The molecular force probe (http://www.asylumresearch.com/) is similar to the AFM but is optimized for accurate measurements and precise control of movement in the Z-direction, enabling pulling on biomolecules to measure inter-molecular and intramolecular interactions. These interactions are visualized as force–distance curves as in Figs. 6 and 10.

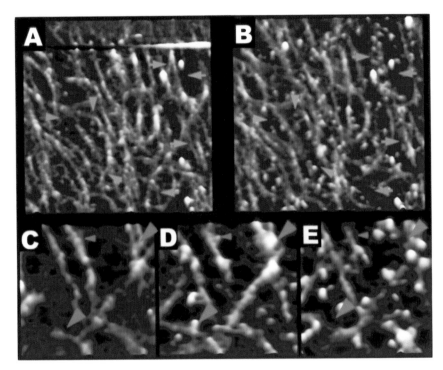

Fig. 7.5 Imaging collagenase digestion of collagen. Height images obtained using the tapping mode are shown that reveal individual collagen I fibers being digested by *C. histolyticum* collagenase. (A, B) Images were collected right after collagenase addition (A) or 4 min after enzyme addition (B). Arrows indicate collagen molecules that are cleaved during this time. The line across the image (A) indicates the end of the collagenase addition. Image area is 1 μm^2. (C–E) Images were collected before (C), immediately after (D), or 4 min after (E) collagenase addition. Arrows indicate sites on collagen that are bound by globular collagenase and then cleaved. Image area = 0.6 μm^2. (From Lin *et al.,* 1999, with permission from the American Chemical Society.)

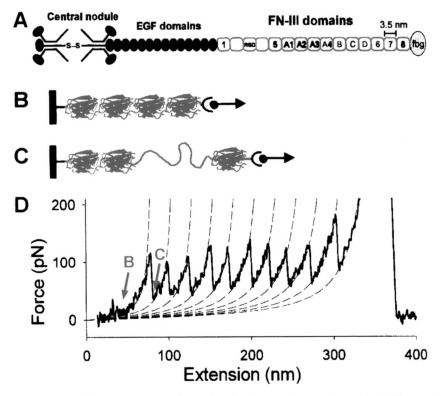

Fig. 7.6 Molecular pulling (force spectroscopy) of tenascin. (A) Diagram of a tenascin arm. (B, C) Diagram of four FN-III domains of tenascin, tethered to surface, and tip (B) at start of pull and (C) after rupture of one FN-III domain. (D) Force–extension spectrum of the unfolding of a molecule containing at least 10 FN-III domains. B and C indicate stages of unfolding as diagramed in (B) and (C) above. Dashed lines show WLC fits. (Figure compiled from Oberhauser *et al.,* 1998, with permission of authors and *Nature*. See also Erickson, 1997; Rief *et al.,* 1997.)

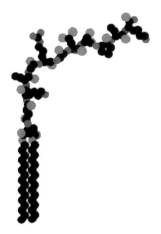

Fig. 7.7 The general structure of the peptide-amphiphile (C_{16})$_2$-Glu-C_2-KAbuGRGDSPAbuK.

Heparin-gold particles bound to Fn dimers

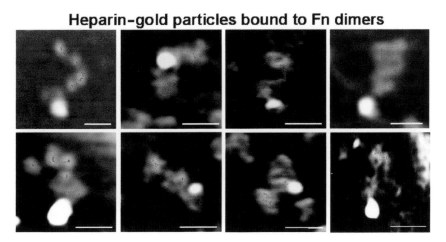

Fig. 7.12 Height mode images of heparin–gold particles bound to Fn dimers. The peptide backbone of Fn is indicated by a dotted line. Scale bars = 50 nm.

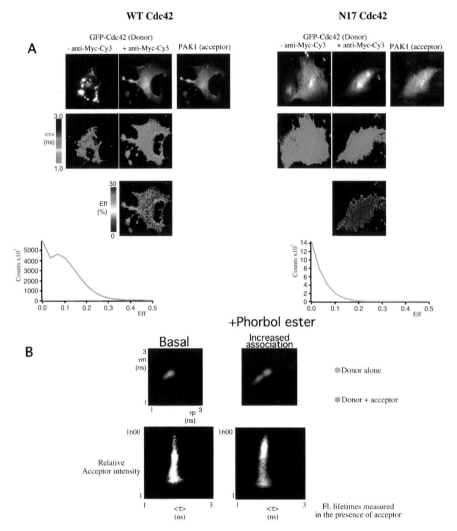

Fig. 12.4 (A) To determine the spatiotemporal distribution of the activated form of Cdc42, FLIM was undertaken to determine the extent of FRET between GFP-Cdc42 (donor) and Myc-tagged PAK1 stained with a Cy3 (acceptor) conjugated antibody 9E10 to the Myc epitope. Porcine aortic endothelial (PAE) cells stably expressing the platelet-derived growth factor (PDGF) PDGF β receptor were dually transfected with both a GFP-Cdc42 (wild-type WT or inactive N17 mutant form) and a Myc-tagged PAK1 construct for 36 h, then stimulated with PDGF (50 ng/ml) for 45 min before fixation. In a third of the WT Cdc42/PAK-cotransfected cells stimulated with PDGF and stained with an anti-Myc-Cy3 post-fixation, it was evident that fluorescence lifetime, <τ>, for GFP was decreased at the cell periphery and filopodial extensions. Because only the activated form of the small GTPases binds PAK, the localization of a reduced <τ> (compared to the GFP lifetime in control cells that were not stained with the Cy3-labeled anti-Myc antibody) indicates the subcellular distribution of the activated form of Cdc42. (B) The detection of a phorbol ester-enhanced association between GFP-PKCα and its binding partner by FLIM. Please refer to text for a detailed description.

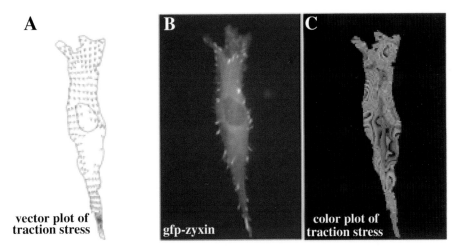

Fig. 16.2 Visualization of traction stress. (A) A vector map of traction stress shows the direction and magnitude of forces per unit area generated by a fish fin fibroblast. (B) GFP-zyxin identifies the focal adhesions. (C) Color rendering of the magnitude of traction stress, where "hot" colors (red) indicate strong forces and "cool" colors weak forces.

INDEX

VOLUMES IN SERIES

Founding Series Editor
DAVID M. PRESCOTT

Volume 1 (1964)
Methods in Cell Physiology
Edited by David M. Prescott

Volume 2 (1966)
Methods in Cell Physiology
Edited by David M. Prescott

Volume 3 (1968)
Methods in Cell Physiology
Edited by David M. Prescott

Volume 4 (1970)
Methods in Cell Physiology
Edited by David M. Prescott

Volume 5 (1972)
Methods in Cell Physiology
Edited by David M. Prescott

Volume 6 (1973)
Methods in Cell Physiology
Edited by David M. Prescott

Volume 7 (1973)
Methods in Cell Biology
Edited by David M. Prescott

Volume 8 (1974)
Methods in Cell Biology
Edited by David M. Prescott

Volume 9 (1975)
Methods in Cell Biology
Edited by David M. Prescott

Volume 10 (1975)
Methods in Cell Biology
Edited by David M. Prescott

Volume 11 (1975)
Yeast Cells
Edited by David M. Prescott

Volume 12 (1975)
Yeast Cells
Edited by David M. Prescott

Volume 13 (1976)
Methods in Cell Biology
Edited by David M. Prescott

Volume 14 (1976)
Methods in Cell Biology
Edited by David M. Prescott

Volume 15 (1977)
Methods in Cell Biology
Edited by David M. Prescott

Volume 16 (1977)
Chromatin and Chromosomal Protein Research I
Edited by Gary Stein, Janet Stein, and
Lewis J. Kleinsmith

Volume 17 (1978)
Chromatin and Chromosomal Protein Research II
Edited by Gary Stein, Janet Stein, and
Lewis J. Kleinsmith

Volume 18 (1978)
Chromatin and Chromosomal Protein Research III
Edited by Gary Stein, Janet Stein, and
Lewis J. Kleinsmith

Volume 19 (1978)
Chromatin and Chromosomal Protein Research IV
Edited by Gary Stein, Janet Stein, and
Lewis J. Kleinsmith

Volume 20 (1978)
Methods in Cell Biology
Edited by David M. Prescott

Advisory Board Chairman
KEITH R. PORTER

Volume 21A (1980)
Normal Human Tissue and Cell Culture, Part A: Respiratory, Cardiovascular, and Integumentary Systems
Edited by Curtis C. Harris, Benjamin F. Trump, and Gary D. Stoner

Volume 21B (1980)
Normal Human Tissue and Cell Culture, Part B: Endocrine, Urogenital, and Gastrointestinal Systems
Edited by Curtis C. Harris, Benjamin F. Trump, and Gray D. Stoner

Volume 22 (1981)
Three-Dimensional Ultrastructure in Biology
Edited by James N. Turner

Volume 23 (1981)
Basic Mechanisms of Cellular Secretion
Edited by Arthur R. Hand and Constance Oliver

Volume 24 (1982)
The Cytoskeleton, Part A: Cytoskeletal Proteins, Isolation and Characterization
Edited by Leslie Wilson

Volume 25 (1982)
The Cytoskeleton, Part B: Biological Systems and *in Vitro* Models
Edited by Leslie Wilson

Volume 26 (1982)
Prenatal Diagnosis: Cell Biological Approaches
Edited by Samuel A. Latt and Gretchen J. Darlington

Series Editor
LESLIE WILSON

Volume 27 (1986)
Echinoderm Gametes and Embryos
Edited by Thomas E. Schroeder

Volume 28 (1987)
***Dictyostelium discoideum:* Molecular Approaches to Cell Biology**
Edited by James A. Spudich

Series Editors
LESLIE WILSON AND PAUL MATSUDAIRA

Volume 50 (1995)
Methods in Plant Cell Biology, Part B
Edited by David W. Galbraith, Don P. Bourque, and Hans J. Bohnert

Volume 51 (1996)
Methods in Avian Embryology
Edited by Marianne Bronner-Fraser

Volume 52 (1997)
Methods in Muscle Biology
Edited by Charles P. Emerson, Jr. and H. Lee Sweeney

Volume 53 (1997)
Nuclear Structure and Function
Edited by Miguel Berrios

Volume 54 (1997)
Cumulative Index

Volume 55 (1997)
Laser Tweezers in Cell Biology
Edited by Michael P. Sheez

Volume 56 (1998)
Video Microscopy
Edited by Greenfield Sluder and David E. Wolf

Volume 57 (1998)
Animal Cell Culture Methods
Edited by Jennie P. Mather and David Barnes

Volume 58 (1998)
Green Fluorescent Protein
Edited by Kevin F. Sullivan and Steve A. Kay

Volume 59 (1998)
The Zebrafish: Biology
Edited by H. William Detrich III, Monte Westerfield, and Leonard I. Zon

Volume 60 (1998)
The Zebrafish: Genetics and Genomics
Edited by H. William Detrich III, Monte Westerfield, and Leonard I. Zon

Volume 61 (1998)
Mitosis and Meiosis
Edited by Conly L. Rieder